OUR PRICE

APPLIED CALCULUS
A GOALS APPROACH

Shirley O. Hockett
Ithaca College

Martin Sternstein
Ithaca College

 D. VAN NOSTRAND COMPANY
New York · Cincinnati · Toronto · London · Melbourne

For Jeanne and Chas

D. Van Nostrand Company Regional Offices:
New York   Cincinnati

D. Van Nostrand Company International Offices:
London   Toronto   Melbourne

Copyright © 1979 by Litton Educational Publishing, Inc.

Library of Congress Catalog Card Number: 78-71027
ISBN: 0-442-23428-7

All rights reserved. No part of this work covered by the copyright hereon may be reproduced or used in any form or by any means—graphic, electronic, or mechanical, including photocopying, recording, taping, or information storage and retrieval systems—without written permission of the publisher. Manufactured in the United States of America.

Published by D. Van Nostrand Company
135 West 50th Street, New York, N.Y.   10020

10 9 8 7 6 5 4 3 2 1

Preface

This book is intended for students in business, economics, social science, or biological science, and any others who take a one-semester or two-quarter course in calculus. Its organization is based on our belief that the objectives of beginning calculus can be made explicit and that most students can achieve these objectives if provided with the proper exposition and reinforcement.

The material is divided into fifty-four units called *Goals*, each focusing on a single topic. The unit begins with the statement of a Goal, followed by a discussion that includes carefully worked out examples, then by a set of exercises of graduated difficulty. The answers to half the exercises, sometimes with hints or complete solutions, are given. Two mini-tests are also included in each Goal, with solutions for the first supplied. We have found that if we give students the opportunity to accomplish specific aims, and if we provide them with immediate feedback on their achievements, we encourage them to go on and we help them overcome their fears and inhibitions about mathematics. Most Goals end with a set of optional Study Questions, which aim to stimulate and provoke the more interested students, thereby leading them to a deeper understanding of the calculus. The Study Questions are *not* prerequisite to subsequent Goals.

The entire set of Goals is grouped into nine chapters. Each chapter begins with a brief outline and ends with a summary of the highlights, a set of comprehensive review exercises (keyed to the pertinent Goals) with answers provided, and a Chapter Test aimed at integrating the material covered up to that point.

A brief review of the basic mathematics that students need on entering into the study of the calculus is included as an appendix.

The presentation of the material is informal and largely intuitive. Although it is not necessary, or even advisable, to offer the intended audience a rigorous treatment of all the mathematics relevant to the topics covered, we believe that what *is* presented should be accurate and complete. A student who pursues the study of mathematics should not at some later date have to unlearn something learned from this course. We have tried to state definitions, theorems, and procedures carefully and illustrate them fully. Numerous graphs and sketches have been included to reinforce the material visually. Special emphasis has been given to word problems and ways of solving them. Our aim throughout has been to maximize student understanding and appreciation of the basic concepts.

The power and usefulness of the calculus are shown through numerous applications from a variety of disciplines; they are used to both motivate and illustrate Goals. These applications are indexed at the back of the book by subject area for easy reference.

The text has been written primarily for a course taught in the traditional way by an instructor. But its format should also make it useful to students in a self-pacing (Keller or PSI) program. In such a course the student would be expected to master each Goal before proceeding to the next. One of the Goal tests can be used for determining the student's mastery of a concept.

Various concepts (for example, concavity, points of inflection, asymptotes) have been introduced intuitively when convenient in an early Goal in anticipation of the more formal definition to come in a later one. We have done this for reinforcement as well as for motivation.

We have also placed considerable emphasis on the encouragement and development of geometric intuition. Although this reflects our preference in teaching the calculus, we are aware that others may choose not to devote as much time to this approach. Many of the abundant geometric examples and exercises can be deleted without loss of continuity, if this is deemed suitable by the individual instructor.

For a one-semester course (meeting 3 or 4 times a week), we recommend the following as a minimum:

Chapter 1 (all); 2 (all); 3 (Goals 14, 15, 17); 4 (Goals 18–21); 5 (Goal 25); 6 (Goals 31–35); and 7 (Goals 37–42). This syllabus includes a total of 32 Goals.

If more time is available (perhaps two quarters or a 5-day-per-week semester course), we suggest the following:

Chapters 1–4 (all); 5 (Goals 25, 28, 29); 6 (all); 7 (37–42); 8 (44, 45); plus additional applications from chapters 5 and 8 or some multivariable calculus (chapter 9), as time permits. This syllabus includes a minimum of 41 Goals.

The *Instructor's Manual* contains Goal Exercise and Chapter Review answers not given in the text; answers to Goal Tests B; a second Chapter Test (B) for each chapter, with answers; and answers to and comments on the Study Questions.

We wish to thank the following reviewers for their helpful suggestions and comments: Albert W. Liberi, Westchester Community College; Stanley Lukawecki, Clemson University; C. David Minda, University of Cincinnati; Richard Semmler, Northern Virginia Community College; Kenneth J. Shabell, Riverside City College; and Ara B. Sullenberger, Tarrant County Junior College.

The authors are happy to acknowledge the help of Ithaca College student Stephen L. Weiman, who proofread much of the manuscript. Finally, we congratulate ourselves on our good fortune in having Charles F. Hockett as our mostly constructive critic, usually friendly arbiter, and always superbly skillful typist.

Ithaca, New York　　　　　　　　　　　　　　　　　　SHIRLEY O. HOCKETT
　　　　　　　　　　　　　　　　　　　　　　　　　　　MARTIN STERNSTEIN

Contents

Chapter 1. Functions and Graphs

GOAL
1. To understand what a function is and to use functional notation — 3
2. To find and graph equations of lines — 12
3. To graph quadratic functions — 22
4. To graph cubic polynomials and other elementary functions — 31
5. To operate on functions — 43
6. To find an equation for a function described in words — 53
   Summary — 65
   Review exercises — 65
   Chapter test A — 68

Chapter 2. Derivatives, Limits, and Continuity

GOAL
7. To approximate the rate of change of a function at a particular instant — 73
8. To find the slope of a curve at a designated point — 81
9. To understand the definition of derivative and to compute derivatives from the definition — 87
10. To evaluate limits — 94
11. To determine whether a function is continuous at a particular point — 106
12. To find derivatives of polynomials — 117
13. To apply the derivative to problems involving rates of change — 124
    Summary — 130
    Review exercises — 130
    Chapter test A — 133

Chapter 3. Differentiation Techniques

GOAL

14.	To use the product and quotient rules	139
15.	To use the chain rule	147
16.	To find derivatives by implicit differentiation	156
17.	To find higher derivatives	164
	Summary	171
	Review exercises	171
	Chapter test A	173

Chapter 4. Curve-Sketching

GOAL

18.	To find equations of tangents and normals	177
19.	To determine where a function is increasing or decreasing	184
20.	To determine local maxima and minima	195
21.	To find absolute maxima and minima	203
22.	To apply the second derivative: concavity, inflection points, testing for maxima and minima	210
23.	To find asymptotes of a curve	220
24.	To graph functions using the calculus	227
	Summary	234
	Review exercises	234
	Chapter test A	237

Chapter 5. Further Applications of the Derivative

GOAL

25.	To solve problems involving maxima and minima	241
26.	To solve related-rate problems	253
27.	To apply the Mean Value Theorem	260
28.	To use differentials for approximation	269
29.	To solve problems involving motion along a line	277
30.	To apply the derivative to problems in economics	283

	Summary	**296**
	Review exercises	**297**
	Chapter test A	**298**

Chapter 6. **Exponential and Logarithmic Functions**

GOAL

31.	To understand exponents and logarithms and functions using them	**303**
32.	To understand the number e and the exponential function e^x	**315**
33.	To understand the natural logarithmic function	**324**
34.	To differentiate $\ln x$ and related functions	**330**
35.	To differentiate e^x and related functions	**335**
36.	To solve problems involving exponential growth or decay	**344**
	Summary	**358**
	Review exercises	**358**
	Chapter test A	**360**

Chapter 7. **The Definite Integral**

GOAL

37.	To approximate limits of sums	**365**
38.	To find the area under a curve	**373**
39.	To understand and use the definite integral	**381**
40.	To use the Fundamental Theorem of the Calculus	**389**
41.	To find antiderivatives	**395**
42.	To integrate by substitution	**404**
43.	To use integration by parts	**413**
	Summary	**419**
	Review exercises	**419**
	Chapter test A	**422**

Chapter 8. **Applications of Integration; Differential Equations**

GOAL

44.	To compute areas between curves	**425**
45.	To analyze motion along a line and related problems	**431**

46.	To understand and apply the average value of a function	439
47.	To use integration in problems in economics	446
48.	To use integration on probability problems	458
49.	To understand and solve simple differential equations	468
50.	To apply separation of variables	474
	Summary	486
	Review exercises	487
	Chapter test A	489

Chapter 9. Multivariable Calculus

GOAL

51.	To understand functions of several variables	493
52.	To find and apply partial derivatives	505
53.	To find extreme values of functions of two variables	520
54.	To find maxima and minima using Lagrange multipliers	532
	Summary	542
	Review exercises	543
	Chapter test A	545

Appendix

Review of basic mathematics	549
Table I Natural logarithms	558
Table II Exponentials	559

Selected Answers

Answers to Goal exercises	563
Answers to Goal tests A	596
Answers to chapter review exercises	624
Answers to chapter tests A	633

Index

641

APPLICATIONS
(for page references consult the index)

BUSINESS AND ECONOMICS

advertising and sales; newspaper vs. TV ads
average cost
Cobb–Douglas formula for productivity
compound interest
competitive vs. complementary products
consumers' and producer's surpluses
cost
demand
depreciation
dropout expectancy
elasticity
equilibrium point
future periodic and continuous incomes
inventory control (optimal order size)
investments
level (*or* contour *or* indifference) curves
maintenance costs
predicting number of subscribers
predicting sales
present and future values of an investment
price
production and productivity
productivities of capital, labor, money
profit
revenue
shipping and storage costs
staffing
stock market
supply and demand
utility of a product (Weber–Fechner law)
wages of skilled vs. unskilled workers

PHYSICAL, BIOLOGICAL, AND SOCIAL SCIENCES

chemistry: autocatalytic reactions; other transformations; decomposition and purification of chemicals
crime: rate, change of
crop yield
demography: estimating growth of world population
diffusion: of disease, of rumor
dolphin mating and water temperature
drugs: determining dosage; concentration and diffusion of in bloodstream; sensitivity to; *see also* medicine, physiology
ecology: predicting population changes; predator–prey relations
fossils, dating of
gene frequency
geology: earthquake magnitude (Richter scale); finding depth of a lake; finding volume of atmosphere
glass blowing
growth: of animal, of tree, of a colony
half-life of radioisotopes
life expectancy in a population
light: change in intensity in various media
line of best fit
linear programming
machining (error tolerance)
mechanics: distance and velocity of rocket; path of missile; tire pressure; density of a cord; optimum speed of vehicle
medicine: change in blood pressure or temperature; sensitivity to drug; growth and treatment of tumor; kidney functioning; *see also* drugs, physiology
meteorology: change in temperature; wind velocity
motion problems involving speed, acceleration, ballistics, limiting velocity
physiology: weight of animal, of baby's brain; laminar flow of blood in artery; coughing and windpipe; absorption of iodine by thyroid; *see also* drugs, medicine
pollution: index of; of air by sulfur dioxide; of river by dumped waste
population change: of bacteria, fruit flies, wildlife; human (demography); unrestricted and restricted growth models; maximum sustainable
probability: using probability-density functions to predict: duration of phone calls, group nature or size, life expectancy in community, SAT scores, income in a group, accidents at intersections, demand for product, time needed to solve puzzle
psychology: learning curves; memorizing nonsense syllables or Ameslan signs; growth of vocabulary; time needed to solve puzzle or learn maze; intensity of response to a stimulus
radioisotope dating
routing problems: most economical route for pipeline, for vehicular transportation
sound: intensity and loudness; depth of well from time splash is heard
spread of disease, of drug, of rumor
statistics: marriage and divorce; reliability tests, including Spearman-Brown
temperature: Newton's law of cooling; human; and length of an iron bar
traffic: interval between cars at intersection; distance between cars on a highway
transportation: finding most economical route

Chapter 1

Functions and Graphs

Since calculus is the branch of mathematics in which we study functions and how they change, we begin by defining a function and describing functional notation. In this chapter our interest is in the so-called elementary functions. Since the behavior of a function is often most conveniently displayed through its graph, we also describe here graphing techniques with which we can sketch curves quickly and which do not require the calculus. Perhaps the most important goal of this chapter is to be able to translate word problems into mathematical terms; that is, to determine functions from verbal descriptions.

Goal 1. To understand what a function is and to use functional notation
Goal 2. To find and graph equations of lines
Goal 3. To graph quadratic functions
Goal 4. To graph cubic polynomials and other elementary functions
Goal 5. To operate on functions
Goal 6. To find an equation for a function that is described in words
Summary
Review Exercises
Chapter Test A

GOAL **1**

To understand what a function is and to use functional notation

One of the most fundamental concepts in mathematics is that of function. All of the applications in this book, to business, science, or social science, will involve functions.

When you complete this goal, you should be able to (1) recognize when a relationship is a function, (2) specify the domain and range of a given function, and (3) evaluate a function at an element in its domain.

DEFINITION AND EXAMPLES

A *function* is a correspondence between two sets such that each element of one set, the *domain*, is associated with one and only one element of the other set, the *range*.

Example 1-1

The distance covered by a jet flying at an average speed of 500 miles per hour (mph) is a function of the time the jet is in transit. If s represents the distance in miles and t the time in hours, then the function is

$$s = 500t$$

If the total time is $7\frac{1}{2}$ hours (h) then the domain is $0 \leq t \leq 7\frac{1}{2}$; the range is $0 \leq s \leq 3750$. This function associates a *unique* distance s with each time t in the domain.

Example 1-2

The association between a particular set of companies on the New York Stock Exchange and their closing prices on a specific day defines a function. The domain is the particular set of companies; the range is the set of prices at which they "closed" on the given day. Each company closed at only *one* price.

Example 1-3

The cost of producing Presto hand calculators is given by the equation

$$C = 500 + 2x$$

where x is the number of calculators and C is the cost in dollars. If Presto manufactures no more than 10,000 calculators, then the domain of this function is the set of positive integers between 1 and 10,000. (Limiting the domain to positive integers says that Presto manufactures only whole calculators!)

Example 1-4

The circumference of a circle is a function of its radius. This is usually expressed by the equation

$$C = 2\pi r$$

where the domain is $r \geq 0$ (with $r = 0$ only for a "point" circle).

Example 1-5

If you earn simple interest on an investment of $4000 at the rate of $8\frac{1}{2}\%$ per year (yr), then the function

$$I = 0.085(4000)t$$

gives the interest I earned in dollars after t yr.

Example 1-6

If A is the set of people in a certain city and B is the set of their mothers, then there is a function from A into B since each person has, or "maps into," a unique mother. However, since many mothers have more than one child, the correspondence between the sets B and A with B as domain is not a function.

This example shows that we can think of a function as mapping its domain into its range. (A map of the United States is a function from the set of cities, towns, villages, and other actual topographical features to the points that identify their relative locations.)

Example 1-7

A function may be defined by a table, such as this one:

Goal 1 | To Understand a Function and Functional Notation

High Temperature (°C)
in Selected United States Cities
on 24 November 1976

Albany	4
Albuquerque	16
Boston	7
Chicago	−1
Denver	16
El Paso	19
Honolulu	27
Juneau	4
Miami Beach	21
Oklahoma City	16
San Francisco	22

The domain of this function is the set of cities in the first column; the range is the set of numbers $\{-1, 4, 7, 16, 19, 21, 22, 27\}$. This correspondence is a function since each city listed maps into a single temperature. Note that two or more cities may (and do!) map into the same temperature.

Example 1-8

A function may be defined by a graph:

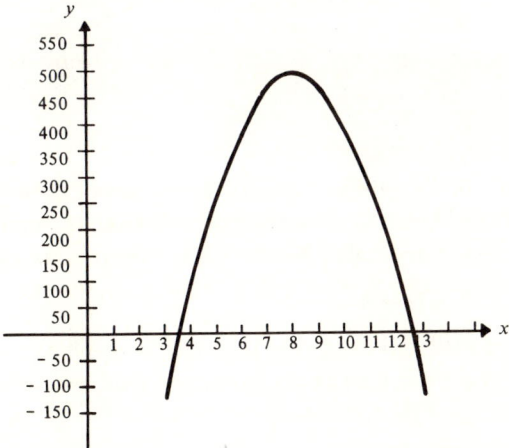

Here the domain is $3 \leq x \leq 13$ and the range is $-125 \leq y \leq 500$. For each x in the domain of the function there is a unique y. But different x's may (and do!) produce the same y; for example, when $x = 6$, $y = 400$, and when $x = 10$, again $y = 400$.

FINDING THE DOMAIN OF A GIVEN FUNCTION

The functions of most interest to us are those whose domains are sets of numbers. *If the domain is not specified, we take it to be the largest subset of the reals for which the function is defined.* (The *reals* are the numbers corresponding to all the points on a number line. We will use the symbol $\mathcal{R}$ for the entire set of reals.)

Example 1-9

Find the domain of each function.

(a) $y = \dfrac{1}{x - 2}$ (b) $y = \dfrac{x + 1}{x^2 - x}$ (c) $y = \sqrt{x - 9}$

(a) Since we cannot divide by zero, no denominator may equal zero. The domain is therefore the set of all reals except 2. This is often written "$x \neq 2$."

(b) The denominator can be factored: $x^2 - x = x(x - 1)$. Since this denominator equals zero if $x = 0$ or 1, the domain is all reals except for these numbers. We can write "$x \neq 0, 1$."

(c) The square root (or any other even root) of a negative number is not a real number. Since $x - 9$ is negative if $x < 9$, the domain of $y = \sqrt{x - 9}$ is $x \geq 9$. Note that $\sqrt{0} = 0$.

Generally, then, a function is not defined at any number that leads to division by zero or to an even root of a negative number. Remember that zero divided by any number except zero is well defined!—it equals zero.

NOTATION

The notation used to represent a function f of a variable x is "$f(x)$," read "f of x." Also, this same notation denotes the element in the range associated by the function with x. If $f(x) = 3x - 1$, then

$$f(-1) = 3(-1) - 1 = -4$$
$$f(0) = 3(0) - 1 = -1$$
$$f(5) = 3(5) - 1 = 14$$
$$f(a) = 3a - 1$$
$$f(-x) = 3(-x) - 1 = -3x - 1$$
$$f(a + h) = 3(a + h) - 1 = 3a + 3h - 1$$

Note above that the symbol "$f(x)$" simply replaces the symbol "y" in the equation

$$y = 3x - 1$$

Also, "$f(5) = 14$" is a precise and brief way of saying, "the value of the function when $x = 5$ is 14."

When y is a function of x we often write "$y = f(x)$" and refer to x as the *independent variable* and to y as the *dependent variable*. Of course, letters other than f are also used to represent functions.

Example 1-10

If $g(x) = x^2 + 1$, find the domain and range of g. Evaluate $g(0)$, $g(-2)$, $g(\sqrt{2})$, $g(1 + h)$, $g(x + h)$.

The domain of g is the set $\mathcal{R}$ of real numbers; its range consists of all the reals that are greater than or equal to 1.

We have

$$g(0) = 0^2 + 1 = 1$$
$$g(-2) = (-2)^2 + 1 = 4 + 1 = 5$$
$$g(\sqrt{2}) = (\sqrt{2})^2 + 1 = 2 + 1 = 3$$
$$g(1 + h) = (1 + h)^2 + 1 = 1 + 2h + h^2 + 1 = 2 + 2h + h^2$$
$$g(x + h) = (x + h)^2 + 1 = x^2 + 2xh + h^2 + 1$$

EXERCISES

Which of the descriptions in Exercises 1 through 3 below define functions (a) with domain S and range T; (b) with domain T and range S?

1. S is a set of 400 freshmen at a college and T is the set of their birthdays.
2. S is the set of all United States citizens alive today and T is the set of all current social security numbers.
3. S is a set of 75 students in a sociology class and T is the set of grades they received on a 50-question multiple-choice examination for which each correct answer earned 2 credits.

Find the domain of each function in Exercises 4 through 13.

4. $f(x) = x^2 - 1$
5. $y = \dfrac{1}{x}$
6. $g(x) = \dfrac{x}{x - 1}$
7. $f(x) = \sqrt{x}$
8. $h(x) = \dfrac{x - 9}{x^2 - 9}$
9. $y = \sqrt{3 - x}$
10. $g(x) = \dfrac{4}{x^2 + 1}$
11. $f(x) = 2x^3 + x - 1$
12. $h(x) = \dfrac{x^2 - 4}{x^2 - 5x}$
13. $y = \sqrt{x - 4}$

14. The cost of producing x calculators in Example 1-3 is given by the function $C(x) = 500 + 2x$. How much does it cost to produce 100 calculators? 500? 10,000? What happens to this cost function as x increases? Find $C(0)$ and suggest an economic interpretation for it.
15. If $f(x) = 5$, find $f(-1)$, $f(3)$, $f(75)$, $f(a)$. What is the domain of this function? What is its range?
16. If $f(x) = 3x^2$, find $f(1), f(2), f(-2), f(0), f(10), f(a), f(a + h), f(x + h)$.
17. Let $f(x) = x^2 + 3$. Find $f(2)$, $f(-2)$, $f(0)$, $f(10)$, $f(x + h)$, $f(x + h) - f(x)$.
18. If $g(x) = (x + 3)^2$, evaluate $g(4)$, $g(-4)$, $g(0)$. For what x does $g(x)$ equal zero?
19. If $g(x) = \dfrac{3}{2x - 1}$, find $g(1), g(-1), g(0), g(1 + h)$.

Find $f(x + h) - f(x)$ and simplify for each of the functions 20 through 24.

20. $f(x) = x^2$
21. $f(x) = \dfrac{1}{x}$
22. $f(x) = \dfrac{2}{3x + 1}$
23. $f(x) = \sqrt{x}$
24. $f(x) = 2x - 1$

In 25 through 29 mark each statement true or false.

25. $f(x)$ means f times x.
26. If $f(x) = 3x$, then $f(x) = 3f(x)$.
27. If $f(x) = x^2$, then $f(-x) = f(x)$.
28. If $f(x) = x^2$, then $f(1) + f(2) = f(1 + 2)$.
29. If $f(x) = 4x - 3$, then $f(-x) = f(x)$.

TEST A

1. Find the domain of each function:
    (a) $f(x) = x^2 - x + 3$
    (b) $f(x) = \dfrac{x + 1}{x - 2}$
    (c) $f(x) = \sqrt{x - 4}$
    (d) $f(x) = \dfrac{1}{x^2 - 3x}$
2. Find $f(1), f(-1), f(0), f(-5)$ if
    (a) $f(x) = 2x + 4$
    (b) $f(x) = 3x^2 - 1$
    (c) $f(x) = \dfrac{2}{x - 3}$
    (d) $f(x) = \sqrt{1 - x}$

3. Find $g(x + h) - g(x)$ and simplify, if
   (a) $g(x) = 3x + 1$
   (b) $g(x) = x^2 - 1$
   (c) $g(x) = 1/x$
   (d) $g(x) = \sqrt{x - 4}$

TEST B

1. Find the domain of each function.
   (a) $f(x) = x^2 + 0.1x + 600$
   (b) $f(x) = \dfrac{x^2 - 1}{x - 3}$
   (c) $f(x) = \sqrt{9 - x}$
   (d) $f(x) = \dfrac{4}{x^2 - 1}$

2. Find $f(0), f(2), f(-2), f(10)$ if
   (a) $f(x) = 1000 + 3x$
   (b) $f(x) = (x - 10)^2 + 20$
   (c) $f(x) = \dfrac{x}{x^2 - 1}$
   (d) $f(x) = x^3 - x - 1$

3. Find $g(x + h) - g(x)$ and simplify, if
   (a) $g(x) = 5 - 2x$
   (b) $g(x) = x^2 + 3$
   (c) $g(x) = \dfrac{1}{x + 2}$
   (d) $g(x) = \dfrac{3}{2x + 1}$

STUDY QUESTIONS

1. Some mathematicians define a function as a set of ordered pairs no two of which have the same first element but different second elements. Which of the following sets define functions?
   (a) $\{(1, 3), (4, 9), (0, 1), (-2, -3), (-10, -19)\}$
   (b) $\{(1, 3), (4, -9), (4, -10), (-2, 1)\}$
   (c) $\{(0, 1), (-1, 1), (3, 1), (9, 1), (5, 1), (8, 1)\}$
   (d) $\{(1, 1), (4, 2), (9, 3), (1, -1), (4, -2), (9, -3)\}$

2. If $f(x) = x^2$, verify that $f(-1) = f(1)$, $f(-3) = f(3)$, and, generally, $f(-x) = f(x)$. Any function that satisfies this condition is called an *even* function. If $f(x) = x^3$, verify that $f(-1) = -f(1)$, $f(-\frac{1}{2}) = -f(\frac{1}{2})$, and, in general, that $f(-x) = -f(x)$. This is an example of an *odd* function.

   Specify which of the following functions are even, which are odd, and which are neither.

(a) $f(x) = x^4$
(b) $h(x) = x^5$
(c) $g(x) = 3$
(d) $F(x) = x^3 + x$
(e) $G(x) = x^2 - 3$
(f) $y = x^3 + 1$
(g) $C(x) = x^2 + 0.01x + 500$
(h) $P(x) = \dfrac{1}{x}$

3. Suppose f is a function with domain S. Let T be the set of all values of f. We know from the definition of function that each element of S maps into a unique element in T. If it is also the case that each element in T is the correspondent under f of only one element in S, then f is said to be *one-to-one*.

   (a) Which of the functions in Examples 1-1 through 1-8 are one-to-one?
   (b) Is the function $f(x) = x^2$ with domain the set of reals ($\mathcal{R}$) one-to-one?
   (c) Is $f(x) = x^3$ with domain $\mathcal{R}$ one-to-one?

4. If $f(x)$ is one-to-one from S into T, then consider the function $g(x)$ from T into S defined by the fact that

$$\text{if } f(a) = b \text{ then } g(b) = a$$

This function is called the *inverse* of f. For example, if $f = \{(1, 3), (2, 7), (3, 11), (8, 23)\}$, then f is one-to-one from $\{1, 2, 3, 8\}$ into $\{3, 7, 11, 23\}$. Its inverse function g has domain $\{3, 7, 11, 23\}$ and is the set of pairs $\{(3, 1), (7, 2), (11, 3), (23, 8)\}$. The domain and range of a one-to-one function are, respectively, the range and domain of its inverse.

In the following table we give the inverse function g for each function f that is one-to-one:

f:	g:
(1) $\{(1, 5), (0, 4), (2, -1), (3, 9), (-7, 8)\}$	$\{(5, 1), (4, 0), (-1, 2), (9, 3), (8, -7)\}$
(2) $\{(1, 3), (2, 3), (5, 3), (5, 3), (12, 3)\}$	f is not one-to-one
(3) $f(x) = 2x + 1$ on domain the set of integers	$g(x) = \dfrac{x - 1}{2}$ on domain the set of odd integers
(4) $f(x) = \sqrt{x}$ on $\{x \mid x \geq 0\}$	$f(x) = x^2$ on $\{x \mid x \geq 0\}$
(5) $f(x) = x^2$ on R	f is not one-to-one
(6) $f(x) = 1/x$ on $\{x \mid x \in R, x \neq 0\}$	$g(x) = 1/x$ on $\{x \mid x \in R, x \neq 0\}$
(7) $f(x) = \dfrac{1}{x - 3}$ $(x \neq 3)$	$g(x) = 3 + 1/x$ $(x \neq 0)$

(a) In (3), (4), (6), and (7) compute $f(g(x))$ and $g(f(x))$.
(b) If $f(x)$ is defined by an equation [as in (3), (4), (6), and (7)], state how you can find its inverse $g(x)$ algebraically.
(c) Find the inverse function for each function below that is one-to-one:
    (i) $f(x) = 3x + 2$
    (ii) $f(x) = 4 - 5x$
    (iii) $f(x) = x^2 - 1$
    (iv) $f(x) = x^3$
    (v) $f(x) = \dfrac{3}{x+1}$
    (vi) $f(x) = x^3 - x$

GOAL 2

To find and graph equations of lines

Many applications of mathematics involve linear functions, each of whose graphs is a straight line. Straight lines are important because in the calculus a small piece of a curve is approximated by a segment of a straight line. In this Goal we aim to find and graph equations of lines and to note properties of linear functions and of lines.

The function

$$f(x) = mx + b \quad \text{or} \quad y = mx + b$$

where m and b are fixed real numbers, is called the *linear function*. Since its graph is a nonvertical straight line, we often call the function the line and vice versa.

The *y-intercept* of any graph is the point where the graph intersects the y-axis. We can therefore find it by letting $x = 0$ in the equation for y. So the y-intercept of the line $y = mx + b$ equals $m \cdot 0 + b$, or b.

If two different points, say (x_1, y_1) and (x_2, y_2), lie on the line $f(x) = mx + b$, then

$$f(x_1) = mx_1 + b = y_1 \tag{1}$$

$$f(x_2) = mx_2 + b = y_2 \tag{2}$$

Subtracting (1) from (2) yields

$$m(x_2 - x_1) = y_2 - y_1$$

and

$$m = \frac{y_2 - y_1}{x_2 - x_1}$$

Goal 2 | To Find and Graph Equations of Lines

The number m is the ratio of the change in y to the change in x. It measures the steepness of the line and is called the *slope* of the line.

Below we graph $y = mx + b$ when $m > 0$, $m = 0$, and $m < 0$.

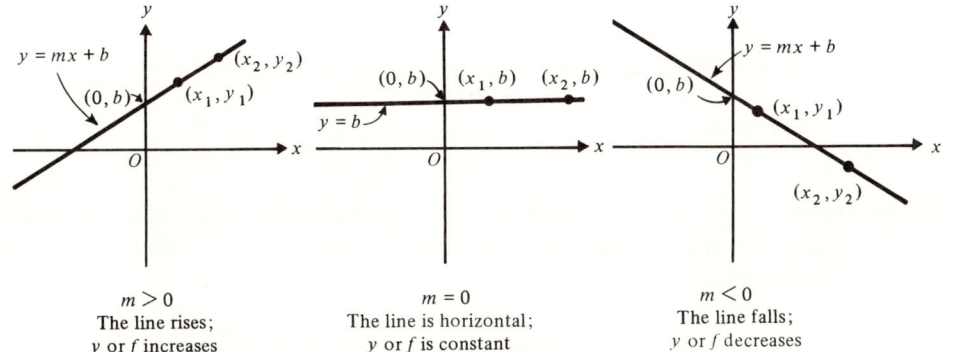

$m > 0$
The line rises;
y or f increases

$m = 0$
The line is horizontal;
y or f is constant

$m < 0$
The line falls;
y or f decreases

In each sketch we choose $x_2 > x_1$, that is, $x_2 - x_1 > 0$; this is equivalent to "moving" along the line from left to right.

When $m > 0$, then $y_2 > y_1$. We say the function is *increasing*.

When $m = 0$, we have the function $f(x) = b$ (or $y = b$), whose graph is a horizontal line. This is called the *constant function*; the value of y is the same for every x and equals b.

When $m < 0$, then $y_2 < y_1$, and the function *decreases*.

Example 2-1

Graph the equation $x = a$, where a is a constant.

The graph is a vertical line through the point $(a, 0)$. But note that this graph does *not* represent a function. (Why?)

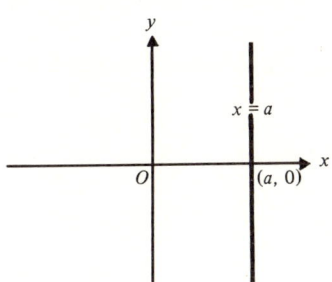

Example 2-2

Graph $y = -3x + 2$.

The easiest way to graph a linear function is to plot both intercepts. When $x = 0$, $y = 2$; when $y = 0$, $0 = -3x + 2$, $3x = 2$, and the x-intercept is $\frac{2}{3}$.

Plot these points, then draw the line through them.

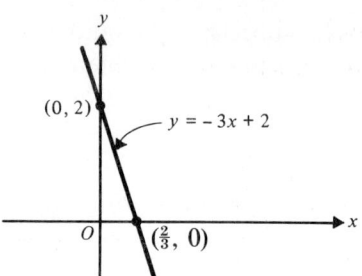

Example 2-3

Graph the line through the points $(-1, 2)$ and $(3, 4)$ and find its equation. The slope of the line is

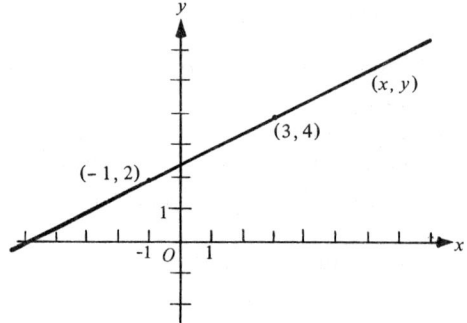

$$m = \frac{4 - 2}{3 - (-1)} = \frac{2}{4} = \frac{1}{2}$$

So we have

$$y = \frac{1}{2}x + b$$

We find b by using the fact that $(-1, 2)$ lies on the line:

$$2 = \frac{1}{2}(-1) + b$$

$$\frac{5}{2} = b$$

So the desired equation is

$$y = \frac{1}{2}x + \frac{5}{2}$$

Goal 2 | **To Find and Graph Equations of Lines**

We could also have used the point (3, 4) to find b; instead, we can verify that (3, 4) satisfies the equation

$$4 \stackrel{?}{=} \frac{1}{2}(3) + \frac{5}{2}$$

$$4 \stackrel{?}{=} \frac{3}{2} + \frac{5}{2}$$

$$4 \stackrel{\checkmark}{=} 4$$

Although we found the equation of the line above by determining b, this is *not* necessary. Since the coordinates of any point on the line are denoted by (x, y), and since we already know that the slope of the line through $(-1, 2)$ and $(3, 4)$ is $\tfrac{1}{2}$, it follows that

$$\frac{y - 4}{x - 3} = \frac{1}{2} \tag{1}$$

which yields

$$y - 4 = \frac{1}{2}(x - 3) \tag{2}$$

or

$$y = \frac{1}{2}x + \frac{5}{2}$$

as before. Note that Equation (1) holds for all points other than (3, 4) but that Equation (2) is satisfied also by (3, 4).

FORMS OF LINEAR EQUATIONS

If a line with slope m contains the point (x_1, y_1), then its equation is

$$y - y_1 = m(x - x_1)$$

This is called the *point-slope form* of the equation for a line, while

$$y = mx + b$$

is called the *slope-intercept form*.

Example 2-4

It costs the Presto Company $(500 + 2x)$ dollars to produce x calculators (see Example 1-3). Draw the graph of the function $C(x) = 500 + 2x$.

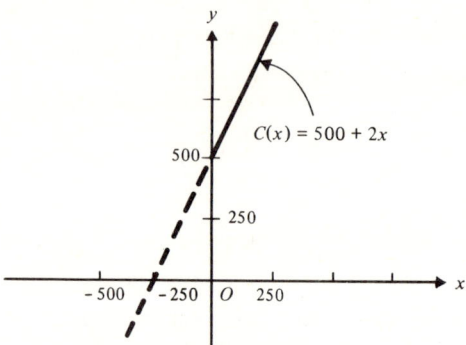

Note that the y-intercept is 500. This is the fixed cost that does not depend on x. The x-intercept is -250. With these points plotted, we can draw the line quickly. We dot the part of the line corresponding to $x < 0$ because the function is meaningless there. By drawing the graph for $x \geq 0$ continuously (without any breaks or holes), we're implying that x may be *any* nonnegative number, not just an integer. Although this does not make sense practically (Presto does not produce a fraction of a calculator!), we will make such assumptions in most problems of this sort so that we can subject them to the techniques of the calculus that we will be developing later.

We see immediately in this example that the cost increases as x does, either from the graph or from the equation itself. The equation tells us that the slope is 2.

ADDITIONAL IMPORTANT FACTS ABOUT LINES

(1) A vertical line has no slope; that is, slope is not defined for it. (As has already been noted, the slope of a horizontal line is zero.)

(2) Nonvertical parallel lines have equal slopes.

(3) If two nonvertical lines are perpendicular, then the product of their slopes is -1; or the slope of one line is the negative reciprocal of that of the other.

(4) The line $y = mx$ goes through the origin. When $m > 0$, then y is said to *vary directly* as x or to be *directly proportional* to x. We call m the *factor of proportionality*.

(5) The equation of any line may be written in the form $Ax + By + C = 0$, where A and B are not both zero.

Example 2-5

Find the equation of the line parallel to $3x + 4y - 5 = 0$ and containing the point $(6, -1)$.

Goal 2 | **To Find and Graph Equations of Lines**

Since the given line equation can be written in the form

$$y = -\frac{3}{4}x + \frac{5}{4}$$

its slope is $-\frac{3}{4}$. Any line parallel to it is therefore of the form

$$y = -\frac{3}{4}x + b$$

because parallel lines have equal slopes. Since the point $(6, -1)$ lies on the required line, we have

$$-1 = -\frac{3}{4}(6) + b$$

so that

$$b = \frac{7}{2}$$

The equation is therefore

$$y = -\frac{3}{4}x + \frac{7}{2} \quad \text{or} \quad 3x + 4y - 14 = 0$$

Example 2-6

Find the equation of the line passing through $(-4, -3)$ and perpendicular to $2x - 5y - 10 = 0$.

We can rewrite the given equation in the form

$$y = \frac{2}{5}x - 2$$

Any line perpendicular to this one has slope $-\frac{5}{2}$ and is of the form

$$y = -\frac{5}{2}x + b$$

Since $(-4, -3)$ lies on the line, we have

$$-3 = -\frac{5}{2}(-4) + b$$

$$-3 = 10 + b$$

$$b = -13$$

The required equation is therefore
$$y = -\frac{5}{2}x - 13 \quad \text{or} \quad 5x + 2y + 26 = 0$$

EXERCISES

1. Find the slope, if it exists, of the line through
   (a) (3, 1) and (6, 4)
   (b) (1, −2) and (−4, 2)
   (c) (−2, 0) and (−2, 5)
   (d) (2, 0) and (0, −3)
   (e) (3, 2) and (3, −4)
   (f) (−4, 1) and (−3, 1)
   (g) (3, 6) and (3 + h, 2(3 + h))
   (h) (x, 4x − 1) and (x + h, 4(x + h) − 1)

2. Find the equation of the line
   (a) whose slope is 3 and whose y-intercept is 2.
   (b) with y-intercept $-\frac{1}{2}$ and slope −1.
   (c) with slope 2 and containing the point (1, 3).
   (d) with slope $-\frac{4}{3}$ and whose x-intercept is −1.
   (e) with y-intercept 3 and whose slope is 0.
   (f) passing through (2, −1) and with slope 0.

3. Find the equation of the line through each pair of points in Exercise 1.

4. Graph:
   (a) $3x - 2y = 6$
   (b) $4x + y + 2 = 0$
   (c) $f(x) = x - 3$
   (d) $3x + 2y = 4$
   (e) $y = 4x$
   (f) $g(x) = -1$
   (g) $x = -2$
   (h) $y = 0$
   (i) $x - y = 0$
   (j) $x = 0$

5. Graph each of the following pairs of lines on a single set of axes, then estimate the coordinates of any point of intersection. Verify algebraically, that is, by solving each pair of equations simultaneously:
   (a) $\begin{cases} 2x + 3y = 11 \\ 3x - 5y = 7 \end{cases}$
   (b) $\begin{cases} 3x - 4y = 12 \\ 6x - 8y = 24 \end{cases}$
   (c) $\begin{cases} 2x - 2y = 9 \\ x - 4y = 6 \end{cases}$
   (d) $\begin{cases} y = 5x - 5 \\ 5x - y + 5 = 0 \end{cases}$

6. Find the equation of a line that
   (a) is horizontal and passes through (−2, −3).
   (b) is parallel to the line $y = x - 7$ and goes through (0, 0).
   (c) is parallel to the line $y = x - 7$ and goes through (−4, 1).
   (d) is perpendicular to the line $5x + y - 2 = 0$ and has y-intercept −1.

(e) is parallel to $3x - 4y = 1$ and has y-intercept 4.

(f) is perpendicular to $3x - 4y = 1$ and goes through $(5, -2)$.

7. Which of the following linear functions is increasing? decreasing? neither?
   (a) $f(x) = x + 2$
   (b) $f(x) = 5$
   (c) $g(x) = -x + 3$
   (d) $C(x) = 1000 + 1.5x$
   (e) $y = -2$
   (f) $y = -0.01x + 850$

8. The Bionic Mouse Company has fixed costs of $200 per month. It costs them $4 to produce a box of bionic mice.
   (a) Find the total monthly cost C of producing x boxes.
   (b) How much does it cost to produce 500 boxes of mice?
   (c) If all the mice produced are sold at $6 a box, express the company's profit P as a function of x, the number of boxes.
   (d) What are the domains of these functions?

9. A function $n(p)$ for the number of items that can be sold at p dollars each is called a *demand* function. If the Tory Company can sell 500 wigs at $20 each but only 300 at $40 each, find the demand function, assuming that it is linear. Graph it. How many wigs can be sold at $25? What is the domain of the function? What happens to the demand for wigs as the unit price increases?

10. Solve the equation of Exercise 9 for p in terms of n and graph this function, with n as independent variable.

11. A function $S(p)$ for the number S of items a seller will supply at unit price p is called a *supply* function. Suppose a wholesale TV dealer is willing to supply 500 TV sets at $150 each but only 350 sets at $110 each. Find and graph the supply function if it is assumed to be linear. How many sets will the dealer supply if he can get $175 for each?

12. Suppose the interest you earn on savings is directly proportional to the amount you invest. If $250 earns $13.75 in 1 yr,
    (a) How much will $350 earn in 1 yr?
    (b) How much must you invest to earn $24.75 interest after one yr?

13. The brain of a baby is proportional to its body weight. If a baby weighing 3.5 kilograms (kg) has a brain weighing 420 grams (g), how much does the brain of a 4-kg baby weigh? Find the equation of variation.

14. Temperatures Fahrenheit (°F) and Celsius (°C) are related as follows:
$$C = \frac{5}{9}(F - 32)$$
    (a) Find C when F = 32°, 68°, 212°.

(b) What is the slope of the line?

(c) Express F as a function of C.

15. *A family* of lines is a set of lines with some common characteristic. For example, the set of parallel lines each with slope 5 is a family. Each has a linear equation of the form $y = 5x + b$. Sketch three different members of each of the following families of lines and describe their common feature.

   (a) $y = x + b$ (c) $y = -2x + k$

   (b) $y = mx + 3$ (d) $y - 2 = k(x + 1)$

16. Write an equation for each of the following families:

   (a) Lines parallel to $2x + y - 7 = 0$.

   (b) Lines with y-intercept -2.

   (c) Lines perpendicular to $3x - 4y = 9$.

   (d) Lines passing through the point $(3, 6)$.

17. Find the particular value of k for which

   (a) $kx - 5y + 3 = 0$ has slope -2.

   (b) $3x + y - k = 0$ has x-intercept 4.

   (c) $y = kx + 3 - k$ passes through $(-1, 1)$.

TEST A

1. Find the slope of the line containing the points.

   (a) $(2, 5)$ and $(3, 1)$

   (b) $(0, -4)$ and $(2, 0)$

   (c) $(-1, 3)$ and $(-2, 3)$

   (d) $(x, 3x + 1)$ and $(x + h, 3(x + h) + 1)$

2. Find the equation of the line

   (a) containing the points $(1, 2)$ and $(-3, -6)$.

   (b) whose slope is -1 and that passes through $(-5, 8)$.

   (c) that is parallel to $x - 3y + 3 = 0$ and that has y-intercept -1.

3. Graph:

   (a) $f(x) = -2x + 3$

   (b) $x - 3y + 2 = 0$

   (c) $f(x) = 0$

   (d) $f(x) = 0.5x + 0.4$

4. Find the linear function $f(x)$ if $f(-1) = 6$ and $f(2) = -3$.

Goal 2 | To Find and Graph Equations of Lines

TEST B

1. Find the slope of the line through the points
   (a) $(1, -2)$ and $(-4, 2)$
   (b) $(0, 0)$ and $(3, -3)$
   (c) $(30, 100)$ and $(50, 150)$
   (d) $(x, -2x + 1)$ and $(x + h, -2(x + h) + 1)$
2. Find the equation of a line if it
   (a) contains the points $(3, -1)$ and $(2, -4)$.
   (b) has slope 0 and passes through $(-2, 3)$.
   (c) is perpendicular to $2x + 4y = 1$ and has y-intercept 1.
3. Graph each of the following and tell which ones are linear functions.
   (a) $y = 2 - 3x$
   (b) $f(x) = \frac{1}{2}x + \frac{1}{2}$
   (c) $x + y = 0$
   (d) $x = 0$
4. Do the points $(-1, 7)$, $(1, 3)$, and $(3, -1)$ lie on the same line? Explain.

STUDY QUESTIONS

1. The equation of a line whose x-intercept is a and whose y-intercept is b is

$$\frac{x}{a} + \frac{y}{b} = 1$$

   Derive this equation. When is it meaningless?

2. Let two lines l_1 and l_2 (neither one vertical) have slopes m_1 and m_2 respectively. If $l_1 \perp l_2$, show that $m_1 m_2 = -1$.
   HINT: We can assume that both lines go through the origin. (Why?) Note from the sketch that

$RS_1 = m_1$
$S_2R = -m_2$
$S_2S_1 = m_1 - m_2$

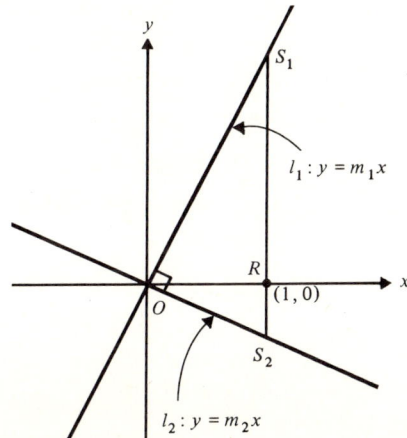

Use the Pythagorean theorem twice.

GOAL 3

To graph quadratic functions

The aim of this goal is to graph quadratic functions of the type $y = ax^2$ ($a \neq 0$) and to use translation of axes to graph a curve whose equation is $y = ax^2 + bx + c$ ($a \neq 0$).

A *quadratic function* is one of the form

$$f(x) = ax^2 + bx + c$$

where a, b, and c are constants and $a \neq 0$. Its graph is called a *parabola*. The domain of x is $\mathcal{R}$, the set of all real numbers.

FUNCTIONS OF THE FORM $y = ax^2$

Example 3-1

Graph $y = x^2$.

We can plot the points that are given in the table of values:

x	-2	-1	0	1	2
y	4	1	0	1	4

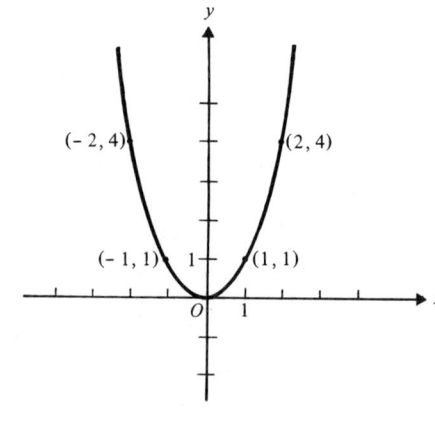

Goal 3 | To Graph Quadratic Functions

We then draw a smooth curve that passes through these points. We note immediately that

(1) The curve goes through the origin.
(2) The curve is symmetric about the y-axis: x and $-x$ produce the same y-value [that is, for all x, $f(-x) = f(x)$].
(3) The curve *opens up*; that is, a line segment joining any two points on the curve lies above the arc of the curve through these points. We also say the curve is *concave upward*.
(4) The lowest point on the curve, called its *vertex*, is the origin.
(5) As x gets larger, either positively or negatively, y increases without bound.

Example 3-2

Use tables of values to graph (a) $y = \frac{1}{2}x^2$; (b) $y = 3x^2$; (c) $y = -x^2$; (d) $y = -2x^2$.

First the tables:

(a)
x	-2	-1	0	1	2
y	2	$\frac{1}{2}$	0	$\frac{1}{2}$	2

(c)
x	-2	-1	0	1	2
y	-4	-1	0	-1	-4

(b)
x	-2	-1	0	1	2
y	12	3	0	3	12

(d)
x	-2	-1	0	1	2
y	-8	-2	0	-2	-8

Below we sketch (a) and (b) on the same set of axes, along with $y = x^2$; we sketch (c) and (d) on another set of axes below at the right.

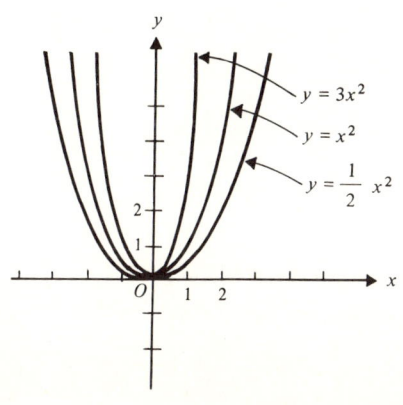
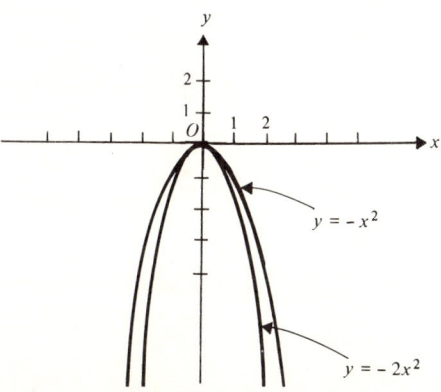

CONCLUSIONS

We see that the differences in the graphs for Examples 3-1 and 3-2 depend only on a in the equation $y = ax^2$:

If a is positive, the parabola opens up; the vertex is the lowest point on the curve and is at the origin.

If a is negative, the parabola opens down (can you define "opening down"?); the vertex is still at the origin but is now the highest or maximum point on the curve.

Also, if a is positive, then the larger a is, the steeper the curve. The curve of $y = 3x^2$ is steeper than that of $y = x^2$ which, in turn, is steeper than that of $y = \frac{1}{2}x^2$. A similar statement can be made when a is negative: $y = -5x^2$ is steeper than $y = -2x^2$, which is steeper than $y = -0.1x^2$.

FUNCTIONS OF THE FORM $y = ax^2 + k$

Example 3-3

Use tables of values to graph (a) $y = x^2 + 4$; (b) $y = x^2 - 3$.

(a)

x	-2	-1	0	1	2
y	8	5	4	5	8

(b)

x	-2	-1	0	1	2
y	1	-2	-3	-2	1

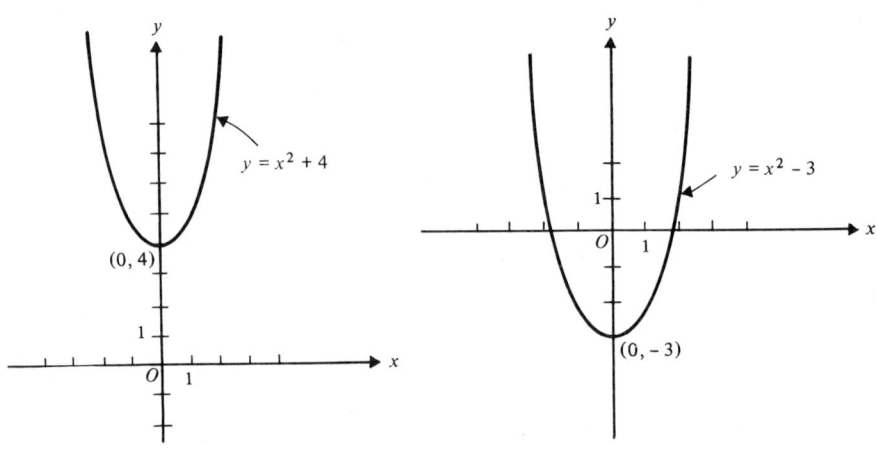

CONCLUSIONS

Tables of values are not necessary for graphing these parabolas. If we can graph $y = x^2$, we can graph *any* parabola of the form $y = x^2 + k$: simply shift the graph of $y = x^2$

up k units if $k > 0$;
down k units if $k < 0$.

Goal 3 | **To Graph Quadratic Functions**

In either case we say we have *translated* the curve vertically k units.
The vertex of $y = ax^2 + k$ is at $(0, k)$; it is a minimum if $a > 0$, a maximum if $a < 0$.

FUNCTIONS OF THE FORM $y = a(x - h)^2$
Example 3-4
Use tables of values to graph (a) $y = (x - 4)^2$, (b) $y = (x + 3)^2$.

(a)
x	2	3	4	5	6
y	4	1	0	1	4

(b)
x	-5	-4	-3	-2	-1
y	4	1	0	1	4

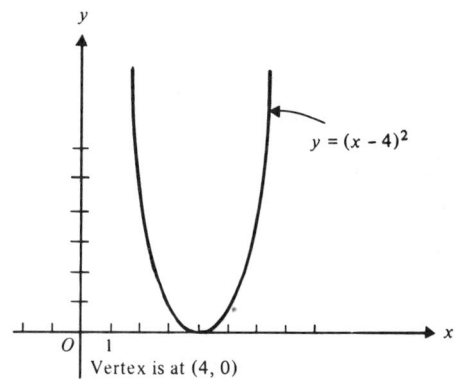
Vertex is at $(4, 0)$

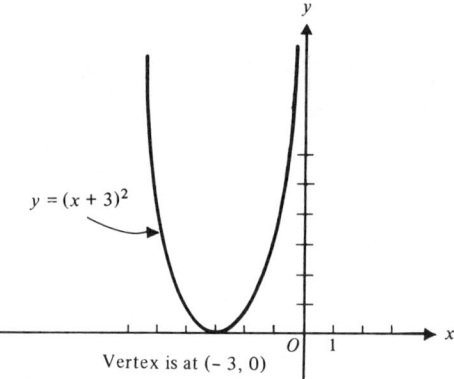
Vertex is at $(-3, 0)$

CONCLUSIONS
The graph of $y = (x - 4)^2$ is just that of $y = x^2$ shifted (or translated) to the right 4 units; the graph of $y = (x + 3)^2$ is the graph of $y = x^2$ translated to the left 3 units.

Generally, the graph of $y = a(x - h)^2$ is that of $y = ax^2$ shifted

to the right h units if h is positive;
to the left h units if h is negative.

In both cases, the vertex of $y = a(x - h)^2$ is at $(h, 0)$; it is a minimum if $a > 0$, a maximum if $a < 0$.

Example 3-5
Graph $y = -\frac{1}{2}(x + 1)^2$ and find its vertex.
The curve is simply that of $y = -\frac{1}{2}x^2$ shifted to the left 1 unit. Plotting its

y-intercept fixes it nicely; when $x = 0$, $y = -\frac{1}{2}$. Its vertex is at $(-1, 0)$. Of course, it opens down.

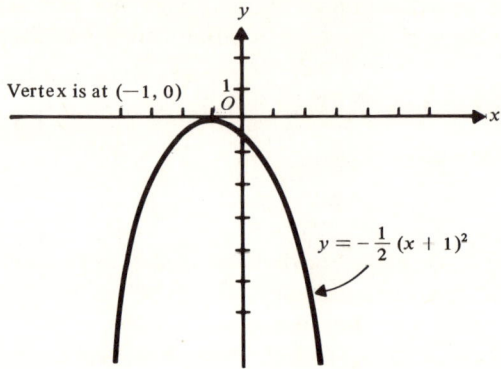

FUNCTIONS OF THE FORM $y = a(x - h)^2 + k$

Example 3-6

Graph $y = (x - 2)^2 - 3$.

This curve is the parabola $y = (x - 2)^2$ shifted down 3 units. But the graph of $y = (x - 2)^2$ is the parabola $y = x^2$ translated to the right 2 units. So the vertex of $y = (x - 2)^2 - 3$ is at $(2, -3)$. Note also that when $x = 0$, $y = 1$.

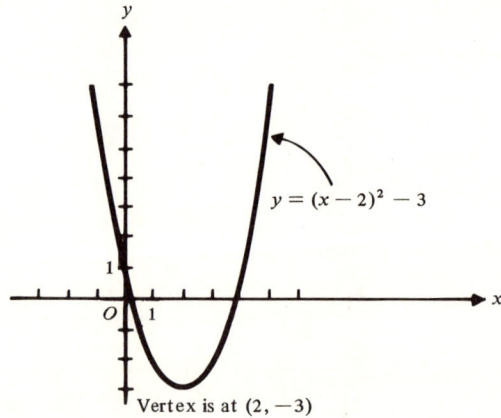

CONCLUSIONS

More generally, we see that the parabola $y = a(x - h)^2 + k$

(1) has vertex at (h, k).

(2) is the parabola $y = ax^2$ translated horizontally h units (to the right if $h > 0$, to the left if $h < 0$) and vertically k units (up if $k > 0$, down if $k < 0$).

(3) has a minimum point if $a > 0$, a maximum if $a < 0$.

(4) opens (or is concave) upward if $a > 0$, downward if $a < 0$.

(5) has a vertical *axis of symmetry* (a line such that the curve to the right of it is a mirror image or reflection of the curve to the left) whose equation is $x = h$. The axis of symmetry is often called just the *axis* of the parabola.

FUNCTIONS OF THE FORM $y = ax^2 + bx + c$

Example 3-7

Graph $y = x^2 - 4x$.

We factor, getting $y = x(x - 4)$, and note that the x-intercepts are 0 and 4. Since this curve has a vertical axis of symmetry, it must be the line midway between $x = 0$ and $x = 4$; that is, the line $x = 2$. When $x = 2$, $y = 2^2 - 4(2) = -4$. So the vertex is at $(2, -4)$. With these three points, we can sketch the parabola quickly.

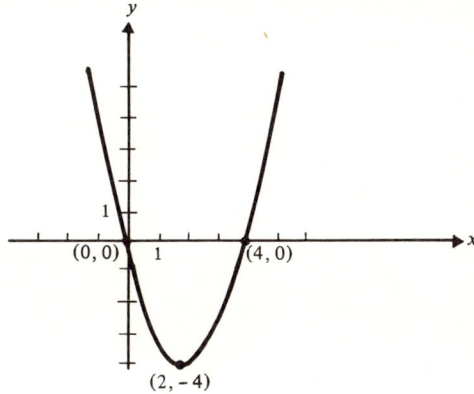

Example 3-8

Sketch $y = -2x^2 - 6x$.

We factor, removing the coefficient of x with its sign: $y = -2x(x + 3)$. Since $y = 0$ if $x = 0$ or $x = -3$, the axis of symmetry is $x = -\frac{3}{2}$. The vertex is $(-\frac{3}{2}, \frac{9}{2})$. The curve opens down. It is shown at left on page 28.

Example 3-9

Sketch $y = -2x^2 - 6x + 3$.

We need only recognize that this is the curve of Example 3-8 shifted up 3 units. (Why?) Plotting the y-intercept helps in sketching: when $x = 0$, $y = 3$. The vertex is $(-\frac{3}{2}, \frac{15}{2})$ and is a maximum. The curve is shown at right on page 28.

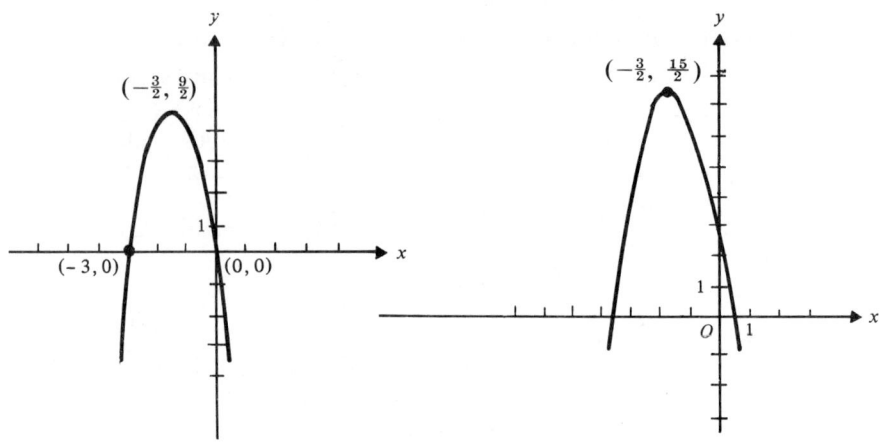

CONCLUSIONS

In summary, we see that the parabola $y = ax^2 + bx + c$ can be easily sketched if we sketch the related parabola $y = ax^2 + bx$ and then shift this curve vertically c units, up if $c > 0$, down if $c < 0$. The easiest way to sketch $y = ax^2 + bx$ is by finding and plotting its x-intercepts, then locating the axis (of symmetry) midway between these intercepts.

EXERCISES

Sketch the graphs of the functions in each single exercise, 1 through 11, on the same set of axes:

1. $y = x^2$, $y = 2x^2$, $y = 0.8x^2$
2. $y = -x^2$, $y = -\frac{1}{2}x^2$, $y = -1.5x^2$
3. $y = x^2$, $y = x^2 + 1$, $y = x^2 - 5$
4. $y = x^2$, $y = (x - 2)^2$, $y = (x + 3)^2$
5. $y = -x^2$, $y = -(x - 3)^2$, $y = -(x + 1)^2$
6. $y = x^2$, $y = (x - 2)^2 + 1$
7. $y = 2x^2$, $y = 2(x + 1)^2 - 1.5$
8. $y = -0.5x^2$, $y = -0.5(x - 4)^2 + 2.5$
9. $y = x^2 - 2x$, $y = x^2 - 2x + 2$
10. $y = -(x^2 + 4x)$, $y = -x^2 - 4x - 5$
11. $y = 2(x^2 - x)$, $y = 2x^2 - 2x - 3$

Sketch graphs of the following functions (12 through 14):

12. $f(x) = 500 + 10x - \dfrac{x^2}{2}$

Goal 3 | To Graph Quadratic Functions

13. $f(x) = 120 + 60x - x^2$
14. $f(x) = 150x - 2.5x^2$

In 15 through 20, find the vertex and axis of symmetry:

15. $y = x^2 + 5$
16. $y = (x - 3)^2 + 9$
17. $y = 4x^2 - 8x$
18. $y = 4x^2 - 8x + 3$
19. $y = 500 + 10x - \dfrac{x^2}{2}$
20. $y = 120 + 40x - x^2$

21. A stone thrown directly upward from the surface of a certain small planet, with an initial velocity 80 feet per second (ft/sec), reaches a height of h ft in t sec, where $h = 80t - t^2$. Graph the function. How long will it take the stone to reach its maximum height? What is the maximum height?

22. A farmer has 24 ft of fencing to enclose a rectangular pigpen. Express the area of the pen in terms of its length x, and sketch the graph of this area function. What length will provide maximum area?

23. Suppose it is determined that the cost in cents per mile of driving a car at speed x mph is given by $C(x) = 0.018x^2 - 1.44x + 40$. Find the most efficient speed for this car. HINT: The most efficient speed is that which costs least per mile; for what x is C least?

24. Sketch on the same set of axes: (a) $x = y^2$; (b) $x = y^2 + 3$; (c) $x = (y - 2)^2 - 1$. Does any of these graphs define y as a function of x?

25. How can you tell whether a graph represents a function?

26. By definition, $y = \sqrt{x}$ is nonnegative for $x \geq 0$. Sketch the graph. Is it a function? How is it related to $y^2 = x$?

27. Sketch on the same set of axes: (a) $y = \sqrt{x}$; (b) $y = \sqrt{x} + 3$; (c) $y = \sqrt{x - 1}$; (d) $y = \sqrt{x - 1} - 2$.

28. A cost analyst decides that the cost of producing x gadgets is given by a function of the form $C(x) = ax^2 + bx + c$, where C is in dollars. Find the function if it costs \$980 to produce 100 gadgets, \$1060 to produce 300 gadgets, and \$1300 to produce 500 gadgets. What production level yields minimum cost?

TEST A

Graph:

1. $y = 2x^2 - 4$
2. $f(x) = -(x - 3)^2 + 3$
3. $g(x) = x^2 - 4x + 1$
4. $y = 1 - 6x - 2x^2$

Without sketching, find the vertex of each of the following parabolas and tell whether it is the highest or lowest point on the curve:

5. $y = 0.01x^2 - 0.4x$

6. $y = 100 + 3x - \dfrac{x^2}{2}$

TEST B

Graph:

1. $y = -x^2 + 2$

2. $y = 4 - (x + 1)^2$

3. $g(x) = 2x^2 - 4x$

4. $f(x) = 3 + 2x - x^2$

Without sketching, find the vertex of each of the following parabolas and tell whether it is the highest or lowest point on the curve:

5. $y = 40 + 1.2x - 0.1x^2$

6. $y = 200x - \dfrac{-x^2}{1000}$

STUDY QUESTIONS

1. Prove that $f(x) = ax^2$ with $a > 0$ is increasing if $x > 0$ but decreasing if $x < 0$. [HINT: Let $x_1 < x_2$ and examine $f(x_2) - f(x_1)$ on the two intervals (a) $x > 0$ and (b) $x < 0$.] State and prove a similar theorem if $f(x) = ax^2$ and $a < 0$.

2. Find the axis of symmetry of the parabola $y = ax^2 + bx + c$. Find the coordinates of the vertex. Show that we can rewrite the equation in the form $y - k = a(x - h)^2$ with $h = -\dfrac{b}{2a}$. What is k? [HINT: $y = ax\left(x + \dfrac{b}{a}\right) + c$ indicates x-intercepts at 0 and at $-\dfrac{b}{a}$. So the vertex lies on the vertical line midway between them, $x = -\dfrac{b}{2a}$. Next complete the square.]

3. The parabola $y = ax^2 + bx + c$ may intersect the x-axis once, twice, or not at all. These intercepts are also called the *zeros* of the function. Show that the number of zeros is one, two, or zero, depending on whether $b^2 - 4ac$ is zero, positive, or negative. (HINT: Let $ax^2 + bx + c$ equal zero and find any roots by using the quadratic formula.)

4. A curve of $y = f(x)$ is said to be symmetric about the origin if, whenever the point (x, y) is on the curve, so is the point $(-x, -y)$. Translate statements about a function being even or odd (see Study Question 2 in Goal 1) into statements about symmetry.

GOAL 4

To graph cubic polynomials and other elementary functions

In analyzing a function to see how it changes, where it rises or falls, where it attains a maximum or minimum, or where it is concave up or down, it is often very helpful to be able to sketch its graph quickly. The calculus will furnish the tools that enable us to get all this information, even for quite complicated functions. Here we will graph cubics, functions of the type $y = 1/x$ or $y = 1/x^2$, and some other special functions, using the techniques so far developed. The translation techniques of Goal 3 will be especially useful.

POLYNOMIAL FUNCTIONS

A *polynomial* function is one of the form

$$f(x) = a_0 x^n + a_1 x^{n-1} + \ldots + a_{n-1} x + a_n$$

where n is a positive integer or zero and where the a's (called the *coefficients*) are constants. If $a_0 \neq 0$, then the *degree* of the polynomial is n. The domain of a polynomial function is the entire set of reals. The linear function $f(x) = mx + b$ (see Goal 2) is a polynomial function of degree 1; the quadratic function $f(x) = ax^2 + bx + c$ (see Goal 3) is a polynomial function of degree 2. Indeed, even a constant (other than 0) is a polynomial; its degree is 0.

In this Goal we graph polynomials only of the third degree; it will be more efficient to graph polynomials of higher degree after we have developed the calculus.

CUBICS

A polynomial function of degree 3 is called a *cubic*; it is of the form

$$f(x) = ax^3 + bx^2 + cx + d$$

We begin by sketching the graph of $y = x^3$, using the table of values given:

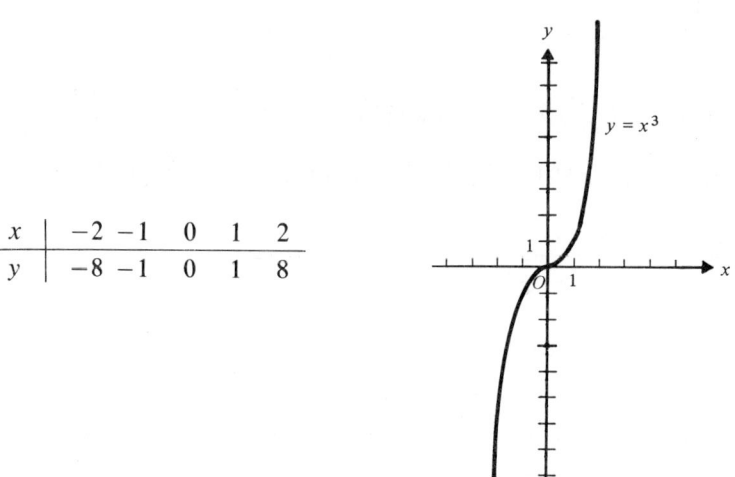

x	-2	-1	0	1	2
y	-8	-1	0	1	8

Note that the curve goes through the origin and that y is positive if x is positive but negative if x is negative. In fact,

$$f(-x) = -f(x)$$

for all x. This means that the graph is *symmetric about the origin*. Also, as x becomes very large positively, so does y; as x becomes very large negatively, so does y. Note further that the curve is concave down if $x < 0$, concave up if $x > 0$ (see p. 23). The point at which the concavity of a curve changes is called a *point of inflection*. The curve $y = x^3$ has a point of inflection at the origin.

Example 4-1

Sketch the graphs of (a) $y = x^3 + 3$; (b) $y = \frac{1}{2}(x + 1)^3 - 2$.

(a) The graph of $y = x^3 + 3$ is that of $y = x^3$ shifted up 3 units. (Why?) Note that the point of inflection is now at $(0, 3)$.

(b) The point of inflection of $y = \frac{1}{2}(x + 1)^3 - 2$ is at $(-1, -2)$, and the curve is half as "steep" as the curve of $y = x^3$. We have dotted in "new" axes here as an aid in sketching.

Example 4-1 shows us that the graph of $y = a(x - h)^3 + k$ has exactly the same shape as that of $y = ax^3$, but with its inflection point at (h, k). Just as for the parabola, the coefficient a determines the steepness of the curve. The curve $y = 2x^3$, for instance, is steeper than that of $y = x^3$ which, in turn, is steeper than that of $y = \frac{1}{2}x^3$.

Goal 4 | **To Graph Cubic Polynomials and Other Elementary Functions**

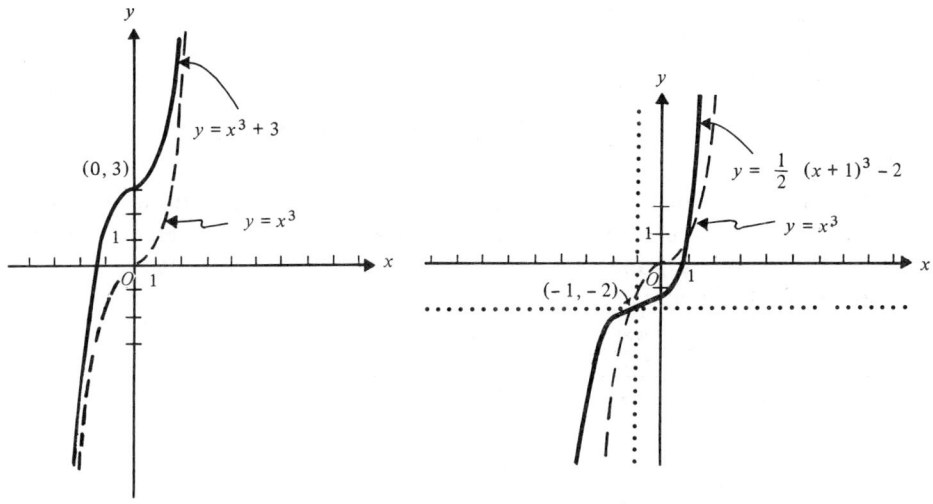

Example 4-2

Sketch (a) $y = -x^3$; (b) $y = -\frac{1}{4}(x - 5)^3 + 1$.

(a) The graph of $y = -x^3$ is the *reflection* of the curve $y = x^3$ through the *x-axis* (its mirror image). The curve $y = -x^3$ is concave up if $x < 0$, down if $x > 0$. Its point of inflection is, of course, $(0, 0)$.

(b) The graph of $y = -\frac{1}{4}(x - 5)^3 + 1$ is that of $y = -\frac{1}{4}x^3$ shifted to the right 5 units, then up one unit. It has exactly the same shape as $y = -\frac{1}{4}x^3$ but its point of inflection is at $(5, 1)$. Both curves are sketched below.

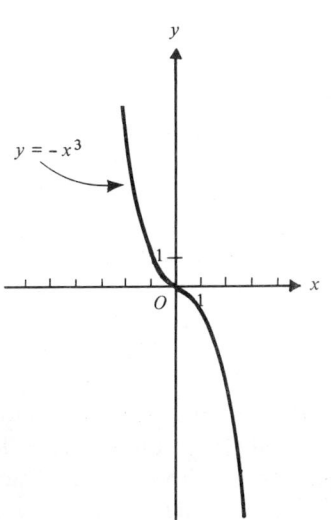 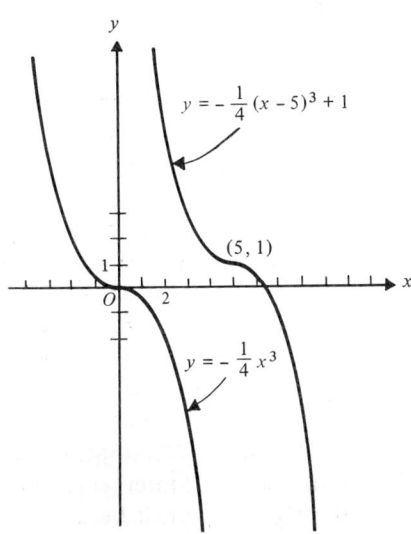

Example 4-3

Sketch $y = x^3 - 4x$.

We note that

$$y = x^3 - 4x = x(x^2 - 4) = x(x-2)(x+2)$$

So the x-intercepts are 0, 2, and −2. Once we've plotted these points, we can sketch the curve easily by plotting just a few more: one between each pair of intercepts, one point to the left of all of them, and one point to the right of all of them:

x	−3	−2	−1	0	1	2	3
y	−15	0	+3	0	−3	0	+15

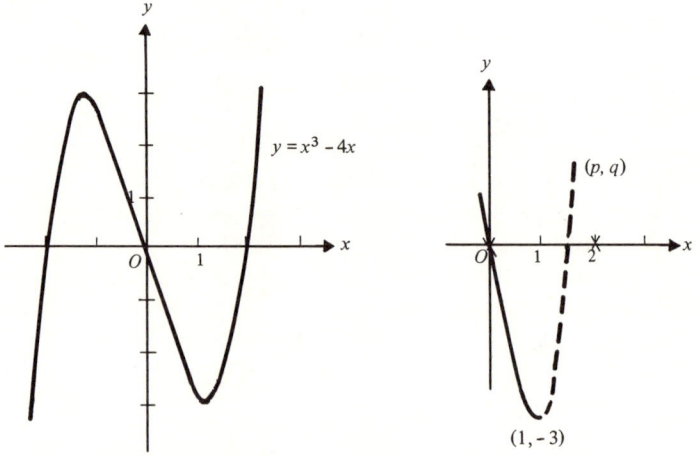

This table tells us that the curve is below the x-axis to the left of $x = -2$, above the x-axis between $x = -2$ and 0, below again between $x = 0$ and 2, and again above the x-axis to the right of $x = 2$. We can assert this because of an important theorem of the calculus (the Intermediate Value Theorem) that leads directly to the following fact: If you can draw a curve without lifting your pencil from the paper, then you cannot get from a point below the x-axis to one above (or vice versa) without crossing the x-axis. The function of this example is zero at $x = 0$ and at $x = 2$ but at no point between. Since y is negative at *one* point between $x = 0$ and $x = 2$ (it is −3 at $x = 1$), it must be negative at *every* point between $x = 0$ and $x = 2$. For if it were positive at any point on this interval, say (p, q), then the curve would have to cross the x-axis somewhere between $x = 1$ and $x = p$. But we have already found *all* the x-intercepts; there is none between $x = 0$ and $x = 2$! Therefore the sign of y must be constant on $0 < x < 2$. A similar statement may be made about the sign of y on the intervals $x < -2$, $-2 < x < 0$, and $x > 2$. The sign of y at any point within an interval is its sign throughout the interval.

Warning! Do not be misled into concluding that the left half of this curve attains its maximum midway between -2 and 0. We will soon see (again we need the calculus) that the maximum actually occurs just to the left of $x = -1$ (and the minimum just to the right of $x = 1$). Our graph-sketching techniques at present do yield the exact intercepts and the general shape of the curve, but not the exact shape.

Example 4-4

Sketch the graph of $f(x) = x^3 - 4x - 1$.

The curve is just that of the preceding example translated down one unit. Again be wary of conclusions about where the high or low points of the graph occur.

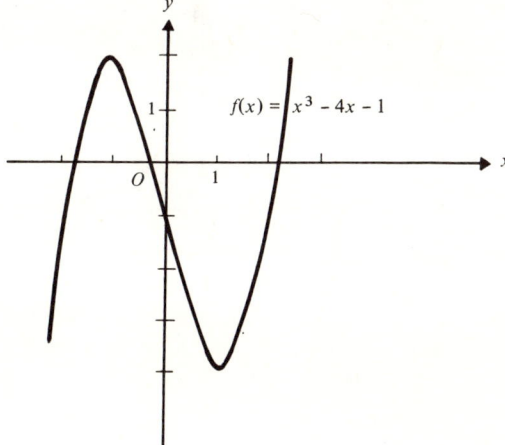

The techniques of Examples 4-3 and 4-4 enable us to sketch roughly any cubic $f(x) = ax^3 + bx^2 + cx + d$ provided we can sketch the "congruent" curve $y = ax^3 + bx^2 + cx$. We then translate this curve up d units if $d > 0$, down if $d < 0$.

RATIONAL FUNCTIONS

A rational function is one of the form

$$f(x) = \frac{P(x)}{Q(x)}$$

where $P(x)$ and $Q(x)$ are polynomials. The domain of f is the subset of $\mathcal{R}$ for which $Q(x) \neq 0$. Each of the following is a rational function:

$$f(x) = \frac{x^2 + 1}{x^2 - 1} \qquad y = \frac{4x}{x^2 + 1} \qquad h(x) = \frac{1}{x}$$

$$g(x) = \frac{1}{x^2} \qquad p(x) = \frac{1}{(x - 2)^2} \qquad q(x) = \frac{2x^2 - x + 7}{x^2 + 4x - 5}$$

(What is the domain of each of these functions?)

Example 4-5

Sketch the graphs of (a) $y = 1/x$, (b) $y = 1/x^2$

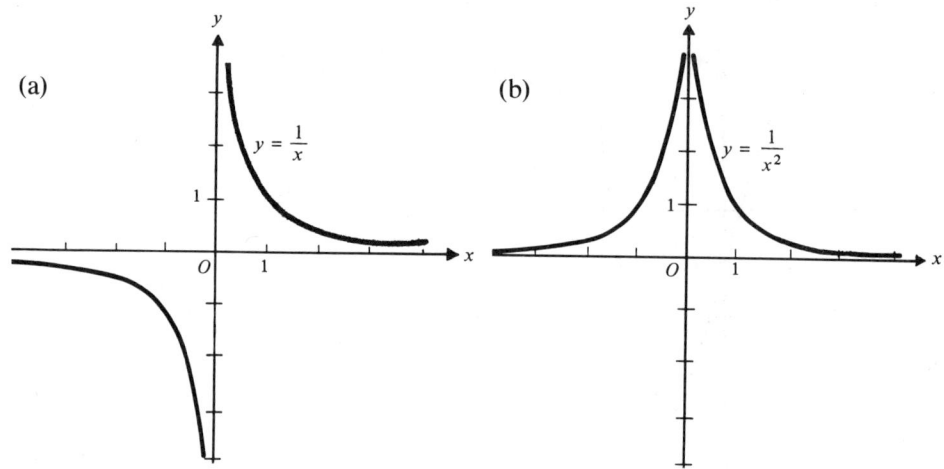

Note the following about the graphs of these functions:

	(a) $y = 1/x$	(b) $y = 1/x^2$
Domain:	$x \neq 0$	$x \neq 0$
Range:	The set of all reals except zero	The set of positive reals
Symmetry:	About the origin: $f(-x) = -f(x)$	About the y-axis: $f(-x) = f(x)$
Behavior for large positive or negative x:		Both curves approach the x-axis
Behavior near zero:	The curve approaches the y-axis. If x is close to zero but positive, y is large and pos-	The curve approaches the y-axis; the closer x is to zero, the larger y is

itive; if x is close to zero and negative, y is a large negative number

Concavity: Concave down if $x < 0$, concave up if $x > 0$. Concave up for all $x \neq 0$

NOTE: The curve with equation $xy = 1$ is called a *hyperbola*.

ODD AND EVEN FUNCTIONS

We know that if $f(-x) = -f(x)$ the graph of f is symmetric about the origin. We also say that such a function is *odd*. If $f(-x) = f(x)$, as is the case for $y = 1/x^2$, then its graph is symmetric about the y-axis. We call a function for which this is true an *even* function.

These functions are odd: $f(x) = x^3$, $g(x) = x^5$, $h(x) = 4x^3 - 3x$.

These functions are even: $f(x) = x^4$, $g(x) = 4x^2 + 1$, $h(x) = -3x^2$, $y = x^6 - 3x^4 - 5x^2 + 7$.

These functions are neither even nor odd: $f(x) = x^3 + 1$, $g(x) = x^2 - 4x$, $h(x) = (x - 2)^2$.

ASYMPTOTES

If a curve gets closer and closer to a horizontal line as x becomes larger and larger either positively or negatively, then this line is called a *(horizontal) asymptote* of the curve. If the curve gets closer and closer to the vertical line $x = c$ as x gets closer and closer to c, then $x = c$ is called a *(vertical) asymptote*. The x-axis is a horizontal asymptote of the curves $y = 1/x$ and $y = 1/x^2$, and both curves approach the y-axis, a vertical asymptote. A curve may also have oblique lines as asymptotes.

Example 4-6

Sketch the graphs of (a) $y = \dfrac{1}{x - 3}$; (b) $y = \dfrac{1}{(x + 1)^2} + 2$.

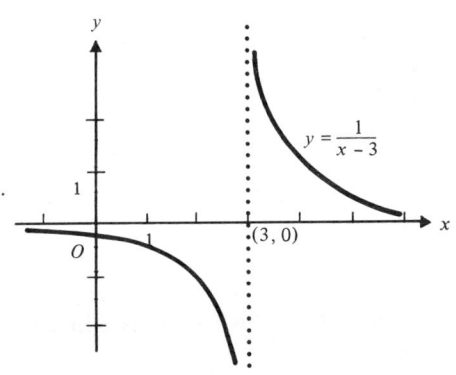
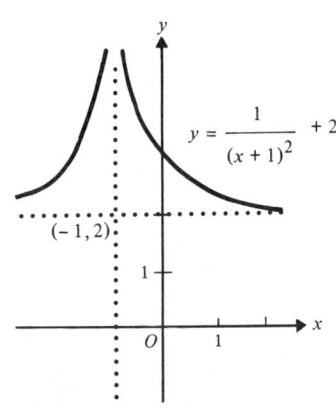

(a) The graph of $y = \dfrac{1}{x - 3}$ is that of $y = \dfrac{1}{x}$ shifted to the right 3 units. (Why?) The x-axis is a horizontal asymptote; the line $x = 3$ (dotted) is a vertical asymptote.

(b) The graph of $y = \dfrac{1}{(x + 1)^2} + 2$ is that of $y = \dfrac{1}{x^2}$ shifted to the left one unit, then up two units. The dotted lines are the asymptotes of the curve; their equations are $x = -1$, $y = 2$.

SPECIAL FUNCTIONS AND THEIR GRAPHS

We complete this goal by defining several types of functions that are not included above but that occur frequently and with which you need to be familiar.

Example 4-7

Graph the function

$$f(x) = \begin{cases} x^2 & \text{if } x \leq 2 \\ x + 2 & \text{if } 2 < x < 5 \\ 7 & \text{if } x \geq 5 \end{cases}$$

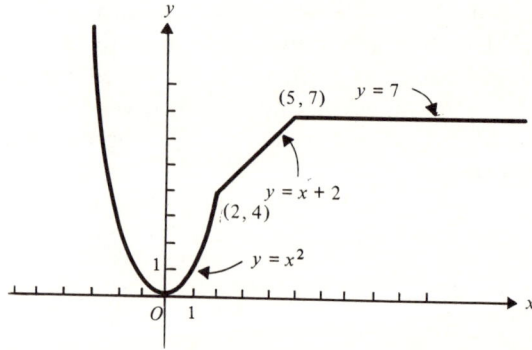

This function is typical of those defined differently on different parts of their domains. Be sure to use the pertinent definition when evaluating a function of this sort. Note that we have sketched only part of the parabola $y = x^2$ (up to $x = 2$), only part of the line $y = x + 2$ (between $x = 2$ and $x = 5$), and only part of the line $y = 7$ (for $x \geq 5$).

Practical applications that give rise to functions of this kind will appear in the exercises.

ABSOLUTE-VALUE FUNCTION: $f(x) = |x|$

The absolute value of a number (see p. 000) is just its magnitude or size:

$|3| = 3$ $|7| = 7$ $|-1| = 1$ $|-40| = 40$ $|0| = 0$

Goal 4 | To Graph Cubic Polynomials and Other Elementary Functions

The absolute value of zero is zero; the absolute value of every other number is positive. The domain of $f(x) = |x|$ is $\mathcal{R}$; its range is the set of nonnegative reals. Its graph consists of two rays meeting at right angles at the origin. Note that $f(x) = |x|$ is *even*, since $f(-x) = f(x)$.

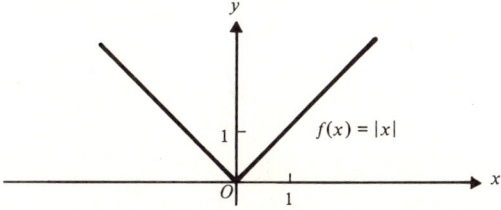

STEP FUNCTIONS

The graph of a step function consists of a finite or infinite number of horizontal line segments ("steps"). The *greatest-integer function* is a step function given by $f(x) = [x]$, where $[x]$ denotes the greatest integer not greater than x (or, the greatest integer that is less than or equal to x). Here is a table for a few selected elements in the domain of this function, which is $\mathcal{R}$:

x	-2	-1.9	-1.1	-1	-0.7	-0.001	0	0.99	0.999	1	1.001	1.999	2	2.9999999	3
$f(x) = [x]$	-2	-2	-2	-1	-1	-1	0	0	0	1	1	1	2	2	3

Carefully check the values of f recorded above, and try to draw the graph of this function. It should look like this (the small circles mean that the end points of the line segments are *not* part of the graph):

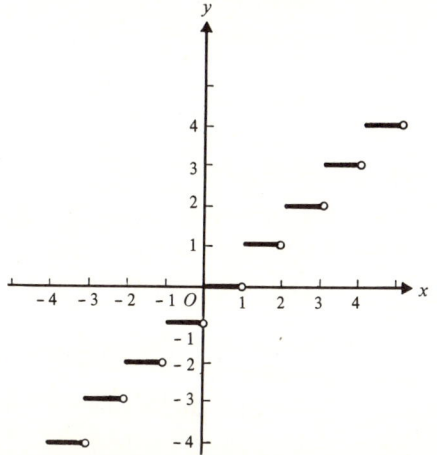

The exercises include a practical application of the greatest-integer function and other examples of step functions.

EXERCISES

Sketch the graphs in each exercise on the same set of axes:

1. $y = -x^3 + 2$, $y = -(x + 2)^3$
2. $y = x^3 - 4$, $y = (x - 4)^3$

Sketch on separate axes:

3. $y = (x - 2)^3 - 1$
4. $y = \frac{1}{2}x^3 + 1$

Find the coordinates of the point of inflection of each cubic and tell where the curve is concave up:

5. $f(x) = x^3$
6. $f(x) = -x^3 - 5$
7. $y = 2x^3 + 6$
8. $g(x) = -4(x - 3)^3$
9. $f(x) = 0.1(x + 2)^3 - 100$
10. $y = 500 - (x + 100)^3$

Sketch each pair on a single set of axes:

11. $y = 4x^3 - 4x$, $y = 4x^3 - 4x + 3$
12. $y = 4x - 4x^3$, $y = 4x - 4x^3 - 2$
13. $y = x^3 - 3x^2$, $y = x^3 - 3x^2 + 2$
14. $y = 3x^2 - x^3$, $y = -3 + 3x^2 - x^3$
15. $y = x(x + 1)(x - 2)$, $y = x(x + 1)(x - 2) + 2$

Sketch on separate sets of axes:

16. $y = 4 + \dfrac{1}{x + 3}$
17. $y = 2 - \dfrac{1}{x^2}$

Find the equations of any asymptotes:

18. $y = \dfrac{1}{(x - 2)^2}$
19. $y = 4 - \dfrac{1}{x}$
20. $y = 2 + \dfrac{1}{x - 3}$
21. $y = 3 - \dfrac{1}{(x - 4)^2}$

22. Which of the following functions are odd? even? neither?
    (a) $f(x) = 5x^2 + 1$
    (b) $g(x) = 3x + 1$
    (c) $y = 2x^3 - 4x$
    (d) $f(x) = \dfrac{x}{x + 1}$
    (e) $h(x) = |x|$
    (f) $y = [x]$
    (g) $f(x) = \sqrt{4 - x^2}$
    (h) $y = \dfrac{4}{x + 1}$

23. Complete the following two tables and plot the points on separate pairs of axes:

(a)

x	-3	-2.5	-2	-1	-0.2	0	0.5	1	1.9	2	2.5
$y = |x|$											

(b)

x	-2	-1.5	-1.1	-1	-0.5	-0.2	-0.001	0.9	1	1.99	2	2.5
$y = [x]$												

24. If h is defined over the reals by
$$h(x) = \begin{cases} 1 & \text{if } x > 0 \\ 0 & \text{if } x = 0 \\ -1 & \text{if } x < 0 \end{cases}$$
   (a) find $h(-3)$, $h(-\frac{1}{2})$, $h(0)$, $h(1)$, $h(\sqrt{2})$, and $h(\pi)$.
   (b) what is the range of h?
   (c) sketch the graph of h and tell why h is a function.

25. Sketch f if
$$f(x) = \begin{cases} x & \text{if } x \geq 0 \\ -x & \text{if } x < 0 \end{cases}$$
   How does your graph compare with that of $f(x) = |x|$?

26. Sketch f if
$$f(x) = \frac{x^2 - 1}{x - 1} \quad \text{if } x \neq 1$$
$$f(1) = 0$$
   [HINT: Simplify $f(x)$ for $x \neq 1$.] How does this graph compare with that of $y = x + 1$?

27. Sketch
$$f(x) = \begin{cases} x^2 - 4 & \text{if } x < 3 \\ \frac{4}{3}x + 1 & \text{if } 3 \leq x < 6 \\ 4 & \text{if } x \geq 6 \end{cases}$$

28. Sketch on the same set of axes: $y = |x|$; $y = |x - 5|$; $y = 3 - |x|$.

29. Sketch the graph of $y = [x] + 2$.

30. In 1978 first-class mail in the United States cost 15¢ for the first ounce or less and 13¢ for each additional ounce or fractions thereof. This defines the so-called *postage function*.
   (a) If c denotes the cost in cents of sending w ounces by first-class mail, complete the following table for the 1978 postage function.

w, ounces	$\frac{1}{4}$	$\frac{1}{2}$	0.9	1	1.1	$1\frac{1}{2}$	2	2.1	2.9	3	3.2
$c(w)$, cents											

   (b) What is the domain of c? Its range?
   (c) Graph the postage function.

31. Suppose a travel agent offers college students a charter tour to Hawaii as follows:

each person in a group of fewer than 60 pays $500;
each person in a group of between 60 and 100 inclusive pays $500 minus $2 for each person in excess of 60;
each person in a group larger than 100 pays $420.

If x denotes the number of students on the tour, express the agent's revenue as a function of x.

TEST A

Sketch each pair of curves on a single set of axes:

1. $y = x^3$, $y = (x - 4)^3 + 3$
2. $y = -\dfrac{1}{x}$, $y = 3 - \dfrac{1}{x+1}$

3. What are the coordinates of the point of inflection of the curve of $y = 4 - (x - 2)^3$? Where is it concave up?
4. Sketch the graph of $f(x) = |x - 2| - 3$.

TEST B

Sketch each pair of curves on a single set of axes:

1. $y = -\dfrac{1}{2}x^3$, $y = -\dfrac{1}{2}x^3 - \dfrac{1}{2}$
2. $y = 4 + \dfrac{1}{(x-1)^2}$, $y = \dfrac{1}{x^2}$

3. What are the coordinates of the point of inflection of the curve of $y = (x + 1)^3 - 5$? Where is it concave down?
4. Sketch the graph of $f(x) = [x]$. What is its domain? its range?

STUDY QUESTIONS

1. Recall that $f(x)$ is even if $f(-x) = f(x)$ but odd if $f(-x) = -f(x)$. Show that the product of any two odd functions is an even function. [HINT: Let $f(x)$ and $g(x)$ be any two odd functions; examine the product $f(-x)g(-x)$.]

2. The inverse of $f(x) = x^3$ is $g(x) = x^{1/3}$ (see Goal 1, Study Question 4). Sketch both curves on the same set of axes and discuss any apparent symmetries. In general, how is the graph of a function related to that of its inverse?

3. Consider the function $f(x) = \dfrac{|x|}{x}$. What is its domain? Sketch its graph and compare with that for Exercise 24 in this Goal.

GOAL 5

To operate on functions

We can add, subtract, multiply, and divide functions much as we do numbers. In this Goal we also define composition of functions and the inverse of a function; both are important in the development of the calculus. Recognizing whether a function is a sum, a product, or a quotient or is composite is useful because we can then break down complicated functions into simpler ones. We will exploit this technique repeatedly in later Goals.

COMBINING FUNCTIONS

If $f(x)$ and $g(x)$ are functions with the same domain, then we can find their sum $f(x) + g(x)$, their difference $f(x) - g(x)$, and their product $f(x)g(x)$. The quotient $f(x)/g(x)$ is defined for all x in the common domain except those for which $g(x) = 0$.

Example 5-1

If $f(x) = x^2 - 1$ and $g(x) = x - 2$, find $f(x) + g(x)$, $f(x) - g(x)$, $f(x)g(x)$, $-4f(x)$, $f(x)/g(x)$, $g(x)/f(x)$, and $-3g^2(x)$.

$f(x) + g(x) = x^2 - 1 + (x-2) = x^2 + x - 3$
$f(x) - g(x) = x^2 - 1 - (x-2) = x^2 - 1 - x + 2 = x^2 - x + 1$
$f(x)g(x) = (x^2-1)(x-2) = x^3 - 2x^2 - x + 2$
$-4f(x) = -4(x^2-1) = -4x^2 + 4$
$\dfrac{f(x)}{g(x)} = \dfrac{x^2 - 1}{x - 2} \qquad (x \neq 2)$
$\dfrac{g(x)}{f(x)} = \dfrac{x - 2}{x^2 - 1} \qquad (x \neq 1, -1)$
$-3g^2(x) = -3[g(x)]^2 = -3(x-2)^2 = -3(x^2 - 4x + 4) = -3x^2 + 12x - 12$

Since both f and g have domain $\mathscr{R}$, so do all of the functions listed above except those for which restrictions appear in parentheses.

Example 5-2

If $f(x) = \dfrac{1}{x+2}$ and $g(x) = \dfrac{1}{x^2-4}$, find $f(x) - g(x)$, $f(x)g(x)$, and $f(x)/g(x)$.

Note that the common domain of f and g is the set of reals different from 2 and from -2.

$$f(x) - g(x) = \frac{1}{x+2} - \frac{1}{x^2-4} = \frac{1 \cdot (x-2)}{(x+2)(x-2)} - \frac{1}{(x+2)(x-2)}$$
$$= \frac{x-2-1}{(x+2)(x-2)} = \frac{x-3}{x^2-4}$$

$$f(x)g(x) = \frac{1}{x+2} \cdot \frac{1}{x^2-4} = \frac{1}{(x+2)(x^2-4)}$$
$$= \frac{1}{(x+2)^2(x-2)} \quad \text{(Why?)}$$

$$\frac{g(x)}{f(x)} = \frac{\dfrac{1}{x^2-4}}{\dfrac{1}{x+2}} = \frac{1}{x^2-4} \cdot \frac{x+2}{1} = \frac{1}{(x+2)(x-2)} \cdot \frac{x+2}{1} = \frac{1}{x-2}$$

COMPOSITION OF FUNCTIONS

We know that the notation "$f(a)$" denotes the value of the function $f(x)$ when $x = a$; we obtain this value by replacing x wherever it occurs in $f(x)$ by a:

$$\text{if } f(x) = x^2 - 1, \text{ then } f(a) = a^2 - 1;$$
$$\text{if } f(x) = -2x^2 + 5x + 3, \text{ then } f(a) = -2a^2 + 5a + 3.$$

If $g(x)$ is a function, then $f(g(x))$ is the function obtained by replacing x by $g(x)$ wherever it occurs in $f(x)$. The function $f(g(x))$ is called the *composition* (or *composite*) *of f with g*. [Some authors denote $f(g(x))$ by $f \circ g$.]

If $f(x) = x^2$ and $g(x) = 3x + 1$, then $f(g(x)) = (g(x))^2 = (3x+1)^2 = 9x^2 + 6x + 1$.

We can also find $g(f(x))$: $g(f(x)) = 3(x^2) + 1 = 3x^2 + 1$.

In general, $f(g(x)) \ne g(f(x))$.

In many applications, y is expressed as a function of one variable, say, $y = f(u)$, and u is expressed as a function of another variable, $u = g(x)$. We can then write

$$y = f(u) = f(g(x))$$

Goal 5 | To Operate on Functions

Given a particular value of x, say x_0, we evaluate $f(g(x_0))$ in two steps:
(1) We first evaluate g at x_0; this is $u_0 = g(x_0)$;
(2) We then evaluate f at u_0; this is $f(g(x_0))$.
We can show the composition of f with g by the following diagram:

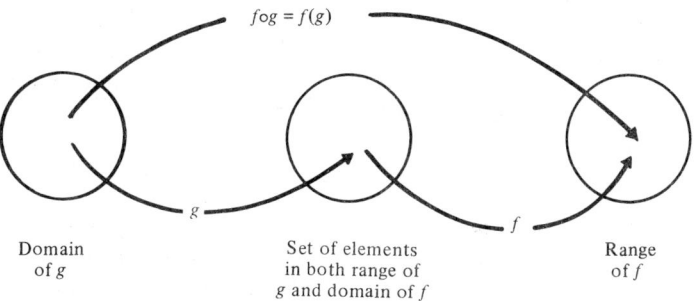

| Domain of g | Set of elements in both range of g and domain of f | Range of f |

This shows that the domain of $f(g(x))$ consists of only those elements x in the domain of g for which $g(x)$ is in the domain of f.

Example 5-3

If $y = f(u) = 1/(u - 1)$ and $u = g(x) = \sqrt{x}$, find $f(g(x))$ and specify its domain.

$$y = \frac{1}{u-1} = \frac{1}{\sqrt{x}-1} = f(g(x))$$

The domain of $f(g(x))$ is the set of all $x \geq 0$ except $x = 1$.

APPLICATIONS OF COMPOSITION

The following are examples of composition of functions.

(1) The revenue of a company is a function of the number of items produced daily. The number of items produced depends on the number of hours that have elapsed since the production run started. The revenue is thus a function of the time of production.

(2) The cost to a farmer of producing eggs is a function of the number of chickens he has. The number of chickens depends, in turn, on the cost of chicken feed. The cost of the eggs is therefore a function of the cost of the chicken feed.

(3) The number of oranges a grower's orchard yields is a function of the number of trees in the orchard. The number of trees planted is a function of the cost of land per unit area. The number of oranges is therefore a function of the cost of land.

(4) The revenue from sales of tickets for home football games during a season is a function of attendance. The attendance is a function of the number of wins of the home team during the previous season. The revenue from ticket sales is thus a function of the team's previous season's wins.

(5) The number of crimes committed daily in a certain city is a function of its population. The population is expressed as a function of the number of years since the founding of the city. The number of crimes is therefore a function of time.

Example 5-4

The Turpen Tine Company estimates that the cost y, in dollars, of producing u forks is $y = 30\sqrt{u} + 1000$. The number of forks produced weekly depends, in turn, on the number x of employees, where $u = 500x - 400$. Express the cost as a function of the number of employees. If Turpen Tine employs 25 workers, how many forks are produced weekly? At what cost?

$$y = 30\sqrt{u} + 1000 = 30\sqrt{500x - 400} + 1000.$$

With 25 workers, $x = 25$ and

$$u = 500(25) - 400 = 12,100;$$

$$y = 30\sqrt{12,100} + 1000 = 30(110) + 1000 = 4300.$$

So 25 workers produce 12,100 forks weekly at a cost of $4300. Note that if we replace x by 25 directly in the composite function $y = 30\sqrt{500x - 400} + 1000$ we also get 4300.

RECOGNIZING COMPOSITE FUNCTIONS

As noted above, it is frequently necessary, when applying the calculus, to recognize a function as a composition of other functions.

Example 5-5

For each of the following functions, find f and g so that $h(x) = f(g(x))$:

(a) $h(x) = (x - 1)^4$; (b) $h(x) = \sqrt{x^2 + 3}$; (c) $h(x) = \dfrac{1}{2x - 5}$;

(d) $h(x) = \dfrac{1}{4(3 - x^2)^3}$.

(a) $h(x)$ first subtracts one from x, then raises the result to the fourth power. Let $u = g(x) = x - 1$, $y = f(u) = u^4$. Then, as a check, $f(g(x)) = f(x - 1) = (x - 1)^4$.

(b) $h(x)$ adds 3 to the square of x, then takes the square root of this sum. Let $u = g(x) = x^2 + 3$, $y = f(u) = \sqrt{u}$. Then $f(g(x)) = f(x^2 + 3) = \sqrt{x^2 + 3}$.

(c) $h(x)$ subtracts 5 from twice x, then takes the reciprocal. Let $u = g(x) = 2x - 5$, $y = f(u) = 1/u$. Then $f(g(x)) = f(2x - 5) = \dfrac{1}{2x - 5}$.

(d) $h(x)$ first subtracts x^2 from 3, then takes the reciprocal of 4 times the cube of this result. Let $u = g(x) = 3 - x^2$, $y = f(u) = 1/4u^3$. We check:
$$f(g(x)) = f(3 - x^2) = \dfrac{1}{4(3 - x^2)^3}.$$

The choice of f and g is not unique. Thus, in (d) we could have let $u = g(x) = (3 - x^2)^3$ and $y = f(x) = 1/4u$. Check: $f(g(x)) = f[(3 - x^2)^3] = \dfrac{1}{4(3 - x^2)^3}$.

INVERSE OF A FUNCTION*

Suppose f is a function with domain X. Let Y be the set of all values of f. If the graph of $y = f(x)$ is such that every horizontal line $y = y_0$, where y_0 is a number in Y, cuts the graph of f at only one point, then f is said to be *one-to-one*.

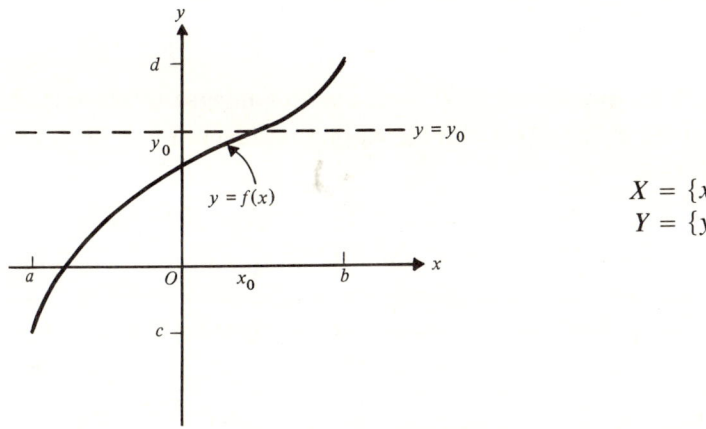

$X = \{x \mid a \leq x \leq b\}$
$Y = \{y \mid c \leq y \leq d\}$

Every horizontal line between $y = c$ and $y = d$ cuts the graph of $y = f(x)$ only once. Since each element in Y is associated with only one x in X, we can define a function g with domain Y and range X such that

$$g(y_0) = x_0 \quad \text{if} \quad f(x_0) = y_0$$

---

*This topic was the subject of Study Questions 4 and 5 of Goal 1.

The function g thus maps y_0 into x_0 if f maps x_0 into y_0.

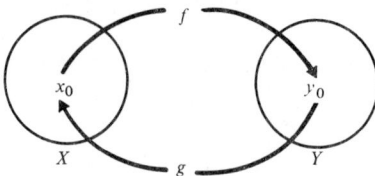

The function g, which in a sense "undoes" f, is called the *inverse* (or *inverse function*) of f. A function has an inverse only if it is one-to-one. NOTE: If g is the inverse of f, then f is the inverse of g.

Example 5-6

Find the inverse of each of the following functions that is one-to-one, and write the inverse as a function of x: (a) $f(x) = x^3$; (b) $f(x) = 3x + 5$; (c) $f(x) = x^2$; (d) $f(x) = 1/x$.

(a) Let $y = x^3$. Then $x = y^{1/3} = g(y)$ is the inverse expressed as a function of y. Therefore $g(x) = x^{1/3}$.

(b) Let $y = 3x + 5$. Then $x = \dfrac{y-5}{3} = g(y)$ is the inverse function, in terms of y. So $g(x) = \dfrac{x-5}{3}$.

(c) Since $f(x) = x^2$ is not one-to-one, it does not have an inverse function.

(d) Let $y = 1/x$. Then $x = 1/y = g(y)$ is the inverse function in terms of y. Thus $g(x) = 1/x$. Note that $f(x) = 1/x$ is its own inverse.

GRAPHICAL RELATION OF A FUNCTION AND ITS INVERSE

If $y = f(x)$ is a one-to-one function and $f(a) = b$, then its inverse $g(x)$ maps b into a. Thus if (a, b) is a point on the graph of $y = f(x)$, then (b, a) is on the graph of $y = g(x)$. This means that the graphs of inverse functions are mirror images, with the line $y = x$ serving as the mirror. If we have the graph of a one-to-one function, we can use this symmetry to draw its inverse.

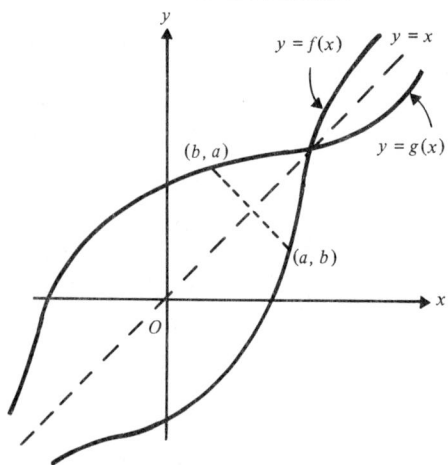

Goal 5 | To Operate on Functions

EXERCISES

In Exercises 1 through 4, find (a) $f(x) + g(x)$, (b) $f(x)g(x)$, (c) $3f(x) - g(x)$, and (d) $g(x)/f(x)$.

1. $f(x) = 3x - 2$, $g(x) = 1 - x$
2. $f(x) = x^2 + 3$, $g(x) = 5x$
3. $f(x) = \dfrac{1}{1 - x}$, $g(x) = 1 - x^2$
4. $f(x) = (2x - 1)^2$, $g(x) = \dfrac{1}{x}$

In Exercises 5 through 10, find $f(g(x))$ and $g(f(x))$.

5. $f(x) = x^3$, $g(x) = x + 2$
6. $f(x) = \dfrac{1}{3x - 1}$, $g(x) = \dfrac{1}{x}$
7. $f(x) = x^3$, $g(x) = x^{1/3}$
8. $f(x) = |x|$, $g(x) = \dfrac{1}{x - 1}$
9. $f(x) = \sqrt{x}$, $g(x) = x^2 + 4$
10. $f(x) = 5x - 2$, $g(x) = \dfrac{x + 2}{5}$

In Exercises 11 through 16, find functions f and g so that $h(x) = f(g(x))$.

11. $h(x) = (3x + 2)^6$
12. $h(x) = \sqrt{x^2 + 5}$
13. $h(x) = \dfrac{4}{\sqrt{2 - x}}$
14. $h(x) = (3x^3 - 0.5x^2 + 14)^5$
15. $h(x) = \dfrac{100}{4 - 3x}$
16. $h(x) = \dfrac{1}{(x^2 - x + 1)^3}$

For each of the functions in Exercises 17 through 26 that is one-to-one, find its inverse g as a function of x.

17. $f(x) = 4x + 1$
18. $f(x) = 7 - 2x$
19. $f(x) = 3x^2$
20. $f(x) = x^3 + 1$
21. $f(x) = \dfrac{1}{x + 2}$
22. $f(x) = |x|$
23. $f(x) = \sqrt{x}$
24. $f(x) = \sqrt{4 - x^2}$
25. $y = x + \dfrac{1}{x}$
26. $y = 1 + \dfrac{1}{x}$

27. Draw the graphs of the inverse of each function f sketched below:

(a)
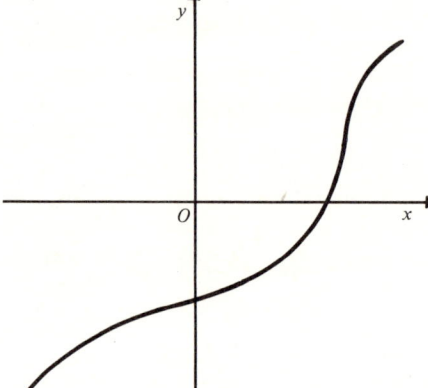

(b)
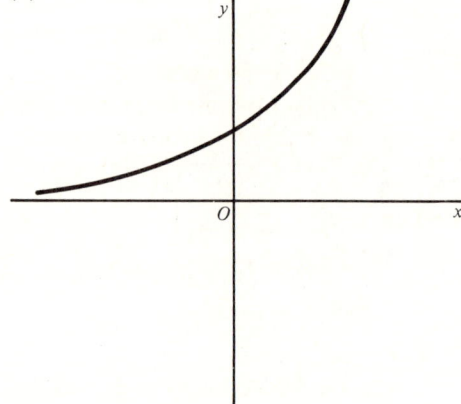

(c)
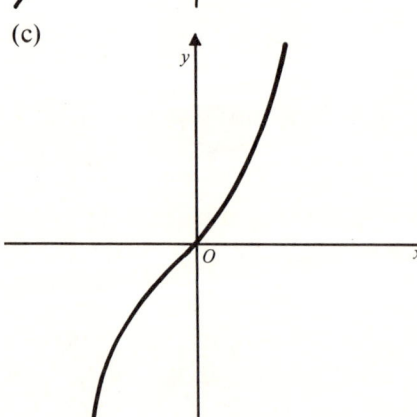

(d)
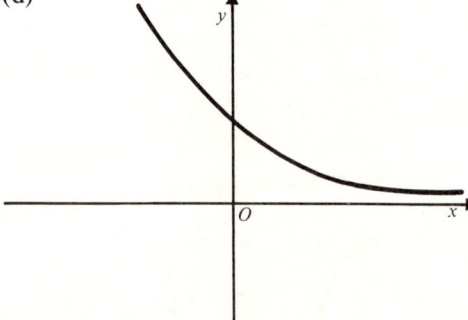

(e)

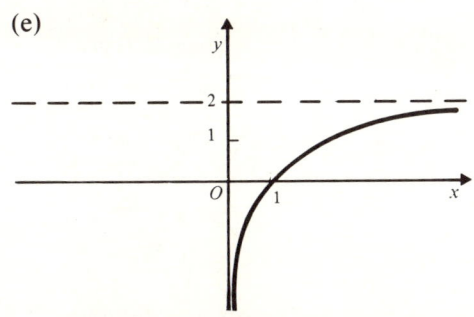

28. It costs the Cupid Doll Company $0.1x^2 - 0.2x + 600$ dollars to produce x dolls a day. The number of dolls produced after t hours is given by $x = 50t - 10$. Express the production cost C as a function of t. Find the production cost of an 8-hour workday.

29. The daily demand D for Poppupp toasters is given by $D(p) = \dfrac{4000}{p + 10}$, where p is the price in dollars. It is estimated that the price in t months will be $0.1t + 25$ dollars. Express the demand for Poppupp toasters as a function of t. How many Poppupp toasters will be bought a year from now?

TEST A

1. Find $f(x) + g(x)$, $f(x)g(x)$, $f(g(x))$, and $g(f(x))$ if $f(x) = 2x^2 - 1$ and $g(x) = \dfrac{1}{x + 1}$.

2. Find functions f and g so that $h(x) = \dfrac{3}{(1 - x - x^2)^2} = f(g(x))$.

3. Which of the following functions are one-to-one?
   (a) $f(x) = 2x^3 + 1$
   (b) $f(x) = 3|x|$
   (c) $f(x) = (x^2 - 1)^3$
   (d) $f(x) = 4x - 2$

4. Find the inverse of $f(x) = x^3 - 4$, express it as a function of x, and sketch both $y = f(x)$ and its inverse on the same set of axes.

TEST B

1. Find $f(x) - g(x)$, $f(x)/g(x)$, $f(g(x))$, and $g(f(x))$ if $f(x) = 3x - x^2$ and $g(x) = \dfrac{1}{x}$.

2. Find functions f and g so that $h(x) = \dfrac{4}{\sqrt{4x^2 - 1}} = f(g(x))$.

3. Which of the following functions are one-to-one?
   (a) $f(x) = 1 - x^2$
   (b) $f(x) = 3 + \dfrac{1}{x}$
   (c) $f(x) = \sqrt{1 + x}$
   (d) $f(x) = \frac{1}{2}x^3 - 4$

4. Find the inverse of $f(x) = 2 - x^3$, express it as a function of x, and sketch both $y = f(x)$ and its inverse on the same set of axes.

STUDY QUESTIONS

1. Let f be a one-to-one function with domain A, and let B be the set of values of f. If g is the inverse of f, show that $g(f(x)) = x$ and that $f(g(x)) = x$. HINT: Recall that $g(b) = a$ if $f(a) = b$. Let x_0 be any element in A such that $f(x_0) = y_0$, and find $g(f(x_0))$.

2. The following definition is often given for a one-to-one function: f is one-to-one if whenever $f(x_1) = f(x_2)$ it follows that $x_1 = x_2$. Show that this definition and the one on page 47 are equivalent.

3. Suppose f is a one-to-one function with domain A and B is the set of values of f. Let g be a one-to-one function with domain B and let C be the set of values of g. It can be proved that $g \circ f$ is a one-to-one function (with domain A and set of values C). Find functions f and g and sets A, B, C that satisfy the hypotheses, and verify that the conclusion holds for $g \circ f$. Provide an argument for the theorem. HINT: Use the definition in Study Question 2; show that if $(g \circ f)(x_1) = (g \circ f)(x_2)$ then $x_1 = x_2$.

GOAL **6**

To find an equation for a function described in words

Applying mathematics to real-life problems involves three steps. First the problem is translated into mathematical terms (this is called finding a *mathematical model*). Then a solution to the mathematical problem is found. Finally, this mathematical answer is interpreted in terms of the original problem.

In this goal we are concerned with only the first step. In particular, we pay special attention here to determining the function (or functions) involved in word problems. We consider a variety of situations from different fields and then summarize the procedure.

Example 6-1

The Original Reproductions Company can produce at most 30 paintings per week. They estimate that they can sell n paintings at p dollars each, where $p = 120 - 2n$. The total cost of producing n paintings is $500 + 8n + n^2$ dollars. Find the profit function.

If $P(n)$ denotes the profit, then we use the fact that

$$\text{profit} = \text{revenue} - \text{cost}$$

Since the revenue is the product of the number sold and the unit price, we have

$$\begin{aligned} P(n) &= n \cdot p - (500 + 8n + n^2) \\ &= n(120-2n) - (500 + 8n + n^2) \\ &= 120n - 2n^2 - 500 - 8n - n^2 \\ &= -3n^2 + 112n - 500 \end{aligned}$$

Note that $0 \leq n \leq 30$.

Example 6-2

Mrs. G. Thumm wants to put a fence around her rectangular garden to keep the rabbits out. She has 250 ft of fence and decides to use all of it. Express the area of her garden as a function of the length of one side of the garden.

Let x and y be the length and width of the garden. Then

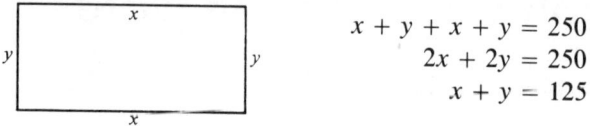

$$x + y + x + y = 250$$
$$2x + 2y = 250$$
$$x + y = 125$$

The area $A = x \cdot y$. To write A as a function of the length of one side, say x, we note that $y = 125 - x$; then

$$A(x) = x(125 - x) = 125x - x^2$$

Example 6-3

Suppose Mrs. Thumm starts a new rectangular garden whose area is 400 square feet (sq ft). If fencing costs $0.30 a foot, express the total cost of fencing the garden as a function of the length of one side.

Using the sketch above, we have that the area is $xy = 400$. The cost C of the fence is the total length (in feet) times the cost per foot. Since the length is $2x + 2y$, the cost, in dollars, at $0.30 per foot, is $0.30(2x + 2y)$ or $0.60x + 0.60y$. To write C as a function of the length of one side, say x, we solve $xy = 400$ for y:

$$y = \frac{400}{x}$$

Therefore

$$C(x) = 0.6x + 0.6\left(\frac{400}{x}\right) = 0.6x + \frac{240}{x}$$

Example 6-4

The owner of an apartment building with 60 units has found that he can rent all his apartments at $200 a month. However, for each $5 increase in the rent per unit he loses one tenant. Express his revenue as a function (a) of the number of $5 increases; (b) of the new rental rate.

We note first that the number of apartments that can be rented depends

Goal 6 | To Find an Equation for a Function Described in Words

on the rental rate. It is often useful, in problems of this sort, to consider one or two particular values of the variable as illustrated below.

(a) We let x be the number of $5 increases:

Number of $5 Increases	Number of Tenants Lost	Number of Occupied Apartments	Total Increase in Rent ($)	New Rent ($)
4	4	60 − 4	4 × 5 = 20	200 + 20
7	7	60 − 7	7 × 5 = 35	200 + 35
x	x	60 − x	x × 5 = 5x	200 + 5x

Therefore, for the revenue $R(x)$ we have

$$R(x) = \text{number of occupied apartments} \times \text{new rent}$$
$$= (60 - x)(200 + 5x)$$

Note how we broke this problem down into smaller parts. And note how easy it was to generalize, after considering specific values of x.

(b) We let r be the new rental rate:

New Rental Rate ($)	New Rate − Old Rate	Number of $5 Increases	Number of Tenants Lost	Number of Occupied Apartments
210	210 − 200 = 10	$\frac{10}{5} = 2$	2	60 − 2
245	245 − 200 = 45	$\frac{45}{5} = 9$	9	60 − 9
r	$r - 200$	$\frac{r - 200}{5}$	$\frac{r - 200}{5}$	$60 - \frac{r - 200}{5}$

Therefore, for the revenue as a function of the rate, $R(r)$, we have

$$R(r) = \text{number of occupied apartments} \times \text{new rent}$$
$$= \left(60 - \frac{r - 200}{5}\right) \times r$$
$$= \frac{r(500 - r)}{5}$$

It is rarely necessary to write out all the above. Working through the given conditions mentally with a specific value of the variable is often sufficient to yield the general expressions sought.

Note that we can easily show that these two functions for the revenue are equivalent: since r, the new rental, equals $200 + 5x$, where x is the number of $5 increases, $R(r)$ becomes, on substituting for r,

$$R(r) = \frac{r(500 - r)}{5} = \frac{(200 + 5x)[500 - (200 + 5x)]}{5}$$

$$= \frac{(200 + 5x)(300 - 5x)}{5}$$

$$= (200 + 5x)(60 - x) \quad \text{or} \quad (60 - x)(200 + 5x)$$

Example 6-5

Suppose it is estimated that the amount of waste dumped into a river is a quadratic function of the time. If 11.5 tons were dumped in a period of 5 days and 20.8 tons had been dumped after 8 days, find the amount of waste dumped in t days.

The first sentence tells us that the waste function is of the following form:

$$w(t) = at^2 + bt + c$$

We need to find a, b, and c, for which three conditions are necessary. That is just what we have:

when $t = 0$, $w = 0$
when $t = 5$, $w = 11.5$
when $t = 8$, $w = 20.8$

Substituting these pairs of values for t and w into the waste function yields

$$0 = a \cdot 0 + b \cdot 0 + c, \quad \text{so} \quad c = 0 \tag{1}$$

$$11.5 = 25a + 5b \tag{2}$$

$$20.8 = 64a + 8b \tag{3}$$

Solving (2) and (3) simultaneously (see Appendix) produces

$$a = 0.1 \quad \text{and} \quad b = 1.8$$

So the waste function is

$$w(t) = 0.1t^2 + 1.8t$$

Goal 6 | **To Find an Equation for a Function Described in Words**

Example 6-6

A sandbox with square base is to have volume 25 cubic feet (cu ft). The bottom is to be made of metal costing $2 per square foot (sq ft); the sides are to be constructed of wood costing $1 per sq ft. Express the cost as a function of the length of a side of the square bottom.

The volume of the box is $l \cdot l \cdot h$, or $l^2 h$. So $l^2 h = 25$. The bottom has $l \cdot l$, or l^2, sq ft, which at $2 per sq ft costs $2l^2$ dollars. In each side there are lh sq ft, which cost $1lh$ dollars; the four sides thus cost $4lh$ dollars. The total cost in dollars is therefore $2l^2 + 4lh$. To express the cost C as a function of l, we use the fact that $l^2 h = 25$, or $h = 25/l^2$. So

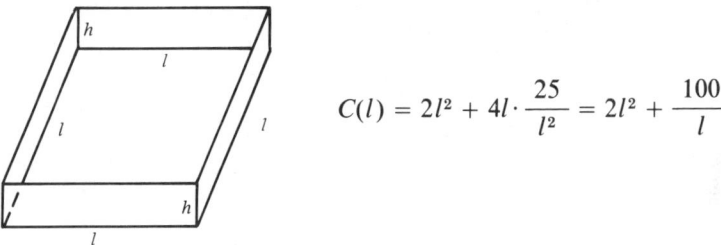

$$C(l) = 2l^2 + 4l \cdot \frac{25}{l^2} = 2l^2 + \frac{100}{l}$$

Example 6-7

A utilities company wants to deliver natural gas from a source S to a plant P located across a straight river 3 miles wide, then downstream 5 miles. It costs $4 per ft to lay the pipe in the river but only $2 per ft to lay it on land. Express the cost of laying the pipe (a) in terms of u; (b) in terms of v.

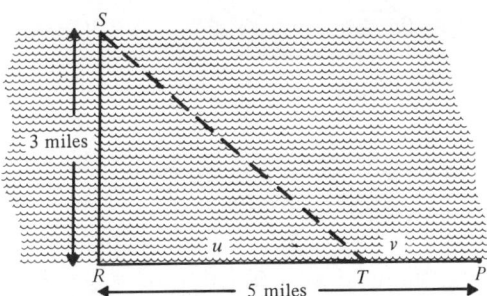

Note that the problem "allows" us to (1) lay all of the pipe in the river, along the line from S to P; (2) lay pipe along SR, in the river, then along RP on land; or (3) lay some pipe in the river, say, along ST, and lay the rest on land along TP. When T coincides with P, we have case (1), with $v = 0$; when T coincides with R, we have case (2), with $u = 0$. So case (3) includes both (1) and (2).

In any event, we need to find the lengths of pipe needed (that is, the distances involved); then we must figure out the cost.

(a) In terms of u.

	In the River	On Land
Distances:		
miles	$ST = \sqrt{9 + u^2}$	$TP = v = 5 - u$
feet	$ST = 5280\sqrt{9 + u^2}$	$TP = 5280(5 - u)$
Costs (dollars):	$4(5280)\sqrt{9 + u^2}$	$2[5280(5 - u)]$

So, if $C(u)$ is the total cost,

$$C(u) = 21{,}120\sqrt{9 + u^2} + 10{,}560(5 - u)$$
$$= 10{,}560(2\sqrt{9 + u^2} + 5 - u)$$

(b) In terms of v.

	In the River	On Land
Distances:		
miles	$ST = \sqrt{9 + u^2}$	$TP = v$
	$= \sqrt{9 + (5 - v)^2}$	
	$= \sqrt{34 - 10v + v^2}$	
feet	$ST = 5280\sqrt{34 - 10v + v^2}$	$TP = 5280v$
Costs ($)	$4(5280)\sqrt{34 - 10v + v^2}$	$2(5280v)$

So, if $C(v)$ is the total cost,

$$C(v) = 21{,}120\sqrt{34 - 10v + v^2} + 10{,}560v$$
$$= 10{,}560(2\sqrt{34 - 10v + v^2} + v)$$

It's apparent that expressing the cost as a function of u is simpler than expressing it as a function of v. Often a judicious choice of variable can minimize tedious computations or cumbersome algebra.

It is of interest to note that it would cost \$116,160 to lay pipe only in the river, and just under \$123,150 to lay it via lines SR and RP. The calculus will enable us to find the exact position of T for which the cost is least.

SUMMARY

The best way to develop skill in handling word problems is to try lots of them. In finding functions described in words, it will often help to consider the following suggestions:

Goal 6 | **To Find an Equation for a Function Described in Words** 59

(1) Read the problem carefully (often more than once); make a list of what information is given and what is sought. Introduce variables where necessary.

(2) Try to decompose the problem into smaller (usually easier!) problems. (For example: profit equals revenue minus total cost; revenue equals number sold times unit cost; total cost equals fixed cost plus variable cost; and so on.)

(3) Draw a sketch if relevant, labeling the parts with variables or with the given data.

(4) Express algebraically any relationships that are given; write down relevant formulas, using any information given.

(5) It may be useful to find the value of the function for one or two particular values of the variable so you can generalize your procedure.

(6) If your choice of variable leads to an unduly complicated function, consider an alternative.

(7) Be careful about units, both monetary and mensurational.

(8) Simplify your results arithmetically and algebraically.

EXERCISES

1. A rectangular garbage dump of area 300 square yards (sq yd) is to be fenced off. One side of the dump is along a river. Find the cost of the fence as a function of the side of the dump along the river, if (a) all four sides are to be fenced at a uniform cost of $4 a yd; (b) the side along the river does not need fencing, but the cost for the remaining three sides is $4 a yd; (c) the side along the river requires more substantial fencing costing $5 a yd, but the rest can be fenced at $4 a yd.

2. A rectangle is to be constructed having fixed perimeter P. Express the area as a function of the length of one side.

3. A rectangle is to be constructed of fixed area A. Express the perimeter as a function of the length of one side.

4. The pollution index in Smogville at 6 A.M. one morning was $P = 20$. If it rose linearly to $P = 45$ at 3 P.M., find P as a function of t, where t is the number of hours since 6 A.M.

5. A rectangular box with square bottom and top is to contain 24 cu ft. The material for the top costs 20¢ per sq ft, for the sides 30¢ per sq ft, and for the bottom 40¢ per sq ft. The labor costs $2. Express the cost of the box in terms of an appropriate variable.

6. Jim Tanwell plans to spend $120 on screening for a solarium with square base and flat roof. Since he can use part of an existing wall and concrete deck of his house, he will need to screen only three sides and the roof of the solarium. If the screening costs 40¢ a sq ft, express

the number of cubic feet (i.e., the volume) in the solarium as a function of the length of one side of the base.

7. If a California grower plants 20 orange trees per acre, then the average yield is 360 oranges per tree. For each additional tree per acre, in excess of 20, the grower gets 15 fewer oranges per tree. Express the total number of oranges per acre as a function (a) of x, where x is the number of trees per acre; (b) of y, where y is the number of trees per acre in excess of 20.

8. The Groovy Record Company estimates that it can sell 4000 albums of a certain rock group at $5 each. For each 10¢ decrease in price it figures it can sell 200 more albums. It costs the company $2 to produce an album, and its fixed overhead is $5000. Find functions for (a) the number of albums the company can sell; (b) its revenue; (c) its total cost; (d) its profit, (i) in terms of x, where x is the number of 10¢ decreases in selling-price and (ii) in terms of y, where y is the new selling price.

9. A charter boat outfit offers a weekend trip at the following rates: Each person in a group of 100 or fewer pays $35. For each person in excess of 100, everyone pays 25¢ less. If a group exceeds 100, express the revenue as a function (a) of the number of passengers on the trip; (b) of the number in excess of 100 on the trip.

10. The owner of a 50-room motel makes a profit of $4 per occupied room per day but loses $1 for each vacant room. Express the daily profit as a function of (a) the number of occupied rooms; (b) the number of vacant rooms.

11. An agent for a famous concert pianist hires a hall for a one-night performance. The agent knows from previous experience that the attendance decreases linearly with the price, that 3000 will attend if admission is $6 but only 2000 will attend if it is $8. Express the admission price as a function of x, the number who attend.

12. Suppose that in Example 11 the agent must pay the pianist $5000 and the owners of the hall $2000 plus 50¢ for each ticket sold, and that other costs come to $300. Find the agent's cost, revenue, and profit functions, each as a function of x.

13. Waikiki U-Drive has 150 compact cars for rent. The owner knows he can rent all the cars at $50 a week but that one car will not be rented for each $2 increase in the weekly rental. Express the owner's revenue as a function of an appropriate variable.

14. The promoter of a preseason professional football game knows that he can fill the 50,000-seat stadium by charging $12 per seat (on the average), but that he will sell only 40,000 tickets if he raises the price to $15. Write the price as a function of the number of tickets sold, assuming that they are related linearly.

15. Suppose that in Exercise 14 the expenses for the game come to $150,000 and on top of this the promoter must pay the stadium owners 60¢ for each ticket sold. Find the promoter's total cost, revenue, and profit functions in terms of the number of tickets sold.

16. A distributor of electric heaters estimates that he can sell 5000 heaters at $30 each. The price drops $4 for each additional 1000 heaters sold. Find the price function in terms of x, the number of heaters he sells.

17. A box with no cover is to be made from a very thin piece of sheet metal, 9 inches (in.) by 12 in., by cutting away a square from each corner and folding up the sides. Find the volume of the box as a function of x.

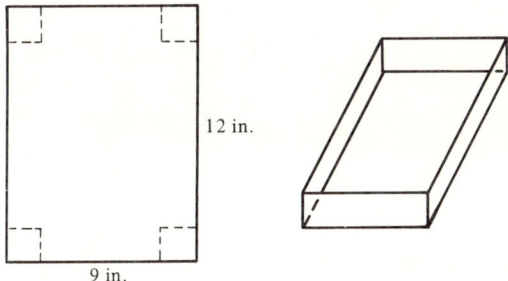

18. As part of his initiation, a fraternity pledge is told to get from point A on one side of a river to point B on the other side (see sketch) within 2 minutes (min). If he can paddle a canoe at 10 ft/sec and run at 20 ft/sec, express the total time it takes him to get from A to B in terms of x.

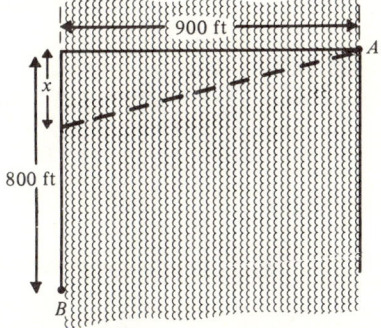

19. A transportation company wants to establish service from town T on one side of a river 2 miles wide to a factory F on the other side and upstream 4 miles. It is considering offering a combination of ferry and

bus service. If the ferry can do 20 mph and the bus 40 mph, express the time required for the entire trip in terms of x. Assume it takes 5 min to change from ferry to bus.

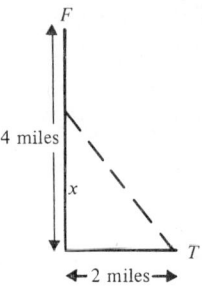

20. A printer is asked to prepare a rectangular poster from tagboard, with 96 sq in. of print. The margins at the top and bottom are to be 3 in.; the margin at each side is to be 2 in. Express the area of tagboard in terms (a) of x, where x is the height of the tagboard; (b) of u, where u is the height of the region containing the printed message.

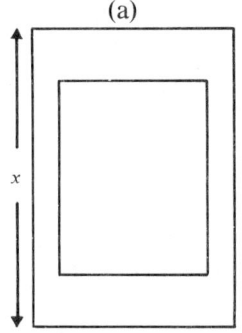 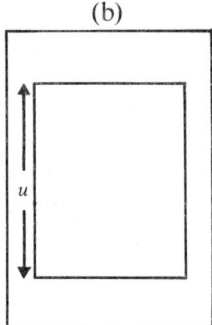

21. At noon on a certain day, two cars A and B are located as shown on perpendicular roads and are traveling toward the intersection. A is traveling at 30 kilometers per hour (km/h), B at 50 km/h. Express the distance between A and B as a function of t, the number of hours past noon.

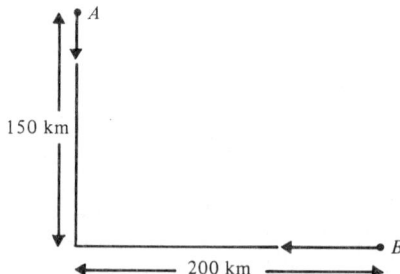

Goal 6 | To Find an Equation for a Function Described in Words

22. In Exercise 21 find the distance between A and B t hours after noon if A is traveling north and B is traveling east.
23. A rectangular box with square ends has combined length and girth of 84 in. (The length is the longest dimension; the girth is twice the sum of the other two dimensions.) Express the volume of the box in terms of an appropriate variable.
24. The volume of a box with cover and square base is V cubic units. Write the total surface area as a function of the length of a side of the base.
25. A piece of wire 20 centimeters (cm) long is cut into two pieces that are then bent to form a square and an equilateral triangle. Express the total area in terms of an appropriate variable.
26. When a substance changes into a new substance so that the rate of change is proportional to the product of the amount of the substance produced and the amount of original substance left, we say we have an *autocatalytic reaction*. Express the rate R of a particular autocatalytic reaction as a function of the amount x of the new substance present, if the factor of proportionality is 0.05% and the original amount of the old substance was 150 units.

TEST A

1. A farmer has 100 ft of fencing to close off a rectangular pasture, one side of which is along the bank of a river. He uses his fencing along the other three sides. Express the area of the pasture as a function of the length of the side along the river.
2. The owner of a small private kindergarten can enroll 25 children at $95 per child per month. She knows that for each $5 increase in the monthly rate she will lose one child. Express the revenue as a function of (a) the new monthly rate; (b) the number of $5 increases in the monthly rate.
3. A watch manufacturer has discovered that he can sell 10,000 watches a month at $15 but only 8000 at $20 each. (a) Express the price as a function of the number he sells, assuming that they are linearly related. (b) If his overhead is $5000 and his manufacturing and labor costs are $8 a watch, find his cost, revenue, and profit functions.

TEST B

1. A manufacturer wants to fill an order for 100,000 boxes. Each box is to have a square base, and its volume is to be 24 cu in. If the material for the bottom costs $1\frac{1}{2}$¢ per sq in., for the sides 1¢ per sq in., and for the top $\frac{1}{2}$¢ per sq in., express the cost as a function of an appropriate variable.

2. A travel agent offers a group tour to the Orient only if the group has at least 60 people. Each person in a group of 60 pays $800. For each person in excess of 60 people, everyone's fare is reduced by $10. Express the agent's revenue as a function of the number of people in the tour.

3. A community theater company knows from experience that there is a linear relationship between the number of people who will buy tickets for any performance and the admission price. At $3.50 each, 1000 will attend, but at $4.50 only 800 will buy tickets. (a) Find the admission price as a function of the number of tickets sold. (b) If it costs $500 to rent the theater per performance and other costs come to 20¢ per ticket sold, find the company's cost, revenue, and profit functions.

STUDY QUESTIONS

1. Show that of all the rectangles of a given perimeter, the one with largest area is a square. HINT: Let the fixed perimeter be P; then express the area of the rectangle as a function of the length of one side. Draw the graph of the area function to see where the function is a maximum.

2. A rumor spreads in a town of 1000 people at a rate proportional to the product of the number who already know it and the number who do not. If x is the number who know it at a particular time and R is the rate at which it spreads, then $R = kx(1000 - x)$. Show that the rumor is being spread fastest when 500 people have already heard it. HINT: The rumor is being spread fastest when the rate R at which it spreads is greatest. Find the vertex of the parabola obtained when the graph of R is sketched.

3. Specify the domains for the functions in the examples and in the odd-numbered exercises in this Goal.

Summary of Goals 1–6

You should now be able to

(1) distinguish functions from nonfunctions;
(2) find the domain of a function;
(3) given $f(x)$, evaluate $f(a)$, $f(-x)$, $f(x+h)$, and so forth;
(4) find the equation of a line, given two points or given its slope and one point;
(5) find the slope and y-intercept of a line from its equation;
(6) graph a line from its equation;
(7) recognize when lines are parallel or perpendicular from their equations;
(8) sketch the parabola of a quadratic function and identify its vertex and axis of symmetry;
(9) sketch the graph of a cubic polynomial;
(10) sketch the graph of $y = 1/x$, $y = 1/x^2$, $y = |x|$, $y = [x]$;
(11) sketch the graph of a function that is defined differently on different parts of its domain;
(12) add, subtract, multiply, or divide functions;
(13) find the composite functions $f(g(x))$ and $g(f(x))$;
(14) recognize a given function as the composition of other functions;
(15) find the inverse of $f(x)$, if the inverse exists;
(16) sketch the inverse of $f(x)$, given the graph of f;
(17) determine a function described in a word problem.

Review Exercises

GOAL 1

1. Which of the following define functions with domain A and range B?
   (a) A is a set of 100 cities in the United States and B is the set of temperatures, to the nearest degree Celsius, in those cities at noon on July 4, 1976.
   (b) $A = \{x \mid x \text{ is an integer}, 0 \leq x \leq 100\}$, and $B = \{y \mid y^2 = x\}$.

2. Find the domain of each function:
   (a) $f(x) = 2x^2 - x + 3$
   (b) $y = \sqrt{x - 1}$
   (c) $y = \dfrac{x - 1}{x^2 + 4}$
   (d) $y = \dfrac{2x - 3}{x^2 - 2x}$

3. For each of the following functions, find $f(1)$, $f(-2)$, $f(q)$, and $f(x + h) - f(x)$:
   (a) $f(x) = 2x + 1$
   (b) $f(x) = \dfrac{3}{x + 1}$
   (c) $f(x) = x^2 + 4$
   (d) $f(x) = (x - 2)^2$

GOAL 2

4. Find the slope, if it exists, of the line through
   (a) $(0, -4)$ and $(3, 0)$
   (b) $(2, 5)$ and $(-3, 5)$
   (c) $(x, 1 - 2x)$ and $(x + h, 1 - 2(x + h))$

5. Find the equation of the line
   (a) whose slope is $\tfrac{1}{2}$ and whose y-intercept is -4.
   (b) containing the points $(-1, 4)$ and $(2, -2)$.
   (c) perpendicular to the line $2x + y - 3 = 0$ and passing through $(4, 1)$.

6. Suppose a bike distributor will supply 500 bikes at $40 each but only 450 at $30 each. Find and graph the supply function if it is assumed to be linear. How many bikes will the distributor supply at $35?

7. For what value of k is the line $kx - 2y + 4 = 0$ parallel to the line $3y = x - 6$?

GOAL 3

8. Sketch graphs of the following on the same set of axes: $y = x^2$, $y = x^2 + 1$, $y = (x - 2)^2 - 3$.

9. A stone thrown directly upward from the ground with initial velocity 96 ft/sec reaches a height of h ft in t sec, where $h = 96t - 16t^2$. How long does it take the stone to reach maximum height? What is the maximum height?

Chap. 1 | Functions and Graphs

10. Without sketching, find the coordinates of the vertex of the parabola $y = 4 + x - x^2$ and tell whether it is the highest or the lowest point on the graph.

GOAL 4

11. Sketch on the same set of axes the graphs of $y = \tfrac{1}{2}x^3$ and $y = \tfrac{1}{2}(x - 1)^3 + 2$.
12. What are the coordinates of the point of inflection of the graph of $f(x) = 5(x + 2)^3 - 1$? Where is the graph concave down?
13. Sketch on separate axes:

    (a) $y = \dfrac{1}{x + 3}$ (b) $y = |2x - 5|$

GOAL 5

14. If $f(x) = x^2 + 3$ and $g(x) = \dfrac{1}{x - 1}$, find $f(x)g(x)$, $f(x)/g(x)$, $f(g(x))$, and $g(f(x))$.
15. Find functions f and g such that $h(x) = f(g(x))$ if

    (a) $h(x) = (0.1x - x^2)^3$ (b) $h(x) = \dfrac{3}{\sqrt{x^2 + 4}}$

16. Find the inverse of each one-to-one function and express the inverse as a function of x:

    (a) $y = |x|$ (b) $y = 2x^3 - 5$ (c) $y = \dfrac{1}{3 - x}$

17. Sketch the inverse of the function graphed below.

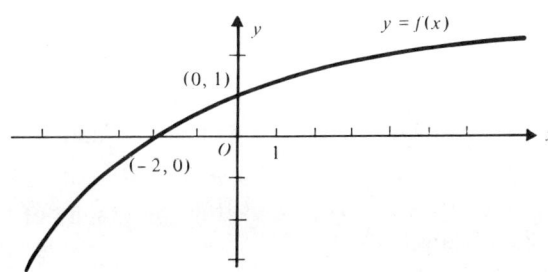

GOAL 6

18. The Dayline Delight offers river cruises to groups of 100 or fewer at $15 per person. For each passenger in excess of 100 the company reduces each person's fare by 5¢. Express the company's revenue as a function of the number of passengers if the ship to be used can accommodate 200 passengers.

19. A two-car carport is to be built with height 8 ft and 200 sq ft of floor space. The materials will cost $1 per sq ft for the floor and $5 per sq ft for the three sides. Other costs, including the roof and labor, will come to $525. Express the total cost as a function of a single variable.

20. A circus manager knows from experience that the number of people who will attend the circus depends linearly on the admission price; that 600 will attend when the price is $5 but only 500 will attend when it costs $6. If the manager figures that the costs per performance average out to $1.50 per ticket and that the fixed costs are $1000, determine the cost, revenue, and profit functions in terms of only one variable.

Chapter Test A

1. Find the domain of $f(x) = \dfrac{3x + 1}{x^2 - x}$.

2. If $f(x) = \dfrac{1}{x - 2}$, find and simplify $f(0)$, $f(-1)$, $f(x + h) - f(x)$.

3. Find the slope of the line through the points $(2, 5)$ and $(2 + h, 3(2 + h) - 1)$.

4. Find the equation of the line through the points $(-4, 0)$ and $(1, 5)$.

5. Find the slope and y-intercept of the line whose equation is $2x - y = 3$. Graph the line.

6. For what value of k is the line $2x + ky = 4$ perpendicular to the line $y = x + 5$?

7. Sketch the graphs of the following functions on separate axes:
   (a) $y = 2x - x^2$ (b) $y = 4x - x^3 + 1$ (c) $y = |x - 2|$

8. If $f(x) = 2x^2 + x - 1$ and $g(x) = 3 - x$, find $f(g(x))$ and $g(f(x))$.

9. If $h(x) = \sqrt{5x + 1}$, find functions f and g such that $h(x) = f(g(x))$.

10. Find the inverse of $f(x) = x^3 + 1$ and sketch the graphs of f and its inverse on the same set of axes.

11. A closed wooden box of volume $\dfrac{64}{3}$ cu ft is to be constructed so that

the length of the base is twice the width. If the wood costs $1.50 per sq ft, express the cost of the wood as a function of the width of the box.

12. A frisbee manufacturer knows that he can sell 100,000 frisbees at $2 each but only 80,000 at $2.50 each. (a) Express the price as a function of the number he sells, assuming that they are linearly related. (b) If the manufacturer's overhead is $3000 and his materials and labor average out to 50¢ a frisbee, determine the cost, revenue, and profit functions.

Chapter 2

Derivatives, Limits, and Continuity

The basic mathematical tool for measuring the rate of change of a function is the derivative. In this chapter we begin by approximating rates of change of specific functions. We then find the exact rate of change and note that we have used the derivative in the process. Since the derivative of a function is defined as a limit, it is important to understand what a limit is and how to evaluate limits. We also see here how limits are used to determine where a function is continuous; that is, where its graph is unbroken. We end the chapter by developing rules that enable us to find derivatives of polynomials quickly; we then apply these rules to practical problems involving rates of change.

Goal 7. To approximate the rate of change of a function at a particular instant
Goal 8. To find the slope of a curve at a designated point
Goal 9. To understand the definition of derivative and to compute derivatives from the definition
Goal 10. To evaluate limits
Goal 11. To determine whether a function is continuous at a particular point
Goal 12. To find derivatives of polynomials
Goal 13. To apply the derivative to problems involving rates of change
Summary
Review Exercises
Chapter Test A

GOAL 7

To approximate the rate of change of a function at a particular instant

One of the major branches of the calculus deals with the rate at which a function changes. In this Goal we consider four seemingly very different problems, each calling for a rate of change at a particular instant. These four examples, together with many others, are actually all disguises for a single basic problem, which is easily solved using a specific tool from the calculus. In this Goal we will not solve the problems but will instead obtain approximations to their solutions. Of special importance is the similar manner in which we attack the different problems. The approach we use here will help us develop the powerful tool needed to find the exact solutions.

THE EXAMPLES

Example 7-1

An object is dropped so that it falls a distance of $16t^2$ ft in t sec. Find its velocity at the end of 3 sec.

From physics we know that the *average velocity* of a moving object over some time interval is equal to the distance traveled divided by the time elapsed. But what is meant by the velocity at a particular instant? Our approach to this problem is to find the average velocity over smaller and smaller time intervals, starting from the point of time in which we are interested. For example, by time $t = 3$ the object has fallen $16(3)^2 = 144$ ft, by time $t = 4$ it has fallen $16(4)^2 = 256$ ft, and during the one-sec time interval from $t = 3$ to $t = 4$ it falls $256 - 144 = 112$ ft. So its average velocity during this one-sec interval is $\dfrac{112 \text{ ft}}{1 \text{ sec}}$, or 112 ft/sec.

73

Elapsed Time (sec)	Distance Covered (ft)	Average Velocity		
3	144	$\dfrac{(256 - 144)\text{ ft}}{(4 - 3)\text{ sec}}$	$= \dfrac{112}{1}$	$= 112$ ft/sec
4	256			

In a similar manner we now find the average velocities in smaller and smaller time intervals, each starting from the 3-sec point:

Elapsed Time (sec)	Distance Covered (ft)	Average Velocity		
3	144	$\dfrac{(196 - 144)\text{ ft}}{(3.5 - 3)\text{ sec}}$	$= \dfrac{52}{0.5}$	$= 104$ ft/sec
3.5	196			
3	144	$\dfrac{(153.76 - 144)\text{ ft}}{(3.1 - 3)\text{ sec}}$	$= \dfrac{9.76}{0.1}$	$= 97.6$ ft/sec
3.1	153.76			
3	144	$\dfrac{(144.9616 - 144)\text{ ft}}{(3.01 - 3)\text{ sec}}$	$= \dfrac{0.9616}{0.01}$	$= 96.16$ ft/sec
3.01	144.9616			
3	144	$\dfrac{(144.096016 - 144)\text{ ft}}{(3.001 - 3)\text{ sec}}$	$= \dfrac{0.096016}{0.001}$	$= 96.016$ ft/sec
3.001	144.096016			

Thus 112 ft/sec, 104 ft/sec, 97.6 ft/sec, 96.16 ft/sec, and 96.016 ft/sec are all approximations to the velocity of the object at the instant when $t = 3$.

Example 7-2

A colony of bacteria grows so that at the end of t hours there are $1000(1 + 0.3t + t^2)$ bacteria present. How fast is the colony growing at the end of 2 hours?

From biology we know that the average rate of growth over some time period is equal to the change in population size divided by the time elapsed. But how do we find the rate of growth at a particular instant? We approximate this by calculating the average rate of growth over smaller and smaller time intervals that start from the point of time in which we are interested. For example, during the time period from 2 to 2.5 hours the population grows from $1000[1 + 0.3(2) + 2^2]$, or 5600, up to $1000[1 + 0.3(2.5) + (2.5)^2]$ or 8000; the average rate of growth is therefore 4800 bacteria/hour:

Goal 7 | To Approximate Rate of Change of a Function at an Instant

Elapsed Time (h)	Number of Bacteria (bac)	Average Rate of Growth		
2	5600	$\dfrac{(8000 - 5600) \text{ bac}}{(2.5 - 2) \text{ h}}$	$= \dfrac{2400}{0.5}$	$= 4800 \text{ bac/h}$
2.5	8000			

In a similar manner we now find the average rates of growth in smaller and smaller time intervals, each beginning at the 2-h point. (Note that in order to apply these techniques we are allowing fractions of bacteria!):

Elapsed Time (h)	Number of Bacteria (bac)	Average Rate of Growth		
2	5600	$\dfrac{(6040 - 5600) \text{ bac}}{(2.1 - 2) \text{ h}}$	$= \dfrac{440}{0.1}$	$= 4400 \text{ bac/h}$
2.1	6040			
2	5600	$\dfrac{(5643.1 - 5600) \text{ bac}}{(2.01 - 2) \text{ h}}$	$= \dfrac{43.1}{0.01}$	$= 4310 \text{ bac/h}$
2.01	5643.1			
2	5600	$\dfrac{(5604.301 - 5600) \text{ bac}}{(2.001 - 2) \text{ h}}$	$= \dfrac{4.301}{0.001}$	$= 4301 \text{ bac/h}$
2.001	5604.301			
2	5600	$\dfrac{(5600.43001 - 5600) \text{ bac}}{(2.0001 - 2) \text{ h}}$	$= \dfrac{0.43001}{0.0001}$	$= 4300.1 \text{ bac/h}$
2.0001	5600.43001			

Thus 4800 bac/h, 4400 bac/h, 4310 bac/h, 4301 bac/h, and 4300.1 bac/h are all approximations to how fast the population is growing at the instant when $t = 2$.

Example 7-3

Suppose a company has determined that the cost of producing x gadgets is $(0.002x^2 - 0.4x + 1000)$ dollars. Find the rate of change of the cost when the production level (given by x) is 500 gadgets.

From economics, we know that the average rate of change of the cost over some change in production level is equal to the change in cost divided by the change in the number of items produced. But how do we find the instantaneous rate of change of the cost at a particular production level? We approximate the answer by finding average rates of change of the cost during small changes in the production level, each starting with the level in which we are interested—in this case, 500 gadgets. For example, going from 500 to 505 units changes the cost from $0.002(500)^2 - 0.4(500) + 1000$, or $1300, to $0.002(505)^2 - 0.4(505) + 1000$, or 1308.05. The average rate of change is

$$\frac{\$(1308.05 - 1300)}{(505 - 500) \text{ items}} = \$1.61/\text{gadget}.$$

In a similar manner we can calculate average rates of change of the cost over smaller and smaller changes in the production level (making the usual assumption that the cost function is defined for all positive x, not just for the positive integers). Here are the calculation already given and two more:

Number of Gadgets	Total Cost ($)	Average Rate of Change of Cost Per Gadget		
500	1300	$\dfrac{1308.05 - 1300}{505 - 500}$	$= \dfrac{8.05}{5}$	$= \$1.61/\text{gadget}$
505	1308.05			
500	1300	$\dfrac{1301.602 - 1300}{501 - 500}$	$= \dfrac{1.602}{1}$	$= \$1.602/\text{gadget}$
501	1301.602			
500	1300	$\dfrac{1300.16002 - 1300}{500.1 - 500}$	$= \dfrac{0.16002}{0.1}$	$= \$1.6002/\text{gadget}$
500.1	1300.16002			

So $1.61/gadget, $1.602/gadget, and $1.6002/gadget are all approximations to the rate of change of the cost per unit gadget when the production level is 500 units.

Example 7-4

Part of the graph of the function $f(x) = x^2 - x + 1$ in the neighborhood of $x = 4$ is sketched below. Find the slope of the curve at the point where $x = 4$.

From Goal 2 we know that the slope of a straight line through two points is equal to the difference in the y-coordinates divided by the difference in the x-coordinates. But what is meant by the slope of a *curve* at a particular point? If we let P be the point $(4, 13)$ on the curve, then our approach to the problem is to find the slopes of straight lines (often called *secant lines*) passing through P and through a second point on the curve, which we choose closer and closer to P. For example, when $x = 5$, $f(5) = 5^2 - 5 + 1 = 21$, and the slope of the line through $(4, 13)$ and $(5, 21)$ is $\dfrac{21 - 13}{5 - 4} = 8$. Here are four successive approximations (note that the sketches and scales are exaggerated and distorted to show the relative positions of the curve and the line):

Goal 7 | **To Approximate Rate of Change of a Function at an Instant**

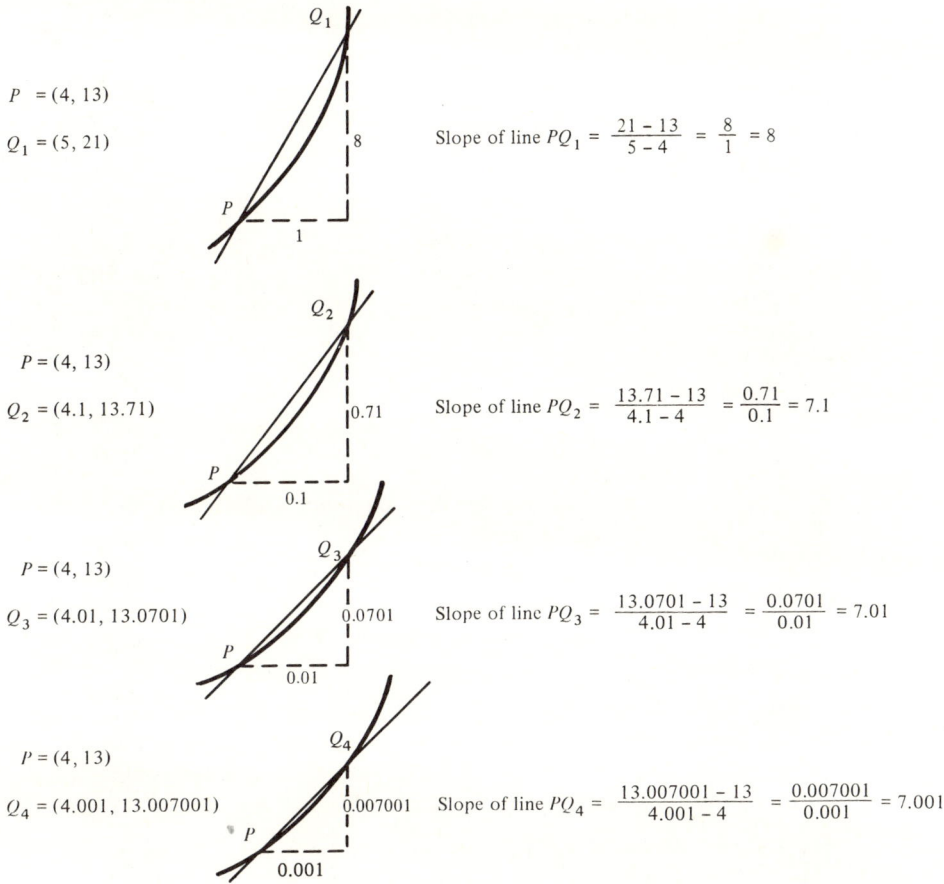

$P = (4, 13)$

$Q_1 = (5, 21)$

Slope of line $PQ_1 = \dfrac{21 - 13}{5 - 4} = \dfrac{8}{1} = 8$

$P = (4, 13)$

$Q_2 = (4.1, 13.71)$

Slope of line $PQ_2 = \dfrac{13.71 - 13}{4.1 - 4} = \dfrac{0.71}{0.1} = 7.1$

$P = (4, 13)$

$Q_3 = (4.01, 13.0701)$

Slope of line $PQ_3 = \dfrac{13.0701 - 13}{4.01 - 4} = \dfrac{0.0701}{0.01} = 7.01$

$P = (4, 13)$

$Q_4 = (4.001, 13.007001)$

Slope of line $PQ_4 = \dfrac{13.007001 - 13}{4.001 - 4} = \dfrac{0.007001}{0.001} = 7.001$

So 8, 7.1, 7.01, and 7.001 are all approximations to the slope of the curve at the point where $x = 4$.

INSTANTANEOUS RATE OF CHANGE

In the preceding discussion, we used the same technique to find approximations in four different problems. In the next few goals we will find exact answers. Note that each example above asks for an *instantaneous rate of change*:

in 7-1, the rate of change of distance with respect to time (called *velocity*) at the instant when $t = 3$;

in 7-2, the rate of change of population with respect to time (called *growth*) at the instant when $t = 2$;

in 7-3, the rate of change of cost with respect to production (called *marginal cost*) (at the instant) when 500 gadgets have been produced;

in 7-4, the rate of change of the y-coordinate of a curve with respect to the x-coordinate (called *slope*) (at the instant) when $x = 4$.

In the discussion of the four examples above we consistently used small intervals *beginning* at the points in which we were interested. If we had computed some average rates of change on intervals *ending* at the points of interest, we would have obtained similar approximations. Verify this statement by working through Example 7-1 on the intervals from 2.9 to 3, from 2.99 to 3, and from 2.999 to 3. (Your answers should be, respectively, 94.4, 95.84, and 95.99 ft/sec.)

Example 7-5

Consider the following four problems:

(1) Suppose $f(x) = x^2$ is the distance an object falls in x sec. Find approximations to its velocity when $x = 2$, using the intervals from (a) $x = 2$ to $x = 2.01$; (b) $x = 1.9$ to $x = 2$.

(2) Suppose $f(x) = x^2$ gives the population of a colony at the end of x h. Find approximations to the rate of growth when two h have elapsed ($x = 2$), using the same intervals as for problem (1).

(3) Suppose $f(x) = x^2$ gives the cost, in dollars, of producing x gadgets, where x is in thousands of gadgets. Find approximations to the rate of change of cost when 2000 gadgets have been produced ($x = 2$), using the same intervals as in (1).

(4) Suppose the graph of $f(x) = x^2$ has been drawn. Find approximations to the slope of this curve at the point (2, 4) (that is, when $x = 2$), using the same intervals as in (1).

How are these four problems related?

ANSWER: We find the approximations for all four different problems at the same time!

(a) On the interval from $x = 2$ to $x = 2.01$ the average rate of change is

$$\frac{f(2.01) - f(2)}{2.01 - 2} = \frac{4.0401 - 4}{0.01} = 4.01$$

(b) On the interval from $x = 1.9$ to $x = 2$, the average rate of change is

$$\frac{f(2) - f(1.9)}{2 - 1.9} = \frac{4 - 3.61}{0.1} = 3.9$$

Both (a) and (b) are approximations for

the velocity in problem (1) when $x = 2$;

Goal 7 | To Approximate Rate of Change of a Function at an Instant

the rate of growth in problem (2) when $x = 2$;
the marginal cost in problem (3) when $x = 2$;
the slope of the curve in problem (4) when $x = 2$.

EXERCISES

NOTE: A hand calculator will be helpful in doing the following exercises.

1. Suppose that t sec after blasting off, a rocket is $t^2/50$ miles above the earth. Find three approximations to the rocket's velocity, in miles per sec, 60 sec after lift-off, using the intervals from $t = 60$ to $t = 61$, from $t = 60$ to $t = 60.1$, and from $t = 59.9$ to $t = 60$.

2. Suppose that a certain animal population is growing at such a rate that after w weeks there are $500(1 + w + 2w^2)$ animals present. Find three approximations to how fast the population is growing at the end of 3 weeks, using the intervals from $w = 3$ to $w = 3.01$, from 3 to 3.001, and from 2.99 to 3.

3. If the cost of producing x gadgets is $0.001x^2 - 0.3x + 500$ dollars, find three approximations to the rate of change of the cost when the production level is 100 units, using the intervals from $x = 100$ to $x = 105$, from 100 to 101, and from 99 to 100.

4. Find three approximations to the slope of the graph of $f(x) = x^2 + x + 2$ when $x = 2$, using the intervals from $x = 2$ to $x = 2.5$, from 2 to 2.05, and from 1.99 to 2.

5. Find two approximations to the slope of the graph of $f(x) = x^2 + x + 1$ when $x = -1$, using the intervals from $x = -1.01$ to $x = -1$ and from -1 to -0.99.

6. Find three approximations to the slope of the graph of $f(x) = \dfrac{1}{x}$ when $x = 2$, using the intervals from $x = 1.9$ to $x = 2$, from 2 to 2.1, and from 2 to 2.001.

7. Suppose we are given a cord of length 5 in., varying in thickness in such a way that the left-hand c in. weighs c^2 oz. Thus the leftmost 1 in. weighs 1 oz, the leftmost 2 in. weigh 4 oz, the second inch from the left weighs $4 - 1$ or 3 oz, and the whole cord weighs 25 oz. From physics we know that the average density of a section of the cord is the weight of that section divided by the length of that section. Find three approximations to the density of the material, in ounces per inch, at a point 3 in. from the left end. Use the intervals from $c = 3$ to $c = 4$, from 3 to 3.1, and from 2.99 to 3.

8. Suppose that the number of gadgets that can be sold at a price p is given by $\dfrac{5000}{p + 1}$, where p is in cents. This is a *demand function*: at

a sales price of 9¢, 500 units will be sold, while at 99¢, only 50 units can be sold. Find three approximations to the rate of change in the demand function when the price is 99¢. The average rate of change over a change in price is found by taking the change in number of units and dividing by the change in the price. Use the intervals from $p = 99$ to $p = 100$, from 99 to 99.1, and from 98.9 to 99.

TEST A

NOTE: A hand calculator will be helpful for both tests.

1. Find three approximations to the slope of the graph of $f(x) = x^2 + 2x - 3$ when $x = 2$, using the intervals from $x = 2$ to $x = 2.5$, from 2 to 2.1, and from 1.99 to 2.

2. Suppose a colony of bacteria is growing at such a rate that when t hours have elapsed there are $100(1 + 0.1t + t^2)$ bacteria present. Find three approximations to how fast the colony is growing when 4 hours have elapsed, using the intervals from $t = 4$ to $t = 4.2$, from 4 to 4.1, and from 3.99 to 4.

TEST B

1. Find three approximations to the slope of the graph of $g(x) = x^2 - 3x + 1$ when $x = 3$, using the intervals from $x = 3$ to $x = 3.1$, from 3 to 3.01, and from 2.99 to 3.

2. Suppose that an object is moving in such a way that at the end of t sec it has moved $2t^2 + t$ ft. Find three approximations to the object's velocity at the end of 2 sec, using the intervals from $t = 1.8$ to $t = 2$, from 2 to 2.02, and from 2 to 2.001.

STUDY QUESTIONS

1. In Example 7-1 the approximations to the velocity of the object at the end of 3 sec were 112, 104, 97.6, 96.16, and 96.016 ft/sec. What would you guess to be the instantaneous velocity when $t = 3$? Similarly, how fast would you guess that the colony of bacteria is growing when $t = 2$; what would you guess to be the rate of change of the cost when the production level is 500 gadgets; and what would you guess to be the slope of the curve when $x = 4$?

    The above questions ask you to find, through your *intuition* rather than by exact methods, four *limits*. The concept of limit will be discussed more fully later.

2. Write down, in your own words, what the similarities and differences are in Examples 7-1 through 7-4 of this Goal.

GOAL 8

To find the slope of a curve at a designated point

As we showed in Goal 7, there are many problems, from a variety of fields, whose solutions can all be approximated using the same mathematical approach. In this Goal we develop a technique for finding the exact solution to one of these problems. We then apply the technique to finding the solutions to various disguised versions of the same problem.

Example 8-1

Find the slope of the curve $y = x^2$ at the point $P(3, 9)$.*

In Goal 7 we computed the slopes of lines through P and a neighboring point as follows:

$P = (3.9)$
$Q_1 = (3.1, (3.1)^2)$

Slope of line $PQ_1 = \dfrac{(3.1)^2 - 9}{3.1 - 3} = \dfrac{9.61 - 9}{3.1 - 3} = \dfrac{0.61}{0.1} = 6.1$

$P = (3, 9)$
$Q_2 = (2.99, (2.99)^2)$

Slope of line $Q_2P = \dfrac{9 - (2.99)^2}{3 - 2.99} = \dfrac{9 - 8.9401}{3 - 2.99} = \dfrac{0.0599}{0.01} = 5.99$

---

*If the point with coordinates 3 and 9 has been labeled "P," then we speak freely of the point $(3, 9)$, the point P, the point $P = (3, 9)$, or the point $P(3, 9)$, whichever is most convenient.

81

So 6.1 and 5.99 are two approximations to the slope of the curve $y = x^2$ at the point (3, 9). If approximations like these to the slope of a curve at a point P become closer and closer to a particular number as we choose the neighboring point closer and closer to P, then we call this number the *slope of the curve at P*. If we let h represent the difference between the x-coordinates of P and a neighboring point Q, then we can generalize our approximation as follows:

$P = (3, 9)$

$Q = (3 + h, (3 + h)^2)$

Slope of line $PQ =$

$$\frac{(3 + h)^2 - 9}{(3 + h) - 3}$$

For appropriate choices of h, this generalization gives us all possible approximations. For example, choosing $h = 0.1$ and then $h = -0.01$ yields the two approximations just above.

What happens to our approximations as Q is chosen closer and closer to P?—that is, as h is chosen closer and closer to zero? The fraction given above for the slope of the line PQ can be simplified as follows:

$$\frac{(3 + h)^2 - 9}{(3 + h) - 3} = \frac{(9 + 6h + h^2) - 9}{h} = \frac{6h + h^2}{h}$$

$$= \frac{h(6 + h)}{h} = 6 + h \qquad \text{if } h \neq 0$$

Now as h is chosen closer and closer to zero, the slope of line PQ, which equals $6 + h$, becomes closer and closer to 6. We say that the slope of the curve $y = x^2$ at the point (3, 9) is 6.

Example 8-2

Find the slope of the curve $y = x^3$ at the point $P(2, 8)$.

We again start with two approximations:

$P = (2, 8)$

$Q_1 = (2.2, (2.2)^3)$

Slope of the line $PQ_1 = \dfrac{(2.2)^3 - 8}{2.2 - 2} = \dfrac{10.648 - 8}{2.2 - 2} = \dfrac{2.648}{0.2} = 13.24$

Goal 8 | **To Find the Slope of a Curve at a Designated Point**

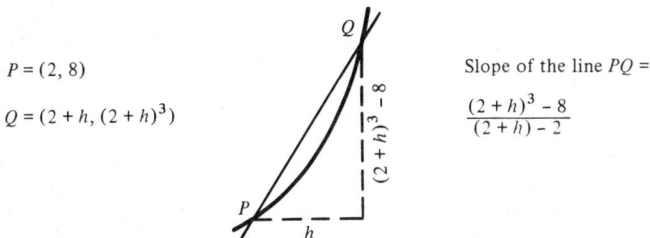

$P = (2, 8)$

$Q_2 = (1.9, (1.9)^3)$

Slope of line $Q_2P =$

$$\frac{8 - (1.9)^3}{2 - 1.9} = \frac{8 - 6.859}{2 - 1.9} = \frac{1.141}{0.1} = 11.41$$

So 13.24 and 11.41 are two approximations to the slope of $y = x^3$ at $(2, 8)$. The generalized approximation is this:

$P = (2, 8)$

$Q = (2 + h, (2 + h)^3)$

Slope of the line $PQ =$

$$\frac{(2 + h)^3 - 8}{(2 + h) - 2}$$

For appropriate choices of h, this generalization gives us all possible approximations. Above we used $h = 0.2$ and $h = -0.1$.

What happens to our approximations as Q is chosen closer and closer to P?—that is, as h is chosen closer and closer to zero? The fraction given above for the slope of line PQ can be simplified as follows:

$$\frac{(2 + h)^3 - 8}{(2 + h) - 2} = \frac{(8 + 12h + 6h^2 + h^3) - 8}{h} = \frac{12h + 6h^2 + h^3}{h}$$

$$= \frac{h(12 + 6h + h^2)}{h} = 12 + 6h + h^2 \qquad \text{if } h \neq 0$$

Now, as h is chosen closer and closer to zero, the slope of PQ, $12 + 6h + h^2$, becomes closer and closer to 12. We conclude that the slope of the curve $y = x^3$ at the point $(2, 8)$ is 12.

Example 8-3

Find the slope of the curve $y = 1/x$ at the point $(2, \frac{1}{2})$.

We go directly to the generalized approximation:

$P = \left(2, \dfrac{1}{2}\right)$

$Q = \left(2 + h, \dfrac{1}{2 + h}\right)$

Slope of line $PQ =$

$$\frac{\dfrac{1}{2 + h} - \dfrac{1}{2}}{(2 + h) - 2}$$

The equation for the slope can be simplified as follows (provided $h \neq 0$):

$$\frac{\dfrac{1}{2+h} - \dfrac{1}{2}}{2+h-2} = \frac{\dfrac{2}{2(2+h)} - \dfrac{2+h}{2(2+h)}}{h}$$

$$= \frac{\dfrac{2-(2+h)}{2(2+h)}}{h} = \frac{\dfrac{-h}{2(2+h)}}{h} = -\frac{1}{2(2+h)}$$

Again we ask: what happens to this slope, of line PQ, as h gets closer and closer to zero? Since $\dfrac{1}{2(2+h)}$ becomes closer and closer to $-\dfrac{1}{2(2)}$, or to $-\dfrac{1}{4}$, we can say that the slope of the curve $y = 1/x$ at the point $(2, \frac{1}{2})$ is $-\frac{1}{4}$.

NOTE: The procedure used in the above examples is called for both in the exercises that follow and in future Goals. It is therefore important that you understand it and any accompanying algebraic simplifications before going on.

EXERCISES

Draw a picture showing the generalized approximation for the slope of each curve at the designated point, then find that slope.

1. $y = 2x^2$ at $(1, 2)$
2. $y = 3x^2$ at $(2, 12)$
3. $y = x^2 + 1$ at $(2, 5)$
4. $y = x^2 + 3$ at $(1, 4)$
5. $y = 3x^3$ at $(2, 24)$
6. $y = 2x^3$ at $(1, 2)$
7. $y = x^3 + x^2$ at $(1, 2)$
8. $y = x^3 - x + 1$ at $(1, 1)$
9. $y = \dfrac{2}{x}$ at $\left(3, \dfrac{2}{3}\right)$
10. $y = \dfrac{5}{x}$ at $(5, 1)$
11. $y = \dfrac{1}{x+1}$ at $(0, 1)$
12. $y = \dfrac{1}{x-1}$ at $(2, 1)$
13. $y = 2x$ at $(5, 10)$
14. $y = 3x$ at $(-1, -3)$
15. $y = 4$ at $(-2, 4)$
16. $y = 19$ at $(3, 19)$
17. $y = x^2 + x + 1$ at $(0, 1)$
18. $y = 2x^2 - x + 2$ at $(-1, 5)$

Goal 8 | To Find the Slope of a Curve at a Designated Point

TEST A

Using the technique developed in this Goal, find the slope of the given curves at the indicated points.

1. $y = \dfrac{x^2}{2}$ at (4, 8)
2. $y = x^3 + 1$ at (2, 9)
3. $y = \dfrac{1}{x}$ at $\left(3, \dfrac{1}{3}\right)$

TEST B

Using the technique developed in this Goal, find the slope of the given curves at the indicated points.

1. $y = x^2 + x$ at (2, 6)
2. $y = 4x^3$ at (1, 4)
3. $y = \dfrac{2}{x}$ at (1, 2)

STUDY QUESTIONS

1. Using the technique developed in this Goal, (a) find the velocity at the end of 3 sec of a falling object whose distance from its starting point equals t^2 ft at the end of t sec; (b) find how fast a bacterial colony is growing at the end of 2 h if after t h there are t^3 bacteria present. HINT: For each of these problems write down several approximations (as in Goal 7), and then write down a generalized approximation (as in this Goal).

2. In Example 8-1 we noted that $6 + h$ becomes closer to 6 as h becomes closer to zero. The mathematical notation for this is

$$\lim_{h \to 0} (6 + h) = 6$$

which can be read "the limit of $6 + h$ as h approaches zero is 6" or "as h goes to zero, the limit of $6 + h$ is 6." Similarly, in Example 8-2 we could have written

$$\lim_{h \to 0} (12 + 6h + h^2) = 12$$

What would be the corresponding notation for Example 8-3?

3. We define the *tangent line to a curve at a designated point P* to be the straight line passing through P whose slope is equal to the slope of the curve at P. So, in Example 8-1, the tangent line to $y = x^2$ at the point (3, 9) is the straight line that passes through (3, 9) and has slope 6. This tangent line is thus given by $y - 9 = 6(x - 3)$, or $y = 6x - 9$.

Similarly, in Example 8-3, the tangent line to the curve $y = \dfrac{1}{x}$ at the point $(2, \tfrac{1}{2})$ is the straight line passing through $(2, \tfrac{1}{2})$ and having slope $-\tfrac{1}{4}$. Its equation is $(y - \tfrac{1}{2}) = -\tfrac{1}{4}(x - 2)$, or $y = -\tfrac{1}{4}x + 1$. What is the equation of the tangent line to the curve $y = x^3$ at the point $(2, 8)$ (Example 8-2)?

4. Find the equation of the tangent line to the curve $y = \sqrt{x}$ at the point $(4, 2)$. HINT: To find the slope of the curve at $(4, 2)$ we use our generalized approximation technique:

$P = (4, 2)$

$Q = (4 + h, \sqrt{4 + h})$

Slope of line $PQ =$

$\dfrac{\sqrt{4 + h} - 2}{(4 + h) - 4}$

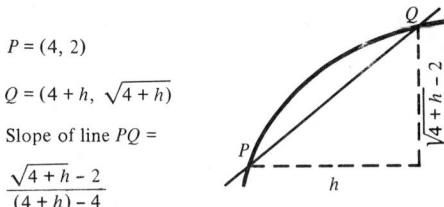

This fraction can be simplified as follows:

$$\dfrac{\sqrt{4 + h} - 2}{h} \cdot \dfrac{\sqrt{4 + h} + 2}{\sqrt{4 + h} + 2} = \dfrac{(\sqrt{4 + h})^2 - 2^2}{h(\sqrt{4 + h} + 2)}$$

$$= \dfrac{4 + h - 4}{h(\sqrt{4 + h} + 2)} = \dfrac{h}{h(\sqrt{4 + h} + 2)}$$

If $h \neq 0$, this equals

$$\dfrac{1}{\sqrt{4 + h} + 2}$$

The slope of the curve is then

$$\lim_{h \to 0} \dfrac{1}{\sqrt{4 + h} + 2} = \dfrac{1}{\sqrt{4} + 2} = \dfrac{1}{4}$$

GOAL 9

To understand the definition of derivative and to compute derivatives from the definition

In Goal 8 we developed a technique for finding the slope of a curve at a point. We noted that this technique would enable us to solve a variety of disguised versions of the same problem. The mathematical tool underlying this technique is called the *derivative*. In this Goal we define the derivative and use the definition to compute derivatives.

Suppose we have a function f and we want to find the slope of its graph at a designated point $P(x, f(x))$. As in earlier discussion, we look at a diagram:

$P = (x, f(x))$
$Q = (x + h, f(x + h))$

Slope of the line through P and Q =

$$\frac{f(x + h) - f(x)}{h}$$

What happens to the quotient $\dfrac{f(x + h) - f(x)}{h}$ as h becomes closer and closer to zero? If this quotient becomes closer and closer to some specific number, then we denote that number by

$$\lim_{h \to 0} \frac{f(x + h) - f(x)}{h}$$

which can be read as "the limit of f of x plus h, minus f of x, all divided by h, as h approaches zero."

The *derivative* of f is a function, denoted by f' (or by $\dfrac{df}{dx}$ or $D_x f$), whose value at x is defined to be the limit above. Thus

$$f'(x) = \lim_{h \to 0} \frac{f(x+h) - f(x)}{h}$$

Of course, it does not matter what symbol is used for the function or for the variable; that is, the following accord with the definition just as much as does the preceding:

$$g'(y) = \lim_{h \to 0} \frac{g(y+h) - g(y)}{h}$$

$$s'(*) = \lim_{h \to 0} \frac{s(*+h) - s(*)}{h}$$

and, for the value of the derivative of the function $j(x)$ at the particular value $x = 3$, we have

$$j'(3) = \lim_{h \to 0} \frac{j(3+h) - j(3)}{h}$$

We note immediately that $f'(x)$ is the slope of the curve of $y = f(x)$ at the point $(x, f(x))$. So, $f'(3)$ is the slope at $x = 3$, $f'(5)$ is the slope at $x = 5$, and so on.

The derivative is one of the most powerful and useful mathematical tools ever invented, and much of this book is concerned with how and when to apply it. In order to use the derivative effectively, you must develop two skills: given an applied problem, you must be able to determine if and in what way the solution calls for the derivative of a function; and then you must be able to find the derivative correctly. In this Goal we concentrate on the latter. In later Goals we will develop formulas that enable us to find derivatives of a wide variety of functions quickly and easily. Since, however, any formula for a derivative depends at bottom on the correct application of the definition, in this Goal we find all derivatives directly from the definition.

Given a function $f(x)$, finding its derivative directly from the definition involves three steps:

Step (1) Write down the quotient $\dfrac{f(x+h) - f(x)}{h}$. In doing this, you must take into consideration whether or not x is a specified value, and you must be able to evaluate f at $x + h$.

Step (2) Simplify the quotient. Without simplification, it can be difficult to calculate what will happen as h is chosen closer and closer to

Goal 9 | To Understand the Definition of Derivative and Compute Derivatives

zero, because h is the denominator and division by zero is not allowed. Usually the simplification involves algebraic manipulation.

Step (3) Find the limit of the simplified quotient as h "goes to zero."

Here are examples illustrating these three steps.

Example 9-1

Find $f'(x)$ if $f(x) = 2x^2 - 3$.

First we write down the quotient $\dfrac{f(x + h) - f(x)}{h}$:

$$\frac{[2(x + h)^2 - 3] - [2x^2 - 3]}{h}$$

Next we simplify:

$$\frac{[2(x^2 + 2xh + h^2) - 3] - [2x^2 - 3]}{h} = \frac{2x^2 + 4xh + 2h^2 - 3 - 2x^2 + 3}{h}$$

$$= \frac{4xh + 2h^2}{h} = 4x + 2h \qquad (h \neq 0)$$

And finally we compute the limit as h becomes closer and closer to zero: as $h \to 0$, $4x + 2h \to 4x$. Thus, if $f(x) = 2x^2 - 3$, then $f'(x) = 4x$.

Example 9-2

Find the slope of the curve of $y = 2x^2 - 3$ at $(2, 5)$.

We do not have to go through an entire calculation because we have already performed the necessary work above. Since $f'(x)$ is the slope of the curve at any point x, the answer is simply $f'(2) = 4(2) = 8$.

Example 9-3

What is $g'(y)$ if $g(y) = y^3 - y$?

First we write down the quotient $\dfrac{g(y + h) - g(y)}{h}$:

$$\frac{[(y + h)^3 - (y + h)] - [y^3 - y]}{h}$$

Next we simplify:

$$\frac{[y^3 + 3y^2h + 3yh^2 + h^3 - y - h] - [y^3 - y]}{h}$$

$$= \frac{3y^2h + 3yh^2 + h^3 - h}{h} = 3y^2 + 3yh + h^2 - 1 \qquad (h \neq 0)$$

And finally we compute the limit as h gets closer and closer to zero:

$$\lim_{h \to 0} (3y^2 + 3yh + h^2 - 1) = 3y^2 - 1$$

Thus, if $g(y) = y^3 - y$, then $g'(y) = 3y^2 - 1$.

Example 9-4

Find $j'(3)$ if $j(x) = 1/x$.

First finding the quotient and then simplifying, we have:

$$\frac{\frac{1}{3+h} - \frac{1}{3}}{h} = \frac{\frac{3}{3(3+h)} - \frac{3+h}{3(3+h)}}{h} = \frac{\frac{3 - (3+h)}{3(3+h)}}{h}$$

$$= \frac{\frac{-h}{3(3+h)}}{h} = \frac{-1}{3(3+h)} \quad (h \neq 0)$$

Finally we take the limit:

$$\lim_{h \to 0} \frac{-1}{3(3+h)} = \frac{-1}{3(3)} = -\frac{1}{9}$$

So $j'(3) = -\frac{1}{9}$.

EXERCISES

From the definition of the derivative as a limit, calculate:

1. $f'(x)$ if $f(x) = 3x^2 + 1$
2. $g'(x)$ if $g(x) = 4x^2 - 2$
3. $j'(x)$ if $j(x) = x^3 - 1$
4. $k'(t)$ if $k(t) = t^3 + t$
5. $s'(2)$ if $s(y) = y^3 - y^2$
6. $r'(3)$ if $r(x) = 2x^3 + x^2$
7. $g'(x)$ if $g(x) = \dfrac{2}{x}$
8. $f'(x)$ if $f(x) = \dfrac{3}{x}$
9. $k'(x)$ if $k(x) = \dfrac{1}{x+1}$
10. $j'(x)$ if $j(x) = \dfrac{1}{x+2}$

TEST A

From the definition of the derivative as a limit, calculate:

1. $f'(x)$ if $f(x) = x^2 + 3$
2. $g'(3)$ if $g(x) = 2x^3 + x$
3. $g'(y)$ if $g(y) = \dfrac{4}{y}$

Goal 9 | **To Understand the Definition of Derivative and Compute Derivatives**

TEST B

From the definition of the derivative as a limit, calculate:

1. $f'(2)$ if $f(x) = x^2 - 1$
2. $g'(x)$ if $g(x) = x^3 + 3x$
3. $f'(x)$ if $f(x) = \dfrac{1}{2x}$

STUDY QUESTIONS

1. It can happen that a function is defined at a point but has no derivative at that point. For example, consider the absolute-value function $f(x) = |x|$ at the point $x = 0$. $f(0) = |0| = 0$, and so f is defined at 0. Then we have

$$f'(0) = \lim_{h \to 0} \frac{f(0+h) - f(0)}{h}$$

$$= \lim_{h \to 0} \frac{|0+h| - |0|}{h}$$

$$= \lim_{h \to 0} \frac{|h|}{h}$$

The question now becomes: is there a number to which $\dfrac{|h|}{h}$ gets closer and closer as h gets closer and closer to zero?

First, we replace h by 1, 0.1, 0.01, and 0.001, and evaluate $\dfrac{|h|}{h}$.

Next replace h by -1, -0.1, -0.01, and -0.001 and evaluate $\dfrac{|h|}{h}$.

What seems to be happening to $\dfrac{|h|}{h}$? Since there is no *one* single value to which $\dfrac{|h|}{h}$ is getting closer and closer, we say that the limit of $\dfrac{|h|}{h}$ as h approaches zero does not exist; it follows that $f(x) = |x|$ has no derivative at $x = 0$.

Interpreting the derivative as the slope of a curve, explain why it is intuitively clear that $f(x) = |x|$ has no derivative at $x = 0$.

In general, if the graph of a function has a "sharp point," the function will have no derivative at that point.

2. What is $f'(x)$ if $f(x) = \sqrt{x}$? What is the value of $f'(4)$? of $f'(2)$? For what values of x does the derivative not exist?

   HINT: The quotient can be simplified by rationalizing the numerator as follows (assuming $h \neq 0$):

$$\frac{f(x+h)-f(x)}{h} = \frac{\sqrt{x+h}-\sqrt{x}}{h}$$

$$= \frac{(\sqrt{x+h}-\sqrt{x})(\sqrt{x+h}+\sqrt{x})}{h(\sqrt{x+h}+\sqrt{x})}$$

$$= \frac{(\sqrt{x+h})^2 - (\sqrt{x})^2}{h(\sqrt{x+h}+\sqrt{x})}$$

$$= \frac{x+h-x}{h(\sqrt{x+h}+\sqrt{x})}$$

$$= \frac{1}{\sqrt{x+h}+\sqrt{x}}$$

3. Suppose we are given a function f with the following two properties. First: $f(x+h) = f(x)f(h)$ for all values of x and h; that is, f evaluated at the sum of two numbers is just the product of f evaluated at each of the numbers individually. Second:

$$\lim_{h \to 0} \frac{f(h)-1}{h} = 1$$

That is, this quotient becomes closer and closer to 1 as h becomes closer and closer to 0.

Show that $f'(x) = f(x)$, that is, that the value of the derivative of f at a point is the same as the value of f itself at that point.

HINT: Use the first property above to simplify the quotient

$$\frac{f(x+h)-f(x)}{h}$$

then use algebra to simplify further; and then use the second property to help determine the limit of the simplified quotient as $h \to 0$.

Later we will encounter a most important function (the exponential function) that has these properties.

4. Suppose we are given two functions f and g with the following three properties:

P1. $f(x+h) = f(x)g(h) + f(h)g(x)$ for all values of x and h

P2. $\lim_{h \to 0} \frac{f(h)}{h} = 1$

P3. $\lim_{h \to 0} \frac{g(h)-1}{h} = 0$

Show that $f'(x) = g(x)$.

Goal 9 | To Understand the Definition of Derivative and Compute Derivatives

HINT: Use P1 to simplify the quotient $\dfrac{f(x+h) - f(x)}{h}$; use algebra to simplify further; and then use P2 and P3 to help determine the limit of the simplified quotient as h approaches 0.

The trigonometric functions $f(x) = \sin x$ and $g(x) = \cos x$ have these properties.

GOAL 10

To evaluate limits

We have seen that the concept of a limit is crucial in the development of the derivative. Later we shall see that it is also crucial in the development of another powerful tool of the calculus, called the *integral*. So far we have dealt with limits intuitively. Here we shall discuss them more carefully and then evaluate some specific limits. We will then be ready to develop techniques for the more efficient recognition and calculation of derivatives.

RIGHT- AND LEFT-HAND LIMITS

Consider the function $f(x) = x^2 + x - 1$. We ask: What happens to $f(x)$ as x is chosen closer and closer to 3? First we choose values for x which are close to but greater than 3:

x	4	3.1	3.01	3.001
$f(x)$	19	11.71	11.0701	11.007001

We see that the values of $f(x)$ get closer and closer to 11. We say that $f(x)$ approaches 11 as x approaches 3 from the right, and we call 11 the *right-hand limit of $f(x)$ at $x = 3$*. Since 11 is a finite number, we also say that the right-hand limit *exists* and we write

$$\lim_{x \to 3^+} f(x) = 11$$

The plus sign above and to the right of the "3" indicates that we have considered values of x *greater* than 3. If we now choose values for x close to but less than 3 we obtain

x	2	2.9	2.99	2.999
$f(x)$	5	10.31	10.9301	10.99301

Again the values of $f(x)$ get closer and closer to 11. But now we say that $f(x)$ approaches 11 as x approaches 3 from the left, call 11 the *left-hand limit of $f(x)$ at $x = 3$*, and write

$$\lim_{x \to 3^-} f(x) = 11$$

where the *minus* sign indicates the investigation of values for x close to but *less* than 3.

In a case such as this, where the left-hand and right-hand limits both exist and are equal, we say that *the limit of $f(x)$ as x approaches 3 exists* and we write

$$\lim_{x \to 3} f(x) = 11$$

The absence of a plus or minus to the right of the "3" indicates explicitly that the right-hand and left-hand limits both equal 11. This notation can be read off as "the limit, as x approaches three, of f of x is eleven." An alternative notation is

$$f(x) \to 11 \text{ as } x \to 3$$

which can be read as "f of x approaches eleven as x approaches 3."

LIMIT OF A POLYNOMIAL FUNCTION

Note, in the example above, that $f(x) = x^2 + x - 1$ has a value at $x = 3$, that this value is 11, and that this value is the same as $\lim f(x)$ as $x \to 3$. It is important to realize that this is not always the case. The limit of a function, say $g(x)$, as x approaches c may exist but be different from $g(c)$; indeed, $g(c)$ may not even be defined and yet $\lim g(x)$ as $x \to c$ exists. We will illustrate this shortly.

However, it is *always* true that, *if $g(x)$ is a polynomial function*, then the limit of $g(x)$ as $x \to c$ is equal to $g(c)$, the value of the function at $x = c$. For example:

$$\lim_{x \to 2} (x^3 - 3x + 1) = 2^3 - 3(2) + 1 = 3$$

$$\lim_{y \to -1} (y^5 + y^2 - 4) = (-1)^5 + (-1)^2 - 4 = -4$$

LIMITS OF NONPOLYNOMIAL FUNCTIONS

In Goal 9 we asked what happens to

$$\frac{4hx + 2h^2}{h}$$

as h is chosen closer and closer to zero. To answer this question, we now choose values for h close to zero, first from the left (that is, less than zero), then from the right (greater than zero). You should verify the results given in the following tables:

h	$\dfrac{4hx + 2h^2}{h}$	h	$\dfrac{4hx + 2h^2}{h}$
-2	$4x - 4$	2	$4x + 4$
-0.1	$4x - 0.2$	0.1	$4x + 0.2$
-0.001	$4x - 0.002$	0.001	$4x + 0.002$
-0.00001	$4x - 0.00002$	0.00001	$4x + 0.00002$

$$\lim_{h \to 0^-} \frac{4hx + 2h^2}{h} = 4x \qquad \lim_{h \to 0^+} \frac{4hx + 2h^2}{h} = 4x$$

We see from the tables that both the right-hand and the left-hand limits exist and that they both equal $4x$. We say, as in the earlier example, that the *limit* exists, this time as h approaches zero, and we write

$$\lim_{h \to 0} \frac{4hx + 2h^2}{h} = 4x$$

Now, what about the value of $\dfrac{4hx + 2h^2}{h}$ when h equals zero? It has none! Division by zero is not allowed; hence the expression is undefined for $h = 0$.

This example demonstrates nicely that $\lim g(t)$ as $t \to c$ does not in any way depend on $g(c)$. The notation

$$\lim_{t \to c} g(t)$$

signals investigation of $g(t)$ as t *becomes closer and closer to* c, but *not at the point where* $t = c$.

Note that

$$\frac{4hx + 2h^2}{h} = (4x + 2h) \cdot \frac{h}{h}$$

Goal 10 | **To Evaluate Limits**

While we cannot divide by h if $h = 0$, as long as $h \neq 0$ we can cancel out the factor h shared by numerator and denominator. Our discussion just above implies, therefore, that

$$\lim_{h \to 0} (4x + 2h) \frac{h}{h} = \lim_{h \to 0} (4x + 2h) = 4x + 2(0) = 4x$$

Note that since $4x$ does not depend on h, it remains fixed as $h \to 0$; in fact, we have treated $4x$ as though it were a constant.

CANCELING COMMON FACTORS

The example just discussed is typical of problems involving limits in which the numerator and the denominator of a function have a common factor. Note that we divided the factor out (canceled it) *before* we took the limit. Here are further examples. In each case below, when the common factor is canceled the resulting function is a polynomial, whose limit we can easily evaluate.

Example 10-1

$$\lim_{x \to 2} \frac{x^2 - 4}{x - 2} = \lim_{x \to 2} (x + 2) \frac{x - 2}{x - 2} = \lim_{x \to 2} (x + 2) = 2 + 2 = 4$$

Example 10-2

$$\lim_{y \to 5} \frac{y + 3y - 10}{y + 5} = \lim_{y \to -5} (y - 2) \frac{y + 5}{y + 5} = \lim_{y \to -5} (y - 2) = -5 - 2 = -7$$

Example 10-3

$$\lim_{h \to 0} \frac{3y^2 h + 3yh^2 + h^3}{h} = \lim_{h \to 0} (3y^2 + 3yh + h^2) \frac{h}{h}$$
$$= \lim_{h \to 0} (3y^2 + 3yh + h^2)$$
$$= 3y^2 + 3y(0) + 0^2 = 3y^2$$

CASES IN WHICH A LIMIT DOES NOT EXIST

If $\lim f(x)$ as $x \to c$ exists, then this limit is a finite number. If the values of $f(x)$ do not become closer and closer to a particular finite number as x becomes closer and closer to c, then the limit does *not* exist. The following examples illustrate this.

A Function Becoming Infinite

Let

$$g(s) = \frac{1}{(s-5)^2}$$

We examine values of g for s close to 5:

s	$g(s)$	s	$g(s)$
4	1	6	1
4.9	100	5.1	100
4.99	10,000	5.01	10,000
4.999	1,000,000	5.001	1,000,000

A sketch is illuminating. It is clear there is no finite number that the values

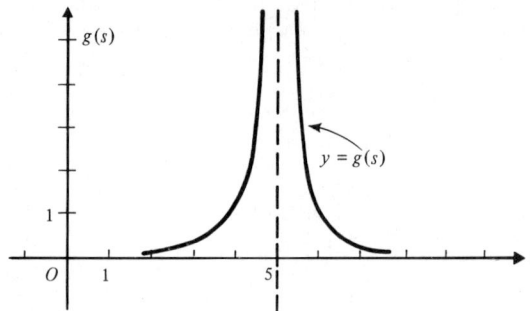

of g approach. We therefore conclude that $\lim_{s \to 5} g(s)$ does not exist. We do see, however, that the values of g increase without bound as s approaches 5 both from the right and from the left. This situation is symbolized mathematically as follows:

$$\lim_{s \to 5^-} g(s) = +\infty \quad \text{and} \quad \lim_{s \to 5^+} g(s) = +\infty$$

Indeed, we also write

$$\lim_{s \to 5} g(s) = +\infty \quad (\text{or just } \lim_{s \to 5} g(s) = \infty)$$

Students often raise questions about these "equations," since they seem to imply that there *is* a limit and that that limit is infinity. Impossible! The statement

$$\lim_{s \to 5} g(s) = \infty$$

or

$$g(s) \to \infty \text{ as } s \to 5$$

implies both of the following: (1) the closer we choose s to 5 (either from the left or from the right) the larger $g(s)$ becomes; (2) $g(s)$ does not have a (finite) limit as s approaches 5. Notice also that in this example neither the left-hand nor the right-hand limit exists.

Below we show the graphs of two functions:

(a) $f(x) = -1/x^2$ (b) $g(x) = 1/x$

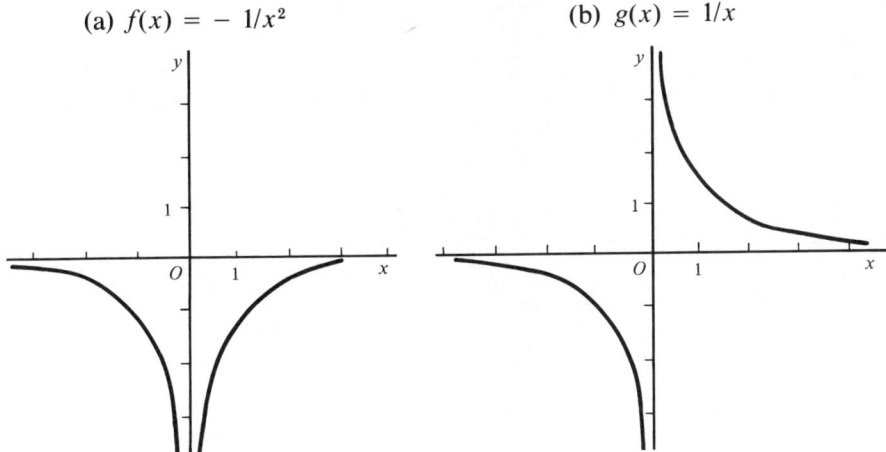

The following statements describe the behavior of these functions in the neighborhood of zero (that is, as x approaches zero):

(a) $\lim\limits_{x \to 0^-} f(x) = \lim\limits_{x \to 0^+} f(x) = \lim\limits_{x \to 0} f(x) = -\infty$

(b) $\lim\limits_{x \to 0^+} g(x) = +\infty$ $\lim\limits_{x \to 0^-} g(x) = -\infty$

Try to say carefully in words what each of these statements means, paying special attention to the difference between the two functions as x gets closer and closer to zero.

VERTICAL ASYMPTOTES

In Goal 4 (page 37) we defined a vertical asymptote of a graph. Note that the line $x = c$ is a vertical asymptote of the graph of $f(x)$ if any one of the following statements is true:

$\lim\limits_{x \to c} f(x) = +\infty$ $\lim\limits_{x \to c^-} f(x) = +\infty$ $\lim\limits_{x \to c^+} f(x) = +\infty$

$\lim\limits_{x \to c} f(x) = -\infty$ $\lim\limits_{x \to c^-} f(x) = -\infty$ $\lim\limits_{x \to c^+} f(x) = -\infty$

Sketch graphs that illustrate each of the above statements.

The graph of $f(s) = \dfrac{1}{(s-5)^2}$ has $s = 5$ as vertical asymptote; $x = 0$ is a vertical asymptote of both $f(x) = -1/x^2$ and $g(x) = 1/x$.

FINITE BUT UNEQUAL LEFT- AND RIGHT-HAND LIMITS

Consider the function

$$f(x) = \begin{cases} -1 & \text{if } x < 0 \\ 1 & \text{if } x \geq 0 \end{cases}$$

The graph of this function is shown below. We are interested in investigating values of $f(x)$ as x approaches zero, first from the left, then from the right:

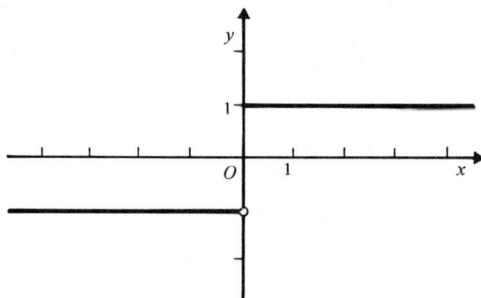

x	-3	-2	-1	-0.5	-0.1	-0.0001
$f(x)$	-1	-1	-1	-1	-1	-1

x	3	2	1	0.5	0.1	0.0001
$f(x)$	1	1	1	1	1	1

We conclude:

$$\lim_{x \to 0^-} f(x) = -1 \qquad \lim_{x \to 0^+} f(x) = +1$$

The left-hand limit is finite (therefore it exists) and is -1; the right-hand limit exists and is $+1$. However, since these numbers are different, $\lim_{x \to 0} f(x)$ does *not* exist.

The so-called postage function is described on page 41 (Exercise 30). The first ounce or fraction thereof costs 15¢, and each additional ounce or fraction thereof costs 13¢. Here is a table of costs for weights close to 1 oz:

w (wt, oz)	0.1	0.8	0.99	1	1.001	1.1	1.2
$c(w)$ (cost, ¢)	15	15	15	15	28	28	28

Goal 10 | To Evaluate Limits

We see that

$$\lim_{w \to 1^-} c(w) = 15 \quad \text{but} \quad \lim_{w \to 1^+} c(w) = 28$$

Although the left-hand and right-hand limits both exist, the limit of $c(w)$ "at" $w = 1$ does not exist. In fact, the limit of the postage function does not exist at any positive integer. (Why?)

PROPERTIES OF LIMITS

In computing limits it is often useful to think of a function in terms of its component parts and consider the limit of each part. In this technique we make use of the following properties of limits; which we state without proof.

If, as x approaches c, $f(x)$ approaches R and $g(x)$ approaches S, then, as x approaches c,

(a) $kf(x)$ approaches kR, where k is any real number;
(b) $f(x) + g(x)$ approaches $R + S$;
(c) $f(x) \cdot g(x)$ approaches RS; and
(d) provided $S \neq 0$, $f(x)/g(x)$ approaches R/S.

For example,

$$\lim_{x \to 2} (3x^2 - 2x + 1) = \left(\lim_{x \to 2} 3\right)\left(\lim_{x \to 2} x\right)\left(\lim_{x \to 2} x\right) + \left(\lim_{x \to 2} -2\right)\left(\lim_{x \to 2} x\right) + \left(\lim_{x \to 2} 1\right)$$

$$= 3(2)(2) + (-2)(2) + 1 = 9$$

In a similar manner, show that

$$\lim_{x \to 1} \frac{2x^2 - 5}{x^3 + 1} = -\frac{3}{2}$$

EXERCISES

In Exercises 1 through 8 find the limit of each polynomial function. (By examining functional values just to the right and just to the left of the point in question, convince yourself that the right-hand and left-hand limits are equal, both to each other and to the answer you've obtained.)

1. $\lim_{x \to 2} (3x + 1)$
2. $\lim_{x \to -3} (x^2 - 4)$
3. $\lim_{x \to 5} 7$
4. $\lim_{x \to 12} -1$
5. $\lim_{x \to 3} (2x^2 + x - 4)$
6. $\lim_{y \to 1} (y^3 - 3y^2 - 2y + 2)$
7. $\lim_{t \to -1} (t^3 + 1)$
8. $\lim_{s \to 0} (2s^2 - 4s - 3)$

In Exercises 9 through 20 find each of the limits.

9. $\lim\limits_{y \to 0} \dfrac{y^2 + 3y}{y}$

10. $\lim\limits_{x \to 2} \dfrac{x^2 - x - 2}{x - 2}$

11. $\lim\limits_{t \to -2} \dfrac{t^3 + t^2}{t + 1}$

12. $\lim\limits_{s \to -2} \dfrac{s^2 + 4s + 4}{s + 2}$

13. $\lim\limits_{h \to 0} \dfrac{2xh + h^2}{h}$

14. $\lim\limits_{h \to 0} \dfrac{5h}{h}$

15. $\lim\limits_{h \to 0} \dfrac{2x^2h - 3xh + h}{h}$

16. $\lim\limits_{h \to 0} \dfrac{2yh + 2h - 4h^2 + h^3}{h}$

17. $\lim\limits_{h \to 0} \dfrac{(2 + h)^2 - 2^2}{h}$

18. $\lim\limits_{h \to 0} \dfrac{(1 + h)^3 - 1}{h}$

19. $\lim\limits_{h \to 0} \dfrac{[5(x + h) + 2] - (5x + 2)}{2}$

20. $\lim\limits_{h \to 0} \dfrac{[3(x + h)^2] - 3x^2}{h}$

In 21 through 26 complete each statement.

21. $\lim\limits_{x \to 4} \dfrac{1}{(x - 4)^2} =$

22. $\lim\limits_{x \to 1} \dfrac{-4}{(x - 1)^2} =$

23. $\lim\limits_{x \to 3^+} \dfrac{1}{x - 3} =$

24. $\lim\limits_{x \to 3^-} \dfrac{1}{x - 3} =$

25. $\lim\limits_{x \to 3^+} \dfrac{1}{3 - x} =$

26. $\lim\limits_{x \to 3^-} \dfrac{1}{3 - x} =$

27. Which of the limits in exercises 21 through 26 exist?

28. Describe any vertical asymptotes of the graph of $y = \dfrac{P(x)}{Q(x)}$, where $P(x)$ and $Q(x)$ are polynomials.

In Exercises 29 through 34 find each of the limits that exists. In these, note that $[x]$ is the greatest integer function defined on page 39.

29. $\lim\limits_{x \to 2^+} (x - 2)$

30. $\lim\limits_{x \to 0} |x|$

31. $\lim\limits_{x \to 2^-} [x]$

32. $\lim\limits_{x \to 2^+} [x]$

33. $\lim\limits_{x \to 2} [x]$

34. $\lim\limits_{x \to 5} [x]$

True or false? (Exercises 35 through 40.)

35. If $P(x)$ is a polynomial, then $\lim\limits_{x \to c} P(x) = P(c)$.

36. If $P(x)$ and $Q(x)$ are polynomials, then

$$\lim\limits_{x \to c} \dfrac{P(x)}{Q(x)} = \dfrac{P(c)}{Q(c)}$$

Goal 10 | To Evaluate Limits

37. If $\lim_{x \to c^+} f(x) = +\infty = \lim_{x \to c^-} f(x)$, then $\lim_{x \to c} f(x)$ exists.

38. If $\lim_{x \to c} f(x)$ does not exist, then $f(x)$ must become (positively or negatively) infinite as x approaches c.

39. It is possible for a function not to be defined at $x = c$, yet for $\lim_{x \to c} f(x)$ to exist.

40. If $\lim_{x \to c} f(x = L$, then $f(c) = L$.

TEST A

Find each of the following limits:

1. $\lim_{y \to 2} (y^2 - y + 3)$

2. $\lim_{x \to 0} (-2)$

3. $\lim_{x \to 3} \dfrac{x^2 - 9}{x - 3}$

4. $\lim_{h \to 0} \dfrac{[3(x + h)^2] - 3x^2}{h}$

Complete each of the following statements:

5. $\lim_{y \to 2^-} \dfrac{3}{y - 2} =$

6. $\lim_{t \to -1} \dfrac{4}{(t + 1)^2} =$

Find any limit that exists.

7. $\lim_{x \to 0^+} [x]$

8. $\lim_{x \to 4} [x]$

TEST B

Find each of the following limits:

1. $\lim_{t \to 0} (t^2 + t - 5)$

2. $\lim_{x \to -1} 4$

3. $\lim_{x \to 2} \dfrac{x^2 - 5x + 6}{x - 2}$

4. $\lim_{h \to 0} \dfrac{2xh + 6h + 2h^2}{h}$

Complete each of the following statements:

5. $\lim_{x \to 1} \dfrac{2}{(1 - x)^2} =$

6. $\lim_{x \to 1^+} \dfrac{2}{1 - x} =$

Find any limit that exists:

7. $\lim_{x \to 2} |x - 2|$

8. $\lim_{x \to 2} [x]$

STUDY QUESTIONS

1. If $100 is deposited in a bank paying 6% compound interest, then the amount on deposit after 1 yr is given by the function

$$A(x) = 100(1 + 0.06x)^{1/x}$$

where x is $\frac{1}{2}$ if the interest is compounded every half year, is $\frac{1}{12}$ if it is compounded every month, and so on. We will determine if it is advantageous to the investor for interest to be compounded more and more often [that is, after shorter and shorter time intervals). Using a calculator and rounding off to the nearest cent, we compute:

yearly compounding: $A(1) = 100(1.06)^1 = \$106.00$

(Calculate the balance if interest is compounded every half year.)

quarterly compounding: $A\left(\dfrac{1}{4}\right) = 100(1.015)^4 = \$106.14$

monthly compounding: $A\left(\dfrac{1}{12}\right) = 100(1.005)^{12} = \$106.17$

daily compounding: $A\left(\dfrac{1}{365}\right) = 100\left(1 + \dfrac{0.06}{365}\right)^{365} = \$106.18$

hourly compounding: $A\left(\dfrac{1}{8760}\right) = 100\left(1 + \dfrac{0.06}{8760}\right)^{8760} = \$106.18$

There is no monetary advantage in having the amount compounded any more often. As x becomes closer and closer to zero, what does $A(x)$ become closer and closer to? We say that $106.18 is the limit of $A(x)$ as x approaches zero, and we write

$$\lim_{x \to 0} A(x) = \$106.18$$

Does $A(x)$ have a value when $x = 0$?

2. Suppose we wish to manufacture square sections of metal whose area is to be 9 sq in. It would be desirable to be able to make squares exactly 3 in. to the side; however, it is physically impossible to be absolutely precise. Whatever the length s of the side, the area will be $A(s) = s^2$, and the closer we desire s^2 to be to 9, the more precise must be our manufacturing machinery. For example, suppose our error tolerance is $\frac{3}{4}$ sq in.; that is, we desire the area of the section to be between $8\frac{1}{4}$ and $9\frac{3}{4}$ sq in. Then it will suffice if our machinery is precise enough to measure s to within $\frac{1}{10}$ in. of 3 in., because if $2\frac{9}{10} < s < 3\frac{1}{10}$, then $8\frac{41}{100} < s^2 < 9\frac{61}{100}$, which is within the range between $8\frac{1}{4}$ and $9\frac{3}{4}$. If our error tolerance is only $\frac{1}{10}$ sq in. (that is, we desire the section

to have an area between $8\frac{9}{10}$ and $9\frac{1}{10}$ sq. in.), then our machinery must be more precise.

Show that in the last case specified it suffices if we can measure to within $\frac{1}{100}$ in. of 3 in.

Similarly, given *any* error tolerance, we can determine how precise our machinery must be for the resulting area to fall within that tolerance.

We say that 9 is the limit of $A(s)$ as s approaches 3, and we write

$$\lim_{s \to 3} A(s) = 9$$

(This study question is actually a disguised version of what is known as the formal epsilon-delta definition of a limit.)

3. Suppose $T(x) = x^2 + 4x$ cubic centimeters (cm³) of medicine is needed to treat a tumor that weighs x g, and suppose a research physiologist is interested in how many cm³ per g are necessary in the treatment of tumors of various sizes.

If the tumor weighs 2 g, then $2^2 + 4(2) = 12$ cm³ is necessary. This is 12 cm³/2g, or 6 cm³/g.

If the tumor weighs $\frac{1}{2}$ g then $(\frac{1}{2})^2 + 4(\frac{1}{2})$ or 2.25 cm³ is necessary. Thus 2.25 cm³/0.5g, or 4.5 cm³/g, is needed.

How many cm³/g are needed if the tumor weighs $\frac{1}{10}$ g?

Is there a limiting value, in cubic centimeters per gram, as the tumor weight is taken to be closer and closer to zero? Yes. We say that 4 cm³/g is the limit of $[T(x)$ cm³$]/[x$ g$]$ as x approaches zero, and we write

$$\lim_{x \to 0} \frac{T(x)}{x} = 4 \text{ cm}^3/\text{g}$$

Does $[T(x)]/x$ have a value when $x = 0$?

GOAL **11**

To determine whether a function is continuous at a particular point

Almost all the functions we consider in this book have the property that the functional values close to a point are close to the functional value at the point. This property, called *continuity*, is related to the idea of a limit defined in Goal 10 and is assumed in the statements of most theorems of calculus. In this Goal we find out how to determine whether or not a function is continuous at a given point.

TWO EXAMPLES OF CONTINUOUS FUNCTIONS
Example 11-1
Suppose we have recorded the temperature readings on a particular thermometer just before and after noon on a given day:

Time	11:55	11:59:50	12:00:20	12:02	12:04
Temperature (°C)	$21\frac{1}{2}$	21	21	21	$20\frac{1}{2}$

A reasonable guess for the temperature exactly at noon is 21°C or some temperature very close to that. This guess is based on our intuition that there can be no great change or jump in the temperature over a short interval of time, such as between 10 sec before noon and 20 sec after noon. We say that temperature is a continuous function of time.

Example 11-2
If we are traveling in a car and the speedometer shows 45 mph at 3 P.M. and 55 mph at 3:10 P.M., then we can assert that the speed of the car must have been exactly 50 mph at some instant between 3 and 3:10 P.M. This conclusion follows from a property of all continuous functions, and the speed

Goal 11 | **To Determine If a Function is Continuous at a Point**

(of a car) is a continuous function of time. (In fact, the property in question assures us that the car must have traveled at *every* speed between 45 and 55 mph during the time interval from 3 to 3:10 P.M.)

TWO EXAMPLES OF DISCONTINUOUS FUNCTIONS

Example 11-3

In 1978 first-class United States postage cost 15¢ for the first ounce or fraction thereof and 13¢ for each additional ounce or fraction thereof. Suppose we arrange a small set of letters in order of descending weight:

Weight (oz)	1.5	1.1	1.01	1.001	1.0001	1
Cost of postage (¢)	28	28	28	28	28	15

We note that there is a sharp jump from the cost for any weight greater than 1 oz to the cost for exactly 1 oz. We say that the cost as a function of weight is *not continuous* or is *discontinuous* at the weight 1 oz, or we say that there is a *discontinuity* in the function at that point.

Example 11-4

A factory, operating for two 8-h shifts a day, has a fixed overhead expense of $5000 for each shift or fraction of a shift, and it costs an additional $1000 per h in salaries during each shift. We thus have the following table of total costs, considering shorter and shorter daily production times:

Production time (h)	10	9	8.1	8.01	8.001	8
Cost for overhead and salaries ($)	20,000	19,000	18,100	18,010	18,001	13,000

We note that there is a sharp jump from the expense for any production time greater than 8 h to that for exactly 8 h. We say that the cost of production as a function of the production time is *discontinuous* at 8 h.

DEFINITION OF CONTINUITY

The functions graphed below are continuous on the indicated domains:

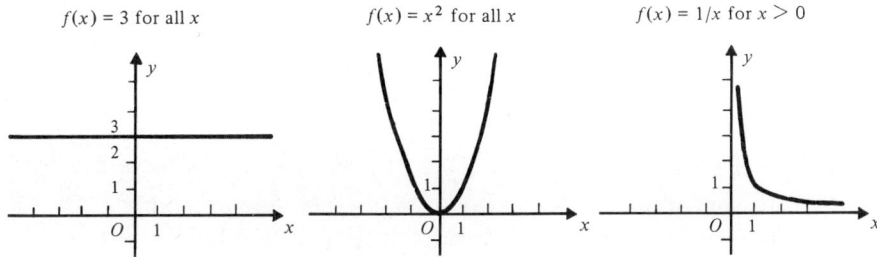

Intuitively, we can say that f is a continuous function if its graph has no breaks or jumps and can be drawn without lifting the pencil from the paper. Each of the functions graphed below has a point of discontinuity at $x = 2$:

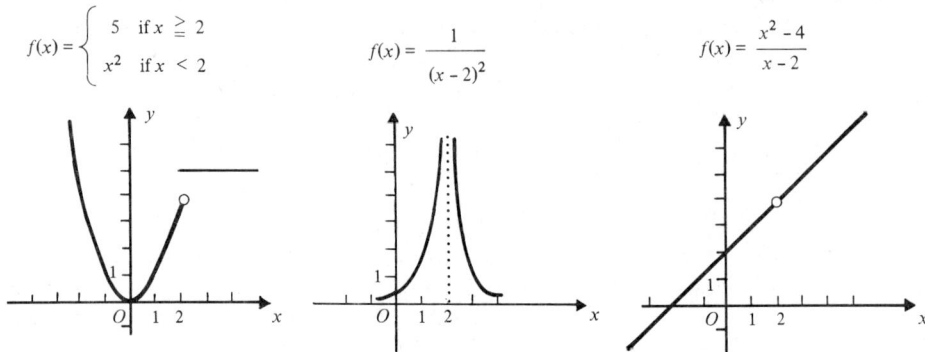

A function $f(x)$ is continuous at a point $x = c$ if f is defined at c and if values of x very close to c produce values of f very close to $f(c)$.

We now give a careful formal definition of continuity which uses the concept of limit from Goal 10.

The function f is said to be *continuous at the point* $x = c$ if the following three conditions hold:

(1) $f(c)$ is defined;
(2) $\lim_{x \to c} f(x)$ exists;
(3) $\lim_{x \to c} f(x) = f(c)$.

The function f is said to be *continuous throughout a specified domain of definition* if it is continuous at every point in the domain. If f is not continuous at some point, then that point is called a *point of discontinuity*. Thus, f will have a point of discontinuity wherever any one of the three above conditions does not hold; that is, at any point where f is not defined, or where f does not have a limit, or where, if it is defined and has a limit, those two values are not equal.

MORE EXAMPLES

In each of the following two examples, f is not continuous at $x = 3$ because $f(3)$ is not defined; that is, condition (1) does not hold:

Goal 11 | **To Determine If a Function is Continuous at a Point**

Example 11-5

$$f(x) = \frac{x^2 - 9}{x - 3}$$

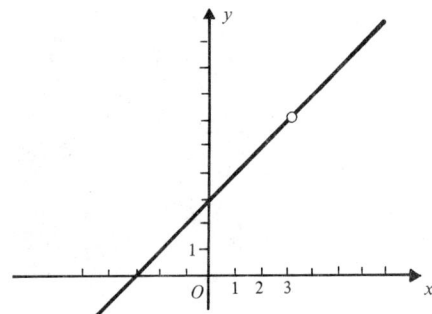

Example 11-6

$$f(x) = \begin{cases} 9 & \text{if } x > 3 \\ x^2 & \text{if } x < 3 \end{cases}$$

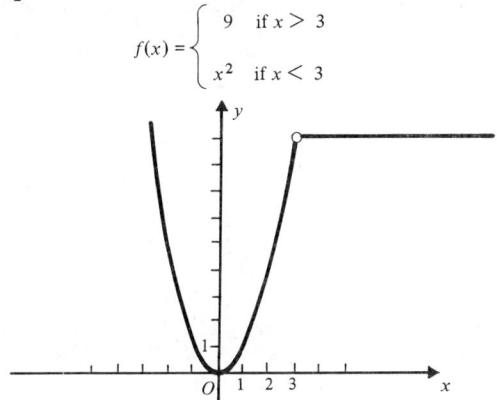

In each of the following two examples, f is not continuous at $x = 2$ because $\lim f(x)$ as $x \to 2$ does not exist; that is, condition (2) does not hold:

Example 11-7

$$f(x) = \frac{1}{(x - 2)^2}$$

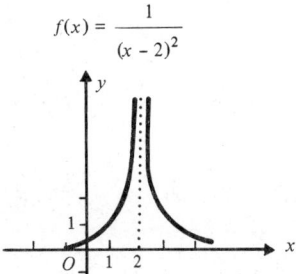

This function is not defined at $x = 2$. Note that

$$\lim_{x \to 2^-} f(x) = +\infty$$

$$\lim_{x \to 2^+} f(x) = +\infty$$

Example 11-8

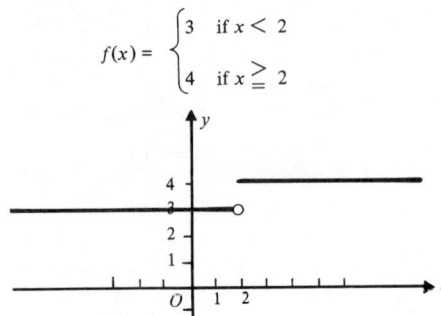

Since $f(2) = 4$, f is defined at 2. Note, however, that

$$\lim_{x \to 2^-} f(x) = 3$$

$$\lim_{x \to 2^+} f(x) = 4$$

In each of the two cases below, f is not continuous at $x = 1$ because $\lim f(x)$ as $x \to 1$ is not equal to $f(1)$; that is, condition (3) does not hold:

Example 11-9

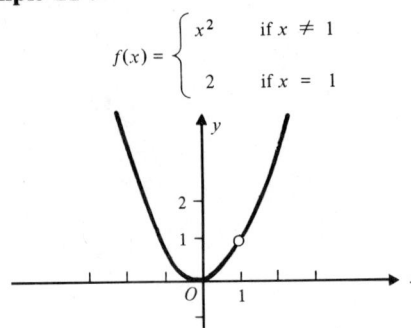

Here

$$\lim_{x \to 1^-} f(x) = \lim_{x \to 1^+} f(x) = 1$$

but $f(1) = 2$.

Example 11-10

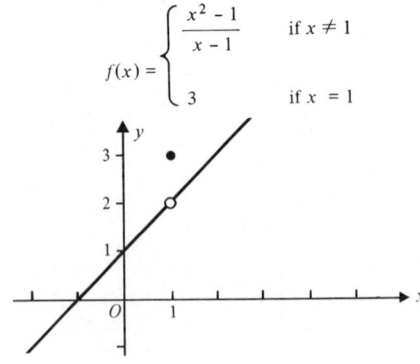

Here

$$\lim_{x \to 1^-} f(x) = \lim_{x \to 1^+} f(x) = 2$$

but $f(1) = 3$.

Goal 11   To Determine If a Function is Continuous at a Point

NOTE. All the functions we consider in this book are either continuous throughout their domain of definition or have discontinuities only at certain isolated points.

In Goal 10 we noted that if $f(x)$ is a polynomial function, then $\lim f(x)$ as $x \to c$ equals $f(c)$. But this says exactly that f is continuous at $x = c$. So a polynomial function is continuous everywhere. Also, if $f(x)$ is a rational function [that is, $f(x) = P(x)/Q(x)$, where P and Q are polynomials] then $\lim f(x)$ as $x \to c$ equals $P(c)/Q(c)$ except where $Q(c) = 0$. We see, then, that a rational function is continuous everywhere except at those points for which the denominator is zero. Thus, for example, the polynomials

$$f(x) = 3x^2 - 4 \qquad f(x) = 9x^4 - 5x + 1 \qquad f(x) = 3$$

are continuous everywhere. The following functions are continuous except where indicated:

$$f(x) = \frac{1}{x} \qquad x \neq 0 \qquad\qquad f(x) = \frac{x^2 + 1}{x^2 - 1} \qquad x \neq 1, -1$$

$$f(x) = \frac{x^2 - 4}{x - 2} \qquad x \neq 2 \qquad\qquad f(x) = \frac{x^3 - 1}{x^2 + x - 6} \qquad x \neq -3, 2$$

$$f(x) = \frac{4}{x^2 + 9} \qquad \text{(continuous everywhere)}$$

EXERCISES

In Exercises 1 through 4 determine if the functions defined are continuous throughout the domain of definition. If not, explain.

1. The cost of a taxi ride, as a function of the distance traveled (in miles) if the taxi company charges 85¢ plus 25¢ for every quarter mile or fraction thereof.
2. The population of the United States as a function of time.
3. The weight of a melting ice cube as a function of time.
4. The number of cars currently on a specified toll road as a function of time.

In 5 through 15, determine if the functions are continuous at the points specified.

5. $f(x) = x^2 + 3x - 5$ at $x = 2$
6. $g(y) = \dfrac{1}{y - 1}$ at $y = 1$

7. $g(y) = \dfrac{1}{y-1}$ at $y = 2$

8. $g(x) = \begin{cases} \dfrac{1}{x-1} & \text{if } x \neq 1 \\ 0 & \text{if } x = 1 \end{cases}$ at $x = 1$

9. $h(z) = 4$ at $z = 0$

10. $f(x) = \dfrac{x^2 - 36}{x - 6}$ at $x = 6$

11. $f(x) = \begin{cases} \dfrac{x^2 - 36}{x - 6} & \text{if } x \neq 6 \\ 0 & \text{if } x = 6 \end{cases}$ at $x = 6$

12. $f(x) = \begin{cases} \dfrac{x^2 - 36}{x - 6} & \text{if } x \neq 6 \\ 12 & \text{if } x = 6 \end{cases}$ at $x = 6$

13. $g(z) = \begin{cases} 3z + 1 & \text{if } z \neq 1 \\ 5 & \text{if } z = 1 \end{cases}$ at $z = 1$

14. $g(z) = \begin{cases} z^2 - z + 1 & \text{if } z \neq 1 \\ 1 & \text{if } z = 1 \end{cases}$ at $z = 1$

15. $f(x) = \dfrac{x - 4}{x + 4}$ at $x = 4$

In 16 through 20 find any points of discontinuity.

16. $f(x) = -1$

17. $g(x) = \dfrac{3x}{(x-1)^2}$

18. $h(y) = \dfrac{y^2 - 2y}{y^2 + 2}$

19. $f(u) = \dfrac{u - 1}{u^2 + u}$

20. $h(x) = |x|$

TEST A

In 1 through 4 determine if the functions are continuous at the points specified; if any function is not continuous at a point specified, tell why.

1. $f(x) = \begin{cases} x^2 + 1 & \text{if } x \neq 3 \\ 8 & \text{if } x = 3 \end{cases}$ at $x = 3$

2. $g(y) = \begin{cases} 1/y & \text{if } y \neq 0 \\ 0 & \text{if } y = 0 \end{cases}$ at $y = 0$

Goal 11   To Determine If a Function is Continuous at a Point

3. $h(z) = 5z^2 - z + 3$ at $z = 2$

4. $g(x) = 3x^2 - 2$ if $x \neq 1$ at $x = 1$

5. For what values of x, if any, is the function

$$f(x) = \frac{3x + 2}{x^2 + x - 12}$$

not continuous?

TEST B

In 1 through 4 determine if the functions are continuous at the points specified; if any function is not continuous at a point specified, tell why.

1. $f(x) = \dfrac{x + 1}{x - 3}$ at $x = 2$

2. $h(y) = \dfrac{y^2 - 49}{y - 7}$ at $y = 7$

3. $g(z) = \begin{cases} z + 5 & \text{if } z < 0 \\ 0 & \text{if } z = 0 \\ 5 - z & \text{if } z > 0 \end{cases}$ at $z = 0$

4. $f(x) = \begin{cases} \dfrac{1}{x + 1} & \text{if } x \neq -1 \\ 0 & \text{if } x = -1 \end{cases}$ at $x = 1$

5. For what values of x, if any, is either of the following functions not continuous?

    (a) $f(x) = \dfrac{x^2 + 1}{x^2 - 1}$ (b) $g(x) = \dfrac{x^2 - 1}{x^2 + 1}$

STUDY QUESTIONS

1. Suppose a function f is discontinuous at point $x = a$. If a function g can be defined which is identical with f at every point except $x = a$ and which is continuous at $x = a$, then f is said to have a *removable discontinuity* at a. For example,

$$f(x) = \begin{cases} 3x + 2 & \text{if } x \neq 1 \\ 6 & \text{if } x = 1 \end{cases}$$

is discontinuous at $x = 1$ because $\lim f(x)$ as $x \to 1$ is 5, but $f(1)$ is 6.

Define $g(x)$ as follows:

$$g(x) = \begin{cases} 3x + 2 & \text{if } x \neq 1 \\ 5 & \text{if } x = 1 \end{cases}$$

We note that the point of discontinuity of f at 1 has been removed. Similarly,

$$f(x) = \frac{x^2 - 25}{x - 5}$$

has a removable discontinuity at $x = 5$ since we can define a function g by

$$g(x) = \begin{cases} \dfrac{x^2 - 25}{x - 5} & \text{if } x \neq 5 \\ 10 & \text{if } x = 5 \end{cases}$$

Which of the six functions in Examples 11-5 through 11-10 have removable discontinuities? Justify your answers.

2. If $f(x)$ and $g(x)$ are continuous at a point $x = c$, then so are $kf(x)$ (where k is a constant), $f(x) + g(x)$, $f(x) \cdot g(x)$, and $\dfrac{f(x)}{g(x)}$, provided in the last case that $g(c) \neq 0$. How can these properties of continuous functions be used to reach conclusions about the continuity of polynomials and of rational functions? [HINT: Start with the assumption that constant functions and the identify function $f(x) = x$ are all continuous.]

3. There is a relationship between continuity and *differentiability*. (A function is said to be differentiable if its derivative exists.) Consider, for example, the function $f(x) = \dfrac{1}{x}$, which is not defined at $x = 0$ and is thus not continuous at that point. Its derivative, found by the methods used in Goal 9, is $f'(x) = \dfrac{-1}{x^2}$, which also has no value at $x = 0$. So the derivative of f does not exist at $x = 0$.

It is true that any function that has a derivative at $x = c$ is continuous at $x = c$. However, if f is continuous at $x = c$, does it automatically follow that it must be differentiable at $x = c$? (HINT: Consider the function $f(x) = |x|$ at the point $x = 0$. Refer to Study Question 1 in Goal 9.)

4. If someone climbs from the bottom to the top of a mountain, at some point in time he must be exactly two-thirds of the way to the top:

Goal 11 | **To Determine If a Function is Continuous at a Point**

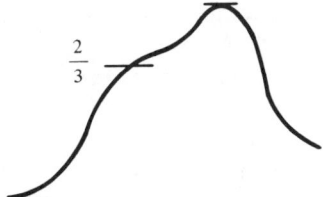

If a rocket accelerates from rest to 10 miles per second (mps) in the first minute after lift-off, then at some time during that first minute the velocity must be 6 mps:

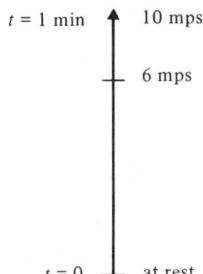

These are further instances of the *intermediate value theorem*, illustrated earlier in Examples 4-3 and 11-2. This theorem concerns a useful property of continuous functions. More precisely, the theorem says that if f is continuous on the interval $a \leq x \leq b$ and Y is some number between $f(a)$ and $f(b)$, then there is at least one number c between a and b such that $f(c) = Y$:

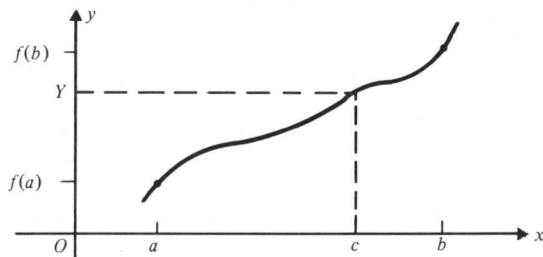

What can be concluded about a continuous function f if $f(a)$ and $f(b)$ have opposite signs?

5. If someone's blood pressure is monitored throughout a 24-h period, then at some point of time the blood pressure must be at the highest value it attains during the 24-h period. Similarly, if the temperature of a computer room is graphically recorded throughout an 8-h work

shift, then at some point of time the temperature must be at the highest value it ever attains during the shift.

These are both examples of the *maximization property* of continuous functions: If f is continuous on the interval $a \leq x \leq b$, then somewhere on the interval f takes on a maximum value [that is, there is some number z such that $a \leq z \leq b$ and such that $f(z) \geq f(x)$ for all values of x in the interval].

Write a statement of, and give some real-life examples of, the corresponding *minimization property* of continuous functions.

GOAL **12**

To find derivatives of polynomials

Since it is often a tedious task to find a derivative directly from the definition of the derivative as a limit, rules have been developed for finding derivatives of the more common types of function quickly and efficiently. Once we recognize a complicated function as being a combination of simpler ones, these rules enable us to obtain derivatives easily. In this Goal we state and illustrate the rules for computing derivatives of polynomials.

NOTATIONS FOR THE DERIVATIVE

When $y = f(x)$, we have written $f'(x)$ for the derivative of f at any point x in the domain. The following notations for the derivative of f with respect to x are also commonly used:

$$D_x y, \quad D_x f, \quad f', \quad y', \quad \frac{dy}{dx}$$

Sometimes one of these notations is more convenient; sometimes another is. Thus, if $y = x^2$, we may write

$$D_x y = 2x \quad \text{or} \quad \frac{dy}{dx} = 2x \quad \text{or} \quad y' = 2x$$

We have been using the notation $f'(a)$ for the derivative of a function $f(x)$ at the point where $x = a$. If $y = f(x)$, this may also be denoted by $y'(a)$.

The process or operation of obtaining a derivative is called *differentiation*. To differentiate a function, then, is to find or "take" its derivative.

DERIVATIVE OF A CONSTANT

If $f(x) = C$, where C is a number, then $f'(x) = 0$.

Examples

If $f(x) = 6$, then $f'(x) = 0$.
$D_x(-3) = 0$.
If $h(y) = 1/\sqrt{2}$, then $D_y h = 0$.

NOTE: This rule says that the derivative of $y = 6$ is 0. We know that the graph of $y = 6$ is a horizontal line, whose slope is 0. We also know that the derivative of a function at a point equals the slope of its graph at that point. This rule confirms that every horizontal line has slope 0 at every point.

DERIVATIVE OF A LINEAR FUNCTION

If $f(x) = mx + b$, where m and b are numbers, then $f'(x) = m$.

PROOF: By the definition of derivative,

$$f(x) = \lim_{h \to 0} \frac{f(x + h) - f(x)}{h}$$

Therefore, if $f(x) = mx + b$,

$$f'(x) = \lim_{h \to 0} \frac{m(x + h) + b - (mx + b)}{h}$$
$$= \lim_{h \to 0} \frac{mx + mh + b - mx - b}{h}$$
$$= \lim_{h \to 0} \frac{mh}{h} = \lim_{h \to 0} m = m$$

Examples

If $f(x = 2x + 3$, then $f'(x) = 2$.
If $y = -5x + 1$, then $dy/dx = -5$.
If $h(t) = 1000 - \frac{1}{2}t$, then $dh/dt = -\frac{1}{2}$.
If $y = 3x$, then $D_x y = 3$.
If $f(x) = x$, then $f'(x) = 1$.
If $f(x) = 4$, then $f'(x) = 0$ (every constant function is a linear function for which $m = 0$).

Goal 12 | To Find Derivatives of Polynomials

NOTE 1: The graph of $f(x) = 2x + 3$ is a straight line whose slope is 2 (at every point). The rule for differentiation of a linear function tells us that the derivative of $f(x) = 2x + 3$ is also 2 (at every point).

NOTE 2: Suppose that the population of an animal colony after t months is given by $P(t) = 4t + 50$. To find the average rate of growth of the population over some time interval, say Δt, we can proceed as follows:

$$\frac{P(t + \Delta t) - P(t)}{\Delta t} = \frac{[4(t + \Delta t) + 50] - (4t + 50)}{\Delta t}$$

$$= \frac{4t + 4\Delta t + 50 - 4t - 50}{\Delta t} = \frac{4\Delta t}{\Delta t} = 4 \qquad \text{if } \Delta t \neq 0$$

Since the average rate of growth is thus constant (and equal to 4), we conclude that the rate of growth at any instant t is 4 (animals per month). If we interpret the derivative of P to be the rate of growth of the (population of the) colony, then the rule above tells us immediately, without any fuss or bother, that the rate of growth is 4 (animals per month) at any instant.

POWER RULE

If $f(x) = x^n$, where n is a positive integer, then $f'(x) = nx^{n-1}$.

Examples

If $f(x) = x^2$, then $f'(x) = 2x$; also $f'(3) = 2(3) = 6$, $f'(-1) = 2(-1) = -2$, $f'(0) = 2(0) = 0$.

If $g(u) = u^3$, then $dg/du = 3u^2$.

If $h(v) = v^7$, then $D_v h = 7v^6$.

If $f(x) = x$ (that is, equals x^1), then $f'(x) = 1 \cdot x^0 = 1 \cdot 1 = 1$.

NOTE: In earlier examples, we found the slope of the graph of $y = x^2$ to be $2x$ and that of $y = x^3$ to be $3x^2$. These functions are now just special cases of $f(x) = x^n$. The rule for differentiating this power function tells us that the slope of the graph of f is nx^{n-1}.

DERIVATIVE OF A CONSTANT TIMES A FUNCTION

If $g(x) = kf(x)$, where k is a constant, then $g'(x) = kf'(x)$.

Examples

If $g(x) = 3x^4$, then $g'(x) = 3(4x^3) = 12x^3$.

If $h(t) = -5t^2$, then $h'(t) = -5(2t) = -10t$.

$D_x 6(x - 1) = 6 D_x(x - 1) = 6 \cdot 1 = 6$.

If $y = x^5/100$, then $dy/dx = \frac{1}{100}(5x^4) = \frac{1}{20}x^4$.

The slope of the graph of $f(x) = 4x^2$ at the point where $x = -1$ is $f'(-1)$. Since $f'(x) = 4(2x) = 8x$, $f'(-1) = 8(-1) = -8$.

NOTE 1: This rule says in words that the derivative of a constant times a function equals the constant times the derivative of the function. In particular,

$$D_x kx^n = k(nx^{n-1}) = knx^{n-1}$$

NOTE 2: Suppose the population of one animal colony at any time is 10 times that of a second colony. If we interpret the derivative of the population function as the rate of growth of the population, then this rule tells us that the rate of growth of the first colony, at any time, is 10 times that of the second.

DERIVATIVE OF THE SUM OF TWO FUNCTIONS

If $w(x) = f(x) + g(x)$, then $w'(x) = f'(x) + g'(x)$.

Examples

If $w(x) = x^3 + 7x$, then $D_x w = D_x(x^3) + D_x(7x) = 3x^2 + 7$.
$D_t(0.1t^2 + 5) = D_t(0.1t^2) + D_t(5) = 0.2t + 0 = 0.2t$.

NOTE: This rule says that the derivative of the sum of two functions is equal to the sum of their individual derivatives. Suppose, for example, that the total cost of producing x gadgets is given by

$$C(x) = f(x) + g(x)$$

where $f(x)$ is the cost of materials and $g(x)$ is the cost of labor. If we interpret the derivative as the (instantaneous) rate of change, then the rule above tells us that the rate of change of the total cost, $C'(x)$, is equal to the rate of change $f'(x)$, of the cost of materials, added to the rate of change, $g'(x)$, of the cost of labor.

PROOF: Every one of the rules we have given can be rigorously proved using the definition of derivative, as we did above for $f(x) = mx + b$. To illustrate this further, we now prove that the derivative of the sum of two functions equals the sum of their derivatives.

Let $w(x) = f(x) + g(x)$. Since

$$w'(x) = \lim_{h \to 0} \frac{w(x+h) - w(x)}{h}$$

Goal 12 | To Find Derivatives of Polynomials

we have

$$w'(x) = \lim_{h \to 0} \frac{[f(x+h) + g(x+h)] - [f(x) + g(x)]}{h}$$

We now rearrange the fraction to "expose" $f'(x)$ and $g'(x)$, as follows:

$$w'(x) = \lim_{h \to 0} \frac{f(x+h) - f(x) + g(x+h) - g(x)}{h}$$

$$= \lim_{h \to 0} \left[\frac{f(x+h) - f(x)}{h} + \frac{g(x+h) - g(x)}{h} \right]$$

But since the limit of a sum of functions is the sum of their limits (p. 101), we have immediately

$$w'(x) = \lim_{h \to 0} \frac{f(x+h) - f(x)}{h} + \lim_{h \to 0} \frac{g(x+h) - g(x)}{h}$$

We note now that the first term in this sum is precisely, by definition, $f'(x)$, the derivative of f, and the second is $g'(x)$, the derivative of g. So

$$w'(x) = f'(x) + g'(x)$$

DERIVATIVE OF A POLYNOMIAL

The rules given above make it easy for us to calculate, almost as quickly as we can write it down, the derivative of any polynomial, since a polynomial is just the sum of functions of the type kx^n plus, perhaps, a constant.

Examples

$$D_x\left(\frac{2}{3}x^5 - 22x^3 + \frac{1}{2}x - 16\right)$$

$$= D_x\left(\frac{2}{3}x^5\right) + D_x(-22x^3) + D_x\left(\frac{1}{2}x\right) + D_x(-16)$$

$$= \frac{10}{3}x^4 + (-66x^2) + \frac{1}{2} + 0 = \frac{10}{3}x^4 - 66x^2 + \frac{1}{2}$$

If $f(x) = 3x^4 - x + 5$, we find $f'(-1)$ by first finding $f'(x)$ and then replacing x by -1, thus:

$$f'(x) = 12x^3 - 1$$
$$f'(-1) = 12(-1)^3 - 1 = -12 - 1 = -13$$

EXERCISES

Calculate the following derivatives:

1. $f'(x)$ if $f(x) = 5$
2. $D_x g$ if $g(y) = -3$
3. $h'(2)$ if $h(t) = -3$
4. $j'(-5)$ if $j(s) = \frac{1}{2}$
5. $\dfrac{dg}{dx}$ if $g(x) = 3x - 2$
6. $f'(y)$ if $f(y) = -\frac{1}{2}y + 2$
7. $k'(4)$ if $k(v) = -5v + \frac{1}{2}$
8. $h'(-2)$ if $h(t) = 3t - 1$
9. $D_x f$ if $f(x) = x^{87}$
10. $g'(t)$ if $g(t) = t^{33}$
11. $g'(y)$ if $g(y) = y^8 - 3y^5 + 2$
12. $\dfrac{dy}{dx}$ if $y = \dfrac{1}{2}x^4 + 3x^2 - 2$
13. $h'(1)$ if $h(t) = \frac{1}{2}t^8 + \frac{2}{3}t - 1$
14. $k'(-1)$ if $k(s) = -\frac{1}{2}s^{10} + 3s - 2$
15. $\dfrac{dy}{dx}$ if $y = x^{12} - x^{10} + x$
16. $\dfrac{dw}{dx}$ if $w = x^9 - 2x + 3$
17. $f'(x)$ if $f(x) = 12(x^3 - x^2 + 3x + 1)$
18. $g'(y)$ if $g(y) = \dfrac{y^3 + 5y^2 - 1}{12}$
19. $h'(-1)$ if $h(x) = \dfrac{x^2 - x + 1}{4} + 3(x^{10} - 2x^2)$
20. $f'\left(\dfrac{1}{2}\right)$ if $f(x) = 4(x^4 + 2) - \dfrac{x^3 - 5x}{2}$

TEST A

Calculate the following derivatives:

1. $f'(x)$ if $f(x) = 5x - 2$
2. $g'(-2)$ if $g(y) = 4$
3. $h'(t)$ if $h(t) = t^{10} - 3t^4 - 2t + 1$
4. $k'(1)$ if $k(s) = \frac{1}{10}(s^2 + 3s - 2)$

TEST B

Calculate the following derivatives:

1. $f'(x)$ if $f(x) = -1$
2. $g'(\frac{1}{2})$ if $g(t) = 3t + 1$

Goal 12 | To Find Derivatives of Polynomials

3. $\dfrac{dy}{dt}$ if $y = 3t^8 + 2t^5 - t - 1$

4. $f'(-1)$ if $f(y) = 4(y^8 - 2y^3 + 3y - 1)$

STUDY QUESTIONS

1. If f is the constant function $f(x) = 5$, what is $f(x + h)$? Calculate f' from the definition of the derivative as a limit.
2. If f is the linear function $f(x) = 3x + 2$, what is $f(x + h)$? Calculate $f'(1)$ from the definition of the derivative as a limit.
3. The binomial expansion of $(x + h)^n$ is equal to

$$x^n + nx^{n-1}h + \{\text{terms all involving } h^i \text{ where } i \geq 2\}$$

Suppose $f(x) = x^n$, where n is a positive integer. Show that

$$f(x + h) - f(x) = nx^{n-1}h + \{\text{terms involving } h^i \text{ where } i \geq 2\}$$

Then show that

$$\dfrac{f(x + h) - f(x)}{h} = nx^{n-1} + \{\text{terms involving } h_i \text{ where } i \geq 1\}$$

Calculate $f'(x)$ from the definition of the derivative as a limit.

4. If $f(x) = cg(x)$, where c is a constant, what is $f(x + h)$? From the definition of the derivative as a limit, show that $f'(x) = cg'(x)$.

GOAL **13**

To apply the derivative to problems involving rates of change

In previous Goals we concentrated on how to find the derivatives of various common functions. Now we shall use the derivative of a function to determine its rate of change. The first step in this process is to recognize when the solution of a problem requires the derivative.

SOME APPLICATIONS OF THE DERIVATIVE

Each of the following calls for the rate of change of a function: how fast one quantity changes relative to, or *with respect to*, another. This rate of change is precisely the derivative of the function. Familiarity with rates of change of the sort given here will help you recognize other such applications or interpretations as they arise:

1. The rate at which the y-coordinate, on the graph of $y = f(x)$, changes with respect to the x-coordinate: $f'(c)$ gives the *slope* of the curve at $x = c$.
2. The rate at which $s(t)$, the distance traveled by a car in t min, changes with respect to time: $s'(c)$ is the (instantaneous) *velocity* of the car at time c.
3. The rate at which the amount of waste $w(t)$ dumped into a river changes (with respect to time): $w'(c)$ tells how fast the amount of waste is increasing at time c.
4. The rate at which the size of a certain population $P(x)$ grows (with respect to time): at time c the rate of growth is $P'(c)$. The population may be composed of people, animals, bacteria, insects, or something else.

Goal 13 | To Apply the Derivative to Problems Involving Rates of Change

5. The rate at which a manufacturer's cost $C(x)$ changes relative to the production level x: $C'(q)$ gives the *marginal cost* when q items are produced.
6. In 5 replace "cost, C" everywhere by "revenue, R."
7. In 5 replace "cost, C" everywhere by "profit, P."
8. The rate at which the volume $V(x)$ of a solid changes with respect to the length of a dimension x: $V'(x_0)$ is the rate of change when $x = x_0$.
9. The rate at which the number of married couples $N(t)$ in a certain community is changing with respect to time, t: $N'(c)$ gives the *marriage rate* at time c.
10. In 9, replace "married couples, N" everywhere by "divorces, S."
11. The rate at which a small animal grows, where $w(t)$ is its weight after t weeks: $w'(10)$ is its *rate of growth* after 10 weeks.
12. The rate at which P dollars grows (with respect to time) if it is reinvested at $6\frac{1}{2}\%$ compounded continuously; $P'(c)$ tells the rate of growth at time c.
13. The rate at which the atmosphere is being polluted (with respect to time) by a factory that releases $H(t)$ tons of a chemical into the air in t h: $H'(p)$ gives the *pollution rate* at the end of p h.
14. The rate at which a disease is spreading (with respect to time) if $N(t)$ people are exposed to the disease in t weeks: $N'(2)$ is the rate of spread of the disease at the end of 2 weeks.
15. The rate at which a tumor grows relative to its radius r, where $V(r)$ is the tumor's volume: $V'(c)$ gives the rate of growth when it has a radius c.
16. The rate at which a person masters a skill, where $P(x)$ gives a measure of his success after x trials: $P'(c)$ is the rate of learning after c trials.
17. The rate at which one substance is being transformed into another in a chemical reaction, where $n(t)$ denotes the number of grams that have changed at time t: $n'(c)$ is the rate of change of the substance at time c.
18. The rate at which radium decays, where $A(t)$ gives the amount of radium present at time t: $A'(c)$ is the rate of decay at time c.
19. The rate at which a product is sold with respect to the amount of advertising, where $A(n)$ is the number of products sold after n dollars has been spent on advertising: $A'(q)$ is the rate of change in sales after an expenditure of q dollars on advertising.
20. The rate at which the noise level $N(a)$ (in decibels) changes at a cocktail party relative to the number of ounces a of alcohol consumed: $N'(q)$ is the rate of change after q oz has been consumed.

21. The rate at which an invalid's temperature $T(t)$ changes with respect to time: $T'(c)$ gives the rate of change at time c.
22. The rate at which a slang expression is adopted by a fixed population, where $N(t)$ is the number of people who have acquired it at time t: $N'(q)$ tells the rate of adoption at time q.
23. The rate at which a rumor spreads in a community, where $N(t)$ is the percentage of the people in the community who have heard it at time t: $N'(c)$ gives the rate of spread at time c.
24. The rate at which a brush fire is spreading if it is moving along a straight line and its distance $s(t)$ is the number of miles it has covered in t h: $s'(t)$ is the velocity at which the fire is spreading.
25. The rate at which the value of a Carson City silver dollar dated 1890 changes if $v(t)$ is its value t yr after it was minted: $v'(80)$ tells the rate of change in its value in 1970.
26. The rate at which the amount of medication in a person's body decreases with time, where $A(t)$ is the amount in the body t h after a dose has been administered: $A'(5)$ is the rate of decrease 5 h after it was administered.

FINDING RATE OF CHANGE

Example 13-1

Suppose a stream is being polluted by chemical waste and the function $w(t) = 0.1t^2 + 1.8t$ gives the quantity of waste, in tons, that has been dumped into the river after t days. How fast is the amount of waste increasing at the end of 3 days?

The solution calls for a derivative, since the question asks how fast one quantity (the amount of waste) is changing relative to a second quantity (time). In fact, we want $w'(t)$ evaluated at $t = 3$. We have

$$w'(t) = 0.1(2t) + 1.8 = 0.2t + 1.8$$

So

$$w'(3) = 0.2(3) + 1.8 = 2.4 \text{ tons/day}$$

Example 13-2

A viral infection spreads through a population so that $V(t) = 130t + 10t^2$ people are exposed to the virus in t weeks. At what rate is exposure to the virus spreading at the end of 4 weeks?

We need to compute $V'(4)$. Since

$$V'(t) = 130 + 20t$$

Goal 13 | To Apply the Derivative to Problems Involving Rates of Change

we have

$$V'(4) = 130 + 20(4) = 210 \text{ people/week}$$

One implication is that during the next day (one-seventh of a week) approximately $\frac{1}{7}(210) = 30$ people will be exposed to the virus.

Example 13-3

The cost of producing x gadgets is $C(x) = 5x + 1000$ dollars, and the revenue from x gadgets is $R(x) = 0.1x^2 + 30x$ dollars. The profit, therefore, is given by

$$P(x) = R(x) - C(x) = 0.1x^2 + 25x - 1000$$

Find the marginal profit when the production level is 500 gadgets.
We want $P'(500)$. Since

$$P'(x) = 0.2x + 25 \quad \text{and} \quad P'(500) = 0.2(500) + 25$$

the answer is \$125 per gadget. One implication is that if the production level is dropped to 495 gadgets (that is, *decreased* from 500 by 5 units) then the profit will change by approximately $(-5) \times 125$, or $-\$625$. That is, the profit will *decrease* by approximately \$625. Compare this answer with the actual decrease in profit when production drops from 500 to 495 gadgets.

Example 13-4

An injection of x g of a drug results in a decrease in blood pressure of $D(x) = 0.5x^3 - 4x$ millimeters (mm) of mercury. Find the sensitivity to 4 g of this drug.

Sensitivity is defined as the rate of change of blood pressure (in millimeters of mercury) with respect to dosage. We want to find $D'(x)$, then evaluate this derivative when $x = 4$.

Since $D'(x) = 1.5x^2 - 4$, $D'(4) = 1.5(16) - 4 = 20$. The sensitivity to an injection of 4 g of the drug is therefore 20 mm of mercury per g. This implies that an additional $\frac{1}{4}$ g of the drug (from 4 up to 4.25 g) will produce a drop in the blood pressure of approximately 0.25×20, or 5 mm (of mercury).

EXERCISES

1. If a car moving down a straight road covers $t^2 + 4t$ miles in t hours, what is its instantaneous velocity at the end of 3 h?
2. Suppose that a particle moves along a straight line so that it goes $t^3 + 8t^2 + 5t$ ft in t sec. What is its (instantaneous) velocity at the end of 5 sec?

3. What is the slope of the curve $y = \frac{1}{2}x^2 - 3x + 2$ at the point $(2, -2)$?
4. What is the slope of the curve $y = -x^3 + x^2 - x + 1$ at the point $(-1, 4)$?
5. If a colony of bacteria grows so that at the end of t h there are $5000(t^2 + 0.01t + 1)$ bacteria present, how fast is the colony growing at the end of 1 hour?
6. Suppose that a small animal is growing so that it weighs, when it is t weeks old, $100(1 + 0.2t)$ g. How fast is the animal growing at the end of 11 weeks?
7. If the cost of producing x units of a particular product is $0.03x^2 - 0.9x + 650$ dollars, what is the marginal cost when the production level is 1000 units?
8. If the cost of producing x units of a particular product is $0.35x^2 - 0.2x + 500$ dollars, what is the marginal cost at a production level of 200 units?
9. If $3000(1 + 3t)$ gallons of waste are dumped into a river in t days, how fast is the amount of waste increasing at the end of 8 days?
10. During a 24-h period, $0.1t^2 + 5t$ tons of a benign chemical are released into the air through a factory chimney in t h. At what rate is the atmosphere being enriched at the end of 20 h?
11. During a flu epidemic, which lasts 20 days, $10t^2 - 2t^3 + 600t$ people are sick at the end of t days. How fast is the flu spreading at the end of 10 days?
12. Rabies is spreading through a squirrel population so that when t weeks have passed there are $t^2 + 3t$ infected animals. How fast is the disease spreading at the end of 5 weeks?
13. The cost of producing x handbicks is $10x + 500$ dollars, and the revenue from x handbicks is $22x$ dollars. What is the marginal profit when the production level is 387 handbicks?
14. The cost of producing x hickbands is $0.01x^2 + 6x + 1000$ dollars and the revenue is $25x$ dollars. What is the marginal profit when the production level is 200 hickbands?
15. If x g of a medication brings about a decrease in blood pressure of $0.3x^2 - x$ mm of mercury, what is the sensitivity to 5 g of the drug?
16. If x cm^3 of a medication result in a decrease in body temperature of $0.1x^2 + 0.6x$ degrees, what is the sensitivity to 4 cm^3 of the drug?
17. If $100t - 0.1t^3$ units are produced on a factory assembly line in t h, at what rate are units being produced at the end of 6 h?
18. Suppose that in a chemical reaction $t^2 + 3t$ g of one substance is changed into another in t sec. How fast is the transformation occurring after 5 sec?

Goal 13 | To Apply the Derivative to Problems Involving Rates of Change

19. Suppose a botanist determines that a particular tree will be $0.1t^2 + t$ ft high in t yr. How fast is the tree growing at the end of 10 yr?
20. A tunnel is dug horizontally through the base of a mountain. At a point x ft from the entrance, the surface of the mountain just above it is at an altitude of $0.3x + 2000$ ft above sea level. What is the slope of the mountain over the point in the tunnel 550 ft from the entrance?

TEST A

1. If $0.7t^2 + 12t$ tons of waste is dumped into the ocean in t days, how fast is the amount of waste increasing at the end of 10 days?
2. If $85t + 6t^2$ people are exposed to a bacterial infection in t weeks, at what rate is exposure to the infection increasing at the end of 5 days?
3. If the cost of producing x units is $9x + 350$ dollars and the revenue from x units is $15x$ dollars, what is the marginal profit when the production level is 900 units?
4. If the daily output of a copper mine is $55t - 0.1t^3$ tons after t h of operation, what is the rate of output at the end of 6 h?

TEST B

1. If $0.4t^2 + 8t$ gallons (gal) of waste is dumped into a lake in t days, how fast is the amount of waste increasing at the end of 5 days?
2. If a bacterial colony has grown to a size of $1000(3t + t^2)$ when t hours have elapsed, how fast is the colony increasing at the end of 3 h?
3. If the cost of producing x units is $0.02x^2 + 25x + 1500$ dollars and the revenue from x units is $85x$ dollars, what is the marginal profit when the production level is 500 units?
4. Suppose that t h after a piece of meat is put into a freezer, its temperature is $72 - 15t$ °F. How fast is the temperature of the meat falling 3 h after it is placed in the freezer?

Summary of Goals 7–13

You should now be able to

(1) obtain approximations to the rate of change of a function by finding the change over a small interval;

(2) find the slope of the graph of $f(x)$ at $x = c$ by evaluating the quotient $\dfrac{f(c + h) - f(c)}{h}$ as h gets closer and closer to zero;

(3) use the definition of the derivative of $f(x)$ to find $f'(c)$ or $f'(x)$;

(4) recognize and use different notations for the derivative;

(5) evaluate the limit of $f(x)$ as x approaches c, if the limit exists;

(6) recognize when a function becomes infinite;

(7) find left- and right-hand limits of a function at a point and use these to decide whether the limit of the function exists at the point;

(8) determine if a function is continuous at a specific point;

(9) find any points of discontinuity of a function;

(10) use formulas to find the derivative of any polynomial;

(11) use the derivative of a function to find its rate of change at a specified instant;

(12) recognize that the following are rates of change: slope of a graph; velocity of a moving object; marginal cost, revenue, or profit; rate of growth, of decay, or of pollution; rate of spread of a disease or of a rumor; sensitivity to a drug; and so on.

Review Exercises

GOAL 7

1. Find three approximations to the slope of the graph of $f(x) = x^2 + x + 1$ when $x = 1$, using the intervals from $x = 1$ to $x = 1.2$, from $x = 1$ to $x = 1.1$, and from $x = 0.9$ to $x = 1$.

2. At the end of t sec a stone thrown directly upward is $80t - 16t^2$ ft above the ground. Find two approximations to its velocity at the end of 2 sec, using the intervals from $t = 2$ to $t = 2.1$ and from $t = 2$ to $t = 2.01$.

GOAL 8

3. Find the slope of the graph of $f(x) = 2x^2 + 1$ at $(2, 9)$ by evaluating the quotient $\dfrac{f(2 + h) - f(2)}{h}$ as h gets closer and closer to 0.

4. Find the slope of the graph of $f(x) = \dfrac{4}{x}$ at $\left(\dfrac{1}{2}, 8\right)$ by evaluating

$$\dfrac{f\left(\dfrac{1}{2} + h\right) - f\left(\dfrac{1}{2}\right)}{h}$$

as h gets closer and closer to zero.

GOAL 9

5. From the definition of the derivative as a limit, find (a) $f'(2)$ if $f(x) = 2x^2 + 1$ (compare Exercise 3 above); (b) $f'(\tfrac{1}{2})$ if $f(x) = \dfrac{4}{x}$ (compare Exercise 4 above); (c) $g'(x)$ if $g(x) = 3 - x^2$; (d) $s'(t)$ if $s(t) = t^3 - t^2$; (e) $f'(1)$ if $f(u) = \dfrac{1}{2 - u}$.

GOAL 10

6. Find each of the following limits:

   (a) $\lim\limits_{x \to 2} (x^3 - x^2 + 4)$

   (b) $\lim\limits_{x \to 1} 5$

   (c) $\lim\limits_{t \to 3} (t - 3)^6$

   (d) $\lim\limits_{x \to 2} \dfrac{x^2 - 4}{x - 2}$

   (e) $\lim\limits_{y \to 4} \dfrac{y^3 - 4y^2}{y - 4}$

   (f) $\lim\limits_{h \to 0} \dfrac{4(1 + h)^2 - 4}{h}$

   (g) $\lim\limits_{h \to 0} \dfrac{[1 - 3(x + h)^2] - (1 - 3x^2)}{h}$

7. Complete each statement:

   (a) $\lim\limits_{x \to 0} \dfrac{1}{x^2} =$

   (b) $\lim\limits_{x \to 2^+} \dfrac{1}{x - 2} =$

   (c) $\lim\limits_{x \to 2^-} [x] =$

   (d) $\lim\limits_{x \to 2^+} [x] =$

(e) $\lim\limits_{x \to 5^+} \dfrac{4}{5-x} =$ (f) $\lim\limits_{x \to 6} |x| =$

8. Find any limit that exists:

   (a) $\lim\limits_{x \to 1} [x]$ (c) $\lim\limits_{x \to 1} \dfrac{1}{(x-1)^2}$

   (b) $\lim\limits_{x \to 0} |x|$ (d) $\lim\limits_{x \to 1^-} [x]$

9. Give an example of a function $f(x)$ for which:

   (a) $f(c)$ is not defined, but $\lim f(x)$ as $x \to c$ exists.

   (b) $\lim f(x)$ as $x \to c$ does not exist, but $f(x)$ does not become infinite as $x \to c$.

GOAL 11

10. If any of the following functions are not continuous at the point specified, tell why:

    (a) $f(x) = x(x-1)(x-2)$ at $x = 1$

    (b) $g(x) = \begin{cases} \dfrac{1}{x^2} & \text{if } x \ne 0 \\ 0 & \text{if } x = 0 \end{cases}$ at $x = 0$

    (c) $f(x) = \dfrac{4}{x^2 + 1}$ at $x = 0$

    (d) $h(x) = \dfrac{x^2 - 9}{x - 3}$ at $x = 3$

    (e) $h(x) = \begin{cases} \dfrac{x^2 - 9}{x - 3} & \text{if } x \ne 3 \\ 6 & \text{if } x = 3 \end{cases}$ at $x = 3$

    (f) $y = \begin{cases} 2x + 1 & \text{if } x \ne 0 \\ 0 & \text{if } x = 0 \end{cases}$ at $x = 0$

11. Find any points of discontinuity:

    (a) $f(x) = 4 - x^2$ (d) $f(x) = \dfrac{3x}{x^2 - 5x + 4}$

    (b) $g(t) = \dfrac{1}{t^2 + 1}$ (e) $f(x) = |x|$

    (c) $h(y) = \dfrac{5}{y^2 - 4}$ (f) $y = [x]$

GOAL 12

12. Find the following derivatives from formulas:
    (a) $f'(x)$ if $f(x) = 4$
    (b) $g'(x)$ if $g(x) = 3x^3 - x + 2$
    (c) $h'(-1)$ if $h(y) = y + 5$
    (d) $\dfrac{dy}{dx}$ if $y = 2x^2 + 7x - 9$
    (e) $D_t x$ if $x = 7t - t^5 - t^6$
    (f) $h'(\tfrac{1}{2})$ if $h(x) = \tfrac{1}{2} x^4 - \tfrac{1}{3} x^3 + 2x - 1$
    (g) $D_x y$ if $y = -\tfrac{1}{10} x^{10}$
    (h) $f'(2)$ if $f(x) = \tfrac{1}{2}$
    (i) $f'(100)$ if $f(x) = 0.01x^2 - 0.2x + 1000$

GOAL 13

13. A ball thrown vertically upward with initial velocity 48 ft/sec reaches a height $h(t) = 48t - 16t^2$ ft after t sec. Find its velocity after 1 sec.
14. Find the slope of the curve $y = x^3 - 4x^2 + 1$ at the point where $x = 2$.
15. At what point on the curve $y = x^3 - 4.32x + 4$ is the slope equal to 0?
16. During a flu epidemic that lasts 22 days, the number of people who have been infected at the end of t days is $N(t) = 288 + 30t^2 - t^3$. How fast is the flu spreading at the end of 5 days?
17. If the Groovy Record Company sells a certain record album at y dollars each, then its revenue is $14{,}000y - 2000y^2$ and its profit is $-33{,}000 + 18{,}000y - 2000y^2$ (both in dollars). Find the company's marginal revenue and marginal profit when the albums sell for $3 each.
18. The volume, in gallons, of oil left in a leaky tank after t sec is given by $V = 1500 - 10t + 0.05t^2$. How fast is the oil leaking out when 20 sec have elapsed?

Chapter Test A

1. Find two approximations to the slope of the graph of $f(x) = 2x^2 - 1$ when $x = 3$, using the intervals from $x = 3$ to $x = 3.1$ and from $x = 2.9$ to $x = 3$.

2. Find the (exact) slope of the graph of $f(x) = 2x^2 - 1$ when $x = 3$ by evaluating the quotient $\dfrac{f(3+h) - f(3)}{h}$ as h gets closer and closer to 0.

3. Use the definition of derivative to find

   (a) $f'(x)$ if $f(x) = 3 - 2x^2$ (b) $g'(-1)$ if $g(y) = \dfrac{3}{y}$

4. Find any of the following limits that exist:

   (a) $\lim\limits_{x \to -1} (3 - x - 2x^2)$ (e) $\lim\limits_{x \to 0} |x|$

   (b) $\lim\limits_{x \to 0} \dfrac{x^2 - 4x}{x}$ (f) $\lim\limits_{x \to 1} [x]$

   (c) $\lim\limits_{x \to 2} \dfrac{4}{x^2 + 4}$ (g) $\lim\limits_{x \to 0^+} [x]$

   (d) $\lim\limits_{x \to 2} \dfrac{4}{x^2 - 4}$ (h) $\lim\limits_{x \to 0} \dfrac{x+1}{x^2}$

5. Give an example of a function $f(x)$ defined at $x = c$ and such that $\lim f(x)$ as $x \to c$ exists but $\lim f(x)$ as $x \to c$ is not equal to $f(c)$.

6. For what values of x, if any, is the function $f(x) = \dfrac{x^2 - 1}{x^2 + 2x}$ not continuous?

7. Is the following function continuous at $x = 2$? Why or why not?

$$f(x) = \begin{cases} x^2 + 1 & \text{if } x < 2 \\ 5 & \text{if } x = 2 \\ 3x - 1 & \text{if } x > 2 \end{cases}$$

8. Use formulas to find the following derivatives:

   (a) $D_x (\tfrac{3}{4} x^4 - 7x^3 + x - 9)$
   (b) $f'(-2)$ if $f(x) = 2x^3 - x^2 + 5$
   (c) $\dfrac{dy}{dx}$ if $y = -2$
   (d) $h'(0)$ if $h(t) = 9 - 9t$

9. At what points on the curve $y = x^4 - 4x^3 + 10$ does the slope of the curve equal zero?

10. A demographer estimates that t yr from now the population of a city will be $P(t) = (\tfrac{1}{6} t^3 - \tfrac{3}{2} t^2 + \tfrac{13}{3} t + 14)$ thousand people ($t \leq 6$). At what rate will the population be growing 3 yr from now?

11. A puka-shell dealer knows he can sell x necklaces at $\left(22.5 - \dfrac{x}{400}\right)$ dollars each. Find his marginal revenue when $x = 2000$.

Chapter 3
Differentiation Techniques

In this chapter we present rules and techniques for differentiating functions which make it unnecessary to resort to using the definition of derivative for such functions. Among these are the rules for finding derivatives of the product and quotient of functions and of functions involving fractional powers. Of great importance is the use of the chain rule in differentiating composite functions. To find a derivative when one variable cannot be expressed explicitly in terms of another, we also introduce in this chapter a technique known as implicit differentiation. Finally, we discuss derivatives of derivatives, called higher derivatives, and present several applications of the second derivative of a function.

Goal 14. To use the product and quotient rules
Goal 15. To use the chain rule
Goal 16. To find derivatives by implicit differentiation
Goal 17. To find higher derivatives
Summary
Review Exercises
Chapter Test A

GOAL 14

To use the product and quotient rules

If we can recognize a given function as the product or the quotient of other functions, then the product and quotient rules enable us to differentiate the given function.

THE DERIVATIVE OF A PRODUCT

Although the derivative of the sum of two functions is equal to the sum of their individual derivatives, the derivative of a product is *not* equal to the product of their derivatives. For example, if

$$f(x) = x^2 \quad \text{and} \quad g(x) = 2x + 1$$

then

$$f(x)g(x) = x^2(2x + 1) = 2x^3 + x^2$$

So the derivative of the product is given by

$$D_x[f(x)g(x)] = D_x[2x^3 + x^2] = 6x^2 + 2x$$

On the other hand, since

$$f'(x) = 2x \quad \text{and} \quad g'(x) = 2$$

the product of the derivatives equals

$$f'(x)g'(x) = (2x)(2) = 4x$$

and clearly, in general (that is, except when $x = 0$ or $\tfrac{1}{3}$),

$$6x^2 + 2x \neq 4x$$

THE PRODUCT RULE

This rule says that, if $f(x)$ and $g(x)$ are functions, then

$$D_x[f(x)g(x)] = f(x) \, D_x g(x) + g(x) \, D_x f(x)$$
$$= f(x)g'(x) + g(x)f'(x)$$

Or in words:

The derivative of the product of two functions is equal to the first function times the derivative of the second function plus the second function times the derivative of the first function.

Example 14-1

If $f(x) = x^4$ and $g(x) = x^3$, then the rule tells us that

$$D_x[x^4 \cdot x^3] = x^4 \cdot D_x x^3 + x^3 \cdot D_x x^4$$
$$= x^4 \cdot 3x^2 + x^3 \cdot 4x^3$$
$$= 3x^6 + 4x^6$$
$$= 7x^6$$

CHECK: Using the power rule (see p. 119), we get $D_x[x^4 \cdot x^3] = D_x(x^7) = 7x^6$, as above.

Example 14-2

If $f(x) = 3x^2 + 2x + 1$ and $g(x) = 5x^3 - x - 2$, then

$$\frac{d}{dx}[f(x)g(x)] = f(x)g'(x) + g(x)f'(x)$$
$$= (3x^2 + 2x + 1)(15x^2 - 1) + (5x^3 - x - 2)(6x + 2)$$

This answer may also be obtained by first multiplying the polynomials $f(x)$ and $g(x)$ together and then differentiating the resulting polynomial. But doing that would be much more cumbersome. Moreover, we will soon encounter examples of the type

$$f(x) = \sqrt{x^2 + 1} \qquad g(x) = x^3 + 5x - 1$$

where it doesn't make sense to multiply the functions out. Use of the product rule will often be unavoidable.

Example 14-3

The owner of an apartment building with 60 units has found that he can rent them all at $200 a month. However, for each $5 increase in the rent per

Goal 14 | To Use the Product and Quotient Rules

unit he loses one tenant. Express his revenue as a function of the number of $5 increases and find his marginal revenue.

This problem is the same as Example 6-4. Letting x be the number of $5 increases, we found that the revenue $R(x) = (60 - x)(200 + 5x)$. Using the product rule to find the marginal revenue yields

$$R'(x) = (60 - x) \cdot 5 + (200 + 5x)(-1)$$
$$= 300 - 5x - 200 - 5x$$
$$= 100 - 10x$$

Note that if we expand $R(x)$ and then find the derivative, we get

$$R(x) = 1200 - 200x + 300x - 5x^2$$
$$= 1200 + 100x - 5x^2$$
$$R'(x) = 100 - 10x$$

THE DERIVATIVE OF A QUOTIENT

The following example shows that the derivative of a quotient of two functions is *not* the quotient of their derivatives.

If $f(x) = 3x^4$ and $g(x) = x^3$, then

$$\frac{f(x)}{g(x)} = \frac{3x^4}{x^3} = 3x$$

and the derivative of this quotient is

$$\frac{d}{dx}\left(\frac{f(x)}{g(x)}\right) = \frac{d}{dx}(3x) = 3$$

On the other hand, since $f'(x) = 12x^3$ and $g'(x) = 3x^2$, the quotient of the derivatives is

$$\frac{f'(x)}{g'(x)} = \frac{12x^3}{3x^2} = 4x$$

and clearly, in general (that is, except just when $x = \frac{3}{4}$),

$$3 \neq 4x$$

The Quotient Rule

This rule says that, if $f(x)$ and $g(x)$ are functions, then, provided $g(x) \neq 0$,

$$D_x \frac{f(x)}{g(x)} = \frac{g(x) D_x f(x) - f(x) D_x g(x)}{[g(x)]^2} = \frac{g(x) f'(x) - f(x) g'(x)}{[g(x)]^2}$$

Or in words:

The derivative of the quotient of two functions is equal to another quotient whose denominator is the old denominator squared and whose numerator is the old denominator times the derivative of the old numerator minus the old numerator times the derivative of the old denominator.*

Example 14-4

Suppose $h(x) = \dfrac{2x^2 + 1}{3x - 5}$. We can let $f(x) = 2x^2 + 1$ and $g(x) = 3x - 5$.
Then

$$h'(x) = \frac{(3x - 5)(4x) - (2x^2 + 1)(3)}{(3x - 5)^2}$$

$$= \frac{12x^2 - 20x - 6x^2 - 3}{(3x - 5)^2} = \frac{6x^2 - 20x - 3}{(3x - 5)^2}$$

Example 14-5

Apply the quotient rule to find the derivatives of (a) $y = 1/x$; (b) $y = 1/x^2$.

(a) Letting $f(x) = 1$ and $g(x) = x$ in the rule, we get

$$D_x \frac{1}{x} = \frac{x \cdot 0 - 1 \cdot 1}{x^2} = -\frac{1}{x^2} = -x^{-2}$$

(b) If we let $f(x) = 1$ and $g(x) = x^2$, we get

$$D_x \frac{1}{x^2} = \frac{x^2 \cdot 0 - 1 \cdot 2x}{x^4} = -\frac{2x}{x^4} = -\frac{2}{x^3} = -2x^{-3}$$

---

*A student pointed out that if we think of a quotient as being in the form $\dfrac{\text{Hi}}{\text{Ho}}$ then the quotient rule tells us that its derivative is $\dfrac{\text{Ho D Hi} - \text{Hi D Ho}}{\text{Ho Ho}}$ (where D stands for an "appropriate" derivative).

Goal 14 | To Use the Product and Quotient Rules

Note that if we allow n to be a negative integer in x^n and apply the power rule

$$\frac{d}{dx} x^n = nx^{n-1}$$

we get

$$\frac{d}{dx}\left(\frac{1}{x}\right) = \frac{d}{dx} x^{-1} = -x^{-2}$$

$$\frac{d}{dx}\left(\frac{1}{x^2}\right) = \frac{d}{dx} x^{-2} = -2x^{-3}$$

Both results are those obtained above by applying the quotient rule.

Example 14-6

The *average cost* $A(x)$ is defined as the quotient $C(x)/x$. If it costs $C(x) = 2x^2 + 10x + 500$ dollars to produce x units, find the marginal average cost, $A'(x)$.

Since

$$A(x) = \frac{2x^2 + 10x + 500}{x}$$

we have

$$A'(x) = \frac{x(4x + 10) - (2x^2 + 10x + 500)(1)}{x^2} = \frac{2x^2 - 500}{x^2}$$

It can be shown that the average cost of production is least when it is equal to the marginal cost. (This point is taken up in Study Question 4.)

EXERCISES

1. Find the derivatives of the following functions:

   (a) $f(x) = (x + 1)(x^2 - 1)$

   (b) $g(y) = y^2(y + 3)$

   (c) $h(z) = \dfrac{z}{z + 1}$

   (d) $f(t) = \dfrac{t^2 + t - 1}{t}$

   (e) $C(x) = (x^3 - 5x^2 + x + 1)(3x^2 - x + 2)$

   (f) $A(y) = \dfrac{y - 2}{y - 3}$

   (g) $f(x) = \dfrac{1}{x^{10}}$

   (h) $g(x) = \dfrac{x + 1}{x^{20}}$

   (i) $y = \dfrac{5}{x^8}$

2. Find the derivatives of the following functions at the indicated points:
   (a) $F(x) = \dfrac{x-1}{x+1}$ at $x = 1$
   (b) $f(x) = (x^3 + 1)(x - 1)$ at $x = 1$
   (c) $g(t) = \dfrac{1}{t^2}$ at $t = -1$
   (d) $R(x) = x(100 - 2x - x^2)$ at $x = 2$
   (e) $g(x) = (x^2 + 5x + 1)(x^3 - x^2 - x - 1)$ at $x = 1$
   (f) $s(t) = \dfrac{t^2 + t + 1}{t^2 - t - 1}$ at $t = 2$

3. What is the slope of the curve $y = \dfrac{x+2}{x-1}$ at the point $x = 3$?

4. Find all points on the curve $y = \dfrac{1 + x^2}{x}$ at which the slope is 0.

5. Find all points on the curve $y = \dfrac{1 - x^2}{x}$ at which the slope is $-\tfrac{5}{4}$.

6. Suppose that the cost of producing x gadgets is $C(x) = 15x + 1000$ dollars.
   (a) What is the marginal cost function?
   (b) What is the average cost function?
   (c) What is the marginal average cost function?
   (d) What are the marginal cost, average cost, and marginal average cost at a production level of 100 gadgets?

7. Suppose that the cost of producing x gadgets is $C(x) = 40 + 20x + 0.1x^2$ dollars.
   (a) What is the marginal cost function?
   (b) What is the average cost function?
   (c) What is the marginal average cost function?
   (d) What are the marginal cost, average cost, and marginal average cost if the production level is 50 gadgets?

8. If a California grower plants 20 orange trees per acre, the average yield is 360 oranges per tree. For each additional tree in excess of 20 per acre, the grower gets 15 fewer oranges per tree.
   (a) Express the total number of oranges per acre as a function of the number of trees per acre in excess of 20. (See Exercise 6–7.)
   (b) Find the derivative of this function.

Goal 14 | To Use the Product and Quotient Rules

9. A box with no cover is to be made from a very thin piece of tin 8 in. by 12 in., by cutting away a square from each corner and folding up the sides. (For a similar problem, see Exercise 6–17.)
   (a) Write the volume of the box as a function of x, the length of an edge of the square removed.
   (b) Find the derivative of this volume function.
10. A printer is asked to prepare a rectangular poster on tagboard, with 96 sq in. of print. The margins at the top and bottom are to be 3 in.; the margin at each side is to be 2 in.
    (a) Express the area of the tagboard in terms of its height.
    (b) Find the derivative of this area function.
11. Suppose we assume that the temperature of a person who has been given x units of a drug is $T(x) = 37 + x^2(a - bx)$, where a and b are constants and where T is measured in degrees Celsius (37°C = 98.6°F).
    (a) Use the product rule to find $T'(x)$.
    (b) For what x is $T'(x)$ equal to zero?

TEST A

1. Find $f'(x)$ if $f(x) = (x^2 + x + 2)(x - 1)$.
2. What is $\dfrac{dy}{dx}$ if $y = \dfrac{1 - x}{2 - x}$?
3. What is the slope of the curve $y = \dfrac{3}{x + 2}$ at the point $(1, 1)$?

TEST B

1. Find $f'(x)$ if $f(x) = \dfrac{x}{1 - x^2}$.
2. What is $\dfrac{dy}{dx}$ if $y = (2 - x)(x^2 - 2x + 1)$?
3. What is the slope of the curve $y = \dfrac{x}{1 - x}$ at the point $(2, -2)$?

STUDY QUESTIONS

1. The following is an outline of a mathematical proof of the product rule. Give a reason for each line (that is, each step) in the proof.

Let $w(x) = f(x)g(x)$. Then

$$w'(x) = \lim_{h \to 0} \frac{w(x+h) - w(x)}{h}$$

$$= \lim_{h \to 0} \frac{f(x+h)g(x+h) - f(x)g(x)}{h}$$

$$= \lim_{h \to 0} \frac{f(x+h)g(x+h) \overbrace{- f(x)g(x+h) + f(x)g(x+h)}^{*} - f(x)g(x)}{h}$$

$$= \lim_{h \to 0} \left(g(x+h) \frac{f(x+h) - f(x)}{h} + f(x) \frac{g(x+h) - g(x)}{h} \right)$$

$$= g(x)f'(x) + f(x)g'(x).$$

*Notice that the proof hinges on the trick, used in this step, of adding zero (the quantity under the brace) to the numerator!

2. Use the product rule to show that the derivative of $cf(x)$, where c is a constant, is $cf'(x)$. (HINT: Interpret c to be the constant function $g(x) = c$.)

3. In Example 14-6 we defined the average cost $A(x)$ to be the quotient $\dfrac{C(x)}{x}$ and calculated the marginal average cost for a specific cost function. Find a general expression for $A'(x)$ by using the quotient rule.

In a similar manner, if revenue is given by $R(p) = p \cdot n(p)$, derive a general expression for the marginal revenue $R'(p)$.

4. Using the definition of average cost, $A(x) = C(x)/x$, show that $A'(x) = 0$ precisely when the average cost equals the marginal cost; that is, when $A(x) = C'(x)$.

5. If $f(x) = g(x)g(x)$ use the product rule to show that $f'(x) = 2g(x)g'(x)$.

6. Use the product rule to show that the derivative of x^2 is $2x$. (HINT: $x^2 = xx$.) Then use this result and the product rule to show that the derivative of x^3 is $3x^2$. One could continue in this manner, to develop the power rule that the derivative of x^n, if n is a positive integer, is nx^{n-1}.

GOAL 15

To use the chain rule

The chain rule is probably the most valuable tool for computing derivatives. With it we can easily differentiate such functions as $(x^2 + 1)^{50}$ or $\sqrt{x^3 - x + 2}$.

THE POWER RULE

In earlier goals we noted that the power rule

$$\frac{d}{dx} x^n = nx^{n-1}$$

is valid for n a negative as well as n a positive integer. In fact, this rule holds if n is any number. Thus

$$\frac{d}{dx} x^{10} = 10x^9 \qquad D_x x^{-3} = -3x^{-4}$$

$$\frac{d}{dx} x^{7/4} = \frac{7}{4} x^{3/4} \qquad D_x x^{-(5/2)} = -\frac{5}{2} x^{-(7/2)}$$

We also have, using the relationship between roots and exponents (see Goal 31):

$$\frac{d}{dx} \sqrt{x} = \frac{d}{dx} x^{1/2} = \frac{1}{2} x^{-1/2} = \frac{1}{2\sqrt{x}}$$

$$D_x \sqrt[3]{x} = \frac{d}{dx} x^{1/3} = \frac{1}{3} x^{-(2/3)} = \frac{1}{3\sqrt[3]{x^2}}$$

$$\frac{d}{dx} \frac{1}{\sqrt{x}} = \frac{d}{dx} x^{-(1/2)} = -\frac{1}{2} x^{-(3/2)} = -\frac{1}{2\sqrt{x^3}}$$

THE EXTENDED POWER RULE

Suppose u is a differentiable function of x (that is, it has a derivative at each point in its domain) and we want to find

$$\frac{d}{dx}(u^2)$$

Regarding u^2 as the product $(u \cdot u)$ and using the product rule, we get

$$\frac{d}{dx}(u^2) = \frac{d}{dx}(u \cdot u) = u\frac{du}{dx} + u\frac{du}{dx} = 2u\frac{du}{dx}$$

To find $(d/dx)(u^3)$, we can proceed as follows:

$$\frac{d}{dx}(u^3) = \frac{d}{dx}(u^2 \cdot u) = u^2\frac{du}{dx} + u\frac{d}{dx}u^2$$

$$= u^2\frac{du}{dx} + u\left(2u\frac{du}{dx}\right) \quad \text{(from above)}$$

$$= u^2\frac{du}{dx} + 2u^2\frac{du}{dx}$$

$$= 3u^2\frac{du}{dx}$$

Similarly, it can be shown that

$$\frac{d}{dx}(u^4) = 4u^3\frac{du}{dx}$$

In general, if n is a positive integer, we have

$$\boxed{\frac{d}{dx}u^n = nu^{n-1}\frac{du}{dx}}$$

If we replace u by $f(x)$, then

$$\boxed{\begin{aligned} D_x[f(x)]^n &= n[f(x)]^{n-1} D_x f \\ &= n[f(x)]^{n-1} f'(x) \end{aligned}}$$

This is called the *extended power rule*. In fact, the power rule $D_x x^n = nx^{n-1}$ is just a special case of the extended power rule, with $u = f(x) = x$ and $du/dx = f'(x) = 1$.

Goal 15 | **To Use the Chain Rule**

The extended power rule, like the special case of it above, holds in fact for any number n, not just for a positive integer. Remember that we restrict ourselves to those x-values in the domain of f for which $[f(x)]^{n-1}$ is defined.

Example 15-1

Find dy/dx if $y = (x^2 + 1)^3$. If we let $u = x^2 + 1$, we get $y = u^3$, $dy/dx = 3u^2 \, du/dx$. Since $du/dx = 2x$, therefore $dy/dx = 3(x^2 + 1)^2 \cdot 2x = 6x(x^2 + 1)^2$.

Note that if we expand y we get $y = x^6 + 3x^4 + 3x^2 + 1$. Then $dy/dx = 6x^5 + 12x^3 + 6x$. This can be rewritten $6x(x^4 + 2x^2 + 1) = 6x(x^2 + 1)^2$, showing that the two answers are identical. But using the extended power rule is both simpler and faster.

Example 15-2

Find $g'(x)$ if $g(x) = (x^2 + 1)^{50}$.

Let $u = x^2 + 1$. Then $g'(x) = 50u^{49}(du/dx) = 50(x^2 + 1)^{49} \cdot 2x = 100x(x^2 + 1)^{49}$.

It would be more awkward now to obtain the derivative by first expanding $(x^2 + 1)^{50}$. But it is just as easy to apply the extended power rule to $(x^2 + 1)^{50}$ as to $(x^2 + 1)^3$. In fact, it is just as painless to differentiate $(x^2 + 1)^{5000}$: $D_x(x^2 + 1)^{5000} = 5000(x^2 + 1)^{4999} \cdot 2x = 10{,}000x(x^2 + 1)^{4999}$. (Note that here we omitted the intermediate step of overtly setting u equal to $x^2 + 1$.)

Example 15-3

Find $D_x\sqrt{x^3 - x + 2}$.

$$D_x\sqrt{x^3 - x + 2} = D_x(x^3 - x + 2)^{1/2}$$
$$= \frac{1}{2}(x^3 - x + 2)^{-(1/2)}(3x^2 - 1) = \frac{3x^2 - 1}{2\sqrt{x^3 - x + 2}}$$

In this example, we thought: let $u = x^3 - x + 2$; then $D_x(x^3 - x + 2)^{1/2}$ equals $D_x u^{1/2}$, which equals $\frac{1}{2}u^{-(1/2)} D_x u$. Usually we do this step mentally instead of writing it out.

NOTE: When finding the derivative of a function raised to a power—that is, of $[f(x)]^n$—observe that

$$D_x[f(x)]^n \text{ does } not \text{ equal } n[f(x)]^{n-1}$$

except in the isolated case when $f(x) = x$. You must be especially careful to avoid this error. The derivative of a function of x to a power n is equal to n times the function to a power one less than n, *times the derivative of the function*.

Example 15-4

(a) $\dfrac{d}{dx}(x^3 + 5x^2 - x + 1)^{123} = 123(x^3 + 5x^2 - x + 1)^{122}(3x^2 + 10x - 1)$

(b) $\dfrac{d}{dx}(x - 3)^4 = 4(x - 3)^3 \cdot 1 = 4(x - 3)^3$

(c) $D_y(2y^4 - y^3 + 2)^{81} = 81(2y^4 - y^3 + 2)^{80}(8y^3 - 3y^2)$

(d) $D_t \dfrac{1}{\sqrt{4 - t^2}} = D_t(4 - t^2)^{-(1/2)} = -\dfrac{1}{2}(4 - t^2)^{-(3/2)}(-2t)$

$= \dfrac{t}{(4 - t^2)^{3/2}} = \dfrac{t}{\sqrt{(4 - t^2)^3}}$

THE CHAIN RULE

The extended power rule is a special case of the *chain rule*. Before stating this important rule we consider some examples.

Example 15-5

A car gets 30 miles per gal of gas. If its speed is such that it uses gas at the rate of $1\frac{1}{2}$ gal per h, what is the car's velocity in miles per hour?

The answer is gotten by multiplying the two given rates of change to obtain the required one:

$$30 \, \dfrac{\text{miles}}{\text{gal}} \times 1\dfrac{1}{2} \, \dfrac{\text{gal}}{\text{h}} = 45 \, \dfrac{\text{miles}}{\text{h}}$$

Example 15-6

Suppose that as a low front moves through a county, the temperature drops at a rate of 18 degrees Fahrenheit per hour. Given that there are $\frac{5}{9}$ of a degree Celsius per degree Fahrenheit, find the rate of drop in the temperature in the county in degrees Celsius per hour.

Again we multiply the two rates of change given to obtain the one sought:

$$\dfrac{5}{9} \, \dfrac{\text{degrees C}}{\text{degrees F}} \times 18 \, \dfrac{\text{degrees F}}{\text{hour}} = 10 \, \dfrac{\text{degrees C}}{\text{hour}}$$

Example 15-7

Suppose a factory makes a profit of $P(x) = 0.01x^2 + 25x - 1000$ dollars on x gadgets and the number of gadgets that can be produced after t h is $x(t) = 10t - 5$. Find the rate at which the profit is changing with respect to time.

There are two ways to do this problem.

Goal 15 | To Use the Chain Rule

Method 1. We can express the profit as a function of time and then find the derivative. Since $P(x) = 0.01x^2 + 25x - 1000$ and $x(t) = 10t - 5$, we have

$$P(x(t)) = 0.01(10t - 5)^2 + 25(10t - 5) - 1000$$
$$= t^2 + 249t - 1125.25 \quad (*)$$
$$\frac{dP}{dt} = 2t + 249$$

[Note that instead of expanding (*) we could have applied the extended power rule to (*) to obtain dP/dt directly:

$$\frac{dP}{dt} = 0.02(10t - 5) \cdot 10 + 25 \cdot 10$$
$$= 2t - 1 + 250 = 2t + 249]$$

Method 2. We can use the method of the preceding two examples, finding the required rate of change of profit with respect to time as the product of two rates of change: the rate of change of profit with respect to number (of gadgets produced) and the rate of change of number (of gadgets) with respect to time:

$$\frac{dP}{dt} = P'(x(t))x'(t) = \frac{dP}{dx} \times \frac{dx}{dt}$$

$$= (0.02x + 25) \frac{\text{dollars}}{\text{gadget}} \times 10 \frac{\text{gadgets}}{\text{hour}}$$

$$= (0.2x + 250) \frac{\text{dollars}}{\text{hour}}$$

$$= [0.2(10t - 5) + 250] \frac{\text{dollars}}{\text{hour}}$$

$$= (2t + 249) \frac{\text{dollars}}{\text{hour}}$$

STATEMENT OF THE CHAIN RULE

The preceding examples lead to the following:

If y is a differentiable function of u and u is a differentiable function of x, then y is a differentiable function of x and

$$\frac{dy}{dx} = \frac{dy}{du} \times \frac{du}{dx}$$

This shows why it is useful to recognize a function as a composite of other functions that are easier to differentiate. (Study Question 2 notes that the extended power rule is a special case of the chain rule.)

Example 15-8

Use the chain rule to evaluate dP/dt in Example 15-7 when $t = 2$. Since $P(x) = 0.01x^2 + 25x - 1000$ and $x(t) = 10t - 5$, we have

$$\frac{dP}{dt}\bigg|_{(t=2)} = \frac{dP}{dx}\bigg|_{(t=2)} \times \frac{dx}{dt}\bigg|_{(t=2)}$$
$$= (0.02x + 25)_{(t=2)} \times 10_{(t=2)}$$

Since $x(2) = 10 \cdot 2 - 5 = 15$,

$$\frac{dP}{dt}\bigg|_{(t=2)} = [0.02(15) + 25]10 = \$253/\text{per h}$$

Note especially that when we use the formula $\dfrac{dy}{dx} = \dfrac{dy}{du} \cdot \dfrac{du}{dx}$ to find $\dfrac{dy}{dx}$ for a particular value of x, say x_0, we must evaluate $\dfrac{dy}{du}$ at $u = u_0$, where $u_0 = u(x_0)$.

Example 15-8

Find $f'(0)$ if $f(x) = (x^3 + 5x - 1)\sqrt{x^2 + 1}$.
We rewrite $\sqrt{x^2 + 1}$, then use the product rule to find $f'(x)$:

$$f(x) = (x^3 + 5x - 1)(x^2 + 1)^{1/2}$$
$$f'(x) = (x^3 + 5x - 1) D_x(x^2 + 1)^{1/2} + (x^2 + 1)^{1/2} D_x(x^3 + 5x - 1)$$

In the next step we apply the chain rule (extended power rule) to $D_x(x^2 + 1)^{1/2}$:

$$f'(x) = (x^3 + 5x - 1)\tfrac{1}{2}(x^2 + 1)^{-(1/2)} \cdot 2x + (x^2 + 1)^{1/2}(3x^2 + 5)$$

At this point it is most efficient to compute $f'(0)$ by replacing x wherever it occurs by 0:

$$f'(0) = (0 + 0 - 1)\tfrac{1}{2}(1)^{-(1/2)} \cdot 0 + (0 + 1)^{1/2} \cdot (0 + 5)$$
$$= 0 + 1 \cdot 5 = 5$$

Goal 15 | To Use the Chain Rule

ANOTHER FORM OF THE CHAIN RULE

If $y = g(u)$ and $u = h(x)$, then it follows that $y = g(h(x))$, which shows that y is actually the composition of g and h (see Goal 5). Now we can rewrite the chain rule $\dfrac{dy}{dx} = \dfrac{dy}{du} \cdot \dfrac{du}{dx}$ as

$$\frac{dy}{dx} = g'(u)h'(x) = g'(h(x))h'(x)$$

A proof of this form of the chain rule is outlined in Study Question 3.

In particular, if we set $y = f(x)$ and if we know that $h(x)$ is differentiable at x_0 and $g(u)$ is differentiable at $u_0 = h(x_0)$, then we have

$$f'(x_0) = g'(h(x_0)) \cdot h'(x_0)$$

Example 15-9

If $f(x) = g(h(x))$, where $g(u) = u^2 + 1$ and $u = h(x) = 3x - 1$, find $f'(2)$.
From the equation for $f'(x_0)$ above, we have, with $x_0 = 2$,

$$f'(2) = g'(h(2)h'(2)$$

Since $h(x) = 3x - 1$, $h(2) = 5$; since $g'(u) = 2u$, $g'(5) = 10$; since $h'(x) = 3$, $h'(2) = 3$. So

$$f'(2) = 10 \cdot 3 = 30$$

EXERCISES

In 1 through 22 find the derivative of the function given:

1. x^{33}
2. x^{-10}
3. $x^{4/3}$
4. $x^{9/7}$
5. $x^{-(3/4)}$
6. $x^{-(7/3)}$
7. $x^{1/4}$
8. $x^{2/5}$
9. $(x + 1)^{10}$
10. $(x - 3)^{-5}$
11. $(5x + 2)^3$
12. $(8x - 4)^{-2}$
13. $\dfrac{1}{x - 1}$
14. $\dfrac{1}{(5x + 1)^2}$
15. $(x^3 + x^2 - 1)^{10}$
16. $(5x^4 - x + 3)^{-2}$
17. $\dfrac{1}{x^3 - 1}$

18. $\dfrac{1}{(x^2 + 1)^5}$

19. $(x^2 + 1)^{3/2}$

20. $(x^3 - x)^{1/2}$

21. $\sqrt{x^4 - 1}$

22. $\dfrac{1}{\sqrt{x^2 + x + 1}}$

In 23 through 32 use the chain rule together with either the product or the quotient rule to find the derivative of the function given:

23. $x(x^2 + 1)^{10}$

24. $x^2(x - 1)^8$

25. $x\sqrt{x + 1}$

26. $x(x^2 - 1)^{-5}$

27. $(x + 1)^5(x - 1)^4$

28. $(x^2 + x)^2(x^3 - 1)^4$

29. $\dfrac{(x + 1)^3}{(x^2 + 1)^4}$

30. $\dfrac{\sqrt{x}}{x}$

31. $\left(\dfrac{x + 1}{x - 1}\right)^5$

32. $\left(\dfrac{x + 3}{x^2 - 1}\right)^{10}$

Find the derivatives at the points indicated:

33. $f'(2)$ if $f(x) = (2x + 1)^3$

34. $f'(1)$ if $f(x) = g(h(x))$ where $g(u) = u^2 + u$ and $u = h(x) = x^5 + 1$

35. $g'(-1)$ if $g(x) = x(x + 2)^{10}$

36. $h'(3)$ if $h(x) = f(g(x))$ where $f(u) = u^3 - 1$ and $u = g(x) = 2x - 3$

TEST A

Find the derivative of each of the following functions:

1. $x^{5/2}$
2. $(3x^2 - x)^{-5}$
3. $\sqrt{x + 1}$
4. $x(2x + 1)^{10}$

TEST B

Find the derivative of each of the following functions:

1. $x^{-(2/3)}$
2. $(5x^3 - 2x)^{15}$
3. $\sqrt{x^2 - 1}$
4. $x(x^2 + 1)^5$

Goal 15 | To Use the Chain Rule

STUDY QUESTIONS

1. x^{12} can be written $(x^2)^6$, $(x^3)^4$, $(x^4)^3$, and $(x^6)^2$. For each composite function, use the chain rule to compute the derivative and show that in every case the result is $12x^{11}$.

2. Show that the extended power rule

$$\frac{d}{dx}[f(x)]^n = n[f(x)]^{n-1} \cdot \frac{df}{dx}$$

is a special case of the chain rule $\frac{dy}{dx} = \frac{dy}{du} \cdot \frac{du}{dx}$. (Examples 15-1 through 15-4 all illustrate this point.)

3. The following provides further justification for the chain rule: Suppose $y = g(u)$ and $u = h(x)$. The diagram shows what happens when x takes on an increment Δx. Give a reason for each equals sign in the computation below the diagram.

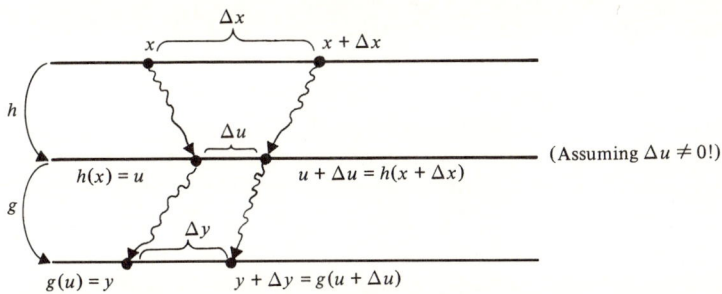

$$\frac{dy}{dx} = \lim_{\Delta x \to 0} \frac{g(h(x + \Delta x)) - g(h(x))}{\Delta x}$$

$$= \lim_{\Delta x \to 0} \frac{g(h(x + \Delta x)) - g(h(x))}{h(x + \Delta x) - h(x)} \times \frac{h(x + \Delta x) - h(x)}{\Delta x}$$

$$= \lim_{\Delta x \to 0} \frac{g(u + \Delta u) - g(u)}{\Delta u} \times \frac{h(x + \Delta x) - h(x)}{\Delta x}$$

$$= \lim_{\Delta x \to 0} \frac{g(u + \Delta u) - g(u)}{\Delta u} \times \lim_{\Delta x \to 0} \frac{h(x + \Delta x) - h(x)}{\Delta x}$$

$$= g'(u)h'(x) = \frac{dy}{du} \cdot \frac{du}{dx}$$

GOAL 16

To find derivatives by implicit differentiation

It is possible to find dy/dx even when y is not expressed explicitly in terms of x. That is, given an equation such as $x^2 - 3xy + 2y^2 = 8$, one can find dy/dx without having to solve first for y.

EXPLICIT VERSUS IMPLICIT FUNCTIONS

If we can write $y = f(x)$, where y stands alone on one side of the equation and the other side does not involve the variable y, then y is said to be an *explicit function* of x, or to be defined *explicitly*. For example,

$$y = x^2 \qquad y = \frac{4}{\sqrt{3-x}} \qquad y = 2x^{3/2} - \frac{1}{x} + 7$$

are all explicit functions of x. But equations such as

$$x^2 + y^2 = 4 \qquad y^2 = x \qquad x^2 - 3xy + 2y^2 = 8 \qquad y^3x - 3y + 4x^2 - 6 = 0$$

are said to define y *implicitly*. No one of these equations has been solved (explicitly) for y in terms of x. In some cases this can be done quite easily. For example, if we solve the equation $y^2 = x$ explicitly for y in terms of x, we get $y = \pm \sqrt{x}$. Thus, two functions of x defined implicitly by the equation $y^2 = x$ are $y_1 = \sqrt{x}$ and $y_2 = -\sqrt{x}$, whose graphs are shown on page 157. However, in many cases it is difficult to solve for y in terms of x; indeed, sometimes it is impossible. Fortunately, the explicit expression is not necessary in order to find the derivative. This is because of the very powerful chain rule.

Goal 16 | **To Find Derivatives by Implicit Differentiation**

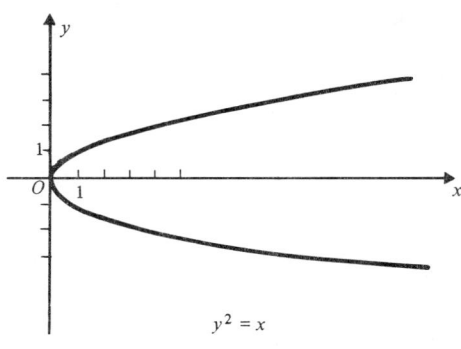
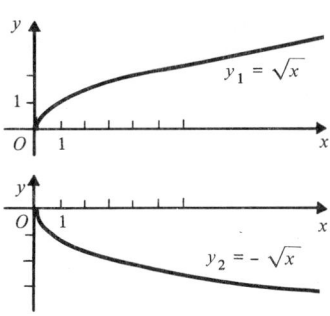

IMPLICIT DIFFERENTIATION

Implicit differentiation is the technique for finding a derivative without solving explicitly for the dependent variable in terms of the independent variable. Since this technique is based on the chain rule, we note in preparation that each of the following is an application of the chain rule:

(1) If $z = (x^2 + 1)^4$, then $dz/dx = 4(x^2 + 1)^3 \cdot 2x$.
(2) If $z = (5 - 2x^3)^7$, then $dz/dx = 7(5 - 2x^3)^6 \cdot (-6x^2)$.
(3) If $z = y^9$, and y is a differentiable function of x, then

$$\frac{dz}{dx} = 9y^8 \cdot \frac{dy}{dx} \quad \text{or} \quad \frac{d}{dx} y^9 = 9y^8 \frac{dy}{dx}$$

(4) If $z = y^2$, and y is a differentiable function of x, then

$$\frac{dz}{dx} = 2y \cdot \frac{dy}{dx} \quad \text{or} \quad \frac{d}{dx} y^2 = 2y \cdot \frac{dy}{dx}$$

Note especially that each of the following statements is *false*. Why?

$$\frac{d}{dx} y^9 = 9y^8 \qquad \frac{d}{dx} y^2 = 2y$$

We are now ready to illustrate implicit differentiation. In each example that follows we assume that there is (at least) one differentiable function y, of x, defined implicitly by the given equation.

Example 16-1
Find dy/dx if $x^2 + y^2 = 4$.

We differentiate term by term with respect to x, remembering that y is a function of x:

$$\frac{d}{dx}(x^2) + \frac{d}{dx}(y^2) = \frac{d}{dx}(4)$$

$$2x + 2y \cdot \frac{dy}{dx} = 0$$

Now we solve for dy/dx:

$$\frac{dy}{dx} = -\frac{2x}{2y} = -\frac{x}{y} \qquad (y \neq 0)$$

NOTE: In the following examples and in the exercises we assume that no operation involves division by zero.

Example 16-2

Find dy/dx if $y^3 x - 3y + 4x^2 - 6 = 0$.

To differentiate term by term we will need $\dfrac{d}{dx}(y^3 x)$. We apply the product rule to this term:

$$\frac{d}{dx}(y^3 x) = y^3 \frac{d}{dx}(x) + x \frac{d}{dx}(y^3)$$

$$= y^3 \cdot 1 + x \cdot 3y^2 \frac{dy}{dx}$$

$$= y^3 + 3xy^2 \frac{dy}{dx}$$

Note also that

$$\frac{d}{dx}(-3y) = -3 \frac{d}{dx} y = -3 \frac{dy}{dx}$$

Therefore we have, differentiating the given equation term by term,

$$\frac{d}{dx}(y^3 x) - \frac{d}{dx}(3y) + \frac{d}{dx}(4x^2) - \frac{d}{dx}(6) = \frac{d}{dx}(0)$$

$$y^3 + 3xy^2 \frac{dy}{dx} - 3 \frac{dy}{dx} + 8x - 0 = 0 \qquad \text{(a)}$$

Combining terms, we get

$$\frac{dy}{dx}(3xy^2 - 3) = -(8x + y^3) \qquad \text{(b)}$$

Goal 16 | To Find Derivatives by Implicit Differentiation

So
$$\frac{dy}{dx} = \frac{8x + y^3}{3 - 3xy^2} \qquad (c)$$

Often the symbol "y'" is used, instead of "dy/dx," when differentiating implicitly. If we use this notation for the equation above, the steps look like this:

$$y^3 x - 3y + 4x^2 - 6 = 0$$
$$y^3 \cdot 1 + x \cdot 3y^2 y' - 3y' + 8x = 0 \qquad (a)$$
$$y'(3xy^2 - 3) = -(8x + y^3) \qquad (b)$$
$$y' = \frac{8x + y^3}{3 - 3xy^2} \qquad (c)$$

This is precisely dy/dx as obtained above.

Example 16-3
Find the slope of the curve whose equation is $y^3 x - 3y + 4x^2 - 6 = 0$ at the point (1, 2).

Since this is the equation of the preceding example, the slope of the curve at the point (1, 2) is the value of y' or dy/dx when $x = 1$ and $y = 2$.

$$\left.\frac{dy}{dx}\right|_{\text{at}(1,2)} = \left.\frac{8x + y^3}{3 - 3xy^2}\right|_{\text{at}(1,2)} = \frac{8 \cdot 1 + 2^3}{3 - 3(1)(2)^2} = -\frac{16}{9}$$

Example 16-4
On a certain island there are two species of animals one of which feeds on the other (this is called a *predator-prey relation*). We assume that the size y of the predator population and the size x of the prey population are related by the equation $10{,}000(y - 500)^2 + (x - 12{,}000)^2 = 100{,}000{,}000$. Find dy/dx when $x = 18{,}000$ and $y = 420$.

Differentiating implicitly, term by term, with respect to x, we get

$$\frac{d}{dx}[10{,}000(y - 500)^2] + \frac{d}{dx}[(x - 12{,}000)^2] = \frac{d}{dx}(100{,}000{,}000)$$

$$10{,}000 \cdot 2(y - 500)\frac{dy}{dx} + 2(x - 12{,}000) = 0$$

So
$$\frac{dy}{dx} = -\frac{(x - 12{,}000)}{10{,}000(y - 500)} = \frac{12{,}000 - x}{10{,}000(y - 500)}$$

When $x = 18{,}000$ and $y = 420$,

$$\frac{dy}{dx} = \frac{-6000}{10{,}000(-80)} = \frac{3}{400}$$

One interpretation of this is that when $x = 18{,}000$ and $y = 420$, an increase of 400 in the prey population produces an increase of approximately 3 in the predator population. (The fact that the answer is an *approximation* will be discussed in Goal 28.)

Example 16-5

A small firm that is offering a new model of table lamp figures that the number n that can be sold at price p dollars is given implicitly by the equation $0.001n^3 + np + 1.3p^2 = 4312.50$. At present the price is $25 per lamp and the demand is for 100 lamps. How will a $1 increase in price affect the demand?

To answer the question we must compute dn/dp when $n = 100$ and $p = 25$. This will give us the rate of change of demand relative to price. We differentiate the given equation implicitly with respect to p:

$$\frac{d}{dp}(0.001n^3) + \frac{d}{dp}(np) + \frac{d}{dp}(1.3p^2) = \frac{d}{dp}(4312.50)$$

$$0.001\left(3n^2 \frac{dn}{dp}\right) + \left(n \frac{d}{dp}p + p \frac{d}{dp}n\right) + 2.6p = 0$$

$$0.003n^2 \frac{dn}{dp} + n + p \frac{dn}{dp} + 2.6p = 0$$

Solve for dn/dp:

$$\frac{dn}{dp}(0.003n^2 + p) = -n - 2.6p$$

$$\frac{dn}{dp} = -\frac{n + 2.6p}{0.003n^2 + p}$$

Substitute $n = 100$ and $p = 25$:

$$\frac{dn}{dp} = -\frac{100 + (2.6)(25)}{0.003(100)^2 + 25} = -\frac{165}{55} = -3$$

So, roughly speaking, a $1 increase in the price decreases the demand by three items. (This is again an approximation, a topic to be taken up in more detail in Goal 28.)

Example 16-6

Suppose the path of a missile is described by the equation $2x^2 - 12x + 3y^2 = 0$, where x gives its horizontal distance and y its vertical distance, both in miles, from the launch pad. At what horizontal distance from the launch pad will the missile be moving horizontally?

The direction of motion of the missile at any instant is the slope of the curve along which the missile travels. It moves horizontally when the slope of the curve (that is, the slope of the tangent line) is equal to zero. (At this point the missile also attains its greatest height.) To solve the problem, therefore, we find dy/dx and set it equal to zero.

Differentiate the given equation implicitly with respect to x:

$$4x - 12 + 6y \frac{dy}{dx} = 0$$

Solve for dy/dx:

$$\frac{dy}{dx} = \frac{12 - 4x}{6y} = \frac{6 - 2x}{3y}$$

Since $dy/dx = 0$ when $6 - 2x = 0$, or when $x = 3$, the missile will be moving horizontally when it is exactly 3 miles horizontally from the launch pad.

EXERCISES

In Exercises 1 through 6, find dy/dx by implicit differentiation.

1. $y^2 = x^2 + 1$
2. $y^3 + x = 3$
3. $\frac{1}{y} + x^2 = 2$
4. $xy + 3 = 0$
5. $x^2y = 6x - 2$
6. $xy^2 = x$

Use implicit differentiation to find y' in Exercises 7 through 9.

7. $x^2y^2 + y = 3$
8. $\frac{x}{y} = 5$
9. $y^2 + xy + x^2 = 3$

In Exercises 10 through 12 find y' at the designated points.

10. $y^4 = x^3 - 11$ at $(3, 2)$
11. $xy = 15$ at $(5, 3)$
12. $y^2 - xy + x^3 = 1$ at $(1, 1)$
13. Find the slope of the circle $x^2 + y^2 = 25$ at the point $(3, 4)$.
14. Find the slope of the graph of $x^2 - 3xy + 2y^2 = 8$ at the point $(0, -2)$.

15. If two populations are related by the equation $y^2 + xy + x^2 = 310{,}000$ and if $y = 100$ when $x = 500$, approximately what change in the population y will result from a unit increase in population x?

16. Suppose that the number n of items that can be sold at a particular price, p dollars, is given implicitly by the equation $0.01n^2 + 2np + p^2 = 12{,}600$, and that at the present time the demand is for 500 items and the price is $10 per item. By approximately how many items will the demand decrease if the price is increased by $1 per item?

17. Suppose that the path of a missile is described by the equation $4x^2 - 8x + 2y^2 = 0$, where x and y are, respectively, the horizontal and vertical distances of the missile, in kilometers, from the firing point. For what x will the missile be moving horizontally?

TEST A

Find dy/dx by implicit differentiation:

1. $3y^2 = x^2 + x - 1$
2. $xy = 5x + 2$
3. $x^2y^2 = y + 1$
4. Find the slope of the ellipse $x^2 + 4y^2 = 13$ at the point $(3, -1)$.

TEST B

Find dy/dx by implicit differentiation:

1. $2y^3 = x^3 - x$
2. $x^2y = 3x^2 + 1$
3. $x^3y^2 = y - 6$
4. Find all points on the circle $x^2 + y^2 - 4x + 2y = 4$ where the tangent to the curve is horizontal.

STUDY QUESTIONS

1. On page 156 we noted that if we are given the equation $y^2 = x$ then two functions of x defined implicitly by this equation are

$$y_1 = \sqrt{x} \quad \text{and} \quad y_2 = -\sqrt{x}$$

Show that the derivative obtained implicitly from $y^2 = x$ is equivalent algebraically to the derivatives of y_1 and y_2 when we differentiate the latter explicitly. Sketch the graphs of y, y_1, and y_2 on separate axes.

2. The technique of implicit differentiation can be approached through

Goal 16 | To Find Derivatives by Implicit Differentiation

consideration of a $\dfrac{\Delta y}{\Delta x}$ limit, applying the basic technique used for explicit functions. For example, consider the equation $x^2 + y^2 = 25$ for a circle. Write down $(x + \Delta x)^2 + (y + \Delta y)^2 = 25$, multiply out the terms, subtract the original equation $x^2 + y^2 = 25$, divide through by Δx, and then take the limit as $\Delta x \to 0$ under the assumption that as $\Delta x \to 0$ both $\Delta y \to 0$ and $\dfrac{\Delta y}{\Delta x} \to \dfrac{dy}{dx}$.

GOAL 17

To find higher derivatives

We first describe how to find the second derivative and other higher derivatives of elementary functions, then suggest a variety of applications that involve the second derivative.

HIGHER DERIVATIVES; NOTATIONS

Since the derivative of a function f is itself a function, it may also have a derivative, called the *second derivative* of f. If, for example, $y = f(x) = x^3 - x^2$, then $f'(x) = 3x^2 - 2x$, and the second derivative of f, denoted by $f''(x)$, is equal to $6x - 2$.

Just as there are various notations for the first derivative, such as

$$f' \qquad y' \qquad \frac{dy}{dx} \qquad \frac{df}{dx} \qquad D_x f$$

so also there are several alternative notations for the second derivative:

$$f'' \qquad y'' \qquad \frac{d^2y}{dx^2} \qquad \frac{d^2f}{dx^2} \qquad D_x^2 f$$

If, after finding the second derivative, we continue to differentiate, we obtain other *higher derivatives*. Thus, starting with $y = f(x) = x^3 - x^2$, we have

$$y' = f'(x) = \frac{dy}{dx} = D_x y = 3x^2 - 2x$$

$$y'' = f''(x) = \frac{d^2y}{dx^2} = D_x^2 y = 6x - 2$$

$$y''' = f'''(x) = \frac{d^3y}{dx^3} = D_x^3 y = 6$$

Goal 17 | **To Find Higher Derivatives**

$$y^{\prime v} = f^{\prime v}(x) = \frac{d^4 y}{dx^4} = D_x^4 y = 0$$

All subsequent higher derivatives of the given function also equal 0.

The following may help to account for the seemingly strange position of the superscript numbers (exponents) in the notation for higher derivatives based on dy/dx. Remember that the symbol $\frac{d}{dx}$ calls for the derivative, with respect to x, of whatever follows. Thus

$$\frac{d}{dx} y = \frac{dy}{dx}$$

$$\frac{d}{dx}\left(\frac{d}{dx} y\right) = \frac{d}{dx}\left(\frac{dy}{dx}\right) = \frac{d^2 y}{dx^2}$$

$$\frac{d}{dx}\left[\frac{d}{dx}\left(\frac{d}{dx} y\right)\right] = \frac{d}{dx}\left(\frac{d^2 y}{dx^2}\right) = \frac{d^3 y}{dx^3}$$

and so on.

Here is another example of successive derivatives:

$$g(y) = y^{10} - 2y^5 + 3y$$
$$g'(y) = 10y^9 - 10y^4 + 3$$
$$g''(y) = 90y^8 - 40y^3$$
$$g'''(y) = 720y^7 - 120y^2$$

and so on.

APPLICATIONS OF THE SECOND DERIVATIVE

The following examples illustrate some of the many applications of the second derivative. They are presented here only to show how important and useful the second derivative is. Many of these applications will be considered more fully in later Goals.

1. If $f(t)$ gives the position of a moving object at time t, then $f'(t)$ gives the instantaneous velocity of the object at time t. It is often important to know how the velocity itself is changing with respect to time. The rate of change of the velocity, that is, the derivative of velocity, is called the *acceleration* of the object. The acceleration, then, is equal to the derivative of $f'(t)$, or to the second derivative of $f(t)$. For example, if the distance of the object from some fixed point after t min is given by $f(t) = t^3 - 2t^2 + 2t$ miles, then the velocity at time t is $f'(t) = 3t^2 - 4t + 2$ miles per min and the acceleration at time t

is $f''(t) = 6t - 4$ miles per min per min. An examination of the signs of the velocity and acceleration at any instant tells whether the object is speeding up or slowing down at that instant. This will be fully discussed in Goal 29.

2. If $P(t)$ is the population of an animal colony at time t, then $P'(t)$ gives the rate of growth of the population. The second derivative $P''(t)$ indicates how this rate of growth is changing; that is, $P''(t)$ is the rate of change of the rate of growth. For example, if $P(t) = 1000(t^2 + 1)$, then $P'(t) = 2000t$ and $P''(t) = 2000$. In Goal 19 we will see that if the derivative of a function is positive, then the function is increasing. Remembering that $P''(t)$ is the derivative of $P'(t)$, we conclude here, since $P''(t) > 0$ for all t, that $P'(t)$ is always increasing. In other words, the rate of growth is always increasing.

3. A psychologist has discovered experimentally that if a particular person is given a string of nonsense syllables, the number N of syllables he can memorize in t min is represented by the curve below at the left (often called a *learning curve*). In the next goal we will define the tangent to a curve at a point as the line that goes through the point and that has slope equal to that of the curve at the point. At the right

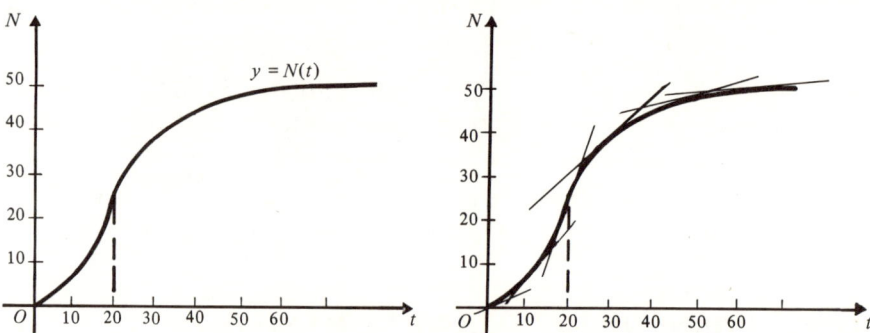

we have drawn segments of tangent lines to the curve as an aid in analyzing how the rate of learning is changing. We see that the slope of the curve is very small for t close to zero; that the slope increases at t does, attaining its maximum at $t = 20$; and that the slope decreases for $t > 20$, tending towards zero. Since the slope of the curve is equal to $N'(t)$, the rate of learning, we have just said that the rate of learning is very slow initially; that it increases with time until $t = 20$, where it reaches its peak; and that it then decreases, eventually becoming very small. The point at which $N'(t)$ is a maximum ($t = 20$) is precisely that at which $N''(t) = 0$.

From Goal 4, page 36, we know that the learning curve has a point of inflection at $t = 20$, that the curve is concave up if $t < 20$, and that it is concave down if $t > 20$. We will see in Goal 23 that $N''(t) > 0$ if $t < 20$, but $N''(t) < 0$ if $t > 20$.

4. It has been found that a contagious disease such as flu spreads in a population so that its curve has the same shape as that of the learning curve above. If $N(t)$ is the number of people who have become infected by time t, then the rate of spread $N'(t)$ increases until it hits a peak, then decreases toward 0. The peak is reached (that is, the disease is being spread fastest) when $N''(t) = 0$.

5. There are other phenomena that exhibit the same pattern as do the preceding two. Some of these are the diffusion of information (as in the spread of a rumor or a bit of news), the growth of a population with limited space or food supply, the growth in sales of a new product, certain chemical reactions in which one substance is transformed into another, the growth of some organisms, and the change in temperature of food that is removed from a freezer. In each of these applications the rate of change (of growth, of spread, or the like) is a maximum when the second derivative (of the population, the sales, the temperature, etc.) is equal to zero.

6. In economics the Weber-Fechner Law claims that the marginal utility (or marginal usefulness) of a product decreases as the quantity of the product used increases. If the function $U(x)$ measures the utility of x units of a product, then $U'(x)$ denotes its marginal utility. Since the derivative of a decreasing function is negative, the Weber-Fechner Law thus says mathematically that $U''(x) < 0$ for all x.

The utility function $U(x) = 1000 \left(1 - \dfrac{1}{x}\right)$ satisfies the Weber-Fechner Law because

$$U'(x) = 1000\left(\dfrac{1}{x^2}\right) = 1000x^{-2}$$

and

$$U''(x) = -2000x^{-3} = -\dfrac{2000}{x^3}$$

We see that $U''(x) < 0$ for all $x > 0$. The graph of $y = U(x)$ is sketched below along with segments of tangents to the curve.

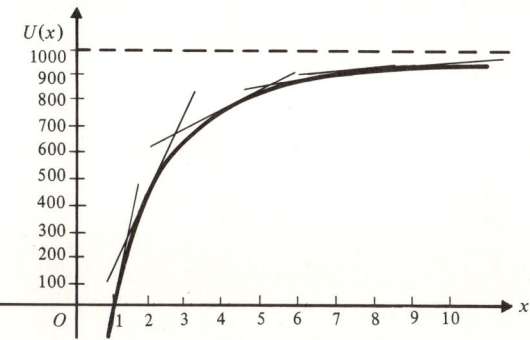

As x increases, the slope of the curve decreases tending toward zero. Since $U''(x)$ is the (first) derivative of $U'(x)$, the slope of the curve, the previous statement implies that $U''(x)$ is negative for all $x > 0$.

7. The graph of a body's response to certain medications looks like the curve in Example 6. While higher doses of medication produce greater response (that is, the derivative or slope of the response function is positive), the rate of change of response (again the first derivative or slope) decreases as the dose increases. Thus the second derivative of the response function is negative.

8. The preceding applications suggest how useful the second derivative is for curve-sketching. If $f''(x_0) > 0$, then the graph of $f(x)$ is concave up at $x = x_0$; if $f''(x_0) < 0$, then the graph is concave down at $x = x_0$. This will be fully discussed in Goal 25. Applications 6 and 7 above imply that the graphs of the relevant functions are everywhere concave down.

EXERCISES

1. Find $f''(x)$ if
   (a) $f(x) = x^8$
   (b) $f(x) = 3x^4 - x + 1$
   (c) $f(x) = x^3 + 2x^2 - 3$
   (d) $f(x) = \dfrac{1}{x}$
   (e) $f(x) = \dfrac{1}{x+1}$
   (f) $f(x) = x^{5/2}$

2. Compute each of the following:
   (a) $f''(9)$ if $f(x) = \sqrt{x}$
   (b) $f''(1)$ if $f(x) = x^4 - x^2 + 7x - 2$
   (c) $f''(0)$ if $f(x) = \dfrac{x-1}{x+2}$
   (d) $\dfrac{d^2y}{dx^2}$ at $x = 1$ if $y = \dfrac{1}{x^2}$

3. Find $\dfrac{d^3y}{dx^3}$ if
   (a) $y = x^6$
   (b) $y = 4x^4 - 3x^3 + 2x^2 - x + 1$
   (c) $y = \dfrac{1}{x}$
   (d) $y = 2\sqrt{x}$

4. Denote by $f^{(n)}(x)$ the nth derivative of $f(x)$. What is the smallest n for which $f^{(n)}(x) = 0$ if $f(x) = x^7$?

5. Find the velocity and acceleration at time t for each of the following functions, where $f(t)$ is the distance from a fixed point of an object moving along a line.
   (a) $f(t) = 16t^2 - 5t - 1$
   (b) $f(t) = 100t - 16t^2$
   (c) $f(t) = t^3 - t^2 + 3$
   (d) $f(t) = \dfrac{1}{t^2}$

Goal 17 | To Find Higher Derivatives

6. For each function in Exercise 5 above, find all t for which the acceleration of the object is positive.
7. Which of the following utility functions satisfy the Weber-Fechner Law? [That is, for which functions is $U''(x) < 0$ for all positive x?]

   (a) $U(x) = 100 - \dfrac{2}{x} + x$ (c) $U(x) = x^2 - \dfrac{1}{x}$

   (b) $U(x) = x^2 - 3x + 25$

8. Suppose a woman can memorize a string of n digits in t min, where $n(t) = -0.002t^3 + 0.05t^2 + 1.8t$ and $t \leq 30$. When is her rate of learning at a peak? [That is, for what t is $n''(t)$ equal to 0?]

TEST A

1. Find the second derivative of $f(x) = 5x^3 - 3x^2 + x + 1$.
2. Find $\dfrac{d^2y}{dx^2}$ if $y = \dfrac{2}{x}$.
3. Compute $f'(1)$, $f''(1)$, and $f'''(1)$ if $f(x) = x^7 - 2x^2 + 5$.
4. (a) Find the velocity and acceleration of an object moving along a line if $f(t) = t^4 - 4t^3$ is its distance from the origin at time t.
   (b) Find all positive t for which the acceleration is positive.

TEST B

1. Find $\dfrac{d^2y}{dx^2}$ if $y = 4x^4 - x^2 + 9$.
2. Find $f''(25)$ if $f(x) = \sqrt{x}$.
3. Find $f'(-1)$, $f''(-1)$, and $f'''(-1)$ if $f(x) = x^8 + x^2 - 3$.
4. Suppose a teen-age boy can memorize a string of n nonsense syllables in t min, where $n(t) = -0.002t^3 + 0.08t^2 + 0.5t$ ($t \leq 25$). When is his rate of learning at a peak? That is, for what t does $n''(t) = 0$?

STUDY QUESTIONS

1. If $f(x) = 8x^2 + 7x + 6$, compute $f(0)$, $f'(0)$, and $f''(0)$. If $f(x) = ax^2 + bx + c$, compute $f(0)$, $f'(0)$, and $f''(0)$. If $f(x) = ax^2 + bx + c$ and if $f(0) = 6$, $f'(0) = 5$, and $f''(0) = 4$, what are a, b, and c?
   Can you generalize this problem to a cubic,
   $$f(x) = ax^3 + bx^2 + cx + d?$$

2. Suppose we wish to find $\dfrac{d^2y}{dx^2}$ where y is given implicitly as a function of x by $x^2 + y^2 = 5$. Using earlier techniques, we get

$$2x + 2y\frac{dy}{dx} = 0, \quad \frac{dy}{dx} = -\frac{x}{y}$$

Differentiating again gives

$$\frac{d^2y}{dx^2} = -\frac{y(1) - x\dfrac{dy}{dx}}{y^2}$$

Substituting from above for $\dfrac{dy}{dx}$ gives

$$\frac{d^2y}{dx^2} = -\frac{y - x\left(-\dfrac{x}{y}\right)}{y^2} = -\frac{y + \dfrac{x^2}{y}}{y^2} = -\frac{y^2 + x^2}{y^3}$$

And since the original equation tells us that $x^2 + y^2 = 5$, we get, finally,

$$\frac{d^2y}{dx^2} = -\frac{5}{y^3}$$

In a similar way, calculate $\dfrac{d^2y}{dx^2}$ if $yx - y = 1$.

Summary of Goals 14-17

You should now be able to

(1) use the rules for finding the derivative of a product or quotient of two functions;
(2) state the chain rule and recognize when it must be used;
(3) find the derivative of $[f(x)]^n$ where n is any real number; for example, $D_x(x^3 - x^2 + 4x + 2)^4$, $\dfrac{d}{dx}\sqrt{2x^2 - x + 1}$, or $f'(x)$ if $f(x) = (x^3 - 1)^{-5}$;
(4) use the technique of implicit differentiation; for example, find $\dfrac{dy}{dx}$ if $x^2y^2 - 3y + x = 5$;
(5) find the second and other higher derivatives of a function.

Review Exercises

GOAL 14

1. Find each of the following derivatives:
   (a) $f'(x)$ if $f(x) = (x^2 + 3x - 2)(1 - x^2)$
   (b) $g'(2)$ if $g(t) = \dfrac{t^2 - 1}{t^2 + 1}$
   (c) $\dfrac{dy}{dx}$ if $y = (4x^4 + x^3 - 2x + 1)(7 - x)$

2. Find any points on the graph of $f(x) = \dfrac{x^2 - 2x + 2}{x - 1}$ at which the slope is zero.

3. Suppose the Beesharp Music Company can produce x music stands at a cost of $3x^2 - 5x + 200$ dollars. (a) Find the average cost function. (b) Compute the marginal average cost when the production level is 20 music stands.

GOAL 15

4. Find $\dfrac{dy}{dx}$ if y equals

   (a) $\dfrac{1}{\sqrt[3]{x}}$

   (b) $(3x^3 + x^2 - 2x + 1)^3$

   (c) $\sqrt{x^2 + 1}$

   (d) $\dfrac{4}{(3-x)^2}$

   (e) $(2x^3 - 1)^{3/2}$

   (f) $\dfrac{x}{\sqrt{x^2 + 4}}$

   (g) $x^2(3 - x^2)^4$

   (h) $\dfrac{(1 - x)^3}{(x + 1)^2}$

5. Evaluate each of the following:

   (a) $g'(-2)$ if $g(x) = \dfrac{x}{\sqrt{x^2 - 1}}$

   (b) $h'(1)$ if $h(x) = f(g(x))$ where $f(u) = 2u^2 + 1$ and $g(x) = 3x + 2$

GOAL 16

6. Find $\dfrac{dy}{dx}$ by implicit differentiation if

   (a) $y^2 - 2xy + x^3 = 4$ (b) $xy = 3x - 1$

7. Find the slope of the curve whose equation is $y^3 - xy + 8 = 0$ at the point $(0, -2)$.

8. In Example 16-5 a lamp manufacturer figures that n lamps can be sold at p dollars each, where n and p are related by the equation

$$0.001n^3 + np + 1.3p^2 = 4312.50$$

At what rate is the price changing with respect to the demand when the demand is for 200 lamps and the price is $20 per lamp?

GOAL 17

9. Find $f''(x)$ if

   (a) $f(x) = x^3 - 3x^2 - 4x + 2$ (b) $f(x) = 4x^{3/2} - 8x^{1/2}$

10. Evaluate

    (a) $f'(4)$ if $f(x) = 2x^{3/2}$

    (b) $f'''(-1)$ if $f(x) = x^5$

    (c) $\dfrac{d^2y}{dx^2}$ at $x = \dfrac{1}{2}$ if $y = \dfrac{1}{x}$

    (d) $g''(1)$ if $g(t) = t^4 - 4t^3$

11. A stone thrown directly upward with initial velocity 60 ft/sec attains a height h equal to $60t - 16t^2$ ft after t sec. For what t is the acceleration of the stone negative?
12. Suppose a college freshman can memorize a string of n signs of Ameslan (American Sign Language for the Deaf) in t min, where $N(t) = -0.003t^3 + 0.1t^2 + t$, for $t \leq 25$. When is the freshman's rate of learning at a peak? That is, when does $N''(t) = 0$?

Chapter Test A

1. Find each of the following derivatives:
   (a) $f'(x)$ if $f(x) = (3x + 2)(4 + x - x^2)$
   (b) $g'(2)$ if $g(x) = \dfrac{1 - x}{2 + x^2}$
2. At what point on the curve $y = \dfrac{x^2 + 4x + 8}{x + 2}$ is the slope equal to zero?
3. Find $\dfrac{dy}{dx}$ if
   (a) $y = \sqrt{4 - x^2}$ (b) $y = \dfrac{x}{(2 - x)^2}$
4. If $xy = x^2 + 1$ find y' (a) by implicit differentiation; (b) by solving for y explicitly. Show that your answers are equivalent.
5. Find the slope of the curve $xy^2 - 2y = 3x^2 - 4$ at the point on the curve where $x = 0$.
6. Find
   (a) $f''(x)$ if $f(x) = 2x^4 - 3x^2 - 3x - 1$
   (b) d^2y/dx^2 if $y = (2x)^{3/2}$
7. An object is moving along a line so that its distance from a fixed point after t sec is $f(t) = 2t^3 - 10t^2 + 7$ ft. When is the acceleration of the object positive?

Chapter 4

Curve-Sketching

The graph of a function is often extremely helpful in understanding how the function behaves. In this chapter we obtain information about various characteristics of a graph that are helpful in sketching. Some of these are where the graph rises or falls, where it has high or low points, where it attains its greatest or least value, where it is concave up or down, and whether it has asymptotes. With this information we can usually sketch the graph of a function quite accurately.

Goal 18. To find equations of tangents and normals
Goal 19. To determine where a function is increasing or decreasing
Goal 20. To determine local maxima and minima
Goal 21. To find absolute maxima and minima
Goal 22. To apply the second derivative: concavity, inflection points, testing for maxima and minima
Goal 23. To find asymptotes of a curve
Goal 24. To graph functions using the calculus
Summary
Review Exercises
Chapter Test A

GOAL **18**

To find equations of tangents and normals

One major application of the derivative of a function is in finding equations of the tangent and normal to the curve at a specified point. The tangent line at a point provides a linear approximation to the behavior of the function in the neighborhood of the point. Tangents to a curve are also extremely useful when one is analyzing where the function increases or decreases and where it has maximum or minimum values. These topics will be fully discussed in the next few Goals.

TANGENT TO A CURVE

The *tangent line* (or just the *tangent*) to the graph of $y = f(x)$ at the point of the curve where $x = c$ is defined to be the straight line passing through $(c, f(c))$ and having slope $f'(c)$.* Thus the slope of the curve at a point and the slope of the tangent to the curve at that point are the same. This is why the tangent line is an approximation to the function near the point. The point $(c, f(c))$ is called the *point of tangency*. (See figure at top of p. 178.)

---

*Unless specified otherwise, we will assume in this Goal that $f'(x)$ exists. If f is continuous at $x = c$ but its derivative becomes infinite at $x = c$, then we speak of the "vertical tangent" to the curve at c; its equation is $x = c$.

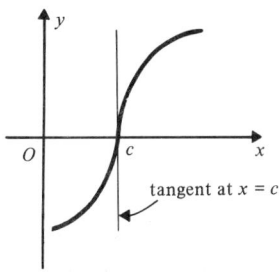

tangent at $x = c$

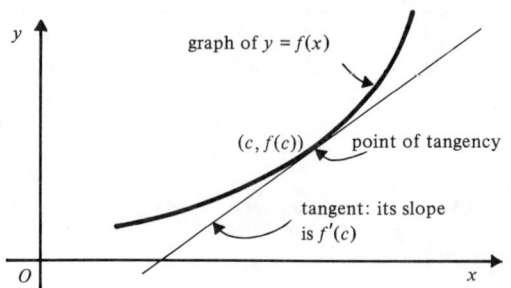

The definition of tangent just above tells us (see p. 15) that the equation of the tangent line to the graph of $y = f(x)$ at $(c, f(c))$ is

$$\frac{y - f(c)}{x - c} = f'(c) \qquad \text{or} \qquad y - f(c) = f'(c)(x - c)$$

Example 18-1

Find the equation of the tangent to the graph of $f(x) = x^2 - x + 2$ at the point $(3, 8)$.

From

$$f'(x) = 2x - 1$$

we have

$$f'(3) = 2(3) - 1 = 5$$

Therefore the tangent has the equation

$$\frac{y - 8}{x - 3} = 5 \qquad \text{or} \qquad y = 5x - 7$$

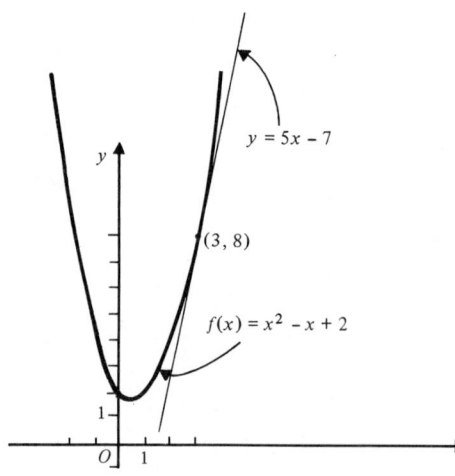

Goal 18 | To Find Equations of Tangents and Normals

Example 18-2

Find the equation of the tangent to the curve $y = 1/x$ at the point where $x = 2$.

When $x = 2$, $y = \frac{1}{2}$. Also, from

$$y = \frac{1}{x} = x^{-1}$$

it follows that

$$\frac{dy}{dx} = -x^{-2} = -\frac{1}{x^2}$$

When $x = 2$, $dy/dx = -\frac{1}{4}$. The equation of the tangent is therefore

$$\frac{y - \frac{1}{2}}{x - 2} = -\frac{1}{4}$$

$$y - \frac{1}{2} = -\frac{1}{4}(x - 2)$$

$$y = -\frac{1}{4}x + 1$$

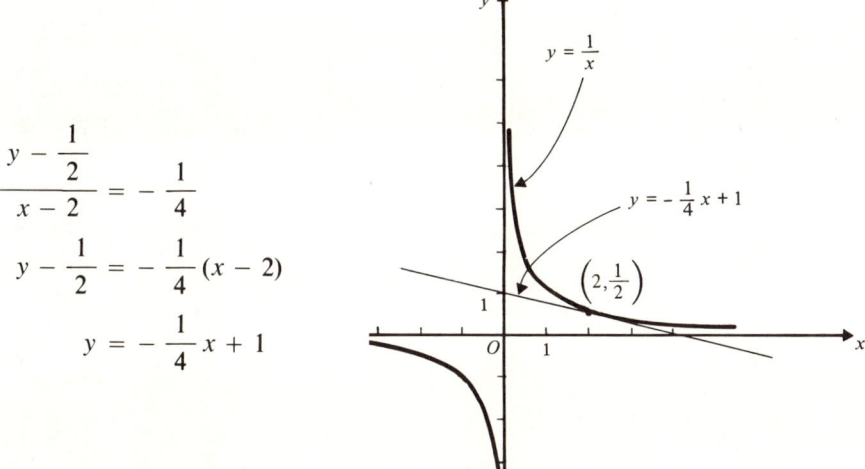

A NOTE OF CAUTION

The following illustrates a common error. Suppose we want to find the equation of the tangent to the graph of $f(x) = x^3$ at $x = 2$. We have $f(2) = 8$ and $f'(x) = 3x^2$. So

$$\frac{y - 8}{x - 2} = 3x^2 \qquad \text{or} \qquad y - 8 = 3x^2(x - 2)$$

But this equation is clearly not the correct one for the tangent line because it is *not linear*. To get the right answer we must use not $f'(x)$ but $f'(2)$; that is, $f'(x)$ evaluated at $x = 2$. The correct equation is then

$$\frac{y - 8}{x - 2} = 3(2^2) \qquad \text{or} \qquad y = 12x - 16$$

This example is given to show how important it is to evaluate the derivative *at the point in question* in order to find the slope of the tangent. The slope of a line (if the line is not vertical) is a constant, and the equation for a tangent must be linear.

Example 18-3

Find the equation of the tangent to the curve $y^2 + 2x^2 = 6$ at the point $(1, 2)$.

We differentiate implicitly (see Goal 16), then solve for dy/dx.

$$2y \frac{dy}{dx} + 4x = 0 \quad \text{and} \quad \frac{dy}{dx} = -\frac{2x}{y}$$

From this we get

$$\left.\frac{dy}{dx}\right|_{(1,\ 2)} = -\frac{2(1)}{2} = -1$$

The equation of the line with slope -1 and passing through the point $(1, 2)$ is therefore

$$\frac{y-2}{x-1} = -1$$

or

$$y - 2 = -(x - 1)$$

or

$$y = -x + 3$$

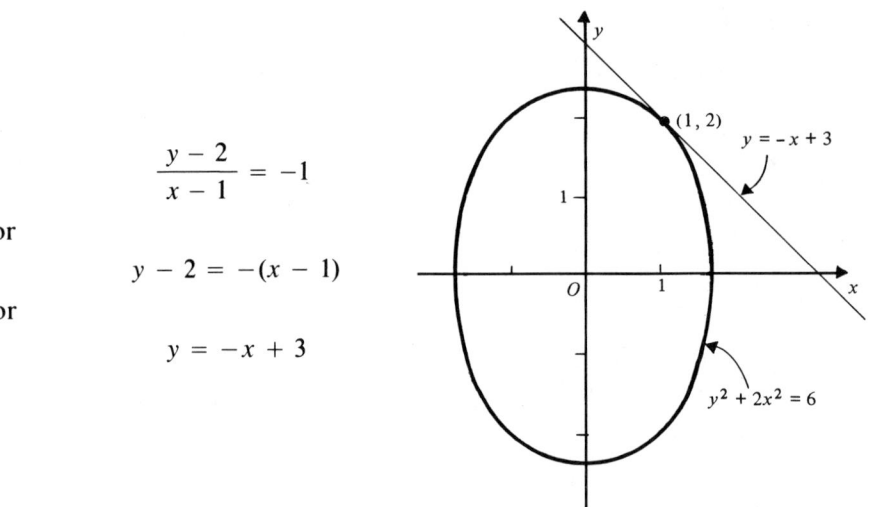

NORMAL TO A CURVE

The *normal line* (or *normal*) to a curve at a point is the straight line that passes through the point and is perpendicular to the tangent at that point. Its slope is therefore the negative reciprocal (see p. 16) of the slope of the tangent at the point.

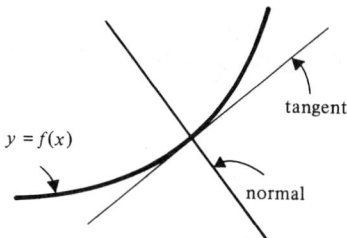

Example 18-4

Find the equations of the tangent and normal to the curve $y = x^2 - 3x + 1$ at the point (3, 1).

$dy/dx = 2x - 3$; at (3, 1), $dy/dx = 2(3) - 3 = 3$. Since the tangent has slope 3, the normal has slope $-\frac{1}{3}$. The equations are therefore

Tangent:	Normal:
$\dfrac{y - 1}{x - 3} = 3$	$\dfrac{y - 1}{x - 3} = -\dfrac{1}{3}$
$y - 1 = 3(x - 3)$	$y - 1 = -\dfrac{1}{3}(x - 3)$
$y = 3x - 8$	$y = -\dfrac{1}{3}x + 2$

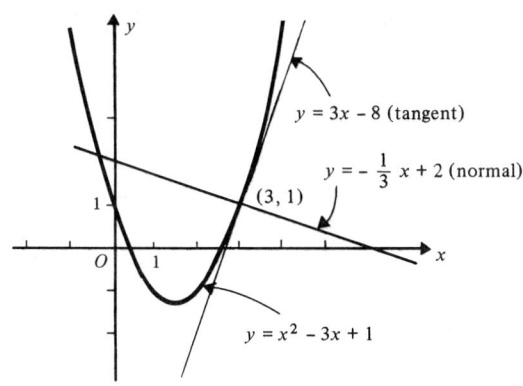

TANGENTS TO CONSTANT OR LINEAR FUNCTIONS

The tangent to a straight line is the line itself. This is the subject of Study Questions 3 and 4.

EXERCISES

Find the equations of the tangent and the normal to each curve at the point specified:

1. $y = x^2 + 2x + 3$ at $(1, 6)$
2. $y = 2x^2 - x + 1$ at $(-1, 4)$
3. $y = -x^2 + 2$ at $(3, -7)$
4. $y = -3x^2 + 6x - 2$ at $(1, 1)$
5. $y = x^3 + 1$ at $(0, 1)$
6. $y = 2x^3 - x$ at $(-1, -1)$
7. $y = \dfrac{1}{x}$ at $\left(3, \dfrac{1}{3}\right)$
8. $y = \dfrac{x}{x+1}$ at $\left(1, \dfrac{1}{2}\right)$
9. $y = \sqrt{x}$ at $(4, 2)$
10. $y = \sqrt{x+1}$ at $(8, 3)$
11. $y^2 + x^3 = 5$ at $(1, 2)$
12. $xy + x + y = 1$ at $(0, 1)$
13. $xy^2 - y = 2$ at $(3, 1)$
14. $\sqrt{y} + y + x = 7$ at $(1, 4)$
15. $y = -2$ at $(1, -2)$
16. $y = x - 5$ at $(8, 3)$

TEST A

Find the equations of the tangent and normal lines to each of the following curves at the specified point:

1. $y = 2x^2 + x - 3$ at $(1, 0)$
2. $y + xy^2 = 3$ at $(2, 1)$

TEST B

Find the equations of the tangent and normal lines to each of the following curves at the point specified:

1. $y = 5x^2 - 3x + 1$ at $(1, 3)$
2. $y^2 - xy = x + 1$ at $(2, -1)$

STUDY QUESTIONS

1. There are many misconceptions about tangent lines. Find and graph examples illustrating why each of the following assertions is *false*:
   (a) The tangent line intersects the curve only at the point of tangency.
   (b) The tangent line does not cross the curve at the point of tangency.
2. Determine the y-intercepts of the normal lines to the circle defined implicitly by $x^2 + y^2 = 25$ at the points $(5, 0)$, $(3, 4)$, and $(4, 3)$. What conclusion can you draw about the normal to a circle at any point? Explain.

3. (a) Show that the equation of the tangent to the line $y = k$ at any point (c, k) is $y = k$; HINT: $f'(c) = 0$. (b) Show that the normal has equation $x = c$.
4. Show that the tangent to the line $y = mx + b$ at any point is the line itself. HINT: If (x_1, y_1) is on the line, then $y_1 = mx_1 + b$.

GOAL **19**

To determine where a function is increasing or decreasing

We have already had frequent occasion to note when a function is increasing over a particular interval or, equivalently, when its graph is rising. In other cases we have observed that a function decreases on some interval (or that its graph falls). We shall now make these important notions more precise and describe how one determines the specific intervals over which a function increases or decreases.

DEFINITIONS

Here are sketches of four different functions on the interval (a, b) (that is, for x between a and b):

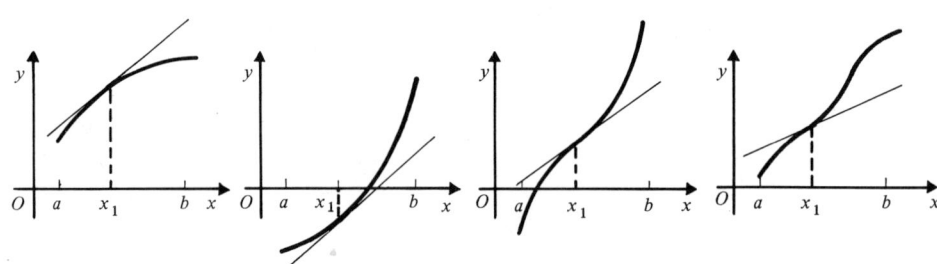

All the following statements are true of all four functions:

(1) The graph rises on the interval (a, b).
(2) Each function is increasing on (a, b).
(3) A tangent to any graph at any point in (a, b), say at x_1, has positive slope.

Here are sketches of four more functions, here on the interval (c, d):

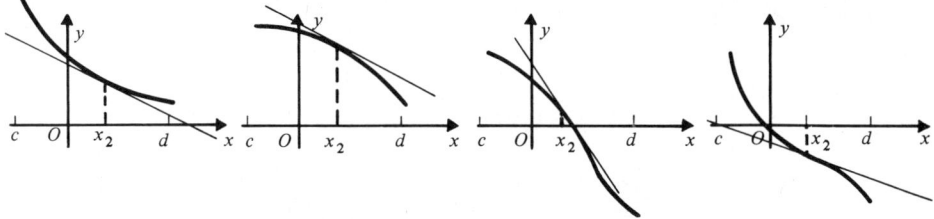

For these we note that

(1) Each graph falls on the interval (c, d).
(2) Each function decreases on (c, d).
(3) A tangent to any graph at any point in (c, d), say at x_2, has negative slope.

More precisely, $f(x)$ is *increasing* on (a, b) if, for any two points x_1 and x_2 in the interval such that $x_1 < x_2$,

$f(x_1) < f(x_2)$

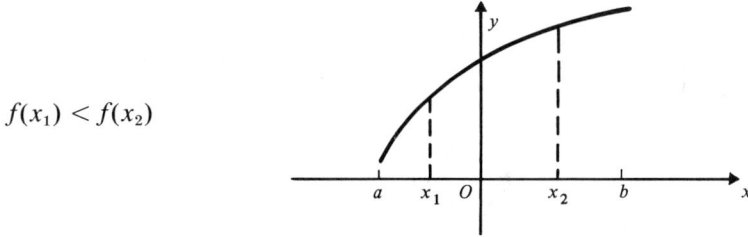

$f(x)$ is *decreasing* on (c, d) if, for any two points x_1 and x_2 in the interval such that $x_1 < x_2$,

$f(x_1) > f(x_2)$

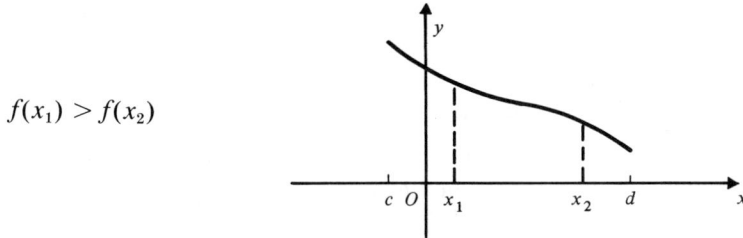

Note especially that these definitions and the preceding statements reflect our habit of discussing the behavior of a function and its graph in the direction of increasing x, that is, from left to right.

Example

For each function sketched below, note where it is increasing and where it is decreasing.

(a)

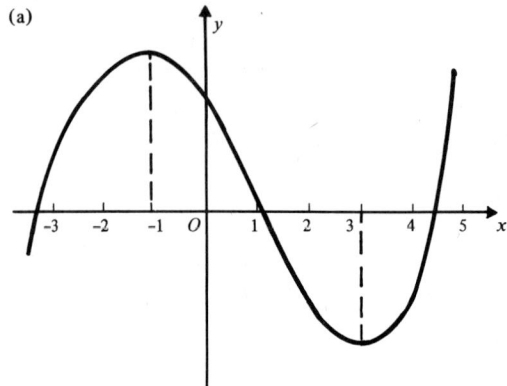

(a) $f(x)$ increases if $x < -1$;
$f(x)$ decreases if $-1 < x < 3$;
$f(x)$ increases if $x > 3$.

(b)

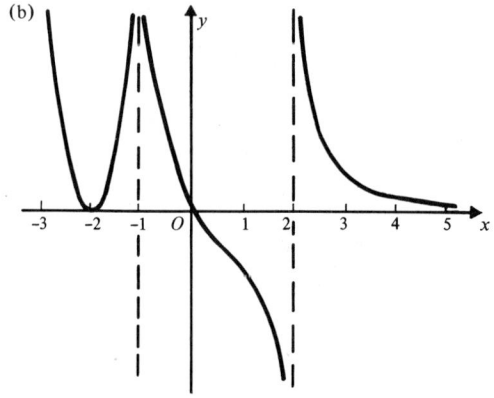

(b) $f(x)$ decreases if $x < -2$;
$f(x)$ increases if $-2 < x < -1$;
$f(x)$ decreases if $-1 < x < 2$;
$f(x)$ decreases if $x > 2$.

In (b) we cannot say that f is decreasing for all $x > -1$. For example, although $1 < 3$, $f(1)$ is *not* greater than $f(3)$. When a function is not defined at one or more points—the function in (b) is not defined either at $x = -1$ or at $x = 2$—then we must examine f on intervals with these points of discontinuity as endpoints or on intervals between such points.

THE SIGN OF THE DERIVATIVE

We have the following important relation between the sign of the derivative of a function and whether the function is increasing or decreasing:

If $f'(x) > 0$ throughout an interval, then $f(x)$ is increasing throughout the interval.

If $f'(x) < 0$ throughout an interval, then $f(x)$ is decreasing throughout the interval.

Goal 19 | To Determine Where a Function is Increasing or Decreasing

Conversely, if f is increasing on an interval we may conclude that $f'(x) \geq 0$ on the interval, and if f is decreasing on an interval we may conclude that $f'(x) \leq 0$ on the interval. [Note that it does *not* follow that $f'(x) > 0$ if f is increasing: for example, $f(x) = x^3$ is increasing on the interval $-1 < x < 1$ but $f'(0) = 0$.]

CRITICAL POINTS OF A FUNCTION

If f has a derivative at $x = a$ then exactly one of the following is true: either $f'(a) > 0$ or $f'(a) < 0$ or $f'(a) = 0$. Any a for which $f'(a) = 0$ is called a *critical point* or *critical value* of f.* If $f'(a) = 0$ then the tangent to the curve at $x = a$ is horizontal.

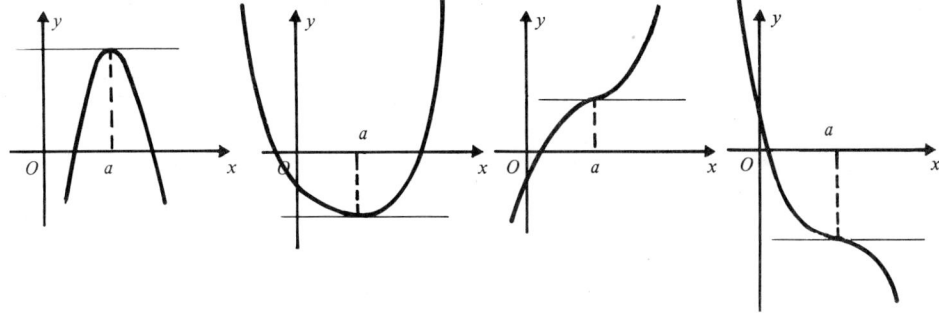

The critical points of a function f are of major importance in sketching its curve. To obtain them, we find $f'(x)$, factor it completely, then solve the equation $f'(x) = 0$.

Example 19-1

Find the critical points of

(a) $f(x) = 2x^3 + 3x^2$
(b) $g(x) = x^2 + 3$
(c) $f(x) = x^3 - 27x$
(d) $y = x^4 - 4x^3 + 4x^2 + 1$

(a) $f'(x) = 6x^2 + 6x = 6x(x + 1)$; this equals zero if $x = 0$ or -1. The critical points are thus 0 and -1.
(b) $g'(x) = 2x$; this is zero if $x = 0$. The only critical point of g is 0.
(c) $f'(x) = 3x^2 - 27 = 3(x^2 - 9) = 3(x - 3)(x + 3)$; this is zero if $x = 3$ or -3. The critical points are ± 3.
(d) $y' = 4x^3 - 12x^2 + 8x = 4x(x^2 - 3x + 2) = 4x(x - 1)(x - 2)$. This is zero if $x = 0$, 1, or 2; these are therefore the critical points.

---

*Often a is also called a critical value of f if $f'(a)$ does not exist. The case in which f is not everywhere differentiable (see definition on p. 114) will be considered shortly.

BEHAVIOR OF A FUNCTION ON THE INTERVAL BETWEEN TWO CRITICAL POINTS

Suppose a function f with a continuous derivative has two critical points, at $x = c_1$ and $x = c_2$, and that $f'(x)$ does *not* equal zero at any x between c_1 and c_2. Then *the sign of $f'(x)$ is the same throughout the interval*:

either $f'(x) > 0$ for all x in (c_1, c_2) and $f(x)$ is increasing throughout the interval;

or $f'(x) < 0$ for all x in (c_1, c_2) and $f(x)$ is decreasing throughout the interval.

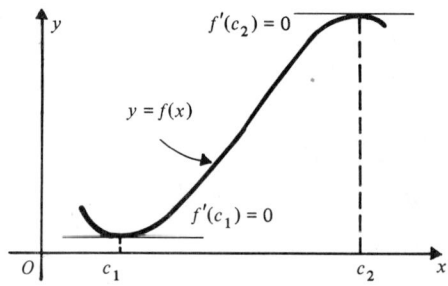

$f'(x) > 0$ for all x in (c_1, c_2)
$f(x)$ is increasing throughout the interval

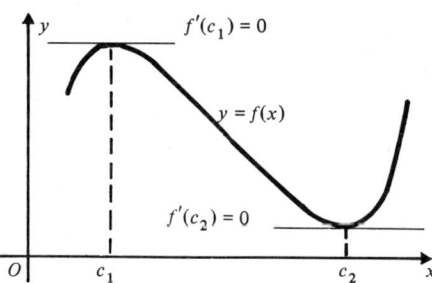

$f'(x) < 0$ for all x in (c_1, c_2)
$f(x)$ is decreasing throughout the interval

Since the derivative is given as continuous, we know that if $f'(x)$ changes sign between c_1 and c_2 then $f'(x)$ *must* equal zero for some x between c_1 and c_2; but this is counter to our assumption about f above. This statement about the constancy of sign of the derivative on the interval between two critical points enables us to determine quite easily the intervals of increase or decrease of a function f. We need only to identify the intervals throughout each of which the sign of the derivative does not change; then for each interval, in turn, we examine the sign of $f'(x)$ at *any* point within the interval.

FINDING WHERE A FUNCTION INCREASES OR DECREASES

There are two cases to consider: that in which the function has a continuous derivative and that in which it does not.

CASE I

The function f has a continuous derivative. We outline the method to be used, then illustrate.

Step (1) Find $f'(x)$.

Step (2) Factor $f'(x)$ completely; then solve the equation $f'(x) = 0$ to obtain the critical points of f.

Goal 19 | To Determine Where a Function is Increasing or Decreasing

Step (3) List the intervals between successive critical points, the interval to the left of the smallest critical point, and the interval to the right of the largest critical point.

Step (4) Let (c_1, c_2) be any of the intervals listed in Step (3). Choose any convenient point, say x_0, in (c_1, c_2); then determine the sign of $f'(x_0)$.

Step (5) If $f'(x_0) > 0$, conclude that f increases on (c_1, c_2); if $f'(x_0) < 0$, conclude that f decreases on (c_1, c_2).

Note that Steps (4) and (5) must be repeated for each interval listed in Step (3).

Example 19-2

If $f(x) = x^3 - 6x^2 + 9x + 1$, where is the function increasing or decreasing?

(1) $f'(x) = 3x^2 - 12x + 9$.

(2) $3x^2 - 12x + 9 = 3(x^2 - 4x + 3) = 3(x - 1)(x - 3)$. This is zero if $x = 1$ or 3; these are the critical points.

(3) The intervals for consideration are $x < 1$, $1 < x < 3$, $3 < x$.

(4) The following shows how we find the sign of $f'(x)$ in each interval:

Interval	x_0	Sign of $f'(x) = 3(x - 1)(x - 3)$ at x_0
$x < 1$	0	$f'(0) = 3(-1)(-3) > 0$
$1 < x < 3$	2	$f'(2) = 3(1)(-1) < 0$
$3 < x$	4	$f'(4) = 3(3)(1) > 0$

Note that we are not at all interested in the value of $f'(x_0)$, only in its sign, which we determine using facts such as the following: the product of an even number of negative numbers is positive, but that of an odd number of negative numbers is negative; a negative number raised to any even power is positive, but raised to an odd power is negative; and so on. For our present purposes it would be a waste of time to evaluate $f'(x_0)$.

(5) The table in Step (4) shows that

if $x < 1$	f increases
if $1 < x < 3$	f decreases
if $3 < x$	f increases

This information, together with the values of f at each critical point (and, for convenience, usually also at zero), enables us to sketch the graph roughly.

Since
$$f(x) = x^3 - 6x^2 + 9x + 1$$
we have

$$f(1) = 1 - 6 + 9 + 1 = 5$$
$$f(3) = 27 - 54 + 27 + 1 = 1$$
$$f(0) = 0 - 0 + 0 + 1 = 1$$

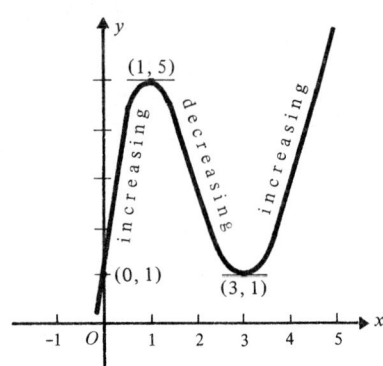

The horizontal tangents to the curve are drawn at the critical points where $f'(x) = 0$. With further information about the concavity of the curve (to be considered in a later Goal), we will be able to sketch curves like these even more accurately.

Example 19-3

Find the intervals over which $f(x) = 4x^5 - 25x^4 + 40x^3$ increases or decreases.

(1) $f'(x) = 20x^4 - 100x^3 + 120x^2$.

(2) $20x^4 - 100x^3 + 120x^2 = 20x^2(x^2 - 5x + 6) = 20x^2(x - 2)(x - 3)$.

This is zero if $x = 0, 2,$ or 3; these are the critical points of f.

(3) The intervals to be examined are $x < 0, 0 < x < 2, 2 < x < 3, 3 < x$.

(4) We now determine the sign of $f'(x)$ at a convenient point in each interval:

Interval	x_0	Sign of $f'(x_0)$	
$x < 0$	-1	$f'(-1) = 20 \cdot (-1)^2(-3)(-4)$	> 0
$0 < x < 2$	1	$f'(1) = 20 \cdot (1)^2(-1)(-2)$	> 0
$2 < x < 3$	$2\frac{1}{2}$	$f'(2\frac{1}{2}) = 20 \cdot (\frac{5}{2})^2(\frac{1}{2})(-\frac{1}{2})$	< 0
$3 < x$	4	$f'(4) = 20 \cdot (4)^2(2)(1)$	> 0

(5) So

if	$x < 0$	f increases
if	$0 < x < 2$	f increases
if	$2 < x < 3$	f decreases
if	$3 < x$	f increases

Once again we offer a rough sketch using the above information and the values of f at the critical points. Since $f(x) = 4x^5 - 25x^4 + 40x^3$,

Goal 19 | To Determine Where a Function is Increasing or Decreasing

$f(0) = 0 - 0 + 0 = 0$
$f(2) = 4(2^5) - 25(2^4) + 40(2^3) = 48$
$f(3) = 4(3^5) - 25(3^4) + 40(3^3) = 27$

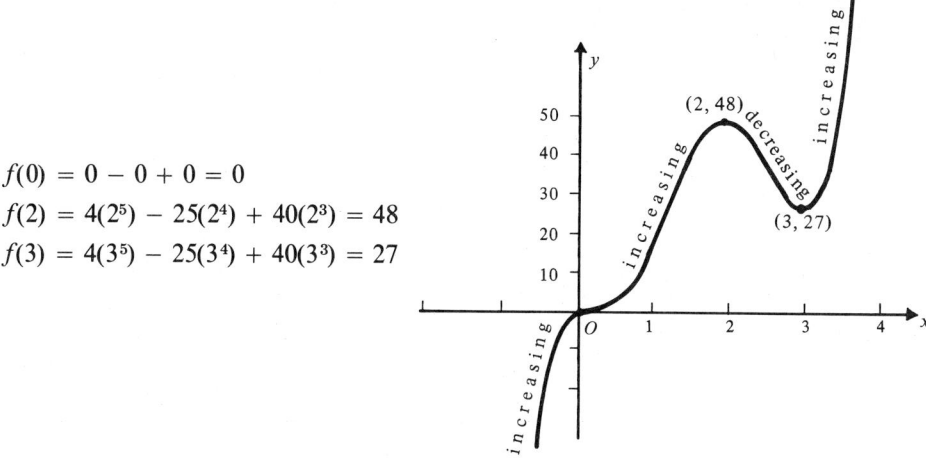

Note the use of different scales on the axes.

CASE II
The function f does not have a continuous derivative. The function may not be defined at one or more points, or it may not be everywhere differentiable. In this case we consider the intervals formed not only by the critical points, as for Case I, but also by these various points of discontinuity of f or f'.

Example 19-4
Use the method discussed above to find where $f(x) = 1/x$ increases or decreases.

We note first that f is not defined at 0. Following the steps of Case I, we have:

(1) $f'(x) = -1/x^2$.
(2) $-1/x^2$ never equals 0.
(3) The intervals to examine are $x < 0$ and $0 < x$.
(4) If $x < 0$, $f'(x) = -(1/x^2) < 0$; if $x < 0$, $f'(x) = -(1/x^2) < 0$.
(5) So f is decreasing if $x < 0$ and f is decreasing if $x > 0$.

Why is it incorrect to say that x is decreasing everywhere (that is, for all x)?

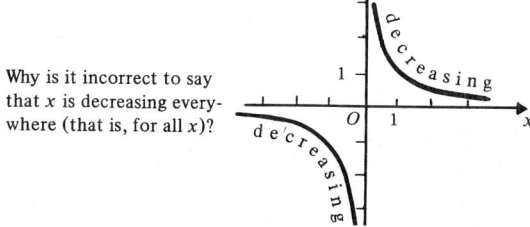

SKETCHING THE GRAPH OF A FUNCTION WITH GIVEN PROPERTIES

Here are two examples.

Example 19-5

Sketch the graph of a function f if $f(1) = 2$, $f(5) = 4$, $f'(x) < 0$ if $x < 1$, $f'(x) > 0$ if $1 < x < 5$, and $f'(x) < 0$ if $x > 5$.

There are many functions that satisfy the conditions given. Here we sketch one possible function.

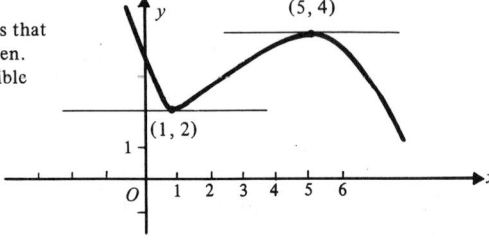

Example 19-6

Sketch the graph of a function that has the following properties: $f(-1) = 3$, $f(2) = 1$; $f'(x) > 0$ if $x < -1$; $f'(x) < 0$ if $-1 < x < 2$; $f'(x) > 0$ if $x > 2$; $f'(-1) = 0$; $f'(x)$ is not defined at $x = 2$.

Again we sketch one possible function.

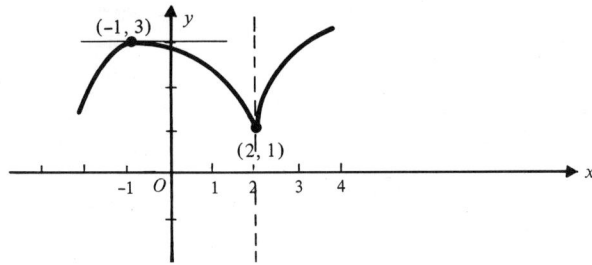

EXERCISES

For each function sketched below, tell where it is increasing or decreasing.

1.   2.

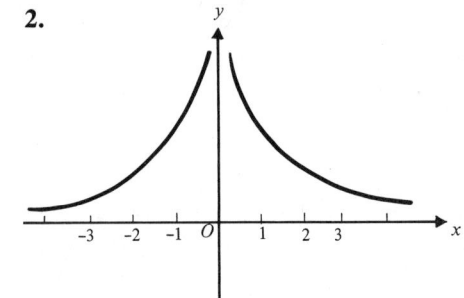

Goal 19 | To Determine Where a Function is Increasing or Decreasing

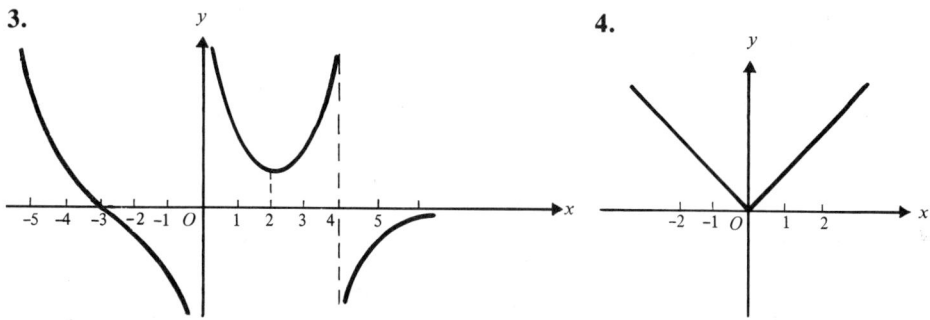

In 5 through 20, find the coordinates of any critical points, and then determine the intervals on which the function is increasing or decreasing.

5. $f(x) = x^2 - 4x + 1$
6. $g(t) = t^2 + 6t + 4$
7. $h(x) = -3x^2 + 4x + 2$
8. $f(x) = -2x^2 - 3x - 2$
9. $f(t) = 2t^3 - 9t^2 + 12t + 1$
10. $g(x) = x^3 - 3x^2 - 9x + 10$
11. $h(x) = -x^3 + 12x$
12. $f(t) = -t^3 - 6t^2 - 9t + 2$
13. $g(x) = x^3 - 3x^2 + 3x + 1$
14. $h(x) = -4x^3 - 18x^2 - 27x$
15. $f(x) = x^4$
16. $g(t) = 3t^5 - 5t^3$
17. $h(x) = 3x^4 + 4x^3 - 12x^2 + 2$
18. $f(x) = x^4 - 4x^3 + 4x^2 + 6$
19. $f(t) = 4t^5 - 25t^4 + 40t^3$
20. $g(x) = x^9$

In 21 and 22, find any discontinuities of the function or its derivative, and then determine where the function increases or decreases.

21. $f(x) = x^{2/3}$
22. $h(t) = \dfrac{1}{t + 1}$

In 23 through 30, sketch the graph of a function with the properties specified.

23. $f(3) = 1$, $f'(x) < 0$ for $x < 3$, $f'(x) > 0$ for $x > 3$
24. $f(2) = -1$, $f'(x) > 0$ for $x < 2$, $f'(x) < 0$ for $x > 2$
25. $f(-2) = 1$, $f(2) = 3$, $f'(x) < 0$ for $x < -2$, $f'(x) > 0$ for $-2 < x < 2$, and $f'(x) < 0$ for $x > 2$
26. $f(1) = 2$, $f(2) = 1$, $f'(x) > 0$ for $x < 1$, $f'(x) < 0$ for $1 < x < 2$, and $f'(x) > 0$ for $x > 2$
27. $f(0) = 1$, $f(1) = 3$, $f(3) = 0$, $f'(x) < 0$ for $x < 0$, $f'(x) > 0$ for $0 < x < 1$, $f'(x) < 0$ for $1 < x < 3$, and $f'(x) > 0$ for $x > 3$
28. $f(-1) = 4$, $f(2) = 1$, $f(3) = 3$, $f'(x) > 0$ for $x < -1$, $f'(x) < 0$ for $-1 < x < 2$, $f'(x) > 0$ for $2 < x < 3$, and $f'(x) < 0$ for $x > 3$

29. $f(2) = 3$, $f'(x)$ is undefined at $x = 2$, $f'(x) > 0$ if $x < 2$, $f'(x) < 0$ if $x > 2$

30. $f(-3) = 2$, $f(0) = 4$, $f(3) = 1$, $f'(x) < 0$ if $x < -3$ or if $0 < x < 3$, $f'(x) > 0$ if $-3 < x < 0$ or if $x > 3$, $f'(-3) = f'(3) = 0$, $f'(x)$ is undefined at $x = 0$

TEST A

1. For each of the following, find any critical points; then determine over what intervals the function is increasing or decreasing:
   (a) $f(x) = 2x^3 + 15x^3 + 36x + 25$
   (b) $g(x) = 3x^4 - 4x^3$

2. Sketch the graph of a function with the following properties: $f(-3) = 2$, $f(1) = 4$, $f'(x) < 0$ for $x < -3$, $f'(x) > 0$ for $-3 < x < 1$, and $f'(x) < 0$ for $x > 1$.

TEST B

1. For each of the following, find any critical values; then determine over what intervals the function is increasing or decreasing:
   (a) $f(x) = -2x^3 + 3x^2 + 36x - 15$
   (b) $g(x) = 3x^4 - 8x^3 + 6x^2$

2. Sketch the graph of a function with the following properties: $f(2) = 2$, $f(3) = 0$, $f'(x) > 0$ for $x < 2$, $f'(x) < 0$ for $2 < x < 3$, and $f'(x) > 0$ for $x > 3$.

STUDY QUESTIONS

1. What general conclusions can be drawn about the intervals on which a constant function may increase or decrease? How about a linear function? a quadratic function? a cubic function?

2. Use the methods of this Goal to confirm that the absolute-value function $f(x) = |x|$ is decreasing if $x < 0$, increasing if $x > 0$. Sketch the graphs of the function and its derivative.

GOAL **20**

To determine local maxima and minima

A great many applications of the calculus involve finding high or low points on a curve. These correspond to what we call local maxima or minima of the function. The first derivative can be used to help identify these points.

TWO PRACTICAL PROBLEMS

1. Suppose a scientist interested in arctic storms near a particular weather station uses apparatus that continuously records the barometric pressure at that location. Since the storms occur at times of low pressure, the crucial points on the graph of pressure as a function of time are the low points, or the local minima of the function. Intuitively, a local minimum of a function f is a point at which the value of f is less than it is at any point in the immediate neighborhood; a local minimum may or may not be the lowest point on the entire graph.

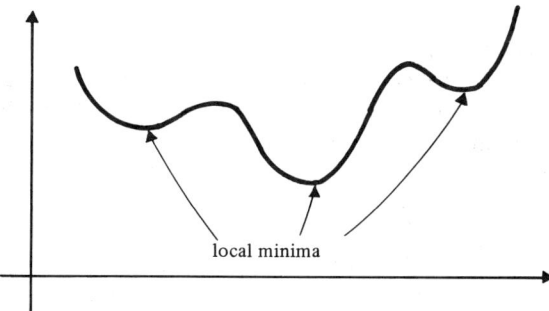

2. Suppose that the temperature of a patient in an intensive-care unit is continuously recorded so that the doctor can know when the peaks occur. On a graph of body temperature as a function of time, the

crucial points for the doctor are the high points, or the local maxima of the function. Intuitively, a local maximum of a function f is a point at which the value of f is greater than it is at any point in the immediate neighborhood; a local maximum may or may not be the highest point on the entire graph.

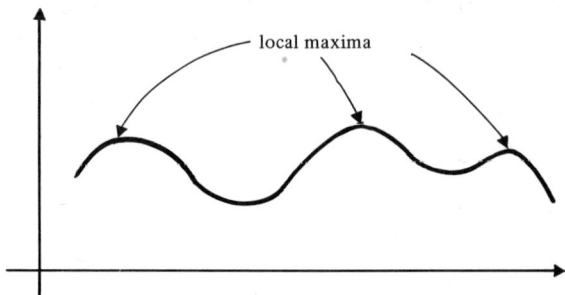

DEFINITIONS

If a function $y = f(x)$ has a *local minimum* at $x = a$ in its domain, then for all x (in its domain) sufficiently close to a, $f(x) \geq f(a)$. We will consider here only those cases in which the function actually changes at $x = a$ from decreasing to increasing, that is, where $f'(x)$ changes from negative to positive. Each of the three functions sketched has a local minimum at $x = a$.

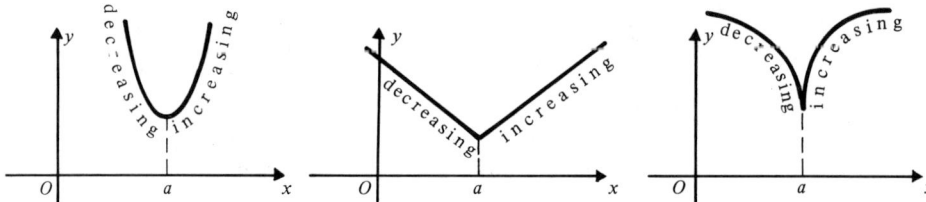

If $y = f(x)$ has a *local maximum* at $x = a$, then for all x sufficiently close to a, $f(x) \leq f(a)$. In our examples, f changes from increasing to decreasing at $x = a$; that is, $f'(x)$ changes from positive to negative. Each function sketched has a local maximum at $x = a$.

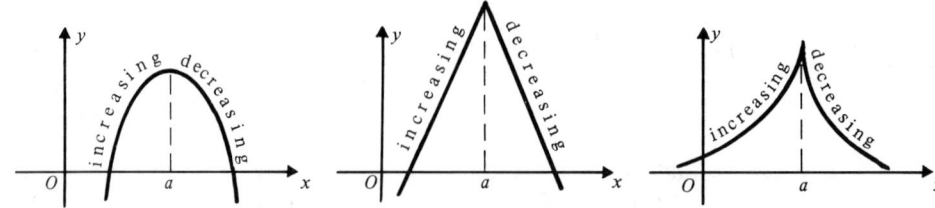

Goal 20 | To Determine Local Maxima and Minima

Note that if a function has a local maximum or minimum at $x = a$, then $x = a$ is a "turning point" of the curve. The definitions given above and the methods of Goal 19 enable us to identify such points easily.

Instead of the term "*local*" maximum (or minimum), some prefer to speak of a "*relative*" maximum (or minimum).

In place of the long words maximum and minimum we shall sometimes use the abbreviations max and min.

FINDING LOCAL MAXIMA OR MINIMA OF A FUNCTION

As in Goal 19, we have to consider two cases: functions that are everywhere differentiable and those that are not.

CASE I

The function f is everywhere differentiable.

We follow the steps outlined under Case I of Goal 19 for finding where a function increases or decreases (p. 188). If $x = a$ is a critical point of f, then this procedure gives us the sign of $f'(x)$ both in the interval to the left of $x = a$ and in the interval to its right. We then conclude:

if $f'(x)$ changes from negative to positive, f has a local minimum at $x = a$;

if $f'(x)$ changes from positive to negative, f has a local maximum at $x = a$;

if $f'(x)$ does not change sign, f has neither a local maximum nor a local minimum at $x = a$.

The following four figures illustrate the possibilities when $x = a$ is a critical point. We sketch tangents at points in intervals to the left and to the right of $x = a$ to show what may happen to the sign of $f'(x)$.

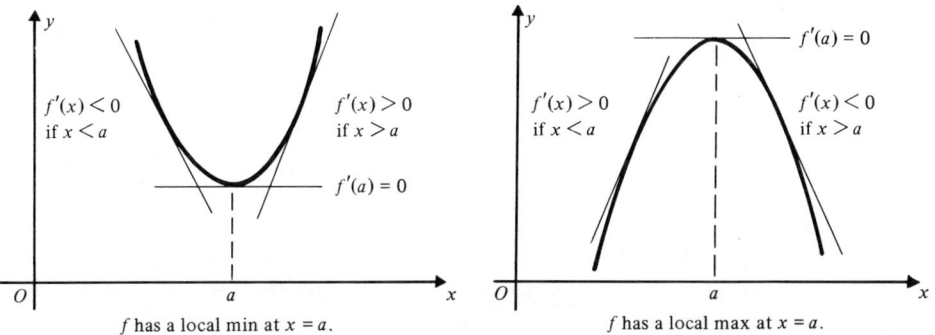

f has a local min at $x = a$. f has a local max at $x = a$.

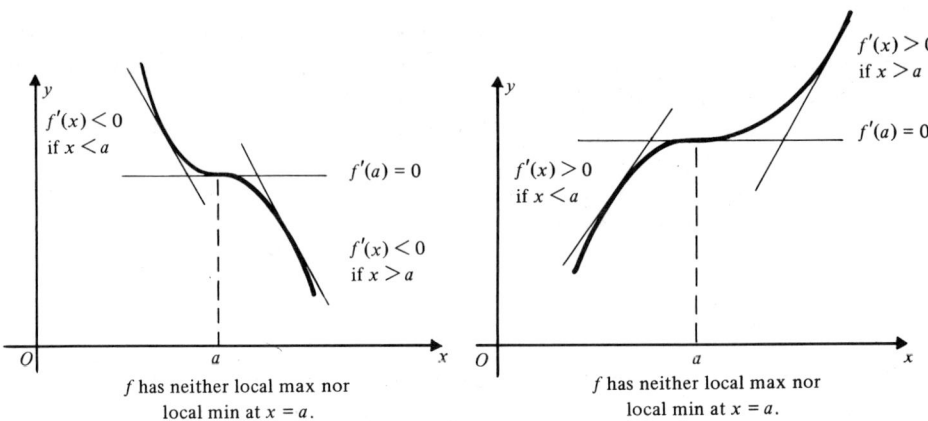

f has neither local max nor
local min at x = a.

f has neither local max nor
local min at x = a.

Example 20-1

Find any points at which $f(x) = x^4 - 4x^3 - 8x^2 + 60$ has a local maximum or minimum.

We follow the steps on page 188.

(1) $f'(x) = 4x^3 - 12x^2 - 16x$.

(2) $4x^3 - 12x^2 - 16x = 4x(x^2 - 3x - 4) = 4x(x + 1)(x - 4)$; this is zero if x is 0, -1, or 4. These are the critical points of f.

(3) The intervals to examine are $x < -1$, $-1 < x < 0$, $0 < x < 4$, and $4 < x$.

(4) We find the sign of $f'(x)$ at a convenient point in each interval.

Interval	x_0	Sign of $f'(x_0)$
$x < -1$	-2	$4(-2)(-1)(-6) < 0$
$-1 < x < 0$	$-\frac{1}{2}$	$4(-\frac{1}{2})(\frac{1}{2})(-\frac{9}{2}) > 0$
$0 < x < 4$	1	$4(1)(2)(-3) < 0$
$4 < x$	5	$4(5)(6)(1) > 0$

(5) So,

if $\quad x < -1 \quad$ f is decreasing
if $\quad -1 < x < 0 \quad$ f is increasing
if $\quad 0 < x < 4 \quad$ f is decreasing
if $\quad 4 < x \quad$ f is increasing

It often helps to show the behavior of f schematically, as on p. 199, ignoring the precise functional values at this stage. It is clear from this sketch that f has local minima at $x = -1$ and $x = 4$, and a local maximum at $x = 0$.

To sketch the graph, we now evaluate f at these points: $f(-1) = 57$, $f(0) = 60$, $f(4) = -68$.

Goal 20 | To Determine Local Maxima and Minima

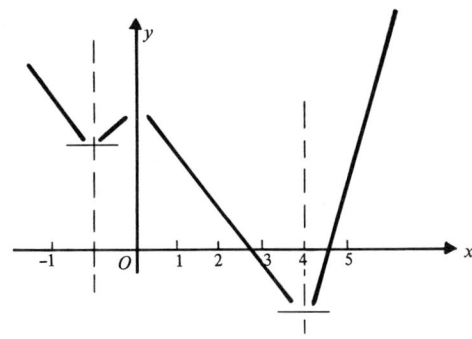

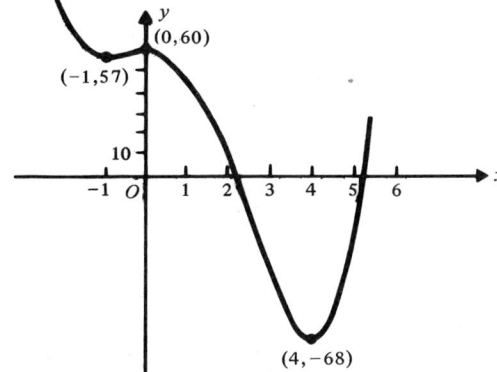

Example 20-2

Find any local maxima and minima of $f(x) = -3x^4 + 8x^3 - 10$.

(1) $f'(x) = -12x^3 + 24x^2$.
(2) $-12x^3 + 24x^2 = -12x^2(x - 2)$; this is zero at $x = 0$ and $x = 2$, which are the critical points.
(3) The intervals are $x < 0$, $0 < x < 2$, and $2 < x$.
(4) We examine the signs of $f'(x)$ in these intervals. For the first, choose $x_0 = -1$; then $f'(x_0) = -12(-1)^2(-3) > 0$. For the second, choose $x_0 = 1$; then $f'(x_0) = -12(1)^2(-1) > 0$. And for $2 < x$, choose $x_0 = 3$; then $f'(x_0) = -12(3)^2(1) < 0$.
(5) f is increasing if $x < 0$ or if $0 < x < 2$; f is decreasing if $x > 2$.

Since f' does not change sign at $x = 0$, f has neither a local max nor a local min at $x = 0$; f does have a local max at $x = 2$.

The behavior of f is shown schematically on p. 200. After noting that $f(0) = -10$ and $f(2) = 6$, we can sketch the curve.

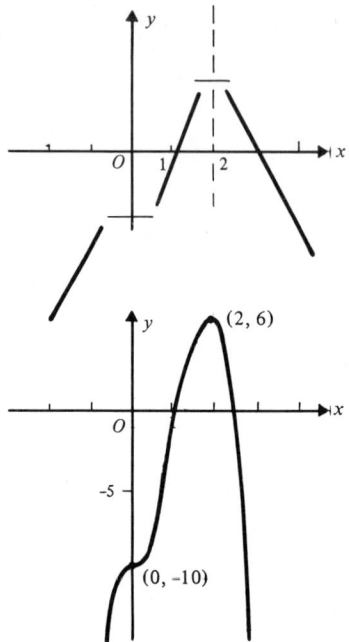

Case II

The function f is not everywhere differentiable. In this situation, we consider the intervals formed by the critical points of f, for which $f'(x) = 0$, together with those bounded by points for which f is defined but $f'(x)$ does not exist.

Example 20-3

Locate any local maximum or minimum points of $f(x) = 1 + x^{2/3}$.

(1) $f'(x) = \dfrac{2}{3x^{1/3}}$

(2) Although no x makes $f'(x)$ equal zero, we note that $f'(x)$ does not exist at $x = 0$.

(3) We therefore look at the intervals $x < 0$ and $x > 0$.

(4) For $x < 0$, choose $x_0 = -1$; then $f'(x_0) = \dfrac{2}{3(-1)^{1/3}} < 0$; for $x > 0$, choose $x_0 = 1$; then $f'(x_0) = \dfrac{2}{3(1)^{1/3}} > 0$.

(5) If $x < 0$, f decreases; if $x > 0$, f increases. On p. 201, we indicate schematically how f behaves, noting that it has a local min at $x = 0$.

In Goal 22 we will see that the graph of this function is concave down for all $x \neq 0$. Note that the domain of f is all reals, that the tangent to the curve at $x = 0$ is vertical, and that $f(0) = 1$.

Goal 20 | To Determine Local Maxima and Minima

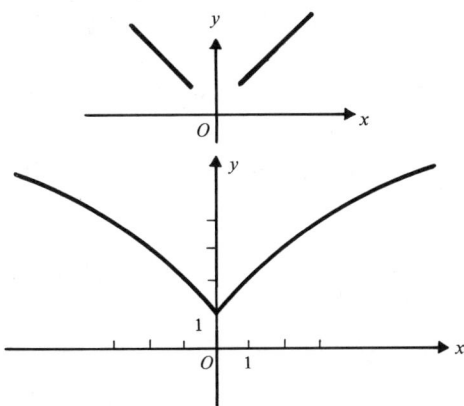

EXERCISES

In Exercises 1 through 4 of Goal 19, list the values of x for which the function has local maxima or minima.

In Exercises 5 through 22 of Goal 19, find the coordinates of each local maximum or minimum.

TEST A

Find the coordinates of any local maxima or minima of

1. $f(x) = 4x^3 + 3x^2 - 6x + 3$
2. $g(x) = x^4 - 4x^3 + 30$

TEST B

Find the coordinates of any local maxima or minima of

1. $f(x) = x^3 - 27x + 50$
2. $h(t) = t^4 + 8t^3 + 18t^2 - 20$

STUDY QUESTIONS

1. Each statement (a) through (h) is false. For each one, find a counterexample; that is, a function, or the graph of a function, that disproves the statement.
   (a) If $f'(x_0) = 0$, then f has a local max or min at $x = x_0$.
   (b) If f is defined for all x on the closed interval $[a, b]$ (that is, $a \leq x \leq b$) and if the only local maximum point of f is at $x = c$ where $a < c < b$, then $f(c) > f(x)$ for all x in $[a, b]$.
   (c) If f has a local max or min at $x = c$, then $f'(c) = 0$.

(d) If $f'(x)$ is not defined at $x = c$, then f does not have a local max or min at $x = c$.

(e) If the tangent to a curve at $x = c$ is horizontal, then the function has a local max or min at $x = c$.

(f) If the tangent to a curve at $x = c$ is vertical, then the function has a local max or min at $x = c$.

(g) If f is continuous and has one local min at $x = a$ and another at $x = b$, then there is some number, say c, between a and b such that $f'(c) = 0$.

(h) If f has a local min at $x = c$ and a local max at $x = d$, then $f(c) < f(d)$.

2. The graph of $f(x) = ax^2 + bx + c$ is a parabola. Use the methods of this Goal to show that if $a > 0$, then at $x = -b/2a$ the function f has a local min, but if $a < 0$ then f has a local max. HINT: Rewrite $f'(x)$ in the form $2a\left(x + \dfrac{b}{2a}\right)$.

3. Show that if $f'(x) = x^2(x - 1)^4(x + 3)^2$, then f has no local max or min. Draw a generalization about the behavior of f at $x = a$ if $(x - a)^n$ is a factor of $f'(x)$.

GOAL **21**

To find absolute maxima and minima

In many problems we are interested in the smallest (or the largest) value of a function over a specified interval. In this Goal we describe how to obtain these absolute extremes of a function.

DEFINITIONS

Recall (p. 550) that "$[a, b]$" denotes the *closed* interval $a \leq x \leq b$, which contains its end points a and b; "(a, b)" is the *open* interval $a < x < b$; $[a, b)$ contains a but not b; $(a, b]$ contains b but not a.

A function f has an *absolute minimum* (abs min) at a point c in its domain if $f(c) \leq f(x)$ for all x in its domain. The function has an *absolute maximum* (abs max) at $x = d$ if $f(d) \geq f(x)$ for all x in its domain. We will be primarily interested here in functions that are continuous on closed intervals, such as those whose graphs are shown below: each function is continuous on the closed interval $[a, b]$.

It is always the case if f is continuous on a closed interval that it has both an absolute maximum and an absolute minimum on that interval. Each ex-

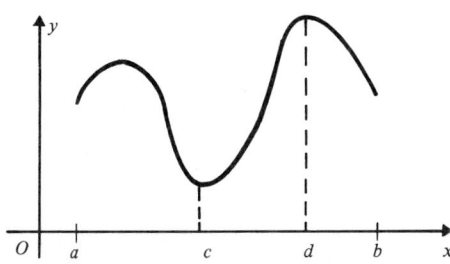

The abs min, at $x = c$, is also a local min; the abs max, at $x = d$, is also a local max.

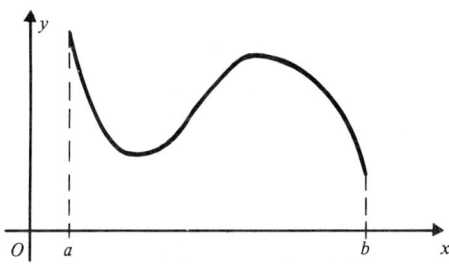

The abs min is at the endpoint $x = b$; the abs max is at the endpoint $x = a$.

203

treme occurs either at a critical point* or at an endpoint of the interval. If a function f is not continuous, or if the interval of continuity is not closed, then f may fail to have an absolute max or min or both.

FINDING ABSOLUTE MAXIMA AND MINIMA

Case I: Differentiable Functions

If f is differentiable on $[a, b]$, then it is continuous at every point in $[a, b]$. We proceed as follows:

(1) Solve $f'(x) = 0$ for critical points in the interval $[a, b]$.

(2) Evaluate f at each critical point in $[a, b]$, at $x = a$, and at $x = b$.

(3) The smallest value obtained in (2) is the absolute minimum of f; the largest is the absolute maximum.

Example 21-1

Find the absolute maximum and minimum of $f(x) = x^4 - 8x^2 + 8$ on the interval $-1 \le x \le 3$.

(1) $f'(x) = 4x^3 - 16x = 4x(x^2 - 4) = 4x(x - 2)(x + 2)$, which is zero if $x = 0, 2$, or -2. Only the critical values $x = 0$ and $x = 2$ are in the interval $[-1, 3]$.

(2) Evaluate f at the critical values 0 and 2 and at the endpoints -1 and 3: $f(-1) = 1$, $f(0) = 8$, $f(2) = -8$, and $f(3) = 17$.

(3) f has an abs min at $x = 2$ and an abs max at $x = 3$.

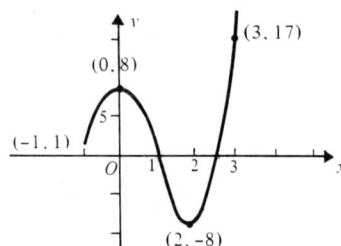

Example 21-2

Find the absolute maximum and minimum of the function in Example 21-1 if the interval is $[-2, 3]$.

The critical value $x = -2$ is also an endpoint of the interval $[-2, 3]$. From Example 21-1, we have $f(0) = 8$, $f(2) = -8$, $f(3) = 17$. Since $f(-2) = -8$,

---

*It is customary to include among the critical points of f not only those for which $f'(x) = 0$ but also any point at which f exists but f' does not.

Goal 21 | To Find Absolute Maxima and Minima

f attains its absolute min *both* at $x = -2$ and at $x = 2$; it has an absolute max at $x = 3$.

CASE II: FUNCTIONS THAT ARE EITHER NOT EVERYWHERE DEFINED OR NOT EVERYWHERE DIFFERENTIABLE

Proceed as for Case I but evaluate f also at each point in $[a, b]$ for which f is defined but f' does not exist. Examine the behavior of f near any point of discontinuity in $[a, b]$.

Example 21-3

Find the absolute max and min, if any, of $f(x) = x + 1/x$.

Since no interval is specified here, we consider f over its entire domain; that is, for all nonzero x. We have

$$f'(x) = 1 - \frac{1}{x^2} = \frac{x^2 - 1}{x^2}$$

and this equals zero if $x = \pm 1$. So -1 and 1 are critical values of f. Also, f' is not defined at 0; but neither is f. We note that

$$f \to +\infty \quad \text{as} \quad x \to 0^+ \qquad f \to -\infty \quad \text{as} \quad x \to 0^-$$

We therefore conclude that f has neither an abs max nor an abs min. It is easy to show that f has a local min at $(1, 2)$ and a local max at $(-1, -2)$.

Note, however, that this function does have both an abs max and an abs min on every closed interval *that does not contain zero*. On the interval $x > 0$, f has an abs min but no abs max. The sketch of f is at the left.

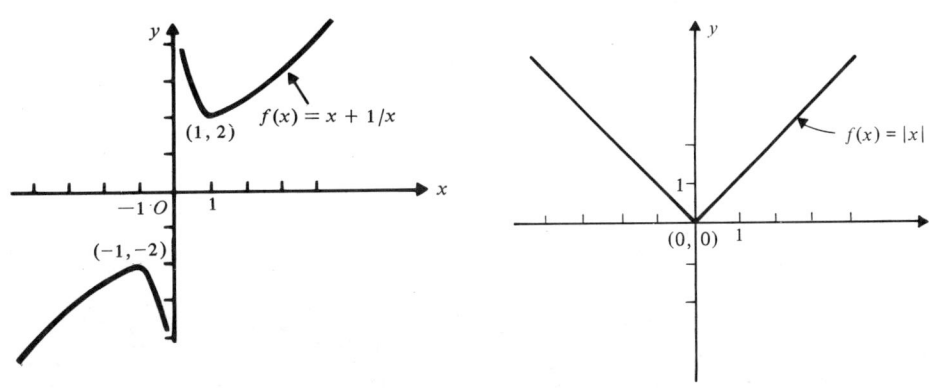

Example 21-4

Find the absolute max and min, if any, of $f(x) = |x|$.

The absolute-value function is defined for all real x. Its derivative exists everywhere except at $x = 0$; indeed, $f'(x) = -1$ if $x < 0$, $f'(x) = 1$ if $x > 0$. Since $f'(x)$ never equals 0, we need note only that f decreases if $x < 0$, increases if $x > 0$. It therefore has an abs min at 0 (equal to 0) but no abs max. The sketch of $f(x) = |x|$ is at the right on p. 205.

WORD PROBLEMS

Example 21-5

The cost of operating a truck at v mph is $C(v) = 0.20 + 0.004v$ dollars per mile. If a driver observes the posted speed limits of 40 to 55 mph, find the most economical speed.

We want the value of v for which the cost is an abs min, over the interval $40 \leq v \leq 55$. Since $C'(v) = 0.004$, the derivative never equals 0 and there are no critical points. We therefore evaluate C at the endpoints: $C(40) = 0.20 + 0.004(40) = \$0.36$/mile; $C(55) = 0.20 + 0.004(55) = \$0.42$/mile. The most economical speed is therefore 40 mph.

Example 21-6

Suppose the driver in Example 21-5 earns $10 an hour. Find the most economical speed for a 1000-mile trip.

We now want the speed v for which $T(v)$ is an abs min on [40, 55], where $T(v)$ is the total cost, in dollars, for 1000 miles of both the truck and the driver. The truck costs, in dollars, $(0.20 + 0.004v) \times 1000 = 200 + 4v$. Since it takes $(1000/v)$ h to complete the trip, the driver's wages, in dollars, are $\left(\dfrac{1000}{v} \times 10\right) = \dfrac{10{,}000}{v}$. So

$$T(v) = 200 + 4v + \frac{10{,}000}{v}$$

$$T'(v) = 4 - \frac{10{,}000}{v^2} = \frac{4v^2 - 10{,}000}{v^2}$$

$$= \frac{4(v^2 - 2500)}{v^2}$$

and this is zero if $v = 50$. (The critical point $v = -50$ is not in the interval $40 \leq v \leq 55$.) We evaluate T at 50 and at each endpoint:

$$T(40) = 200 + 4(40) + \frac{10{,}000}{40} = \$610$$

$$T(50) = 200 + 4(50) + \frac{10,000}{50} = \$600$$

$$T(55) = 200 + 4(55) + \frac{10,000}{55} = \$601.82$$

The total cost is therefore least (an absolute minimum) when the speed is 50 mph.

EXERCISES

In 1 through 20, find any absolute maxima or minima of the functions over the specified intervals.

1. $f(x) = x^2$ on $-1 \leq x \leq 2$
2. $f(x) = -x^2 + 2$ on $[-3, 1]$
3. $f(x) = x^2 + 2x - 1$ on $[0, 4]$
4. $f(x) = -x^2 + 3x + 1$ on $-1 \leq x \leq 1$
5. $f(x) = 2x^3 - 3x^2 - 12x$ on $-2 \leq x \leq 3$
6. $f(x) = x^3 - 3x + 2$ on $0 \leq x \leq 3$
7. $f(x) = 2x^3 - 3x^2 + 3$ on $[-1, 3]$
8. $f(x) = x^3 - 12x + 5$ on $-1 \leq x \leq 1$
9. $f(x) = 2x^3 + 1$ on $-1 \leq x \leq 2$
10. $f(x) = -x^3 - 6x^2 + 15x$ on $[0, 2]$
11a. $f(x) = \dfrac{1}{x}$ on $[1, 5]$
11b. $f(x) = \dfrac{1}{x}$ on $[0, 1]$
12a. $f(x) = \dfrac{2}{x - 1}$ on $-3 \leq x \leq 0$
12b. $f(x) = \dfrac{2}{x - 1}$ on $0 \leq x < 1$
13. $f(x) = \dfrac{1}{x^2 + 1}$ on $[-2, 1]$
14. $f(x) = x^4 - 2x^2 + 1$ on $[0, 2]$
15a. $f(x) = x + \dfrac{1}{x}$ on $[2, 5]$
15b. $f(x) = x + \dfrac{1}{x}$ on $[1, 3]$

16. $f(x) = \sqrt{x}$ on $4 \leq x \leq 9$
17. $f(x) = x^5 - 5x^4$ on $-1 \leq x \leq 2$
18. $f(x) = 3x^4 - 4x^3 - 12x^2$ on $[-1, 3]$
19. $f(x) = \dfrac{x-1}{x+1}$ on $0 \leq x \leq 5$
20. $f(x) = x^4 - 4x^3 + 4x^2 + 1$ on $[-1, 3]$
21. A carpenter who wants to build a box, of fixed volume, with a square base has figured that it will cost him $C(x) = x^2 + \dfrac{432}{x}$ dollars, where x is the length in feet of the side of the base. Find the length of the side of the base for which the box will cost least to build, if
    (a) there are no limitations on the dimensions of the box.
    (b) the length of the base must be at least 2 ft but not greater than 5 ft.
22. An agent for a concert pianist knows that the profit, after all expenses, on a one-night performance will be

$$P(x) = 11.50x - \frac{x^2}{500} - 7300$$

dollars, where x is the number of tickets sold (compare with Exercises 11 and 12 on p. 60). Find the number of tickets that should be sold for maximum profit if
    (a) the theater seats 3000 people.
    (b) only 2500 people can be seated.

TEST A

For each function find the coordinates of the absolute maximum and minimum, if any, on the designated interval:

1. $f(x) = x^2 - 4x + 2$ on $\leq x \leq 3$
2. $f(x) = -x^2 + 2x + 1$ on $[2, 4]$
3. $f(x) = x^3 - 6x^2 + 9x - 1$ on $[0, 2]$
4. $f(x) = x^2 + \dfrac{2}{x}$ on $\dfrac{1}{2} \leq x \leq 2$

TEST B

For each function find the coordinates of the absolute maximum and minimum, if any, on the designated interval:

1. $f(x) = 3x^3 + 6x - 5$ on $0 \leq x \leq 2$

Goal 21 | To Find Absolute Maxima and Minima

2. $f(x) = -2x^2 + 4x + 2$ on $[0, 3]$
3. $f(x) = x^3 + 3x^2 - 3$ on $-1 \leq x \leq 0$
4. $f(x) = \dfrac{1}{x^2}$ on (a) $1 \leq x \leq 4$, (b) $-1 \leq x \leq 1$

STUDY QUESTIONS

1. Find the locations of any absolute maxima or minima of $f(x) = x^2 - 4x + 5$ on each of the following intervals:
   (a) [0, 3]   (d) [2, 3)
   (b) [0, 3)   (e) (2, 3)
   (c) [2, 3]   (f) (0, ∞)

   What do the answers show about the existence of absolute extremes of a continuous function on a specified interval?

2. Find the locations of any absolute maximum or minimum of $f(x) = \dfrac{1}{x}$ on each of the following intervals:
   (a) [1, 4]   (d) $[\tfrac{1}{100}, 2]$
   (b) $[\tfrac{1}{5}, 3]$ (e) (0, 2)
   (c) $[\tfrac{1}{10}, 2]$

GOAL **22**

To apply the second derivative: concavity, inflection points, testing for maxima and minima

The shape of a curve depends on many factors, such as where the corresponding function is positive or negative, where it is increasing or decreasing, where the local maxima and minima occur, where the curve is concave up and concave down, and where the concavity changes. The second derivative of a function can be used to determine concavity; it can also be used to test whether a critical point of a function yields a maximum or a minimum.

DEFINITIONS AND PRELIMINARIES

The graph of $f(x)$ is *concave up* over an interval (a, b) if throughout the interval the curve lies above the tangents to it (see sketches). If we examine the slopes of the tangent lines, moving along the curve from left to right, we note that the slope—that is, $f'(x)$—increases. In the first figure $f'(x)$ is negative to the left of the critical point, positive to the right; in the second figure $f'(x)$ is always negative yet increases; in the third figure, $f'(x)$ is always

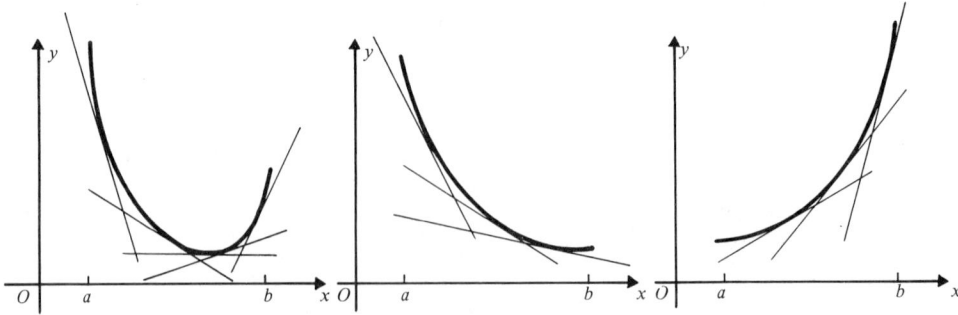

positive yet increases. In each figure, the tangents are "turning" or "rotating" counterclockwise.

From Goal 19 we know that if a function is increasing over an interval then its derivative (if it exists) is nonnegative on that interval. Since $f''(x)$ is the derivative of $f'(x)$, we can conclude that

If $f'(x)$ increases on (a, b), then $f''(x) \geq 0$ on (a, b).

Sometimes we say that a curve or arc that is concave up "holds water."

If the graph of f is *concave down* on (a, b), then throughout the interval the curve lies below the tangents to it (see sketches). Now the slope $f'(x)$

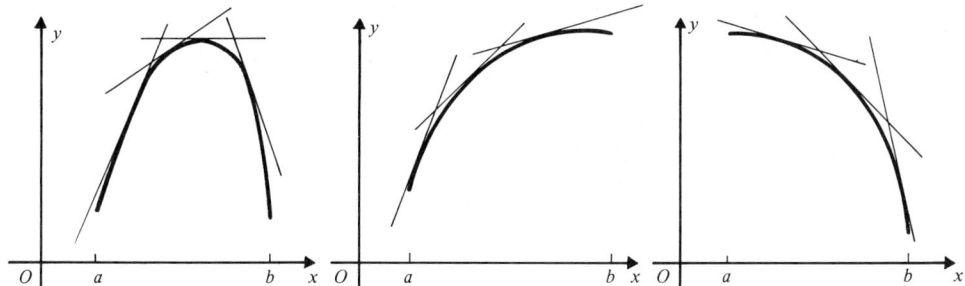

decreases as we move along the curve from left to right: in the first figure, $f'(x)$ changes from positive to negative; in the second, $f'(x)$ is always positive yet decreases; on the third, $f'(x)$ is always negative yet decreases. Now the tangents are "turning" or "rotating" clockwise. Once again we have, from Goal 19:

If $f'(x)$ decreases on (a, b) then $f''(x) \leq 0$ on (a, b).

The crucial facts about concavity can be summarized this way:

If $f''(x) > 0$ on (a, b), then the graph of f is concave up on (a, b); if $f''(x) < 0$ on (a, b), then the graph of f is concave down on (a, b).

POINTS OF INFLECTION

A *point of inflection* is a point I on the curve at which the concavity changes, from up to down or from down to up (see sketches on p. 212). At a point of inflection the tangent line cuts through the curve; if the tangent is not

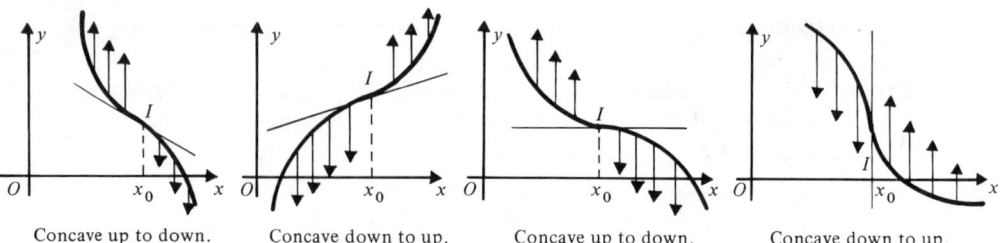

Concave up to down.   Concave down to up.   Concave up to down.   Concave down to up.

vertical, it lies below the graph to the left of $x = x_0$ but above to the right, or vice versa.

Since the concavity changes at a point of inflection with x-coordinate x_0, this implies that $f''(x)$ must change sign at $x = x_0$, from positive to negative or from negative to positive. That can happen only if $f''(x_0) = 0$ or if $f''(x_0)$ is not defined. If $f''(x)$ does not change sign at $x = x_0$, then there is not a point of inflection at $x = x_0$.

FINDING POINTS OF INFLECTION AND DETERMINING CONCAVITY

CASE I: $f''(x)$ IS CONTINUOUS

(1) Find $f''(x)$.
(2) Solve $f''(x) = 0$. If $f''(x_0) = 0$, then x_0 yields a possible point of inflection.
(3) Examine the signs of $f''(x)$ in the intervals to the left and right of $x = x_0$. If $f''(x)$ changes sign at $x = x_0$, then x_0 is the abscissa of a point of inflection; if $f''(x)$ does not change sign at $x = x_0$ then there is no point of inflection there.
(4) Determine the intervals on which $f''(x) > 0$ to find where the curve is concave up, and those on which $f''(x) < 0$ to find where the curve is concave down.

Example 22-1

Find any points of inflection and determine the concavity of the graph of $f(x) = -x^3$.

(1) $f'(x) = -3x^2$; $f''(x) = -6x$.
(2) $-6x = 0$ at $x = 0$, the abscissa of a possible point of inflection.
(3) If $x < 0$, $-6x > 0$; if $x > 0$, $-6x < 0$. So there is a point of inflection at $x = 0$.
(4) The curve is concave up if $x < 0$, down if $x > 0$.

Goal 22 | To Apply the Second Derivative

For completeness we show the curve here (Goal 24 will be devoted to curve-sketching). Note that $f(0) = 0, f'(0) = 0$, and the tangent at $(0, 0)$ is horizontal.

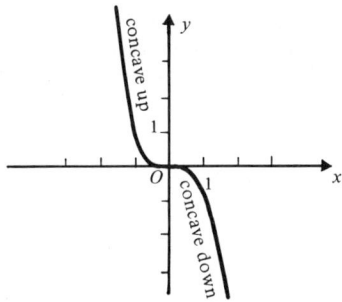

Example 22-2

Find any points of inflection and determine the concavity of the graph of $f(x) = 3x^5 - 5x^4 + 4$.

(1) $f'(x) = 15x^4 - 20x^3$; $f''(x) = 60x^3 - 60x^2 = 60x^2(x - 1)$.

(2) $f''(x) = 0$ if $x = 0$ or 1; these yield possible points of inflection.

(3) We find the signs of $f''(x)$ at a convenient point q on each of the intervals $x < 0$, $0 < x < 1$, and $1 < x$:

Interval	q	Sign of $f''(q)$
$x < 0$	-1	$60(-1)^2(-2) < 0$
$0 < x < 1$	$\frac{1}{2}$	$60(\frac{1}{2})^2(-\frac{1}{2}) < 0$
$1 < x$	2	$60(2)^2(1) > 0$

Since $f''(x)$ does not change sign at $x = 0$, there is no point of inflection at $x = 0$; there is at $x = 1$.

(4) The curve is concave down if $x < 0$ or $0 < x < 1$, up if $x > 1$. Since the graph lies below the tangents on the interval $x < 1$, the curve is concave down on this interval.

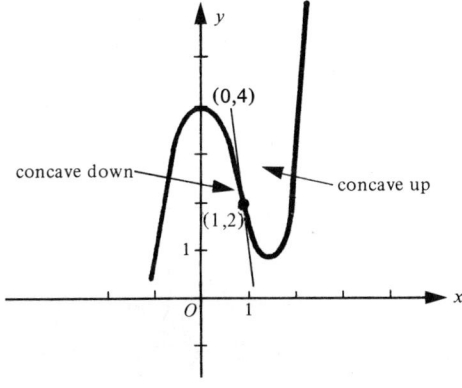

Note that $f(0) = 4$ and $f(1) = 2$; also $f'(0) = 0$ and $f'(1) = -5$. It can easily be verified that f has a local max (a very "flat" one) at $x = 0$ and a local min at $(\frac{4}{3}, \approx 0.8)$.

Example 22-3

Find any points of inflection and determine the concavity of $f(x) = x^4 - 8x^3 + 18x^2 - 10$.

(1) $f'(x) = 4x^3 - 24x^2 + 36x;$
$f''(x) = 12x^2 - 48x + 36 = 12(x - 1)(x - 3)$.

(2) $f''(x) = 0$ if $x = 1$ or 3, x-coordinates of possible points of inflection.

(3) The intervals to consider are $x < 1$, $1 < x < 3$, and $3 < x$.

Interval	q	Sign of $f''(q)$
$x < 1$	0	$12(-1)(-3) > 0$
$1 < x < 3$	2	$12(1)(-1) < 0$
$3 < x$	4	$12(3)(1) > 0$

So there are two points of inflection, at $x = 1$ and $x = 3$.

(4) The curve is concave up if $x < 1$, down if $1 < x < 3$, up if $3 < x$. To sketch the curve, we note that $f(1) = 1$ and $f(3) = 17$; also, $f'(1) = 16$, $f'(3) = 0$.

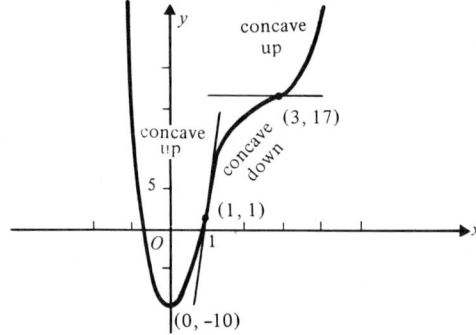

CASE II: $f''(x)$ HAS DISCONTINUITIES

As in earlier Goals, we must also consider any points of discontinuity, examining the signs of f'' both to the left and to the right of such points.

Example 22-4

Find any points of inflection and determine the concavity of the graph of $f(x) = x^{1/3}$.

Goal 22 | To Apply the Second Derivative

(1) $f'(x) = \frac{1}{3}x^{-2/3}$; $f''(x) = -\frac{2}{9}x^{-5/3} = -\dfrac{2}{9x^{5/3}}$.
(2) $f''(x)$ never equals 0; $f''(0)$ is not defined.
(3) Since $f''(x) > 0$ if $x < 0$ and $f''(x) < 0$ if $x > 0$, there is a point of inflection at $x = 0$.
(4) The curve is concave up if $x < 0$, down if $x > 0$. Note that $f(0) = 0$, that $f'(0)$ is not defined, and that the tangent at $(0, 0)$ is vertical.

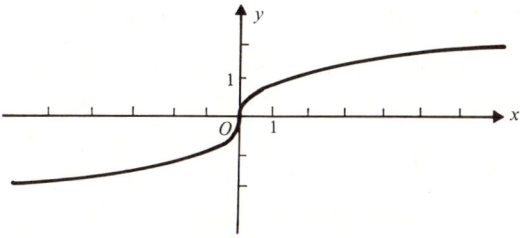

THE SECOND-DERIVATIVE TEST FOR LOCAL MAXIMA AND MINIMA

If f is differentiable and if $f'(c) = 0$, we know that the tangent to the graph of f is horizontal at $x = c$. The graph may look like one of these:

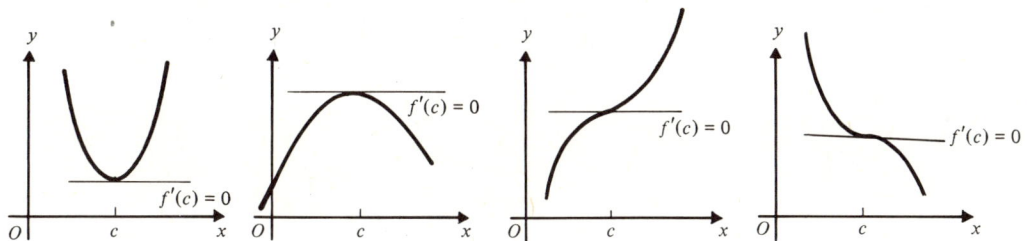

Since $f'(c) = 0$, f may have a local max or a local min at $x = c$. Instead of examining the signs of f' to the left and right of $x = c$, as in Goal 20, it is often more convenient to use the following second-derivative test:

(1) If $f''(c) > 0$, f has a local min at $x = c$.
(2) If $f''(c) < 0$, f has a local max at $x = c$.

The first figure above illustrates (1); the second figure illustrates (2). In the third and fourth figures, we see that $f''(c) = 0$ and note that $x = c$ yields a point of inflection. But we have already seen that it is possible for both $f'(c)$ and $f''(c)$ to be zero, yet for f to have a local max or min at $x = c$. In

Example 22-2, for instance, both $f'(0)$ and $f''(0)$ are zero and f has a local max at $x = 0$. Compare this with Example 22-1, where once again $f'(0)$ and $f''(0)$ both equal zero but the graph has an inflection point at $x = 0$. These examples show that the second-derivative test fails to settle the question if $f''(c) = 0$. Indeed, if both $f'(c)$ and $f''(c)$ are zero, we must fall back on the first-derivative test of Goal 20 to determine if f has a local max or min, or neither, at $x = c$.

Using the Second-Derivative Test If $f''(x)$ Is Continuous

(1) Find $f'(x)$ and $f''(x)$.
(2) Find any critical points of f by solving $f'(x) = 0$.
(3) If $f'(c) = 0$, examine the sign of $f''(c)$:
  if $f''(c) > 0$, f has a local min at $x = c$;
  if $f''(c) < 0$, f has a local max at $x = c$;
  if $f''(c) = 0$, the second-derivative test is inconclusive;
    resort to the first-derivative test of Goal 20.

Sometimes when the second derivative is complicated or difficult to obtain, it is simpler to resort to the first-derivative test altogether.

The sketches for the following three examples appear on page 217.

Example 22-5
Use the second-derivative test to locate any local max or min of $f(x) = x^2$.

(1) $f'(x) = 2x$; $f''(x) = 2$.
(2) The only critical value is $x = 0$.
(3) Since $f''(0) = 2$, a positive number, f has a local min at $x = 0$.

Example 22-6
Use the second-derivative test to locate any local max or min of $g(x) = x^3$.
$g'(x) = 3x^2$; $g''(x) = 6x$. $g'(x) = 0$ only at $x = 0$. Since $g''(0) = 6 \cdot 0 = 0$, the test fails. Since $g'(x) > 0$ if $x < 0$, $g'(x) > 0$ if $x > 0$, f has neither a local max nor a local min at $x = 0$.

Example 22-7
Use the second-derivative test to locate any local max or min of $f(x) = x^4$.
$h'(x) = 4x^3$; $h''(x) = 12x^2$. $h'(x) = 0$ only at $x = 0$. Since $h''(0) = 12 \cdot 0^2 = 0$, the test fails. Since $h'(x) < 0$ if $x < 0$ and $h'(x) > 0$ if $x > 0$, f has a local minimum (a very flat one) at $x = 0$.

Goal 22 | To Apply the Second Derivative

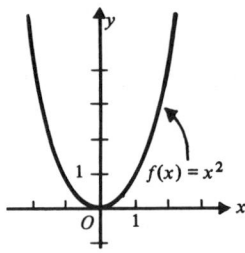
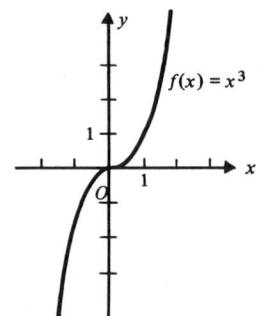
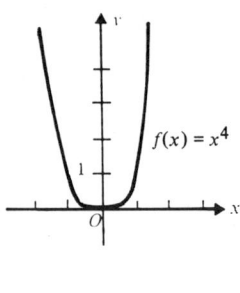

EXERCISES

1. Find any points of inflection and determine the concavity of the graph of each of the following:

    (a) $f(x) = x^3 - x^2 + 2x - 1$
    (b) $f(x) = 2x^3 + x^2 + x - 3$
    (c) $f(x) = x^3 - 6x^2 + 5x - 1$
    (d) $f(x) = 3x^3 + 9x^2 + 8x$
    (e) $f(x) = x^4 - 6x^2 + 4$
    (f) $f(x) = x^4 - 6x^3 + x + 10$
    (g) $f(x) = \dfrac{1}{1 + x^2}$
    (h) $f(x) = |x^2 - 1|$
    (i) $f(x) = x + \dfrac{1}{x}$
    (j) $f(x) = x^{2/3}$

In 2 through 5, for each sketch determine any intervals on which (a) $f'(x) > 0$; (b) $f''(x) > 0$; (c) both $f'(x) > 0$ and $f''(x) > 0$:

2.
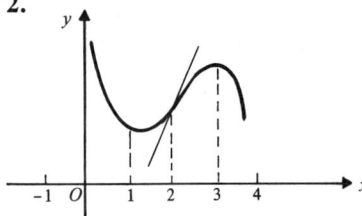

3.
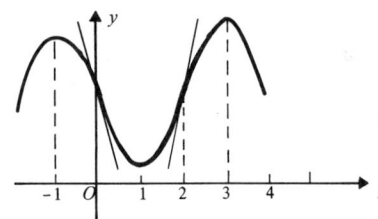

4.
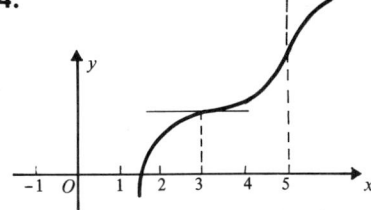

5.
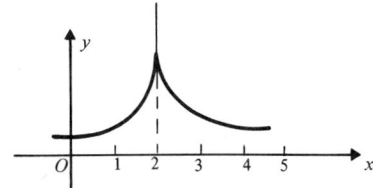

6. For each of the following sketch a curve with the properties specified:
   (a) $f''(x) > 0$ for $x > 2$, $f''(x) < 0$ for $x < 2$, $f'(2) = -2$
   (b) $f''(x) > 0$ for $x < 1$, $f''(x) < 0$ for $x > 1$, $f'(1) = -1$
   (c) $f''(x) > 0$ for $x < -1$ and for $x > 1$, $f''(x) < 0$ for $-1 \leq x \leq 1$, $f'(-1) = 1$, $f'(1) = 0$
   (d) $f''(x) < 0$ for $x < 3$ and for $x > 3$, $f'(3)$ not defined
   (e) $f''(x) > 0$ for $x < -1$, $f''(x) < 0$ for $x > -1$, $f'(-1)$ not defined
7. Use the second-derivative test to find any local max or min of each function in Exercises 5 through 20 of Goal 19.

TEST A

1. Find any points of inflection and determine the concavity of the graph of $f(x) = -x^4 + 2x^3 + 12x^2 - 10$.
2. Determine the intervals on which (a) $f'(x) > 0$, (b) $f''(x) > 0$, and (c) both $f'(x) > 0$ and $f''(x) > 0$, if the graph of f is

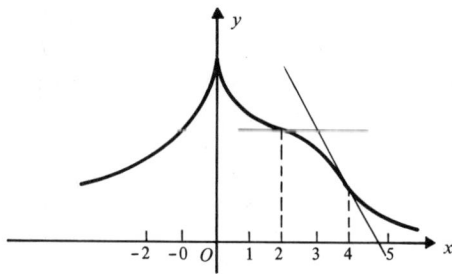

3. Use the second-derivative test to find the coordinates of any local maxima or minima of $f(x) = (x^2 - 1)^2$.

TEST B

1. Find any points of inflection and determine the concavity of the graph of $f(x) = 3x^5 - 10x^3$.
2. Determine the intervals on which (a) $f'(x) < 0$, (b) $f''(x) < 0$, and (c) both $f'(x) < 0$ and $f''(x) < 0$, if the graph of f is

Goal 22 | To Apply the Second Derivative

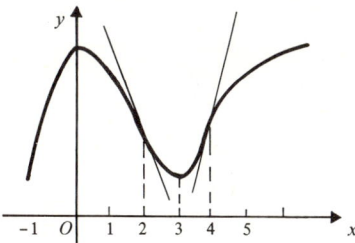

3. Use the second-derivative test to find the coordinates of any local maxima or minima of $f(x) = 3x^4 - 4x^3 + 5$.

STUDY QUESTIONS

1. Suppose $f''(x) = x(x - 1)^2(x - 2)^3(x - 3)^4$. For what values of x does f have points of inflection? If $f''(x) = (x - c)^m(x - d)^n$, then f may have points of inflection at c and d. If m and n are positive integers, what values for m and n will guarantee points of inflection both at c and at d?

2. Suppose $f(x) > 0$ for all x and the graph of f is concave upward everywhere. Show that the graph of $g(x) = [f(x)]^2$ is concave upward everywhere. HINT: Use the chain rule and the product rule to calculate $g''(x) = 2[f(x)f''(x) + f'(x)f'(x)]$, and then show that this is positive.

GOAL 23

To find asymptotes of a curve

In Goal 10 we considered the limit of $f(x)$ as x approaches c, where c is a finite number. Here we shall investigate limits as the variable becomes infinite. We shall also describe how to find the values of x for which $f(x)$ becomes infinite. This discussion involves asymptotes of a curve, which are very helpful when one is sketching it.

HORIZONTAL ASYMPTOTES

We begin with an application.

Example 23-1

Suppose that the cost of producing lawn chairs is $\dfrac{2x + 1600}{x + 200}$ dollars per chair when the production level is x chairs. Thus when 100 chairs are produced the cost per chair is $\dfrac{2(100) + 1600}{100 + 200}$, or \$6. When 1000 chairs are produced, the cost per chair is $\dfrac{2(1000) + 1600}{1000 + 200}$, or only \$3. It is easy to show that the higher the production level, the lower the unit cost [in fact, $f'(x) < 0$ if $x > 0$]. But as the production level rises (that is, as x increases) does the unit cost decrease toward zero, or is the lower bound greater than zero? Indeed, is there a lower bound? If there is, then it is the limit of the cost-per-chair function as x becomes arbitrarily large, a limit "at infinity," denoted by

$$\lim_{x \to \infty} \frac{2x + 1600}{x + 200}$$

As x becomes larger and larger, does the function get closer and closer to some number? To investigate, we evaluate the function for some large values of x and sketch the graph:

Goal 23 | To Find Asymptotes of a Curve

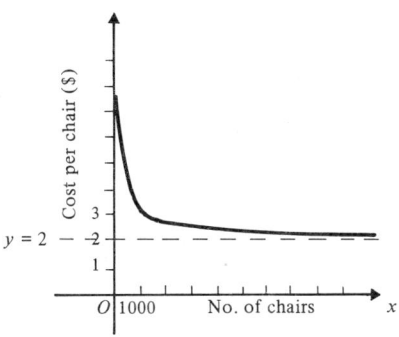

x	$\dfrac{2x + 1600}{x + 200}$
1,000	3.000
5,000	2.231
10,000	2.118
100,000	2.012
500,000	2.002

The table shows that the line $y = 2$ is a horizontal asymptote of the curve. Generally, the line $y = b$ is a *horizontal asymptote* of the graph of f if either

$$\lim_{x \to \infty} f(x) = b \quad \text{or} \quad \lim_{x \to -\infty} f(x) = b$$

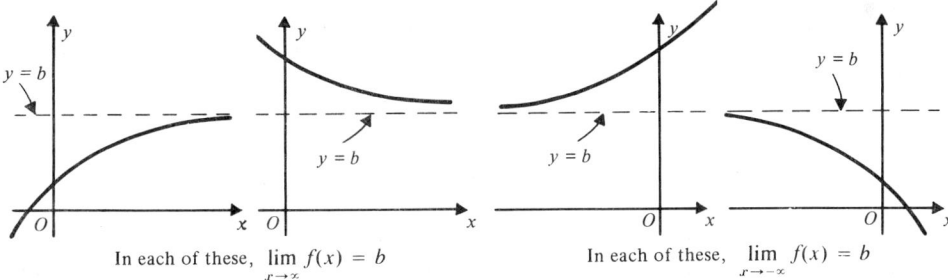

In each of these, $\lim_{x \to \infty} f(x) = b$ In each of these, $\lim_{x \to -\infty} f(x) = b$

Compare this definition with that given in Goal 4 (p. 37): a line to which the curve gets closer and closer as x becomes larger and larger either positively or negatively.

To find horizontal asymptotes of the graph of f, we find the limit of $f(x)$ as $x \to \infty$, as $x \to -\infty$, or both.

Example 23-2

Find any horizontal asymptotes of (a) $y = 1/x$; (b) $y = x + 1/x$.

(a) $\lim\limits_{x \to \infty} (1/x) = 0$ and $\lim\limits_{x \to -\infty} (1/x) = 0$. So $y = 0$ is a horizontal asymptote.

(b) $\lim\limits_{x \to \infty} (x + 1/x) = \infty$ and $\lim\limits_{x \to -\infty} (x + 1/x) = -\infty$. Since neither limit exists, the curve has no horizontal asymptotes. (The graph is on p. 205.)

FINDING HORIZONTAL ASYMPTOTES: THE LIMIT OF A RATIONAL FUNCTION "AT INFINITY"

Recall that a rational function is the quotient of two polynomials. To find the limit of the quotient as $x \to \infty$, divide both numerator and denominator by the highest power of x that occurs and use the fact that $\lim\limits_{x \to \infty} 1/x = 0$.

Example 23-3

$$\lim_{x \to \infty} \frac{x+5}{x^2+2x+3} = \lim_{x \to \infty} \frac{\frac{x}{x^2}+\frac{5}{x^2}}{\frac{x^2}{x^2}+\frac{2x}{x^2}+\frac{3}{x^2}} = \frac{0+0}{1+0+0} = \frac{0}{1} = 0$$

So $y = 0$ is a horizontal asymptote of the graph of this function.

Example 23-4

$$\lim_{x \to \infty} \frac{2x^4-5x+3}{15x^3-8} = \lim_{x \to \infty} \frac{\frac{2x^4}{x^4}-\frac{5x}{x^4}+\frac{3}{x^4}}{\frac{15x^3}{x^4}-\frac{8}{x^4}} = \frac{2-0+0}{0-0} = \infty$$

In this case there is no limit and the curve has no horizontal asymptote. (The limit of the function as $x \to -\infty$ is $-\infty$; again no limit exists.)

Example 23-5

$$\lim_{x \to \infty} \frac{x^3+2x^2+3}{4-7x-3x^3} = \lim_{x \to \infty} \frac{\frac{x^3}{x^3}+\frac{2x^2}{x^3}+\frac{3}{x^3}}{\frac{4}{x^3}-\frac{7x}{x^3}-\frac{3x^3}{x^3}} = \frac{1+0+0}{0-0-3} = -\frac{1}{3}$$

Therefore $y = 1/3$ is a horizontal asymptote.

RULE

The above examples lead us to the following rule:
If $P(x)$ and $Q(x)$ are polynomials, there are three possibilities for

$$\lim_{x \to \infty} \frac{P(x)}{Q(x)}$$

(1) If $Q(x)$ is of higher degree than $P(x)$, then the limit is 0.
(2) If $P(x)$ is of higher degree than $Q(x)$, then the limit is ∞ or $-\infty$; that is, the limit does not exist.
(3) If $P(x)$ and $Q(x)$ are of the same degree, then the limit is equal to a/b, where a and b are, respectively, the coefficients of the terms of highest degree of $P(x)$ and $Q(x)$.

This rule tells us immediately, for the cost-per-chair function of Example 23-1, that

Goal 23 | To Find Asymptotes of a Curve

$$\lim_{x \to \infty} \frac{2x + 1600}{x + 200} = \frac{2}{1} = 2$$

since the numerator and denominator are of the same degree; this confirms that $y = 2$ is a horizontal asymptote.

VERTICAL ASYMPTOTES

As noted in Goal 10, the line $x = c$ is a *vertical asymptote* if any one of the following is true:

$$\lim_{x \to c^+} f(x) = \infty \qquad \lim_{x \to c^+} f(x) = -\infty \qquad \lim_{x \to c^-} f(x) = \infty \qquad \lim_{x \to c^-} f(x) = -\infty$$

Thus, as x gets closer and closer to c, from either the right or the left, the values of f either increase or decrease without bound.

Example 23-6

Find $\lim_{x \to 2^+} y$ and $\lim_{x \to 2^-} y$ if $y = \dfrac{1}{x - 2}$.

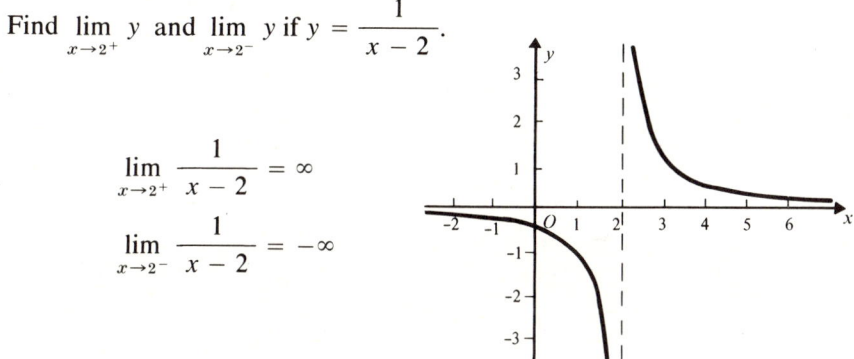

$$\lim_{x \to 2^+} \frac{1}{x - 2} = \infty$$

$$\lim_{x \to 2^-} \frac{1}{x - 2} = -\infty$$

So $x = 2$ is a vertical asymptote. Also, $y = 0$ is a horizontal asymptote.

FINDING VERTICAL ASYMPTOTES

The rational function $P(x)/Q(x)$ has a vertical asymptote at any x for which $Q(x) = 0$, provided $P(x)$ is not also zero there. So to find vertical asymptotes of this function we set the denominator equal to zero and solve.

Example 23-7

Find any vertical or horizontal asymptotes of

$$y = \frac{2x^2}{x^2 - 4x + 3}$$

The denominator $x^2 - 4x + 3 = (x - 1)(x - 3)$, and this equals zero if $x = 1$ or 3. Since the numerator is different from zero for both values, the graph has vertical asymptotes at $x = 1$ and $x = 3$. Noting that

$$\lim_{x \to \infty} \frac{2x^2}{x^2 - 4x + 3} = \frac{2}{1} = 2 \quad \left(\text{and also } \lim_{x \to -\infty} y = 2\right)$$

we conclude that the line $y = 2$ is a horizontal asymptote.

NOTE: The graph of a nonconstant polynomial function has no horizontal or vertical asymptotes.

EXERCISES

1. For each of the following, find the limit of $f(x)$ as $x \to \infty$.

(a) $f(x) = \dfrac{1}{x - 2}$

(b) $f(x) = \dfrac{x + 2}{x}$

(c) $f(x) = \dfrac{-x^2}{1 + 2x^2}$

(d) $f(x) = \dfrac{x^2}{1 - x^3}$

(e) $f(x) = \dfrac{x + 1}{1 - x + x^2}$

(f) $f(x) = \dfrac{x^3 + x - 1}{2x^2 + 1}$

(g) $f(x) = \dfrac{x^2 - x + 1}{3x^2 - 5}$

(h) $f(x) = \dfrac{1 - x^3}{x^5 - x^3}$

(i) $f(x) = \dfrac{5x^4 - x^2 + 1}{x^4 + x^3 - x}$

(j) $f(x) = \dfrac{x - x^2}{3x^2 + 2}$

(k) $f(x) = \dfrac{x^{10} - 9x^3 - 1}{3 + x^5 - 2x^{10}}$

(l) $f(x) = \dfrac{x^8 - 3}{5 - x^6}$

2. For each of the following find any horizontal or vertical asymptotes.

(a) $y = \dfrac{1}{x}$

(b) $y = \dfrac{3}{x^2 + 2x - 3}$

(c) $y = \dfrac{x + 5}{x - 1}$

(d) $y = \dfrac{3x - 5}{2x + 4}$

(e) $y = \dfrac{x^2 + 1}{(x + 1)(x - 2)}$

(f) $y = \dfrac{2x^3 + 1}{x^3 - x^2}$

(g) $y = \dfrac{5x^4 + 3x^3 - 1}{x^2(x^2 - 1)}$

(h) $y = \dfrac{3x^2 - 1}{x + 1}$

(i) $y = \dfrac{x}{3x - x^2}$

(j) $y = \dfrac{x^2 - 1}{x^2 - x}$

Goal 23 | To Find Asymptotes of a Curve

3. The quantity of drug in the bloodstream t h after being injected into a muscle is $\dfrac{10t}{t^2+1}$ g. As time passes, what is the lower limit to the quantity of the drug in the blood?

4. Using a particular processing technique, a chemical can be made $\dfrac{95t^2}{t^2+25}$ % pure after t h of refining. What is the upper limit to how pure a sample this technique will produce?

5. A family of rabbits is released in a game preserve. After t yr the number of rabbits has grown to $\dfrac{5500t^3 - 330t^2 - 582}{25 + 36t^2 + 2t^3}$. If this function continues to be accurate, is there a limit to the size of the rabbit population?

TEST A

1. For each of the following, find the limit as $x \to \infty$.

   (a) $\dfrac{3x^2 - x + 2}{5 + x - 2x^2}$

   (b) $\dfrac{x^4 - 2}{5 + x^5}$

   (c) $\dfrac{x^3 + x - 1}{x^2 + 10x}$

   (d) $\dfrac{14x^4 + x^2 - 1}{2x^4 - 10x^2 + 1}$

2. For each of the following, list all horizontal and vertical asymptotes.

   (a) $y = \dfrac{1}{x^2}$

   (b) $y = \dfrac{5}{x^2 + 2x - 3}$

   (c) $y = \dfrac{3x + 4}{5 - 2x}$

   (d) $y = \dfrac{(2+x)(1-x)}{(2-x)(1+x)}$

TEST B

1. For each of the following, find the limit as $x \to \infty$.

   (a) $\dfrac{x^6 - x}{x^7 + 2}$

   (b) $\dfrac{13x^4 - 2x + 1}{4x^4 + x^3 - 1}$

   (c) $\dfrac{3x^7 - x}{4x^4 + x^7}$

   (d) $\dfrac{x^9 - x^7 + 2}{x + 5x^5}$

2. For each of these, list all horizontal and vertical asymptotes.

   (a) $y = \dfrac{1}{x + 2}$

   (b) $y = \dfrac{x - 8}{x + 5}$

   (c) $y = \dfrac{3x^2}{x^2 - 1}$

   (d) $y = \dfrac{8x^3 + x}{x^2(x + 3)}$

STUDY QUESTIONS

1. Sometimes a curve has an *oblique asymptote*: one which is neither horizontal nor vertical. For example, if we examine the function $f(x) = x + \dfrac{1}{x}$ as x becomes infinite, we see that f takes on values closer and closer to the corresponding ones of the linear function $y = x$. As the sketch on page 205 shows, the line $y = x$ is an asymptote of the curve.
   Similarly, since

   $$f(x) = \frac{2x^2 + 3x + 1}{x - 5} = \frac{2x + 3 + \dfrac{1}{x}}{1 - \dfrac{5}{x}} \qquad (x > 5)$$

   we see that as x becomes infinite f approaches the function $y = 2x + 3$, showing that this line is an asymptote of the graph of f.
   In a similar manner, find an asymptote of the graph of
   (a) $f(x) = \dfrac{6x^2 + x}{2x + 1}$ (b) $f(x) = \dfrac{4x^3 - 3x^2 + x - 1}{2x^2 - x + 3}$

2. Use the methods of this Goal to show that a nonconstant polynomial function (defined on p. 31) has no horizontal or vertical asymptotes.

GOAL **24**

To graph functions using the calculus

The Chinese proverb says a picture is worth ten thousand words. Perhaps that is an exaggeration when the picture is that of a function; nonetheless, the graph of a function is a powerful tool for displaying and understanding the function's behavior. In earlier Goals we have seen how invaluable the calculus is in sketching curves. Here we integrate the ideas from those earlier Goals and systematically describe how to graph a function.

GRAPHING A FUNCTION

The following is a rather lengthy and detailed statement of the use of the calculus in graphing. It is presented here for reference and for completeness. In the first two examples we follow the steps carefully, but in Example 3, as is the general practice, we condense the procedure considerably. It should also be noted that for some functions all the steps listed are routine. However, for others one or more steps may be too tedious or complicated to be worthwhile. Fortunately, this does not usually cause difficulty, since the procedure has built-in redundancy: that is, it includes alternative steps for obtaining the same information.

CASE I: THE FUNCTION f IS A POLYNOMINAL

(1) Find the intercepts. For the x-intercept, evaluate $f(0)$; to determine any y-intercepts, factor $f(x)$ and solve $f(x) = 0$. Plot these points.
(2) Find $f'(x)$ and $f''(x)$ and factor each completely.
(3) Solve $f'(x) = 0$ for critical points; evaluate f at each one. Plot these and draw dotted horizontal tangents at these critical points.
(4) Examine the sign of $f''(x)$ at each critical point. If c is a critical point, and

if $f''(c) > 0$, then f has a local min at $(c, f(c))$

if $f''(c) < 0$, then f has a local max at $(c, f(c))$

But if $f''(c) = 0$, examine the signs of $f'(c)$ on both sides of $x = c$; if $f'(c)$ changes sign at $x = c$, f has a local max or min there, otherwise not.

Sketch the curve roughly in the neighborhood of each critical point.

(5) Solve $f''(x) = 0$ for possible points of inflection. If $f''(x_0) = 0$, examine the signs of f'' on both sides of x. If $f''(x)$ changes sign at $x = x_0$, then the graph has an inflection point at x_0, otherwise not.

Evaluate f at each inflection point and plot it. This step also gives information about the intervals of concavity: up where $f''(x) > 0$, down where $f''(x) < 0$.

(6) Complete the sketch of the curve. Although often unnecessary, it is sometimes helpful to examine.

  (a) The signs of f: where $f(x) > 0$ the graph is above the x-axis; where $f(x) < 0$, the graph is below the x-axis;

  (b) The signs of f': where $f'(x) > 0$, the graph rises; where $f'(x) < 0$, the graph falls;

  (c) The slope of the tangent at each inflection point, since drawing these tangents improves the accuracy of the graph;

  (d) The behavior of f as $x \to \infty$ or as $x \to -\infty$; f behaves as does its leading term.

Example 24-1

Sketch the graph of $f(x) = x^3 + 3x^2 - 4$.

The figures below show successive steps in the sketching process.

(1) $f(0) = -4$. Since $f(x) = (x - 1)(x^2 + 4x + 4) = (x - 1)(x + 2)^2$, the x-intercepts are 1 and -2.

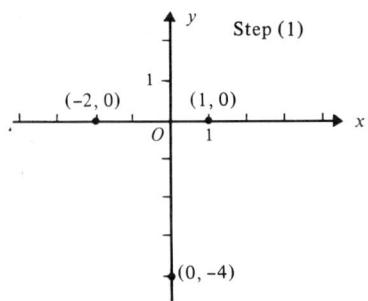

Step (1)

(2) $f'(x) = 3x^2 + 6x = 3x(x + 2)$; $f''(x) = 6x + 6 = 6(x + 1)$.

(3) $f'(x) = 0$ if $x = 0$ or -2, the critical points; $f(0) = -4$, $f(-2) = (-2)^3 + 3(-2)^2 - 4 = 0$.
(4) Since $f''(0) = 6(0 + 1) > 0$, f is concave up and has a local min at $(0, -4)$; since $f''(-2) = 6(-1) < 0$, f is concave down and has a local max at $(-2, 0)$.

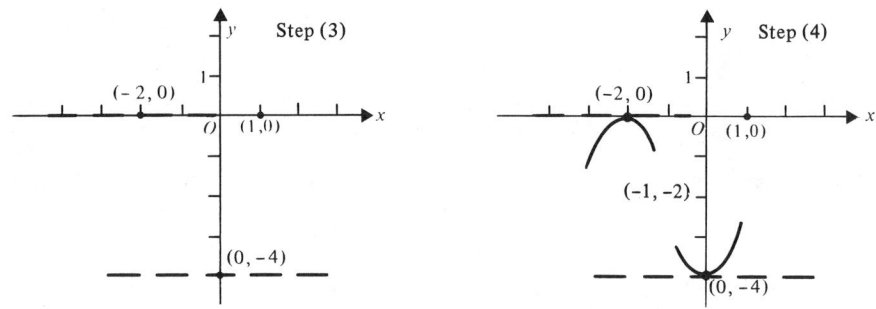

(5) $f''(x) = 0$ if $x = -1$, which may yield an inflection point; since $f''(x) < 0$ if $x < -1$, $f''(x) > 0$ if $x > -1$, there is an inflection point at $x = -1$. $f(-1) = -2$. The curve is concave down if $x < -1$, up if $x > -1$.
(6) Miscellaneous information:
   (a) From the factored form of $f(x)$ in Step (1) we see that $f(x) > 0$ if $x > 1$, $f(x) < 0$ if $x < 1$ (except at $x = -2$).
   (b) From $f'(x)$ in Step (2) we see that $f(x)$ increases if $x < -2$ or if $x > 0$, decreases if $-2 < x < 0$.
   (c) $f'(-1) = 3(-1)(-1 + 2) = -3$.
   (d) As $x \to \infty$, $f(x) \to \infty$; as $x \to -\infty$, $f(x) \to -\infty$.

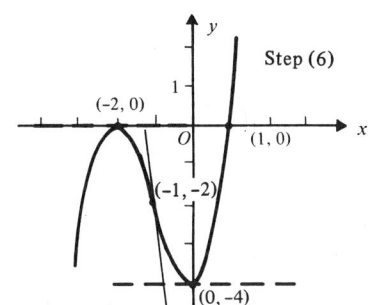

CASE II: THE FUNCTION f IS NOT A POLYNOMIAL
At the outset

(i) Find the domain of f.
(ii) Find asymptotes, if any: if $f(x) = P(x)/Q(x)$, find where $Q(x)$ equals zero to get the vertical asymptotes; let $x \to \pm\infty$ to find any horizontal asymptotes. Draw dotted lines for the asymptotes.
(iii) If $x = c$ is a vertical asymptote, examine f as $x \to c^-$ and as $x \to c^+$; sketch the graph roughly near each vertical asymptote.

Now proceed as in Case I, but also investigate, if necessary, the function and its derivative near any point of discontinuity.

Example 24-2

Sketch $f(x) = \dfrac{2x - 3}{x + 1}$.

(i) f is defined everywhere except at $x = -1$.
(ii) $x = -1$ is a vertical asymptote. Since as $x \to \infty$ and as $x \to -\infty$, $f(x) \to 2$, $y = 2$ is a horizontal asymptote.
(iii) To see how f behaves near $x = -1$, we can evaluate f at a number just less than -1, say -1.1, and at a number just greater than -1, say -0.9. Since $f(-1.1) = \dfrac{-5.2}{-0.1}$, we see $f(x) \to +\infty$ as $x \to -1^-$. Since $f(-0.9) = \dfrac{-4.8}{+0.1}$, we see that $f(x) \to -\infty$ as $x \to -1^+$.

Goal 24 | To Graph Functions Using the Calculus

(1) $f(0) = -3$; also $f(x) = 0$ if $2x - 3 = 0$, i.e., if $x = \frac{3}{2}$.

(2) $f'(x) = \dfrac{(x + 1)2 - (2x - 3) \cdot 1}{(x + 1)^2} = \dfrac{5}{(x + 1)^2}$

$f''(x) = -\dfrac{10}{(x + 1)^3}$

(3) $f'(x)$ never equals zero; f has no critical points. Skip Step 4.

(5) $f''(x)$ never equals zero; f has no inflection points. Since $f''(x) > 0$ if $x < -1$, the curve is concave up there; since $f''(x) < 0$ if $x > -1$, the curve is concave down there.

(6) (a) f can change sign only where the numerator or denominator does, i.e., at $x = -1$ or $\frac{3}{2}$. If $x < -1$, $f(x) > 0$; if $-1 < x < \frac{3}{2}$, $f(x) < 0$; if $x > \frac{3}{2}$, $f(x) > 0$.

(b) Since $f'(x) > 0$ both if $x < -1$ and if $x > -1$, f is increasing on both intervals.

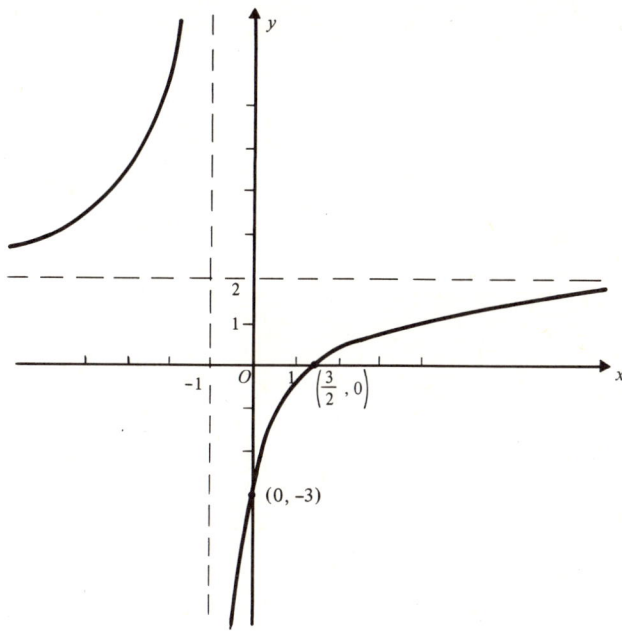

Example 24-3

Sketch the graph of $f(x) = x^4 - 4x^3$.

As noted earlier, we will condense the procedure here. Usually, we apply information to a single graph as we gather it, when appropriate, then "firm up" the curve as the final step.

Since f is a polynomial, there are no asymptotes; f becomes positively infinite as $x \to \pm\infty$.

$f(0) = 0$; since $x^4 - 4x^3 = x^3(x - 4)$, f has x-intercepts at $x = 0$ and $x = 4$.

$f'(x) = 4x^3 - 12x^2 = 4x^2(x - 3)$; $f''(x) = 12x^2 - 24x = 12x(x - 2)$. The critical values are $x = 0$ and $x = 3$. Since $f''(3) > 0$, f has a local min at $x = 3$, where $f(x) = -27$. Since $f''(0) = 0$, we inspect the signs of $f'(x)$: $f'(x) < 0$ both if $x < 0$ and if $0 < x < 3$, so f has neither a local max nor a local min at $x = 0$. $f''(x) = 0$ if $x = 0$ or 2; since f'' changes sign at each, both are inflection points. The curve is concave up if $x < 0$ or if $x > 2$. Note also that $f > 0$ if $x < 0$ or if $x > 4$ and that f decreases if $x < 3$ (except at $x = 0$), increases if $x > 3$. Note the different scales.

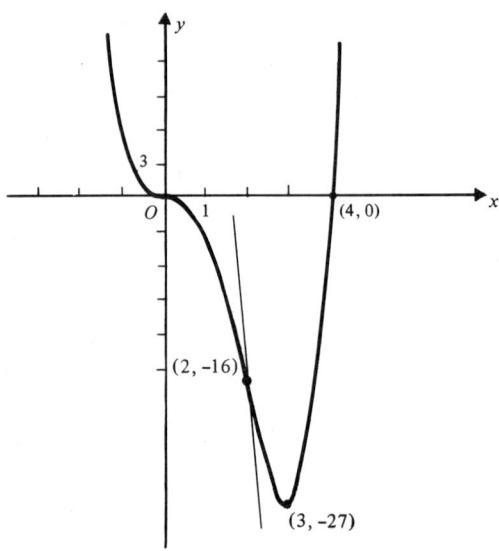

EXERCISES

Sketch the graph of each of the following functions:

1. $f(x) = x^2 - x - 6$
2. $f(x) = -x^2 + 6x - 5$
3. $f(x) = x^3 + 6x^2$
4. $f(x) = x^3 - 3x + 2$
5. $f(x) = \dfrac{3x - 12}{x + 2}$
6. $f(x) = \dfrac{x + 5}{2x - 4}$
7. $f(x) = x^4 - 8x^3$
8. $f(x) = x^4 - 8x^2 + 7$
9. $f(x) = \dfrac{x^2 + 1}{x}$
10. $f(x) = \dfrac{x^2}{x^2 - 1}$

Goal 24 | To Graph Functions Using the Calculus

TEST A

Sketch the graph of each of the following functions:

1. $f(x) = -x^2 + 2x + 3$
2. $f(x) = x^3 + 2x + 2$
3. $f(x) = \dfrac{4x - 4}{2x + 3}$

TEST B

Sketch the graph of each of the following functions:

1. $f(x) = (x - 3)(2 - x)$
2. $f(x) = x^4 - 2x^2 + 1$
3. $f(x) = \dfrac{2x - 6}{x + 3}$

Summary of Goals 18–24

You should now be able to

(1) Find the equation of the tangent and normal to a curve at a specified point;
(2) Find the critical points of a function: the values of x for which $f'(x) = 0$;
(3) Find any x for which $f(x)$ or $f'(x)$ does not exist;
(4) Determine the specific intervals over which a function increases or decreases;
(5) Determine, if $f'(c)$ equals zero or does not exist, whether f has a local max or min at $x = c$;
(6) Find the absolute max and min of a function that is continuous on a closed interval;
(7) Recognize when a function does not attain an absolute max or min;
(8) Determine the specific intervals over which a graph is concave up or down;
(9) Find any inflection points of a graph;
(10) Use the second-derivative test, if $f'(c) = 0$, to determine whether f has a local max or min at $x = c$;
(11) Sketch the graph of a function given sufficient information about its first and/or second derivatives;
(12) Determine from a given graph of a function $f(x)$ the intervals over which $f'(x)$ or $f''(x)$ is positive or negative;
(13) Find any horizontal or vertical asymptotes of a curve;
(14) Graph a function after obtaining information about intercepts, local max or min, inflection points, concavity, behavior for large x, and asymptotes if any.

Review Exercises

GOAL 18

1. Find the equations of the tangent and normal to each curve at the specified point:
    (a) $y = -x^3 + 3x^2 - x + 4$ at $(2, 6)$
    (b) $x^2 - xy + 4y^2 = 4$ at $(1, -1)$

GOAL 19

2. Find the coordinates of any critical points:
   (a) $f(x) = 4 - 2x - x^2$
   (b) $y = 2x^3 - 9x^2 + 12x - 4$
   (c) $y = \frac{1}{3}x^3 + 2x$
   (d) $g(t) = -t^3 + 3t + 1$
   (e) $f(x) = 2x^4 - 3$
   (f) $y = -(x + 3)^4$
   (g) $x^2 + y^2 = 25$

3. For each function (a) through (f) in Exercise 2, find the intervals over which it is increasing or decreasing.

4. Find the intervals over which f is increasing or decreasing if
   (a) $f(x) = \dfrac{4}{x - 2}$
   (b) $f(x) = \sqrt[3]{x - 1}$
   (c) $f(x) = x^{2/3}$

5. Sketch the graph of a function f such that
   (a) $f(0) = 2$; $f(3) = -2$; $f'(x) < 0$ if $x < 0$ or if $0 < x < 3$; $f'(x) > 0$ if $x > 3$; $f'(0) = f'(3) = 0$
   (b) $f(0) = 3$; $f(2) = -4$; $f(4) = 5$; if $'(x) > 0$ if $x < 0$ or if $2 < x < 4$; $f'(x) < 0$ if $0 < x < 2$ or if $x > 4$; $f'(2) = f'(4) = 0$; $f'(0)$ is not defined

GOAL 20

6. Find the coordinates of any local max or min of
   (a) $f(x) = x^3 - 3x^2 - 9x - 1$
   (b) $f(x) = 3x^4 - 16x^3 + 10$
   (c) $y = 4 - x^{2/3}$
   (d) $y = \dfrac{x^2 + 2x + 2}{x + 1}$

GOAL 21

7. Find the coordinates of the abs max or min, if any, of the following functions on the intervals specified:
   (a) $y = 2 + 5x - x^2$ on $-2 \leq x \leq 1$
   (b) $f(x) = x^3 - 3x^2 + 3x + 1$ on $[0, 3]$
   (c) $g(x) = 2x^3 + 4x - 5$ on $[-1, 1]$
   (d) $y = \dfrac{x^2 - 4x + 8}{x - 2}$ on $-1 \leq x \leq 1$
   (e) the function in (d) on $(2, 5]$

8. An excursion boat can accommodate 35 passengers. The owner figures that his revenue on a weekend excursion is $x(120 - 2x)$ dollars if he sells x tickets. How many tickets should he sell for maximum revenue?

9. Suppose, in Exercise 8, that the owner uses a boat that can accommodate only 25 passengers.

GOAL 22

10. Find any points of inflection and determine the concavity of the graphs of the functions in Exercise 6 above.

11. For each of the following, sketch a curve with the specified properties:
    (a) $f'(x) > 0$ if $x < 2$; $f'(x) < 0$ if $x > 2$; $f'(2)$ is undefined; $f''(x) > 0$ if $x < 2$ or if $x > 2$.
    (b) $f'(x) < 0$ if $x < 0$; $f'(x) > 0$ if $x > 0$; $f'(0) = f'(3) = 0$; $f''(x) > 0$ if $x < 1$ or if $x > 3$; $f''(x) < 0$ if $1 < x < 3$

12. Use the second-derivative test to show that
    (a) $f(x) = 3 + 8x - x^2$ has a local max but no local min
    (b) $g(x) = (1 - x)^4$ has a local min but no local max
    (c) $h(x) = x^3 + x$ has neither a local max nor a local min

GOAL 23

13. For each of the following find $\lim f(x)$ as $x \to \infty$:
    (a) $f(x) = \dfrac{4}{x + 1}$
    (c) $f(x) = \dfrac{x^2 + 1}{x - 1}$
    (b) $f(x) = \dfrac{2x + 1}{x - 3}$
    (d) $f(x) = \dfrac{3x^2 + x - 1}{2 - x^2}$

14. Find any horizontal or vertical asymptotes:
    (a) $y = \dfrac{x + 4}{x^2 - 4}$
    (d) $y = \dfrac{x^2 + 4x + 8}{x + 2}$
    (b) $y = \dfrac{x + 1}{x - 1}$
    (e) $y = x^3 - 3x^2 + 3x + 1$
    (c) $y = \dfrac{8x}{x^2 + 4}$
    (f) $y = x + \dfrac{1}{x}$

15. Suppose it is assumed that a person has learned N words by the time he is t years old, where

$$N(t) = \frac{50{,}000 t^2}{300 + t^2} \qquad (t > 2)$$

What can be said about the number of words he knows as he gets older and older?

GOAL 24

16. After obtaining information about intercepts, max or min, inflection points, concavity, and asymptotes, sketch the graph of each of the following functions:
    (a) $f(x) = -\frac{1}{2}x^2 + 4x - 5\frac{1}{2}$
    (b) $f(x) = x^3 - 3x + 2$
    (c) $f(x) = \dfrac{4x + 13}{x + 3}$
    (d) $f(x) = \dfrac{x^2}{x^2 - 1}$

Chapter Test A

1. Find the equations of the tangent and normal to the curve $y = (2 - x)^3 + 1$ at its point of inflection.
2. Find any local max, min, or inflection points of the graph of
    (a) $f(x) = x^3 - 4x - 1$ (b) $f(x) = x^4 - 2x^2 + 4$
3. Sketch the graphs for the functions in Exercise 2.
4. Sketch the graph of a function f such that $f'(x) > 0$ if $x < 0$ or if $0 < x < 2$; $f'(x) < 0$ if $x > 2$; $f''(x) < 0$ if $x < 0$ or if $x > \frac{4}{3}$; $f''(x) > 0$ if $0 < x < \frac{4}{3}$; $f'(0) = f'(2) = 0$.
5. Find the coordinates of the absolute max or min, if any, of the following functions on the intervals specified:
    (a) $f(x) = 4x^2 - 8x + 2$ on $[0, 3]$
    (b) $f(x) = \dfrac{1}{x}$ on $\left[\dfrac{1}{2}, 4\right]$ (c) $f(x) = \dfrac{1}{x}$ on $0 < x \leq 8$
6. Find any asymptotes of
    (a) $y = x^5 - 5x^3$ (c) $f(x) = \sqrt[3]{x - 1}$
    (b) $f(x) = \dfrac{x^2}{4 - x^2}$ (d) $y = \dfrac{x - 1}{1 + x^2}$
7. Sketch the graph of $y = \dfrac{2x^2 - 1}{x^2}$.

8. Suppose a 12-year-old can memorize a string of N nonsense syllables in t min, where

$$N(t) = \frac{30t^2}{9 + t^2}$$

(a) What happens as t increases indefinitely?

(b) Is there a length of time in which the maximum number of syllables is memorized?

Chapter 5

Further Applications of the Derivative

In this chapter we apply the methods of the calculus developed so far to practical problems. We begin by solving a wide variety of problems dealing with maxima and minima. We then consider rate-of-change problems involving relationships among three or more variables. Both the chain rule and the technique of implicit differentiation are used to solve these related-rate problems. Velocity and acceleration are discussed more fully in this chapter to provide further information about the motion of an object along a line. We will also see how the derivative of a function can be used to approximate a function at a specific point when we know its value at a point nearby.

Another important application of the derivative to be presented here is the so-called Mean Value theorem, with corollaries that we will need in later chapters. Finally, we turn our attention to problems in business and economics that can be solved using the techniques and methods of the calculus now available to us.

Goal 25. To solve problems involving maxima and minima
Goal 26. To solve related-rate problems
Goal 27. To apply the Mean Value Theorem
Goal 28. To use differentials for approximation
Goal 29. To solve problems involving motion along a line
Goal 30. To apply the derivative to problems in economics
Summary
Review Exercises
Chapter Test A

GOAL **25**

To solve problems involving maxima and minima

We shall now apply the techniques for finding the maximum or minimum values of a function to a variety of practical problems. We noted in Goal 6 that applying mathematics to real-life problems involves three steps: first, we find a mathematical model—that is, we translate the problem into mathematical terms; next, we find a solution to the mathematical problem; finally, we interpret this mathematical answer back into the terms of the original problem. In Goal 6 we were concerned with only the first step; here we deal with the entire process. Many of the problems to be considered here are based on those presented in Goal 6.

PRELIMINARY EXAMPLES

Example 25-1

Mrs. G. Thumm wants to put a fence around her rectangular garden to keep the rabbits out. She has 250 ft of fencing and decides to use all of it. What should the dimensions be for the largest garden? (See Example 6-2, p. 54.)

The largest garden is the one with biggest, or maximum, area. To pinpoint what is involved, note, for example, that the garden might have either of the shapes shown in the figures.

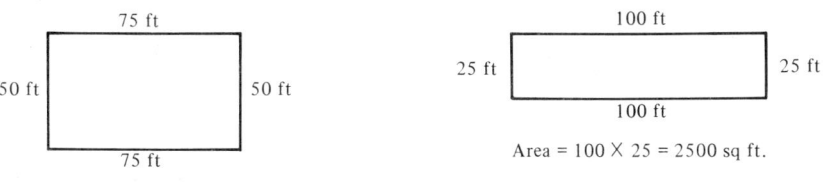

Of course, many other shapes are also possible. It is clear that the area depends on the particular dimensions of the rectangle. If we let x be the length and y the width, then the area of the rectangle is given by $A = xy$.

To express A as a function of only one variable, say x, we use the fact that there are 250 ft of fencing: $2x + 2y = 250$, or $y = 125 - x$. Then $A(x) = x(125 - x) = 125x - x^2$. We want to find the value of x for which A is a maximum, on the interval $[0,125]$. Since $A'(x) = 125 - 2x$ and this equals zero if $x = 125/2 = 62\frac{1}{2}$, there is only one critical point. We evaluate A at it and at the endpoints:

$A(62\frac{1}{2}) = 3906\frac{1}{4}$
$A(0) = 0$
$A(125) = 0$

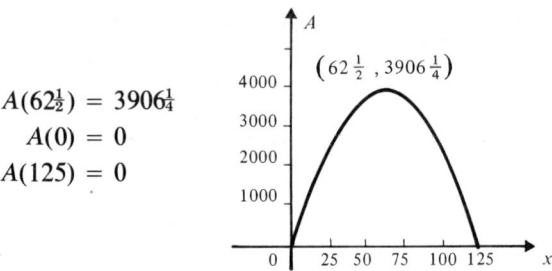

The area is therefore a maximum when $x = 62\frac{1}{2}$. Also, $y = 62\frac{1}{2}$, which shows that the largest garden is a square. We have graphed the area function for completeness.

Example 25-2

Suppose Mrs. Thumm starts a new rectangular garden whose area is to be 400 sq ft. If fencing costs 30¢ per ft, what is the least amount it will cost her to fence it? (See Example 6-3, p. 54.)

To minimize the cost, we first determine the cost function, which is equal to the total length of the fence, $2x + 2y$, times the cost per foot:

$$C = (2x + 2y)(0.30) = 0.6x + 0.6y = 0.6(x + y)$$

To express C in terms of one variable, say x, we note that, since the area is 400 sq ft, $xy = 400$ and $y = 400/x$. So

$$C(x) = 0.6\left(x + \frac{400}{x}\right) \qquad \text{on the interval } x > 0$$

Then

$$C'(x) = 0.6\left(1 - \frac{400}{x^2}\right) = 0.6\left(\frac{x^2 - 400}{x^2}\right)$$

This equals zero if $x = 20$. We discard the other critical value, -20, as meaningless in this problem. Since the domain of C is an open interval, we find $C''(x)$, then examine the sign of $C''(20)$ to determine whether $x = 20$ gives a local max or a local min:

Goal 25 | To Solve Problems Involving Maxima and Minima

$$C''(x) = 0.6\left(\frac{800}{x^3}\right) > 0 \quad \text{for all } x > 0$$

Therefore the curve is concave up when $x > 0$ and C has a local (and absolute) minimum at $x = 20$. The minimum cost is equal to

$$C(20) = 0.6\left(20 + \frac{400}{20}\right) = \$24$$

Example 25-3

The Original Reproductions Company can produce at most 30 paintings per week. They estimate that they can sell n paintings at p dollars each, where $p = 120 - 2n$. The total cost of producing n paintings is $500 + 8n + n^2$ dollars. How many pictures should be sold to maximize the company's profit?

As noted in Example 6-1, page 53, the profit function here is $P(n) = -3n^2 + 112n - 500$. Since

$$P'(n) = -6n + 112$$

and this equals zero when $n = 112/6 = 18\tfrac{2}{3}$, and since

$$P''(n) = -6$$

we know that $P(n)$ reaches its maximum when $n = 18\tfrac{2}{3}$. But this cannot be the answer to the question in the original problem, since the answer must

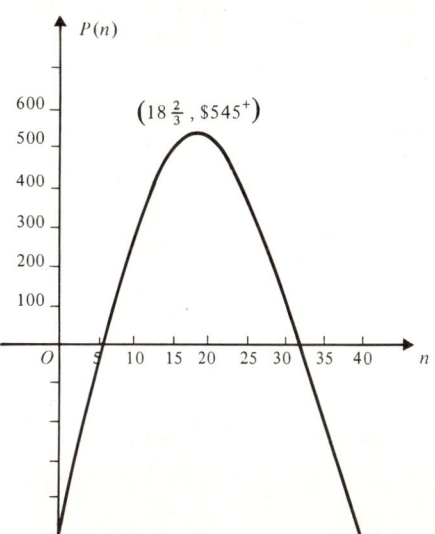

be an integer. Only a whole number of paintings (but no more than 30) can be sold. Still, from our work above and a rough sketch of the graph we can tell immediately that the integer that yields maximum profit is either 18 or 19. Since $P(18) = 544$ and $P(19) = 545$, the answer to the question in the original problem is 19 paintings.

This situation is typical of many in which the domain of a function is a set of integers. To apply the calculus we must assume that the function is continuous over all the positive numbers. We must therefore allow for this assumption when we return to the original problem, by interpreting the mathematical solution so as to yield answers meaningful in a practical sense—in this case, answers that are integers.

SOLVING PRACTICAL PROBLEMS INVOLVING MAXIMA OR MINIMA

Our preliminary examples suggest the following procedure:

(1) After reading the problem carefully, draw a figure (if relevant). Label the parts with variables or with the data given.
(2) Decide what quantity is to be maximized or minimized. Then determine an equation involving this quantity. (It may help to review the specific suggestions on this topic on p. 58, Goal 6.)
(3) Express the quantity in (2) as a function of only one variable: say, $f(x)$. If your choice of variable leads to an unduly complicated function, try an alternative.
(4) Find the critical points at which $f'(x) = 0$. (None of the problems in this Goal involve extreme values at points where $f'(x)$ is not defined.)
(5) If the domain of f is the closed interval $[a, b]$, evaluate f at each critical point, at a, and at b, to see which yields the desired max or min of f. Sometimes it is more convenient to use the second- or first-derivative test. If the domain of f is an open interval, then the latter is necessary to determine if a critical point yields a max or a min.
(6) Interpret the mathematical solution so that it makes sense in terms of the original problem.

Example 25-4

The owner of an apartment house with 60 units has found that he can rent them all at $200 a month. However, for each $5 increase in the rent he loses one tenant. By how much should the owner raise the rent to obtain maximum revenue? (See Example 6-4, p. 54.)

Let x be the number of $5 increases. Then, as in Example 6-4, the revenue

Goal 25 | To Solve Problems Involving Maxima and Minima

$$R(x) = \text{number of occupied apartments} \times \text{new rent}$$
$$= (60 - x)(200 + 5x)$$
$$R'(x) = (60 - x)(5) + (-1)(200 + 5x)$$
$$= 300 - 5x - 200 - 5x = 100 - 10x$$

and this is zero if $x = 10$.

The domain of R is $[0, 60]$ (0 corresponds to no increase; 60 five-dollar increases would lead to no tenants). We note then that

$$R(10) = (50)(250) = \$12{,}500$$
$$R(0) = (60)(200) = \$12{,}000$$
$$R(60) = \$0$$

Therefore 10 five-dollar increases or a \$50 increase in the rent yields maximum revenue.

It is instructive to do this problem with the revenue as a function of the new rental rate, say r. In Exercise 1 you will be asked to show that a new rental rate equal to \$250 produces maximum revenue. This is equivalent, of course, to a \$50 increase.

Example 25-5

A sandbox with square base is to have volume 25 cu ft. The bottom is to be made of metal costing 50¢ per ft; the sides are to be made of wood costing \$1.25 per sq ft. What dimensions will minimize the cost? What is the minimum cost? (Compare Example 6-6, p. 57.)

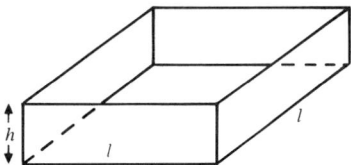

Let l and h be the dimensions, as shown in the figure. The total cost in dollars is given by

$$C = \quad \text{cost of bottom} + \text{cost of four sides}$$

$$C = (\text{no. of sq. ft. in bottom}) \times \frac{1}{2} + (\text{no. of sq. ft. in 4 sides}) \times \frac{5}{4}$$

$$= \quad l^2 \times \frac{1}{2} + lh \times 4 \times \frac{5}{4}$$

$$= \quad \frac{1}{2} l^2 + 5lh$$

To express C as a function of one variable, we use the fact that the volume is given as 25 cu ft. Therefore

$$l^2 h = 25 \quad \text{and} \quad h = \frac{25}{l^2}$$

(It's easier to solve for h than for l.) So

$$C(l) = \frac{1}{2} l^2 + 5l\left(\frac{25}{l^2}\right) = \frac{1}{2} l^2 + \frac{125}{l}$$

and the domain of C is $l > 0$. Then

$$C'(l) = l - \frac{125}{l^2} = \frac{l^3 - 125}{l^2}$$

And this is zero when $l^3 - 125 = 0$, or $l = \sqrt[3]{125} = 5$. Rewriting $C'(l)$ as $l - 125 l^{-2}$, we get

$$C''(l) = 1 + 250 l^{-3} = 1 + \frac{250}{l^3}$$

Since $C''(l) > 0$ for all positive l, the graph of C is everywhere concave up and C has an absolute min at $l = 5$. The dimensions for minimum cost are therefore 5 ft × 5 ft × 1 ft; the minimum cost is

$$C(5) = \frac{1}{2}(5)^2 + \frac{125}{5} = \$37.50$$

Example 25-6

A utilities company wants to deliver natural gas from a source S to a plant P located across a river 3 miles wide then down the road along the river 5 miles. If it costs $4 per ft to lay the pipe in the river but only $2 per ft on land, how can the pipe be laid most economically? (See Example 6-7, page 57.)

As pointed out earlier, we can (1) lay all of the pipe in the river (along line segment SP), (2) lay some of it in the river along SR and the rest of it on land along RP, or (3) lay some in the river along ST and the rest on land along TP. In case (1) $v = 0$, $u = 5$, and $T = P$; in case (2) $u = 0$, $v = 5$, and $T = R$. Case (3) includes both (1) and (2).

In Example 6-7 we found the cost as a function first of u, then of v:

$$C(u) = 10{,}560(2\sqrt{9 + u^2} + 5 - u)$$
$$C(v) = 10{,}560(2\sqrt{34 - 10v + v^2} + v)$$

Goal 25 | To Solve Problems Involving Maxima and Minima

Since the former is simpler, we now minimize $C(u)$:

$$C'(u) = 10{,}560\left(2 \cdot \frac{1}{2} \frac{2u}{\sqrt{9+u^2}} - 1\right) = 10{,}560\left(\frac{2u}{\sqrt{9+u^2}} - 1\right)$$

We now set $C'(u)$ equal to zero and solve for u:

$$\frac{2u}{\sqrt{9+u^2}} - 1 = 0 \quad \rightarrow \quad \frac{2u}{\sqrt{9+u^2}} = 1 \quad \rightarrow \quad \frac{4u^2}{9+u^2} = 1$$

where, in the last step, we squared both sides; then

$$4u^2 = 9 + u^2 \qquad 3u^2 = 9 \qquad u^2 = 3 \qquad u = \sqrt{3}$$

where we discard $u = -\sqrt{3}$ as meaningless for this problem.

Our earlier discussion tells us that the domain of $C(u)$ is $[0, 5]$. Since

$$C(0) = 10{,}560(2\sqrt{9} + 5) = \$116{,}160$$
$$C(5) = 10{,}560(2\sqrt{34}) \approx \$123{,}150$$
$$C(\sqrt{3}) = 10{,}560(2\sqrt{12} + 5 - \sqrt{3}) = \$107{,}670$$

we see that $u = \sqrt{3}$ yields minimum cost. Thus, the pipe can be laid most economically if some of it is laid in the river from the source S to a point T that is $\sqrt{3}$ miles toward the plant P from R, and the rest is laid along the road from T to P.

EXERCISES

1. Express the revenue R in Example 25-4 as a function of the new rental rate r, then show that the rent that produces maximum revenue is precisely the rent resulting from 10 five-dollar increases.
2. A rectangular garbage dump of area 400 sq miles is to be fenced off. Find the dimensions for which the fence will cost least if a meter of fence costs \$4.
3. Suppose the dump in Exercise 2 has area 200 sq miles and one side along the river requires more substantial fencing, costing \$5 a meter. What dimensions yield minimum cost?
4. In Exercise 6 on page 59, what shape solarium (that is, what dimensions) will yield maximum volume?
5. In Exercise 7 on page 60, how many trees per acre will yield a maximum harvest?
6. In Exercise 8 on page 60, determine the optimum price, that is, the

price that will result in maximum profit. How many albums can be sold at this price?

7. In Exercise 9 on page 60, determine the number of passengers for maximum revenue. Find the maximum revenue.

8. In Exercise 12 on page 60, how many tickets should be sold for maximum profit?

9. In Exercise 15 on page 61, how many tickets should be sold for maximum profit?

10. In Exercise 17 on page 61, what size square should be cut away to yield a box of maximum volume?

11. As part of his initiation, a fraternity pledge is told to get from point A on one side of a river to point B on the opposite side (see sketch) within 2 min. If he can paddle a canoe at 10 ft/sec and run at 20 ft/sec, can he make it? (For sketch see Exercise 18 on p. 61.)

12. In Exercise 19 on page 61, determine what route the transportation company should follow to minimize the time it will take to get from town T to factory F.

13. In Exercise 20 on page 62, what are the dimensions of the tagboard that will minimize the area of the tagboard? HINT: The computation is easier if the area of the tagboard is expressed in terms of the dimensions of the printed message.

14. How does the preceding exercise change if the printer receives an order for 100 posters, knows that the tagboard costs 8¢ per sq ft, and wants to minimize the cost of the material?

15. In Exercise 21 on page 62, when is the distance between the cars least? HINT: Minimize the square of the distance rather than the distance.

16. In Exercise 23 on page 63, find the dimensions of the largest possible box; i.e., the box of maximum volume.

17. A group interested in chartering a 25-passenger boat is told that it will cost $80 per person if 20 or fewer sign up, but that the price will drop $3 per ticket for each passenger over 20. How many tickets should be sold for maximum revenue?

18. Do the preceding exercise with the boat company reducing the fare by only $2 for each passenger in excess of 20.

19. The sum of the squares of two nonnegative numbers equals 200. Find their maximum and minimum products, if they exist.

20. How should the wire in Exercise 25 on page 63 be cut to maximize the sum of the areas of the square and the triangle?

21. A rectangular box with square bottom and top is to contain 27 cu ft. What dimensions will yield least surface area?

Goal 25 | To Solve Problems Involving Maxima and Minima

22. A rectangular box with square bottom and top is to be constructed to have volume 48 cu ft. The material for the sides and bottom costs 40¢ per sq ft, but that for the top costs only 20¢ per sq ft. If the labor costs $2.50 a box, what are the dimensions of the most economical box?

23. What are the dimensions of the most economical 1-liter closed cylindrical can? HINT: If r and h are the radius and height, respectively, of the can, then its volume $V = \pi r^2 h$ and its surface area $S = 2\pi r^2 + 2\pi r h$. Here V is fixed at 1 (liter) and we want to minimize S.

24. A stone thrown directly upward with initial velocity 80 ft/sec reaches a height h ft in t sec, where $h(t) = 80t - 16t^2$. How long does it take the stone to reach maximum height? What is the maximum height?

25. If it costs $(0.018x^2 - 1.44x + 40)$ cents per mile to drive a car at speed x mph, what is the most efficient speed for the car? HINT: We want the speed that yields least cost per mile.

26. A cost analyst determines that it costs $(0.002x^2 - 0.4x + 1000)$ dollars to produce x wristwatches. What production level yields minimum cost per watch? HINT: Divide the cost function by x and minimize the result.

27. During a flu epidemic that lasts 22 days, the number of people who have been infected at the end of t days is $N(t) = 288 + 30t^2 - t^3$. When is the disease being spread fastest? HINT: Maximize $N'(t)$.

28. The daily output of a copper mine that can be operated for not more than 14 hr per day is $54t - 0.09t^3$ tons at the end of t hr. How many hours per day should the mine be worked for maximum output?

29. No-Con-U-Drive has 150 cars for rent. The owner can rent them all at $55 a week but knows from experience that he loses two customers for each $1 increase in the weekly rental. How much should he charge to maximize his revenue? What is the maximum revenue?

30. A rumor spreads in a town of 1000 people at a rate proportional to the product of the number who already know it and the number who do not. If x is the number who know it at a particular time and R is the rate at which it is spreading, then $R = kx(1000 - x)$. Show that the rumor is being spread fastest when 500 people know it. (Compare Study Question 6-3.)

31. It has been shown experimentally that the velocity v of air forced through the windpipe when a person coughs is given by $v(r) = kr^2(r_0 - r)$; here r_0 is the radius of the windpipe when the person is not coughing, r is the radius during a cough, and k is an appropriate constant. For what r is the velocity of the airflow greatest?

32. In an autocatalytic reaction (see Exercise 26, p. 63), the rate R at

which a new substance is formed from 150 units of a given substance is equal to $R(x) = 10x(150 - x)$, where x is the amount of the new substance present. For what x is the reaction rate a maximum?

33. Suppose an injection of x grams of a drug produces a change in blood pressure given by $P(x) = 1.5x^2 - 0.2x^3$, where P is measured in millimeters of mercury. What dose produces maximum sensitivity? [In Example 13-4, p. 127, the sensitivity to a drug is defined to be $P'(x)$.]

TEST A

1. A farmer has 100 ft of fencing to close off a rectangular pasture, one side of which is along the bank of a river. If he uses all his fencing along the other three sides, what are the dimensions of the biggest pasture he can close off?

2. A travel agent offers a group tour to the Orient only if the group has at least 60 people. Each member of a group of 60 pays $800. For each person in excess of 60 people, everyone's fare is reduced by $8. How many participants will maximize the agent's revenue, if at most 75 people can be accommodated?

3. A manufacturer wants to fill an order for 100,000 boxes. Each is to have a square base and to contain 24 cu in. If the material for the bottom costs $1\frac{1}{2}$¢ per sq in., for the top, $\frac{1}{2}$¢ per sq in., and for the sides, $1\frac{1}{8}$¢ per sq in., for what shape box will the materials cost least?

TEST B

1. A rectangular play area of 600 sq ft is to be fenced off. One side, which is along an existing wall, does not require fencing. The other three sides are to be fenced at a uniform cost of $4 a yard. What dimensions will minimize the cost of the fencing?

2. The owner of a small private kindergarten can enroll 25 children at $95 per child per month. She knows that for each $5 increase in the monthly rate, she will lose one child. What fee will maximize her revenue, if she can accept no more than 35 children?

3. A watch manufacturer can sell x watches at

$$40 - \frac{x}{400}$$

dollars each. If his overhead is $5000 and his manufacturing and labor costs come to $8 a watch, how many watches should he sell to maximize his profit?

Goal 25 | To Solve Problems Involving Maxima and Minima 251

STUDY QUESTIONS

1. We usually try, in max-min problems, to express the quantity to be maximized or minimized as a function of a single variable. Sometimes this is inconvenient; indeed, sometimes it is impossible. The following problem illustrates an alternative method:

   Find the shape of the largest rectangle of perimeter 27 ft.
   If A is the area, then

   $$A = xy$$

   We know that the perimeter is 27 ft, so

   $$2x + 2y = 27$$

   Now, instead of solving for y, say, in terms of x in order to express A explicitly as a function just of x, we differentiate each equation above implicitly *with respect to one variable*, say x, regarding y as a function of x and using the chain rule:

   $$A = xy \qquad\qquad 2x + 2y = 27$$
   $$\frac{dA}{dx} = x\frac{dy}{dx} + y \cdot 1 \qquad 2 \cdot 1 + 2\frac{dy}{dx} = 0$$

   Next we solve the equation on the right for dy/dx:

   $$2 + 2\frac{dy}{dx} = 0, \qquad \frac{dy}{dx} = -\frac{2}{2} = -1$$

   Now we substitute this value for $\frac{dy}{dx}$ into the equation on the left:

   $$\frac{dA}{dx} = x(-1) + y \cdot 1$$

   To maximize A, we now set $\frac{dA}{dx}$ equal to zero:

   $$-x + y = 0 \qquad \text{if} \qquad x = y$$

To verify that $x = y$ yields maximum area, we find d^2A/dx^2:

$$\frac{d^2A}{dx^2} = -1 + \frac{dy}{dx} = -1 + (-1) = -2$$

And -2 is a negative number. Therefore the rectangle of maximum area is a square.

Do this problem by expressing the area as a function of only one variable, to show the equivalence of the methods.

2. Use the alternative method of Study Question 1 on Exercises 19 and 22 of this Goal.
3. In Exercise 15 we suggested minimizing the square of the distance between two cars, rather than the distance, because doing the former is considerably simpler. Justify this mathematically.

GOAL 26

To solve related-rate problems

Frequently a problem involves two or more variables that are related and each of which is a function of time. A problem that gives information about all but one of the variables and asks for the rate of change of that one is referred to as a *related-rate* problem.

INTRODUCTION

When variables are related, so are their rates of change. If we have an equation involving several variables, each of which is a function of time, we can determine how their rates of change are related by differentiating the equation with respect to time. This often involves both the chain rule (Goal 15) and implicit differentiation (Goal 16). We illustrate with examples.

Example 26-1

The number of bacteria in a culture, growing in a circular shape on a Petri dish, is proportional to the area occupied by the culture. Suppose that $x = 1000A$, where x is the number of bacteria and A is the area occupied, in square inches. Suppose also that a lab technician determines, at the instant when the radius of the culture is 2 in., that the radius is increasing at the rate of 0.1 in. per h. At what rate is the culture growing (i.e., at what rate is the number of bacteria increasing) at that instant?

We can proceed as follows:

(1) Note that the crucial variables are x, the number of bacteria, and r, the radius of the circular region. The problem asks for dx/dt at the instant when $dr/dt = 0.1$ and $r = 2$.

(2) Find an equation involving x and r. Since $x = 1000A$ and $A = \pi r^2$, we derive the equation $x = 1000\pi r^2$.

(3) Differentiate the equation in (2) implicitly with respect to time t:

$$\frac{d}{dt}(x) = \frac{d}{dt}(1000\pi r^2)$$

$$\frac{dx}{dt} = 1000 \cdot 2\pi r \frac{dr}{dt}$$

This gives us the rate of growth of the culture at *any* time.

(4) Find the rate of growth at the specified instant, that is, when $r = 2$ and $dr/dt = 0.1$:

$$\frac{dx}{dt} = 1000 \cdot 2\pi(2)(0.1) = 400\pi$$

or about 1256 bacteria per h.

NOTE: We could have omitted step (2) as follows. Since

$$x = 1000A \quad \text{and} \quad A = \pi r^2$$

by the chain rule

$$\frac{dx}{dt} = \frac{dx}{dA} \cdot \frac{dA}{dt}$$

so at any time

$$\frac{dx}{dt} = 1000 \cdot 2\pi r \frac{dr}{dt}$$

and at the specified instant

$$\frac{dx}{dt} = 1000 \cdot 2\pi(2)(0.1) = 400\pi$$

as above.

Example 26-2

A painter is standing at the top of a 13-ft ladder that is leaning against a wall. If the foot of the ladder begins sliding away from the wall at the constant rate of 1 ft per sec, how fast will the painter be falling when he is 5 ft above the ground?

(1) We note that the crucial variables here are x, the distance the painter is above the ground, and y, the distance of the foot of the ladder from the wall. We are told that dy/dt is equal to the constant 1; we want to find dx/dt (at the instant) when $x = 5$.

Goal 26 | To Solve Related-Rate Problems

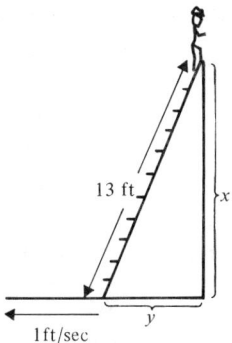

(2) The variables are related by the equation

$$x^2 + y^2 = 13^2$$

(3) We differentiate implicitly with respect to t, obtaining

$$2x\frac{dx}{dt} + 2y\frac{dy}{dt} = 0 \quad \text{or} \quad x\frac{dx}{dt} + y\frac{dy}{dt} = 0$$

Then we solve for dx/dt:

$$\frac{dx}{dt} = -\frac{y}{x}\frac{dy}{dt}$$

Since dy/dt is constant (= 1), we can substitute for it:

$$\frac{dx}{dt} = -\frac{y}{x}(1) = -\frac{y}{x}$$

This gives us the rate of change (of x with respect to t) at *any* time. It is clear that this rate depends on both x and y.

(4) At the instant specified, $x = 5$. At that instant, we note from the equation $x^2 + y^2 = 169$ that $y = 12$. Thus

$$\frac{dx}{dt} = -\frac{12}{5} = -2\frac{2}{5}$$

So when the painter is 5 ft above the ground, he is *falling* (indicated by the minus sign) at the rate of $2\frac{2}{5}$ ft/sec.

NOTE: It is important not to substitute values given for rates of change or for other variables prematurely. It is usually safer to find a required rate of change at any time, as above, then evaluate it at a specified instant using the data given.

GENERAL PROCEDURE

The above examples illustrate the general procedure to be used in solving related-rate problems:

(1) Note the crucial variables, drawing a figure or graph, if relevant, that shows the situation at *any* time. Note what is given and what is called for.

(2) Write an equation involving the crucial variables.

(3) Differentiate this equation implicitly with respect to time, then solve for the desired rate of change at *any* time.

(4) Evaluate the rate of change in (3) at the specified instant, using any given information.

Example 26-3

Two airplanes have left an airport. At a given instant, airplane A is 300 miles north of the airport and is heading due north at 200 mph; airplane B is 400 miles east of the airport and is heading due east at 250 mph. How fast are the planes separating at that instant?

(1) x is the distance of A from the airport (at any time); y is the distance of B from the airport; z is the distance between the planes.

$$\frac{dx}{dt} = 200; \qquad \frac{dy}{dt} = 250$$

We want dz/dt at the instant when $x = 300$ and $y = 400$.

(2) $x^2 + y^2 = z^2$

(3) $2x \dfrac{dx}{dt} + 2y \dfrac{dy}{dt} = 2z \dfrac{dz}{dt}$; $\qquad x \dfrac{dx}{dt} + y \dfrac{dy}{dt} = z \dfrac{dz}{dt}$;

$$\frac{dz}{dt} = \frac{x \dfrac{dx}{dt} + y \dfrac{dy}{dt}}{z}$$

Since dx/dt and dy/dt are fixed throughout, we can substitute their constant values here:

Goal 26 | To Solve Related-Rate Problems

$$\frac{dz}{dt} = \frac{x(200) + y(250)}{z} = \frac{200x + 250y}{z}$$

This is the rate at which the planes are separating at any time.

(4) At the specified instant $x = 300$, $y = 400$, and the equation in (2) yields $z = 500$. So

$$\frac{dz}{dt} = \frac{200(300) + 250(400)}{500} = 320$$

The planes are therefore separating at the given instant at the rate of 320 mph.

EXERCISES

1. Suppose that the number of bacteria in a culture on a Petri dish is $500A$, where A is the area, in square centimeters, of the circular region occupied. Under certain experimental conditions a technician determines that the population grows at a constant rate equal approximately to 4800 more bacteria per hour. How fast is the radius of the culture increasing when the radius is 3 cm?

2. A man is standing at the top of a 15-ft ladder that is leaning against a wall. The foot of the ladder begins sliding away from the wall so that it is moving at 2 ft/per sec when it is 9 ft from the wall. How fast is the man falling at that moment?

3. One ship is heading toward a port from the north at 25 knots, while a second ship is heading east away from the port at 15 knots. At a certain instant, the first ship is 30 nautical miles from the port while the second is 40 nautical miles from port. At this instant are the ships separating or approaching each other and at what speed? (NOTE: one knot is 1 nautical mile per h.)

4. Oil is being stored in a cylindrical tank of radius 3 meters (m). If it is poured into the tank at the rate of $28\frac{1}{4}$ cu m per min, how fast is the oil level rising? HINT: The volume V of a cylinder is $\pi r^2 h$, where r is the radius and h is the height.

5. Suppose that two species of animals are in a predator-prey relationship as given in Example 16-4 (p. 159). Then if the prey population is x and the predator population is y, we have

$$10{,}000(y - 500)^2 + (x - 12{,}000)^2 = 100{,}000{,}000$$

Suppose that at a specified time there are 4000 prey and 560 predators and that the prey population is increasing at the rate of 75 animals per month. How is the predator population changing at that time?

6. Consider the problem of blowing a spherical glass bottle. How fast is the radius increasing when the radius is 2 cm if air is being blown in at a rate of 10 cm³ per min? ($V = \frac{4}{3}\pi r^3$).

7. Suppose the best price at which to sell x thingamajigs is $30 + 0.02x - 0.00001x^2$ dollars. If the production level is now 400 thingamajigs and it is anticipated that demand will necessitate increasing production at the rate of 100 thingamajigs per year, at what rate is the price increasing now?

8. A water storage tank is shaped like a cone with radius 5 ft and height 10 ft. If the current depth of water is 7 ft and water is running out of the bottom at 77 cu ft per h, how fast is the water level dropping? ($V = \frac{1}{3}\pi r^2 h$).

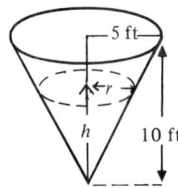

9. A rocket is sent straight up in such a way that when it attains a height of x km its velocity is $0.02x^2 + 10x$ km per h. What is its acceleration at a height of 100 km?

10. The Do-It-Yourshelf Bookcase Company figures that its profit P, in dollars, is related to x, the number of units sold, by the equation $xP - 8x^2 = 1{,}000{,}000$. If the level of production at present is 1000 units per year, and this number is increasing at the rate of 50 units per year, at what rate is the profit changing at present?

TEST A

1. A car is speeding north toward a city at 60 km/hr while another is speeding west away from the city at 50 km/h. If at a certain instant the first car is 30 km south of the city and the second is 40 km west of the city, are the cars approaching each other or separating, and at what speed?

2. In a controlled setting, biologists note that the population x of one organism is related to the population y of another by the equation $y^2 = x^2 - 750x$. If the x-population at a particular time is found to be 1000 and to be increasing at the rate of four organisms per hour, how is the y-population changing at that time?

Goal 26 | **To Solve Related-Rate Problems**

TEST B

1. The length L in centimeters of a particular iron bar is related to its temperature T in degrees Celsius by the formula $L = 25 + 0.0005T$. If the temperature of the bar is lowered at the rate of 2 degrees Celsius per minute, at what rate will its length be changing?

2. A tractor distributor figures that the number N of salesmen he needs depends on the number x of tractors he sells in one season, the relationship being given by $N = 25 + 0.001x - 0.00001x^2$. Is the distributor hiring or firing salesmen, and at what rate, when 500 tractors are sold in one season and sales are increasing at the rate of 100 tractors per season?

GOAL **27**

To apply the Mean Value Theorem

The Mean Value Theorem (which we will sometimes abbreviate "MVT") is important because of its usefulness in establishing other theorems of the calculus. Of special interest to us is a pair of corollaries that we will need in later chapters.

We start with a preliminary example.

Example 27-1

Below we show the graph of $f(x) = 2x^2 - 3x - 1$ between the points $(-1, 4)$ and $(3, 8)$. We note that the slope of chord AB equals

$$\frac{8-4}{3-(-1)} = \frac{4}{4} = 1$$

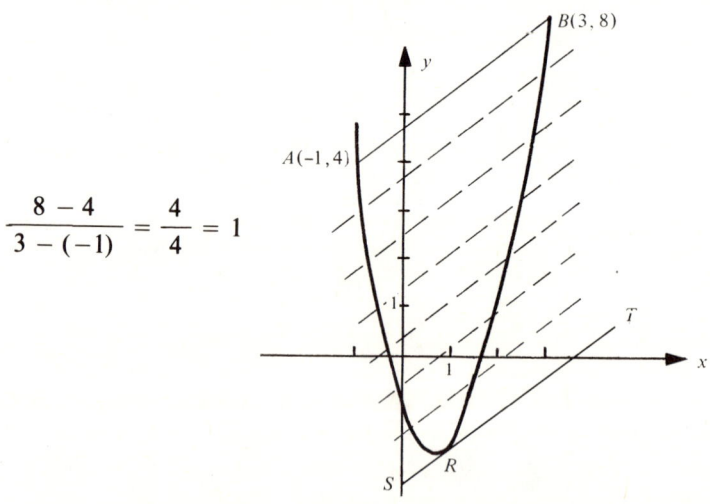

We now ask: Is there a point on the curve between $x = -1$ and $x = 3$ where the slope of the curve is equal to 1? That is, is there a point on the arc connecting A and B where the tangent to the curve is parallel to chord AB?

Goal 27 | To Apply the Mean Value Theorem

Our intuition says yes. If, for example, we place a straightedge along AB and slide it downward so that it is always parallel to AB, we will surely "hit" a point on the curve where the straightedge is tangent to the curve. We've labeled this point R, and the tangent to the graph at R, ST. We can easily find the x-coordinate of R by noting that the slope of ST is to equal 1, the slope of chord AB. Since $f'(x) = 4x - 3$, we solve the equation $4x - 3 = 1$, getting $x = 1$. We now have

$$f'(1) = \frac{f(3) - f(-1)}{3 - (-1)}$$

where $-1 < 1 < 3$; that is, point R on the curve is between A and B.

STATEMENT OF THE THEOREM

Example 27-1 is an application of the Mean Value Theorem. If we examine the graph of any differentiable function f between points $A(a, f(a))$ and $B(b, f(b))$, it is clear that at some point, say at $x = c$, between a and b the tangent to the curve is parallel to chord AB; that is, the tangent at $x = c$ and the chord AB have equal slopes. Since the slope of the tangent at $x = c$ is $f'(c)$, this means that

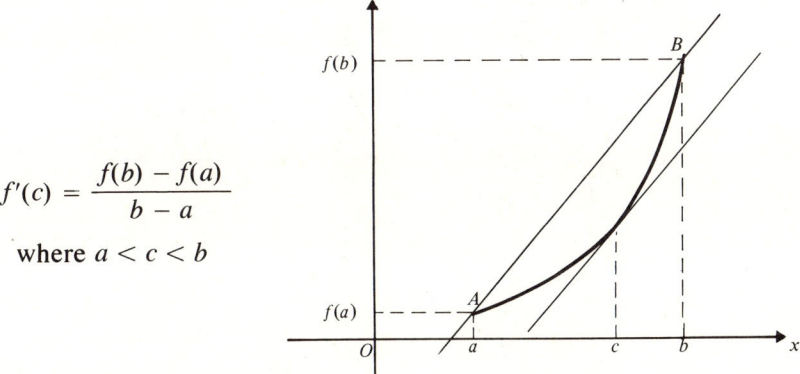

$$f'(c) = \frac{f(b) - f(a)}{b - a}$$

where $a < c < b$

The MVT tells us that this statement holds true for any function that is both continuous on the closed interval $a \leq x \leq b$ and differentiable at every point on the open interval $a < x < b$. If a function satisfies these conditions, then the theorem guarantees the existence of at least one c in the specified interval. For some functions there are two or more values of c in a given interval. Examples of the latter sort will be found in the exercises.

Example 27-2

Suppose that when you drive onto a toll road you receive a ticket stamped "2 P.M.—150 miles," indicating the time of entry and the distance of the tollbooth from one end of the toll road. When you are ready to leave the road, you hand your ticket to a collector, who stamps it "5 P.M.—330 miles." He then calls over a patrolman, who gives you another kind of ticket. When you protest, the officer offers the following explanation. Since average speed is obtained by dividing the total distance covered by the total time, your average speed for the toll-road trip, he notes, was

$$\frac{(330 - 150) \text{ miles}}{(5 - 2) \text{ hours}} \text{ or } 60 \text{ mph}$$

He then states that at some instant between 2 and 5 P.M. your instantaneous speed must also have been 60 mph. Since the legal speed limit is 55 mph and you exceeded that, he has now concluded that you are guilty of an infraction. (At present—fortunately?—not all states require their patrolmen to have completed a course in calculus.)

This example is, of course, another application of the "inescapable" MVT. In general, if $f(t)$ is the distance traveled in time t and if $t = a$ and $t = b$ are two points in time, then

the distance covered between $t = a$ and $t = b$ is $f(b) - f(a)$
the time elapsed is $b - a$
the average speed is $\dfrac{f(b) - f(a)}{b - a}$

The claim made in Example 27-2 is that at some time c between a and b, the instantaneous speed must be equal exactly to the average speed. That is, there is at least one c between a and b such that

$$f'(c) = \frac{f(b) - f(a)}{b - a}$$

And this is precisely what the MVT states.

Example 27-3

Verify the MVT for the function $f(x) = x^3 + x^2 - 15x - 12$ on the interval $-1 \leq x \leq 4$.

We must find at least one value of c promised by the theorem to verify it. In this example, $a = -1$, $b = 4$, $f(-1) = 3$, and $f(4) = 8$. We seek a c such that

$$f'(c) = \frac{f(b) - f(a)}{b - a} = \frac{8 - 3}{4 - (-1)} = \frac{5}{5} = 1$$

Goal 27 — To Apply the Mean Value Theorem

Since $f'(x) = 3x^2 + 2x - 15$, we solve

$$3x^2 + 2x - 15 = 1$$
$$3x^2 + 2x - 16 = 0$$
$$(3x + 8)(x - 2) = 0$$

The solutions are $x = -8/3$ and $x = 2$. Since only 2 is between -1 and 4, 2 serves as the promised c.

CHECKING THE HYPOTHESES OF THE MVT

Two examples suffice to make an important point.

Example 27-4

Find any values of c promised by the MVT for $f(x) = 1/x$ on the interval $-1 \leq x \leq 2$.

We note that

$$\frac{f(2) - f(-1)}{2 - (-1)} = \frac{\frac{1}{2} - (-1)}{3} = \frac{\frac{3}{2}}{3} = \frac{1}{2}$$

Since $f'(x) = -1/x^2$, we seek an x such that

$$-\frac{1}{x^2} = \frac{1}{2}$$

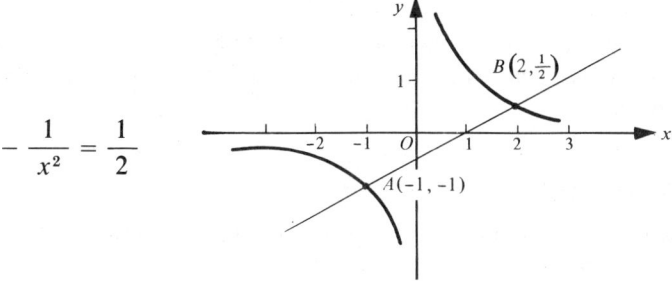

But there is none. Indeed, $-1/x^2 < 0$ for all $x \neq 0$. Does this contradict the MVT? Not at all! Since f is not defined at $x = 0$, this function does not satisfy the hypotheses of the MVT, which require f to be continuous on the given interval. The figure affirms that there is no point on the graph between $x = -1$ and $x = 2$ where the tangent to the graph is parallel to chord AB.

Example 27-5

Does the MVT apply to the function $f(x) = x^{2/3}$ on the interval $-8 \leq x \leq 8$?

Once again we look for a number c between a and b, with $a = -8$ and $b = 8$, such that

$$f'(c) = \frac{f(b) - f(a)}{b - a} = \frac{4 - 4}{8 - (-8)} = 0$$

Since

$$f'(x) = \frac{2}{3x^{1/3}}$$

we seek x here such that

$$\frac{2}{3x^{1/3}} = 0$$

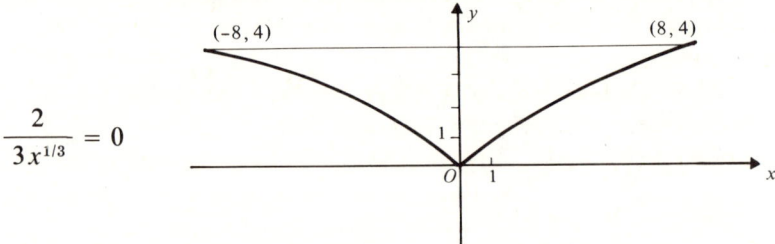

But since the left side of this equation never equals 0, we conclude that the required c does not exist. Although this function is continuous on the interval $-8 \leqq x \leqq 8$, it has no derivative at $x = 0$. The MVT does not apply to this function because its hypotheses are not satisfied.

These examples show the importance of checking to be sure that a given function meets the hypotheses of the Mean Value Theorem before trying to apply the theorem.

COROLLARIES OF THE MVT

The following two corollaries will be used frequently when we discuss the topic of antiderivatives.

COROLLARY (I): If $f'(x) = 0$ throughout the interval $a \leqq x \leqq b$, then $f(x)$ is constant throughout the interval.

The proof is easy: If x is any point in the given interval, then by the MVT applied to the interval (a, x), there exists at least one value of c such that $a < c < x$ and such that

$$\frac{f(x) - f(a)}{x - a} = f'(c)$$

But by hypothesis $f'(c) = 0$. Therefore

$$\frac{f(x) - f(a)}{x - a} = 0$$
$$f(x) - f(a) = 0$$
$$f(x) = f(a)$$

Since x is *any* point in the interval $a < x < b$, the very last equation implies that the value of f is the same at every point in the interval [and that value is $f(a)$]; so f is constant throughout the interval.

COROLLARY (II): If two functions have the same derivative throughout an interval, then the functions differ by a constant.

This corollary says that if $f'(x) = g'(x)$ on the interval $a < x < b$, then $f(x) = g(x) + C$, where C is a constant.*

The proof uses Corollary (I). If we let $h(x) = f(x) - g(x)$, then $h'(x) = f'(x) - g'(x) = 0$ throughout the interval $a < x < c$. By Corollary (I), then, $h(x)$ is a constant function; say, $h(x) = C$. So $f(x) - g(x) = C$ or $f(x) = g(x) + C$.

Example 27-6

Find all the functions whose derivatives are equal to $2x$.

From experience we know that x^2 is one such function. Corollary (II) tells us that any other function with derivative $2x$ *must* be of the form $x^2 + C$, where C is a constant. Some candidates are $x^2 + 1$, $x^2 - 4$, $x^2 + \sqrt{2}$, $x^2 - \pi$. The graph of any member of this family of functions $y = x^2 + C$ is a parabola "parallel" to $y = x^2$. (Two graphs are parallel if tangents to the graphs at any points with the same abscissa are parallel; thus one graph is merely the other shifted up or down. See the figure on p. 266.)

In Chapter 7 we will define x^2 to be an *antiderivative* of $2x$.

Example 27-7

Find a function f such that $f'(x) = 3x^2$ and $f(2) = 5$.

Since the derivative of x^3 is $3x^2$, we know that $f(x) = x^3 + C$, where C is a constant. Since it is given further here that $f(2) = 5$, it follows that $2^3 + C$ must equal 5 and that $C = 5 - 8 = -3$. Thus there is only one function satisfying the conditions given: it is $f(x) = x^3 - 3$.

---

*Note that the arbitrary constant C is totally different from the c of the MVT.

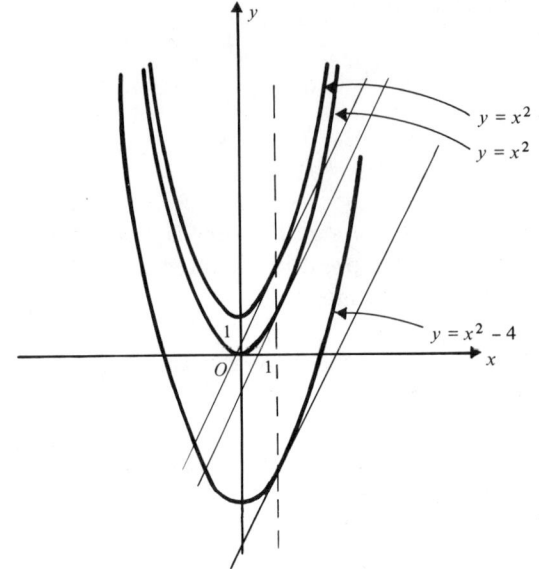

Graph of functions whose derivatives equal $2x$

EXERCISES

For each of the following functions (1 through 11) and the specified intervals, graph the function, then either find any points promised by the Mean Value Theorem or explain why the theorem does not apply.

1. $f(x) = x^2 + 2x + 3$ on $1 \leq x \leq 5$
2. $f(x) = x^2 - 3x + 4$ on $3 \leq x \leq 4$
3. $f(x) = -x^2 + x - 1$ on $-1 \leq x \leq 3$
4. $f(x) = -2x^2 - 3x + 4$ on $0 \leq x \leq 1$
5. $f(x) = x^3 + x^2 - 2x + 3$ on $-1 \leq x \leq 4$
6. $f(x) = x^3 - 3x^2 + 2x - 1$ on $0 \leq x \leq 3$
7. $f(x) = x^3 - \frac{1}{2}x^2 + x + 1$ on $-2 \leq x \leq 2$
8. $f(x) = \dfrac{1}{x-1}$ on $0 \leq x \leq 2$
9. $f(x) = \dfrac{1}{x-1}$ on $2 \leq x \leq 5$
10. $f(x) = x^{2/3}$ on $0 \leq x \leq 1$
11. $f(x) = |x|$ on $-1 \leq x \leq 3$

Goal 27 | **To Apply the Mean Value Theorem**

In each of the following (12 through 17), find a function f that has the specified properties:

12. $f'(x) = 2x$, $f(0) = 5$
13. $f'(x) = 3x^2$, $f(1) = 2$
14. $f'(x) = 5x^4$, $f(-1) = 2$
15. $f'(x) = -1/x^2$, $f(1) = 0$
16. $f'(x) = x$, $f(2) = 1$
17. $f'(x) = x^2$, $f(0) = -1$

TEST A

1. For each of the following functions and specified intervals, either find any values of c promised by the Mean Value Theorem or explain why that theorem does not apply.

    (a) $f(x) = -x^2 + 3x - 2$ on $-1 \leq x \leq 2$
    (b) $f(x) = x^3 - 2x^2 + x - 1$ on $0 \leq x \leq 5$
    (c) $f(x) = \dfrac{1}{x}$ on $2 \leq x \leq 8$

2. In each of the following, find a function that has the properties specified.

    (a) $f'(x) = 4x^3$, $f(0) = 3$
    (b) $f'(x) = 1$, $f(1) = -1$

TEST B

1. For each of the following functions and specified intervals, either find any values of c promised by the Mean Value Theorem or explain why that theorem does not apply.

    (a) $f(x) = 2x^2 - x + 3$ on $0 \leq x \leq 2$
    (b) $f(x) = x^3 + \frac{1}{4}x^2 - x + \frac{3}{4}$ on $-1 \leq x \leq 1$
    (c) $f(x) = \dfrac{1}{x+1}$ on $-3 \leq x \leq 0$

2. In each of the following, find a function that has the properties specified.

    (a) $f'(x) = \dfrac{1}{x^2}$, $f(-1) = -1$
    (b) $f'(x) = x$, $f(0) = -\dfrac{1}{2}$

STUDY QUESTIONS

1. Show that for a parabola on the interval $x_1 \leq x \leq x_2$ the c promised by the MVT is equal to $\dfrac{x_1 + x_2}{2}$. HINT: Given that $f(x) = Ax^2 + Bx + C$, show that

$$\frac{f(x_2) - f(x_1)}{x_2 - x_1} = f'\left(\frac{x_1 + x_2}{2}\right)$$

2. Give an example in which more than one point within the specified interval satisfies the MVT.

3. The MVT is a generalization of *Rolle's Theorem*, which states that if f is continuous on $a \leq x \leq b$ and differentiable on $a < x < b$, and if, furthermore, $f(a) = f(b)$, then for some c between a and b, $f'(c) = 0$. Show that the MVT follows from Rolle's Theorem. HINT: If f satisfies the hypotheses of the MVT and if L is the function whose graph is the straight line passing through the end points, then show that the function $f - L$ satisfies the hypotheses of Rolle's Theorem.

4. Suppose that $f'(x) = \dfrac{1}{x}$ for all $x > 0$ and $f(1) = 0$. Show that $\dfrac{1}{2} < f(2) < 1$. HINT: Apply the MVT using the points $x = 1$ and $x = 2$.

5. Suppose $-1 < f'(x) < 1$ for all x. Show that $|f(x_2) - f(x_1)| < |x_2 - x_1|$ for all x_1 and x_2.

6. Suppose $-2 < f'(x) < 2$ for all x. Show that $|f(x_2) - f(x_1)| < 2|x_2 - x_1|$ for all x_1 and x_2.

7. In Goal 19 we stated that if $f'(x) > 0$ throughout an interval, then f is increasing throughout that interval. Show that this follows from the MVT. HINT: You must show that if $b > a$ then $f(b) > f(a)$.

GOAL **28**

To use differentials for approximation

There are many contexts in which we know the value of a function at a specified point and are interested in its value at a nearby point. For example:

A plant manager knows the cost associated with the current level of production and wants to know what the cost will be if there is a slight increase in production.

A scientist knows the length of an iron bar at a given temperature and is interested in the length of the bar if he decreases the temperature slightly.

An engineer knows the amount of light that will pass through glass of specified thickness and is concerned with the possible change in light intensity if there is a small error in the thickness of the glass delivered by the manufacturer.

In this Goal we develop a technique for arriving quickly at approximate answers to problems like these, when we know not only the value of a function at a given point but also the value of its derivative at that point. Since the technique consists of approximating the function by the tangent line to the curve at the given point, the method is often called "linear approximation."

AN APPROXIMATION FOR Δy

Suppose we know the values of both $f(c)$ and $f'(c)$ at a point P on the curve and want to approximate the value of f when x is close to c, say, Δx units away. On page 270, the segment RQ, which is equal to the change in y resulting from the change Δx, is Δy. The value of f at $(c + \Delta x)$ is therefore equal to $f(c) + \Delta y$. When Δx is small, segment RT, labeled dy, is a reasonable

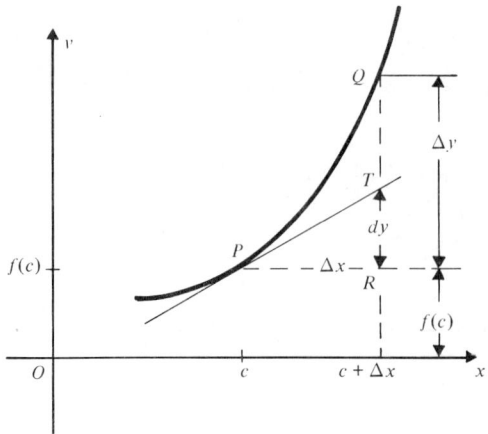

approximation for Δy. Noting that $dy/\Delta x =$ the slope of the tangent at $P = f'(c)$, we see that

$$dy = f'(c)\Delta x$$

We call the quantity dy a *differential*. We now approximate $f(c) + \Delta y$ by $f(c) + dy$; that is, by $f(c) + f'(c) \Delta x$. We will find approximations both for Δy and for $f(c + \Delta x)$ in the examples and exercises that follow.

Example 28-1

(a) Find the actual change Δy in the function $y = f(x) = x^3 + 1$ when x changes from 2 to 2.01. (b) Find dy.

(a) $\Delta y = f(2.01) - f(2) = [(2.01)^3 + 1] - [2^3 + 1] = 0.120601$.

(b) $dy = f'(c) \Delta x$, where $c = 2$ and $\Delta x = 0.01$. Since $f'(x) = 3x^2$, $f'(c) = 3(2)^2(0.01) = 0.12$. For most purposes this is a fine approximation.

Example 28-2

Use differentials to approximate $\sqrt[3]{26.7}$.

Since we are approximating a cube root, we focus on a number that is close to 26.7 and that is also a perfect cube. We will first find by how much (approximately) $\sqrt[3]{26.7}$ differs from $\sqrt[3]{27}$ (this difference is dy), then add that to $\sqrt[3]{27}$. Choosing a function is easy. We let $f(x) = \sqrt[3]{x}$. Then, with $c = 27$ and $\Delta x = -0.3$, we have

$$dy = f'(c) \Delta x = f'(27)(-0.3)$$

Goal 28 | To Use Differentials for Approximation

Since $f(x) = \sqrt[3]{x} = x^{1/3}$,

$$f'(x) = \frac{1}{3}x^{-2/3} = \frac{1}{3x^{2/3}}$$

So

$$dy = \frac{1}{3(27)^{2/3}}(-0.3) = \frac{1}{3(9)}(-0.3) = -0.011111$$

Thus $\sqrt[3]{26.7} \approx \sqrt[3]{27} - 0.011111 = 3 - 0.011111 = 2.988889$. In a table of cube roots, $\sqrt[3]{26.7}$ begins 2.988847; obviously we were able quickly to obtain an excellent approximation. With little additional work we can obtain related approximations:

$$\sqrt[3]{27.1} \approx 3 + \frac{1}{3(9)}(0.1) = 3.003704 \text{ (actual answer begins 3.003699)}$$

$$\sqrt[3]{26.99} \approx 3 + \frac{1}{3(9)}(-0.01) = 2.999630 \text{ (actual answer begins 2.999629)}$$

Note that the smaller Δx is, the better is the approximation dy for Δy.

Example 28-3

The profit of Genuine Imported Products (GIP) is $P(x) = -8000 + 6000x - 500x^2$, where x is the wholesale price per product in dollars. Find the percentage change in profit when x increases from 4 to 4.25.

The percentage change is defined to be $100 \cdot \frac{\Delta P}{P}$. We approximate ΔP by dP, which equals $P'(c) \Delta x$, with $c = 4$ and $\Delta x = \frac{1}{4}$. Since $P'(x) = 6000 - 1000x$, we have

$$dP = P'(4)\left(\frac{1}{4}\right) = (6000 - 4000)\left(\frac{1}{4}\right) = 500$$

and

$$100 \cdot \frac{\Delta P}{P} \approx 100 \cdot \frac{500}{P(4)} = \frac{50,000}{8000} = 6.25\%$$

Example 28-4

What is the approximate volume of rubber used in manufacturing a hollow rubber ball with diameter 4 cm if the thickness of the rubber is 1 mm?

The exact volume of rubber is ΔV, where V is a function of the radius r;

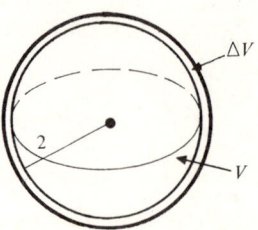

here $V(r) = \frac{4}{3}\pi r^3$, $c = 2$, and $\Delta r = 0.1$ (cm). We approximate ΔV by dV, which equals $V'(c) \Delta r$ or $V'(2) (0.1)$. Since $V'(r) = 4\pi r^2$, $dV = 4\pi (2)^2(0.1) = 1.6\pi \approx 5.03$; so the volume of rubber used is approximately 5.03 cm³.

USING DIFFERENTIALS TO ESTIMATE ERROR TOLERANCE

Suppose y is a function of x. We can use differentials to estimate the effect on y of an error in x (caused perhaps by imprecise measurement, production, judgment, prediction, or the like). We can also approximate with differentials the error allowable in x if the error in y is not to exceed a given amount (tolerance).

Example 28-5

Suppose a cubical steel block of volume 125 cu in. is to be manufactured, and an error of 0.01 cu in. can be tolerated. What is the allowable error in the length of an edge of the block?

We note that if the edge can be constructed to be exactly 5 in., then the volume will be exactly 125 cu in. If we let $y = f(x) = x^3$ be the volume function, then we are told that the magnitude of Δy cannot exceed 0.01 when $x = 5$; that is, the volume must be within 0.01 of 125. Since $\Delta y \approx dy = f'(5) \Delta x$ and $f'(x) = 3x^2$, we have $dy = 3(5)^2 \Delta x = 75 \Delta x$. We are given that $-0.01 < \Delta y < 0.01$; so our approximation yields $-0.01 < 75 \Delta x < 0.01$, and (dividing through by 75),

$$-0.00013 < \Delta x < 0.00013$$

So the allowable error in the length of the edge is 0.00013 in.

DEFINITION OF DIFFERENTIALS

If $y = f(x)$, then in our discussion above we called

$$dy \quad \text{or} \quad f'(x) \Delta x$$

Goal 28 | To Use Differentials for Approximation

the *differential* of y. If we now define dx to be equal to Δx, then we can define the differential of y as follows:

$$dy = f'(x)\, dx$$

Thus if $y = f(x)$ we can find dy by multiplying the derivative $f'(x)$ by dx. So

$$\begin{aligned}
&\text{if } y = x^3 &&\text{then } dy = 3x^2\, dx \\
&\text{if } y = x^4 - 2x^2 + 1 &&\text{then } dy = (4x^3 - 4x)\, dx \\
&\text{if } y = (x^2 + 8)^6 &&\text{then } dy = 6(x^2 + 8)^5 \cdot 2x\, dx \\
&&&\quad\quad\quad\;\; = 12x(x^2 + 8)^5\, dx
\end{aligned}$$

Of course, other variables can also be used:

$$\begin{aligned}
&\text{if } s = 4t^2 + 3t &&\text{then } ds = (8t + 3)\, dt \\
&\text{if } A = \pi r^2 &&\text{then } dA = 2\pi r\, dr \\
&\text{if } w = \sqrt{u^3 - 4} &&\text{then } dw = \frac{3u^2}{2\sqrt{u^3 - 4}}\, du \\
&\text{if } u = (3x^2 - 1)^3 &&\text{then } du = 3(3x^2 - 1)^2 \cdot 6x\, dx \\
&&&\quad\quad\quad\;\; = 18x(3x^2 - 1)^2\, dx
\end{aligned}$$

Under the definition just given, it turns out that the symbol dy/dx is a quotient. This definition will prove very useful when we consider the inverse of differentiation in Chapter 7.

EXERCISES

1. Let $f(x) = x^2$ and $c = 3$. For each Δx given below, compute both Δy and dy. Draw a picture illustrating the results.
   (a) $\Delta x = 1$
   (b) $\Delta x = 0.5$
   (c) $\Delta x = -0.1$
   (d) $\Delta x = 0.01$
   (e) $\Delta x = -0.0001$

2. Use the technique of this goal to approximate each of the following:
   (a) $\sqrt{65}$
   (b) $\sqrt{63}$
   (c) $\sqrt[3]{8.2}$
   (d) $\sqrt[3]{0.96}$
   (e) $\sqrt[3]{9.04} + \dfrac{1}{\sqrt{9.04}}$

3. What is the approximate additional cost of manufacturing 10 more units if the cost of making x units is $5000 + 3.2x + 0.001x^2$ dollars and the current production level is 550 units?

4. The sides of a hollow cubical metal box are $\frac{1}{3}$ cm thick. If the interior volume of the box is 125 cm³, what is the approximate volume of metal used in the construction of the box?

5. Suppose an airplane flies around the Earth at the equator. If a second plane flies the same path but at an altitude 100 ft higher, approximately how much farther must it fly than the first? (Note that in solving this problem it is not necessary to know Earth's radius.)

6. What change in profit results from selling 25 more units of a product if the marginal profit (derivative of the profit) is $5 per unit at the current production level of 54,325 units?

7. A virus is spreading at such a rate that after t days $\dfrac{10{,}000t}{t + 100}$ people have come in contact with it. Thus after 100 days, 5000 people have been infected. Approximately how many people will have been infected after 101 days? 102 days? 103 days? 104 days? 105 days?

8. If the radius of Earth is about 4000 miles and the depth of the atmosphere is about 20 miles, approximately how much volume does the atmosphere occupy (in cubic miles)?

9. Approximate the allowable error in measuring the side of a square sheet of plastic if the area is to be 36 ± 0.1 sq in.

10. Suppose x g of a drug will lower a person's blood pressure by $10\sqrt{x}$ mm of mercury. Approximate the allowable error in measuring the dosage if the doctor wishes the blood pressure to be lowered by 20 ± 1 mm of mercury.

11. Suppose a company figures that its profit will be $P(x) = -4900 + 8.8x - 0.0011x^2$ dollars when x units are sold. If the company estimates that it will sell 3000 units with a possible 5% error, find the company's estimated profit and the error and percentage error in that estimated profit.

12. Refer to Example 16-4, page 159. Use differentials to show, for the predator-prey relation given, that when $x = 18{,}000$ and $y = 420$ an increase of 400 in the prey population produces an increase of approximately 3 in the predator population.

13. Refer to Example 16-5, page 160. For the demand-price equation given, use differentials to approximate the change in demand caused by a $1 increase in price when the price is $25 per lamp and the demand is for 100 lamps.

14. Find the indicated differential:

    (a) $y = 2x^3 - 5x + 2$; dy (d) $y = 4x$; dy

    (b) $z = \dfrac{1}{u}$; dz (e) $u = (3 + 7x)^4$; du

    (c) $s = \sqrt{4 - t^2}$; ds (f) $u = (x^2 + 3x + 8)^3$; du

Goal 28 | To Use Differentials for Approximation

TEST A

1. Let $f(x) = x^3 - x$, $c = 2$, and $\Delta x = 0.1$. Find Δy and dy.
2. Approximate $\sqrt{4.2}$.
3. Suppose that at a wholesale price of x dollars per unit, a company's total profit is $-10{,}000 + 5000x - 400x^2$. Approximate the change in profit that will result from an increase in wholesale price from \$6.50 to \$6.75.
4. A colony of bacteria is growing in a circular pattern on a Petri dish. The relationship between the number x of bacteria and the area A in square inches is $x = 700A$. A bacteriologist measures the radius of the occupied region and finds it to be $2\frac{1}{4}$ in. Approximately how accurate must that measurement be for the number of bacteria to be accurate to within an error of 100?

TEST B

1. Let $f(x) = x^2 + 2x + 1$, $c = 1$, and $\Delta x = -0.2$. Find Δy and dy.
2. Approximate $\sqrt[3]{7.8}$.
3. A company's total profit after t yr of production is $-100{,}000 + 25{,}000t + 42{,}000\sqrt{t}$ dollars. Approximate the profit during the tenth year, that is, between $t = 9$ and $t = 10$.
4. Suppose x pounds (lb) of a certain fissionable material will explode with a force equivalent to $500x$ tons of TNT. What is the allowable error in measuring 2 lb of the material if the error in the resulting explosion is not to exceed the equivalent of 1 ton of TNT?

STUDY QUESTIONS

1. Suppose f has the property that for a certain fixed x, df and Δf are equal for all Δx. Describe the graph of f.
2. Suppose $y = uv$. Let Δu and Δv, changes in u and v respectively, result in Δy, a change in y. Since this new point must also satisfy the original equation, we have

$$y + \Delta y = (u + \Delta u)(v + \Delta v) = uv + v\,\Delta u + u\,\Delta v + \Delta u\,\Delta v$$
$$= y + v\,\Delta u + u\,\Delta v + \Delta u\,\Delta v$$

So $\Delta y = u\,\Delta v + v\,\Delta u + \Delta u\,\Delta v$, and when Δu and Δv are both small, a reasonable approximation to Δy is $dy = u\,dv + v\,du = u\,\Delta v + v\,\Delta u$. For example, a company's yearly revenue for a particular product is equal to the unit price times the number of units sold. Suppose the current price is \$5.50 and 10,000 units are being sold yearly. Approximate the change in revenue resulting from a drop in price to \$5.48 and a rise in units sold to 10,100.

3. Show that if $y = \dfrac{u}{v}$ and Δu and Δv are both small, then a reasonable approximation to Δy is $dy = \dfrac{v\,du - u\,dv}{v^2} = \dfrac{v\,\Delta u - u\,\Delta v}{v^2}$

   HINT: $\Delta y = \dfrac{u + \Delta u}{v + \Delta v} - \dfrac{u}{v}$.

GOAL **29**

To solve problems involving motion along a line

Analysis of motion along a line is one of the traditional applications of the calculus. It is interesting to discover how much detailed information can be gleaned simply from a position function and its derivatives. In Goal 45 we will use the integral, a tool of the calculus as valuable as the derivative, to further our understanding of motion along a line. Differential equations, to be considered in Chapter 8, afford additional insight.

VELOCITY AND ACCELERATION

As noted in Goal 17, if $f(t)$ is the position of a particle P (say, on a coordinate axis) at time t, then the velocity and acceleration of P at time t are, respectively,

$$v(t) = f'(t) \quad \text{and} \quad a(t) = f''(t)$$

In this Goal all position functions and their derivatives are continuous everywhere.

The following statements tell what information the derivatives furnish about the motion of a particle.

(1) If $v(t) > 0$ over some time interval, then the particle P is moving to the right over that interval; if $v(t) < 0$, then P is moving to the left. Since $v(t)$ is the derivative of $f(t)$, therefore when $v(t) > 0$, $f(t)$ is increasing. If, for example, $t_1 < t_2$ and $f(t_1) = 3$, then $f(t_2) > 3$. Or if $f(t_1) = -4$, then $f(t_2) > -4$.

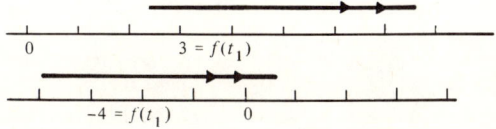

(2) When $t = t_0$, the speed of P is $|v(t_0)|$. For example, if $v(t) = t^3 - 4t^2 + 5$, then the speed when $t = 1$ is $|(1)^3 - 4(1)^2 + 5|$, or 2; the speed when $t = 2$ is $|(2)^3 - 4(2)^2 + 5| = |-3| = 3$.

(3) If $v(t_0) = 0$, then P is at rest when $t = t_0$.

(4) If P reverses direction at $t = t_0$, then v changes sign at t_0; if v changes sign at $t = t_0$, then P reverses direction at t_0. If v changes sign (and P reverses direction) at $t = t_0$, then $v(t_0) = 0$. But the converse of this last statement is not true. That is, there can be a time t_0 such that $v(t_0) = 0$ but v does not change sign (and P does not reverse direction). Suppose, for example, that $v(t) = (t - 1)^2(t - 4)$. Then $v(1)$ and $v(4)$ both equal zero; v changes sign, and P reverses direction, at $t - 4$, but v does not change sign and P does not reverse direction at $t = 1$.

(5) If $v(t_0)$ and $a(t_0)$ have the same sign, then P is *speeding up* at time t_0; if they have opposite signs, then P is slowing down. There are four cases to consider:

(i) $v > 0$ and $a > 0$: P is moving to the right with increasing speed.

(ii) $v > 0$ and $a < 0$: P is moving to the right with decreasing speed.

(iii) $v < 0$ and $a > 0$: P is moving to the left with increasing velocity but with decreasing speed.

(iv) $v < 0$ and $a < 0$: P is moving to the left with decreasing velocity but with increasing speed.

In cases (iii) and (iv), remember that speed is the *magnitude* of velocity. So, for example, if $v(t_0) = -5$ and $a(t_0) > 0$, as in case (iii), then P is slowing down, while if $v(t_0) = -5$ and $a(t_0) < 0$, as in case (iv), then P is speeding up.

Example 29-1

A particle P is propelled along a coordinate axis so that its position at time t is $f(t) = t^3 - 6t^2 + 9t + 5$, $t \geq 0$. (a) Express the velocity, acceleration, and speed of P as functions of time. (b) Where is P when $t = 0$? (c) When is $f(t)$ increasing? decreasing? (d) When does P reverse direction? (e) When is v increasing? decreasing? (f) When is P speeding up? slowing down? (g) Show the motion of P on a graph.

(a) $v(t) = 3t^2 - 12t + 9$; $a(t) = 6t - 12$; the speed is $|v(t)| = |3t^2 - 12t + 9|$.

(b) When $t = 0$, $f(t) = 5$.

(c) $f(t)$ increases when $v(t) > 0$, decreases when $v(t) < 0$. Since $v(t) = 3(t^2 - 4t + 3) = 3(t - 1)(t - 3)$, we can use the techniques of Goal 19, page 189, to determine the sign of v in the relevant intervals:

$$\begin{array}{ll} t < 1 & v > 0 \\ 1 < t < 3 & v < 0 \\ 3 < t & v > 0 \end{array}$$

Goal 29 | To Solve Problems Involving Motion Along a Line

So f increases if $0 < t < 1$ or if $t > 3$, decreases if $1 < t < 3$.

(d) Since $v = 0$ when $t = 1$ or 3, and v changes sign at both times, P reverses direction at both times.

(e) Since v increases when $a(t) > 0$, decreases when $a(t) < 0$, and $a(t) = 6(t - 2)$, v increases if $t > 2$, decreases if $t < 2$.

(f) Since $a(t)$ changes sign at $t = 2$ while $v(t)$ changes sign both at $t = 1$ and at $t = 3$, we consider the following intervals:

Interval	Sign of v	Sign of a	Behavior of Particle
$0 < t < 1$	+	−	slowing down
$1 < t < 2$	−	−	speeding up
$2 < t < 3$	−	+	slowing down
$3 < t$	+	+	speeding up

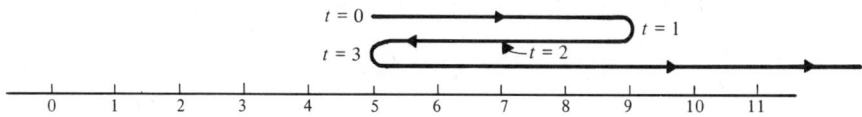

(g) Note that $f(1) = 9$, $f(2) = 7$, $f(3) = 5$.

So the particle starts at position 5 on the axis, moving to the right and slowing down until it gets to position 9, where it reverses direction and speeds up until it reaches position 7. It then slows down again until it gets to position 5, when it once again reverses direction, then speeds up forever.

Example 29-2

From the top of an 80-ft-high flagpole a stone is thrown directly up with

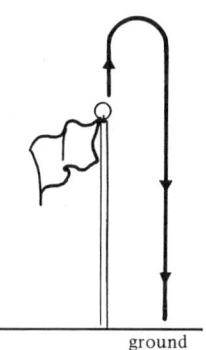
ground

such a velocity that its position after t sec is $f(t) = -16t^2 + 64t + 80$ ft above the ground. (a) Express the velocity, acceleration, and speed of the stone as functions of time. (b) What is the initial velocity? (c) When is the stone rising? falling? (d) What is the maximum height the stone attains? (e) At what instant does it pass the top of the flagpole on its way down? with what velocity? (f) When does it strike the ground? with what velocity? (g) Exhibit the motion of the stone pictorially.

(a) $v(t) = -32t + 64$; $a(t) = -32$ ft/sec² (the constant of gravity); the speed is $|-32t + 64|$.

(b) $v(0) = -32(0) + 64 = 64$ ft/sec, the initial velocity with which the stone is thrown upwards.

(c) The stone rises when $v > 0$, falls when $v < 0$. Since $v(t) = -32(t - 2)$, the stone rises when $t < 2$, falls when $t > 2$. It therefore rises during the first two seconds, then reverses its direction.

(d) From (c) we see that the stone reverses its direction when $t = 2$. At this time it attains its maximum height, $f(2)$, which equals 144 ft.

(e) It passes the top of the flagpole when its height above ground is precisely 80 ft. We therefore solve the equation $f(t) = 80$:

$$-16t^2 + 64t + 80 = 80$$
$$-16t^2 + 64t = 0$$
$$-16t(t - 4) = 0$$

The solution $t = 0$ appears because initially the stone was at a height of 80 ft. The stone is again at a height of 80 ft, this time on its way down, when 4 sec have elapsed. Its velocity then is $v(4) = -64$ ft/sec. Note that its *speed* at this height is the same on the way up and on the way down.

(f) The stone strikes the ground when its height (above ground) is zero; so we solve the equation $f(t) = 0$:

$$-16t^2 + 64t + 80 = 0$$
$$-16(t^2 - 4t - 5) = 0$$
$$-16(t + 1)(t - 5) = 0$$

So $f(t) = 0$ when $t = -1$ or 5. Since the domain of f here is nonnegative t, the stone hits the ground after 5 sec. Its velocity at the moment of impact is $v(5) = -96$ ft/sec.

(g)

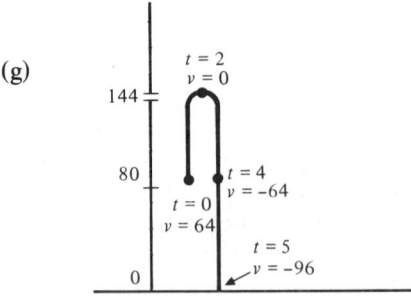

Goal 29 | To Solve Problems Involving Motion Along a Line

EXERCISES

1. For each of the following position functions ($t \geq 0$), find the initial position, when and where the velocity is zero, and when and where the object changes direction. Exhibit these results on an appropriate picture.
   (a) $f(t) = t^3 - 12t^2 + 45t + 6$
   (b) $f(t) = 2t^3 - 15t^2 + 24t + 10$
   (c) $f(t) = t^2 - 6t + 1$
   (d) $f(t) = 2t^3 + 3t^2 - 36t + 20$
   (e) $f(t) = 3t^4 - 20t^3 + 48t^2 - 48t + 10$

2. In parts (b) and (c) of Exercise 1, find where the object is speeding up and where it is slowing down.

3. A stone is thrown straight up (or down) from the top of a building. Its height above the ground in feet after t sec is given in turn by each of the following functions. For each, find the height of the building, the initial velocity of the stone, the maximum height it attains, and the velocity with which it strikes the ground.
   (a) $f(t) = -16t^2 + 128t + 320$ (d) $f(t) = -16t^2 + 160t + 3200$
   (b) $f(t) = -16t^2 + 96t + 880$ (e) $f(t) = -16t^2 - 5t + 425$
   (c) $f(t) = -16t^2 + 16t + 32$

4. For each of the following pairs of position functions (with $t \geq 0$), determine at which times the particles have the same position and at which times they have the same velocity.
   (a) $f(t) = t^2 + 5t - 4$
   $g(t) = 2t^2 - 3t + 3$
   (b) $f(t) = 3t^2 - t + 1$
   $g(t) = t^2 - 2t + 2$

5. Mr. Wright is pushed from the top of a 150-ft building; his height above ground t sec later is $150 - 16t^2$ ft. Two sec after he starts falling, Superman arrives and leaps from the top of the building at such a speed that his height above ground m sec after leaping is $150 - 128m - 16m^2$ ft. Will Superman catch Mr. Wright in time; and if so, how far above the ground will the rescue take place? (HINT: What is the relation between t and m?)

TEST A

1. For the position function $f(t) = 2t^3 - 21t^2 + 60t + 5$, determine the initial position, when and where the object changes direction, and when it is speeding up. Exhibit its motion in an appropriate picture.

2. From the top of a 96-ft-high building a ball is thrown straight up with

such a velocity that its height above ground t sec later is $f(t) = -16t^2 + 80t + 96$ ft. What is the maximum height achieved, and with what velocity does the ball strike the ground?

TEST B

1. For the position function $f(t) = t^3 - 3t + 1$, determine the initial position, when and where the object changes direction, and when it is slowing down. Exhibit its motion in an appropriate picture.

2. From the edge of a 160-ft-high cliff a stone is thrown straight up with such a velocity that its height t sec later is $f(t) = -16t^2 + 48t + 160$ ft. What is the maximum height achieved, and with what velocity does the stone strike the ground?

STUDY QUESTIONS

1. In Example 29-2 (p. 279) suppose that we set up a coordinate axis with the top of the flagpole at the origin. The position function then becomes $f(t) = -16t^2 + 64t$. Rework the various parts of the example and show that the same conclusions are reached.

2. The motion of a ball thrown upward from ground level with an initial velocity v_0 is described by the position function $f(t) = -16t^2 + v_0 t$. Show that the ball takes the same length of time going up as it does coming down.

3. With what initial velocity must an object be thrown straight up in order to achieve a height of 1 mile? (HINT: The position function is $f(t) = -16t^2 + v_0 t$; when the velocity is zero, the height must be 5280 ft.)

GOAL **30**

To apply the derivative to problems in economics

Economists and business administrators have been turning more and more to the use of mathematical techniques in a variety of decision-making situations. Numerous applications involve rates of change; for these, the methods of the differential calculus naturally are especially useful. Here we shall trace through a simplified version of a typical decision-making problem. We will use the techniques of the calculus developed so far to provide further insight into the economic concepts of revenue, cost, profit, and demand and to help in reaching managerial decisions.

APPLYING THE CALCULUS TO REAL PROBLEMS

As has been noted before, the functions arising in business and economics are usually defined only for positive integers (companies manufacture and sell only whole numbers of units!). To apply the calculus, however, we must assume that the various functions are defined continuously over the positive numbers. We must then allow for this assumption, when we return to the real-life situation, by interpreting the mathematical solution so as to yield answers that are integers.

In a problem to be considered shortly, involving optimal order size (what quantity should be ordered per shipment to minimize total costs?), we will make several assumptions of somewhat dubious validity in some practical applications but of unquestionable value in simplifying the mathematical solution. For example, one assumption is that the demand for a product does not fluctuate with the season: the number sold in January is the same as in August. What if the product is a surfboard? a toboggan? an air-conditioner? A second simplifying assumption is that a new shipment arrives at the warehouse just as the preceding one has been sold out. (Not only convenient, but ideal! No supplier would ever run out, nor would any dealer ever have to stockpile.) Still another assumption is that costs (for materials, machinery, labor, insurance, storage, etc.) do not vary during the period of time involved in the problem. (This has overtones of wishful thinking.)

Despite such assumptions, the simplified presentation can be very instructive to students of elementary calculus. Moreover, understanding the simplified techniques is a necessary step toward mastering the more complicated ones that are found in real business problems.

In addition to these observations, we note that students often wonder how functions for cost, revenue, profit, and so on are "cooked up." A mathematics consultant may examine all the data a company or business can gather on fixed and variable costs incurred in producing a product and on the demand for and sales of the product. The consultant will then try to fit these data as closely as possible with approximating functions—functions that make it easier (since the efficient methods of the calculus can now be applied) to obtain various items of information: how an increase in wages might affect the price of the product, or how a change in costs, price, or demand might affect the company's revenue and profit. Of course, the better the approximating functions fit the data, the better can be the predictions of future activity.

In this text the functions we present are truly contrived; indeed, we invent them. Since the companies involved are fictitious, the functions we assign to their cost, revenue, profit, and so on are, at best, reasonable guesses. Perhaps of even more significance, the functions are sufficiently simplified so as not to obscure the mathematics being expounded. Nonetheless, as mentioned above, problems of the sort we are about to discuss can provide a basic and worthwhile introduction on how mathematics may be applied.

REVENUE

The Lectrue Company is faced with production-level decisions in a small subsidiary plant manufacturing electronic units. Current market conditions are such that the price, in dollars per unit, at which x units can be sold is given by the function

$$p(x) = 10 - 0.001x \qquad (0 < x \leq 10{,}000)$$

As expected, to sell more units the price must be lowered. [Note that $p(x)$ is linear and has negative slope.] The revenue R from this plant is the product of the number of units sold and the price per unit; so

$$R(x) = xp(x) = x(10 - 0.001x) = 10x - 0.001x^2$$

What production level will maximize Lectrue's revenue?

Using the techniques of Goal 21, we find where $R'(x)$ is zero:

$$R'(x) = 10 - 0.002x$$

and this is zero when $x = 5000$. We note that at production levels under

5000 units, $R'(x) > 0$ and an increase in production results in increased revenue; at production levels over 5000 units, $R'(x) < 0$ and an increase in production decreases revenue. The revenue is therefore maximized when 5000 units are produced.

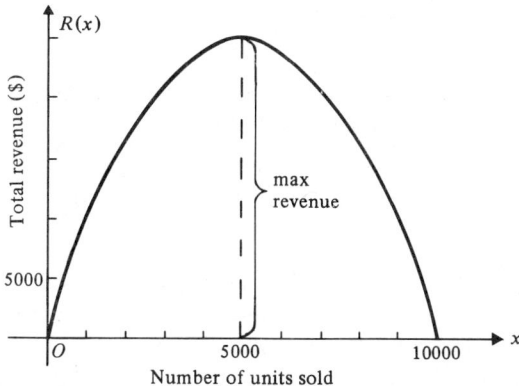

COST

A decision about level of production is usually not based only on considerations of revenue. It is also important to note the cost of production. Suppose, for the Lectrue plant, that the total cost in dollars of producing x units a year is given by the function

$$C(x) = 4900 + 1.2x + 0.0001x^2$$

The first term in the expression for $C(x)$ represents the fixed yearly investment, $4900, in equipment and buildings. There is also a basic production cost of $1.20 per unit for salaries and materials. The last term, $0.0001x^2$, is a measurement, in dollars, of the expenditures for repairs resulting from more frequent breakdowns of equipment when larger quantities are produced. In Goal 28, on the use of differentials for approximation, we noted that $C'(x)$, the marginal cost,* indicates how small changes in the number of items produced at a given production level, affect the total cost. Since

$$C'(x) = 1.2 + 0.0002x$$

it follows that

$$C'(5000) = 1.2 + 0.0002(5000) = \$2.20$$

So, when the production level is 5000 units, it costs approximately $2.20 to produce an additional unit.

---

*The term "marginal," which has already appeared very frequently, seems most appropriate when interpreted as "additional." Economists regard the marginal cost as the additional cost of an additional unit (at a specified production level), the marginal revenue as the additional revenue from the sale of an additional unit, and so on.

Average Cost

In Example 14-6, p. 143, we defined the average cost per unit, A, as the total cost divided by the number of units. Thus, for the Lectrue plant,

$$A(x) = \frac{C(x)}{x} = \frac{4900 + 1.2x + 0.0001x^2}{x} = \frac{4900}{x} + 1.2 + 0.0001x$$

What level of production minimizes average cost? To answer this, we find where $A'(x)$, the marginal average cost, equals zero:

$$A'(x) = -\frac{4900}{x^2} + 0.0001 = \frac{-4900 + 0.0001x^2}{x^2}$$

which equals zero when $x = 7000$ (we discard the negative critical value). When production is under 7000 units, $A'(x) < 0$ and an increase in production decreases the average cost per unit. When production is over 7000 units, $A'(x) > 0$ and an increase in production increases average cost. The average cost is therefore minimized when 7000 units are produced.

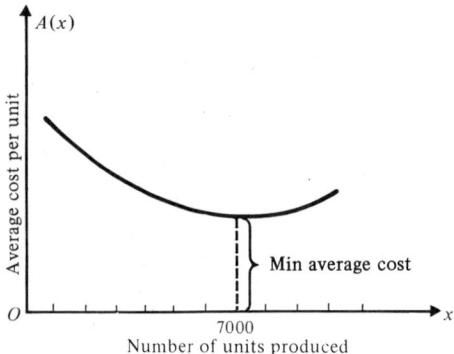

PROFIT

What decision should Lectrue now make? Should the production level be 5000 units (where the revenue is maximized) or 7000 units (where the average cost is minimized)? Or should it be something else? What is ultimately of interest to management, of course, is the profit. Lectrue wants to maximize its profit. The profit, the difference between revenue and cost, is given by

$$P(x) = R(x) - C(x) = (10x - 0.001x^2) - (4900 + 1.2x + 0.0001x^2)$$
$$= -4900 + 8.8x - 0.0011x^2$$

What production level will maximize Lectrue's profit? We must determine when $P'(x)$, the marginal profit, equals zero. We have

$$P'(x) = 8.8 - 0.0022x$$

Goal 30 | To Apply the Derivative to Problems in Economics

which is zero when $x = 4000$. When fewer than 4000 units are produced, $P'(x) > 0$ and an increase in production yields increased profit; when more than 4000 units are produced, $P'(x) < 0$ and an increase in production results in decreased profit. The profit for Lectrue is thus maximized when 4000 units are produced. This is the best production-level decision Lectrue can make, based on the information given above.

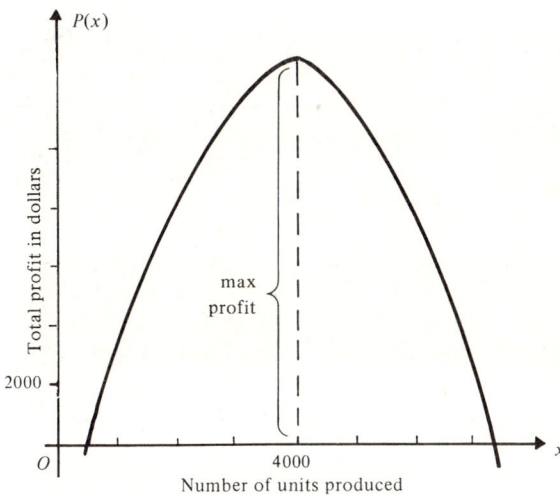

Relevant Data

For completeness we calculate here the values of the economic functions at the production-level which maximizes profit, namely $x = 4000$:

price: $p(4000) = 10 - 0.001(4000) = \$6$

revenue: $R(4000) = 10(4000) - 0.001(4000)^2 = \$24{,}000$

marginal revenue: $R'(4000) = 10 - 0.002(4000) = \$2$ per unit

cost: $C(4000) = 4900 + 1.2(4000) + 0.0001(4000)^2 = \$11{,}300$

marginal cost: $C'(4000) = 1.2 + 0.0002(4000) = \$2$ per unit

average cost: $A(4000) = \dfrac{4900}{4000} + 1.2 + 0.0001(4000) = \$2.825$ per unit

$\left[\text{this is the same as } \dfrac{C(4000)}{4000} \right]$

profit: $P(4000) = R(4000) - C(4000) = -4900 + 8.8(4000) - 0.0011(4000)^2 = \$12{,}700$

marginal profit: $P'(4000) = 8.8 - 0.0022(4000) = \$0$

Note that when the profit is maximized, the marginal revenue $R'(x)$ is equal to the marginal cost $C'(x)$. This is to be expected, since to determine the level for maximum profit we set $P'(x)$, which equals $R'(x) - C'(x)$, equal to zero.

DEMAND: ELASTICITY

In the discussion so far, the price has been given as a function $p(x)$ of the number of units. We can also regard the number of units to be a function $D(p)$ of the price (see the discussion of inverse functions in Goal 5). Recall that $D(p)$ is the *demand function*. Revenue can now also be expressed as a function of price:

$$R(p) = \text{(price per unit)} \times \text{(number of units)} = pD(p)$$

If the graph of $R(p)$ is of the form ⌢ then we can maximize the revenue by determining when $R'(p) = 0$. By the product rule (Goal 14),

$$R'(p) = D(p) + pD'(p)$$

We now rewrite $R'(p)$ by engaging in some algebraic manipulation:

$$R'(p) = D(p) + D(p)\frac{p}{D(p)} D'(p)$$

$$= D(p)\left[1 + \frac{p}{D(p)} D'(p)\right] = D(p)\left\{1 - \left[-\frac{p}{D(p)} D'(p)\right]\right\}$$

Note that $R'(p) = 0$ when $-\dfrac{p}{D(p)} D'(p) = 1$. The quantity

$$E(p) = -\frac{p}{D(p)} D'(p)$$

is called the *elasticity of demand*. It is a measure of the percentage change in demand with respect to the percentage change in price (see Study Question 6). The minus sign in the expression for $E(p)$ above is just a convenience: it guarantees that $E(p)$ is never negative. This is because the quotient $p/D(p)$ is never negative and $D'(p)$ is never positive [in fact, $D'(p)$ is usually negative because demand normally decreases as the price increases].

We can now rewrite $R'(p)$:

$$R'(p) = D(p)[1 - E(p)]$$

There are three cases to be considered.

(1) $E(p) > 1$. In this case $R'(p) < 0$, implying that revenue decreases with small increases in price. Since the decrease in demand is proportionately *greater* than the increase in price, the demand is significantly affected by price changes. The demand is said to be *elastic*.

Goal 30 | To Apply the Derivative to Problems in Economics

(2) $E(p) < 1$. So $R'(p) > 0$, which implies that revenue increases with small increases in price. In this case the decrease in demand is proportionately *less* than the increase in price. Since the demand is relatively insensitive to changes in price, economists say that the demand is *inelastic*.

(3) $E(p) = 1$.

(3a) Suppose $E(p)$ equals 1 for only one value of p, say p_0. Then $R'(p_0) = 0$ and p_0 is a critical value of R. If $R''(p_0) < 0$, then $R(p_0)$ is the maximum revenue.

(3b) If $E(p)$ is the constant function 1, then $R'(p) = 0$ for all p and $R(p)$ is constant, say k. Then $R(p) = pD(p) = k$ and

$$D(p) = \frac{k}{p}$$

The demand in this case is thus inversely proportional to the price.

BACK TO LECTRUE

We recall that for Lectrue the price function is

$$p(x) = 10 - 0.001x$$

Solving for x in terms of p yields the demand function

$$x = D(p) = \frac{10 - p}{0.001} = 1000(10 - p)$$

Note that the domain of D is $p < 10$. (Why?) The elasticity of demand can now be determined:

$$E(p) = -\frac{p}{D(p)} D'(p) = -\frac{p \cdot 1000(-1)}{1000(10 - p)} = \frac{p}{10 - p}$$

We now consider the three cases above for this particular elasticity function.

(1) $E(p) > 1$: $\dfrac{p}{10 - p} > 1$ if $p > 10 - p$, that is, if $p > 5$.

(2) $E(p) < 1$: if $p < 5$.

(3a) $E(p) = 1$: if $\dfrac{p}{10 - p} = 1$, that is, if $p = 10 - p$, $p = 5$.

We conclude:

When the price is less than $5, as in case (2), small increases in price will increase the revenue (the demand is inelastic).

When the price is exactly $5 the revenue is maximized.

When the price is greater than $5, as in case (3), small increases in price will result in decreased revenue (the demand is elastic).

Note also that when the price is $5, the demand is

$$D(5) = 1000(10 - 5) = 5000 \text{ units}$$

This is, of course, precisely the answer we obtained earlier when we sought the production level x that would yield maximum revenue.

INVENTORY CONTROL

Suppose that the Eastern regional distributor of Lectrue products knows that she can sell 10,000 electronic units a year in her district and that she can sell them at a constant rate throughout the year. (It takes several small subsidiary plants to meet her demand.) Aside from the cost per unit that Lectrue charges, the distributor must also take into account both storage and shipping costs. The *storage costs* include those for warehouse space, for fire and theft insurance, and so on; the *shipping costs* include delivery and labor charges that accompany each order, paperwork, and so on. The distributor is now faced with a decision. If she orders all 10,000 units at one time, then her storage costs will be unnecessarily large; if, at the other extreme, she places very many small orders, then her shipping costs will be unduly great. How many units should she order each time to minimize her total annual cost?

Suppose that Lectrue charges $5 for each electronic unit, that it costs the distributor 20¢ to store one unit for a year, and that each shipment of units costs $10.

In addition to the assumption above that the merchandise can be sold at a constant rate, we impose the following (simplifying!) assumptions:*

(1) All costs involved are constant.

(2) The shipping cost is the same for every order, no matter what the size of the order may be.

(3) A new order arrives just when the preceding order has been sold out.

(4) All the orders are of the same size.

Average Inventory

If there are x units in each shipment, then, as the figure shows, assumptions (2) and (3) imply that the average inventory over the year is equal to $x/2$ units. In computing our distributor's annual storage costs we will therefore assume that $x/2$ electronic units are in storage for the whole year.

---

*See comments on these assumptions on page 283.

Goal 30 | To Apply the Derivative to Problems in Economics

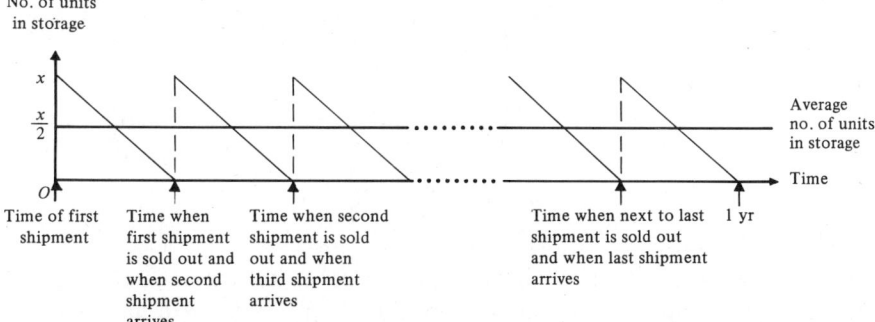

Minimizing Annual Cost

The total yearly cost $C(x)$ is the sum of three costs:

$$C(x) = \underset{\text{10,000 units}}{\underset{(1)}{\text{the cost of}}} + \underset{\text{storage costs}}{\underset{(2)}{\text{the yearly}}} + \underset{\text{shipping costs}}{\underset{(3)}{\text{the yearly}}}$$

(1) At $5 a unit, 10,000 units cost $50,000.

(2) At 20¢ per unit for a year, the storage costs for $x/2$ units is (20¢)($x/2$), or $10x$¢.

(3) Since there are x units in each shipment and a total of 10,000 units annually, there are $10,000/x$ shipments per yr. At $10 per shipment, the yearly shipping costs total $10(10,000/x)$.

In dollars, then,

$$C(x) = 50,000 + 0.10x + \frac{100,000}{x}$$

where the domain of C is $0 < x \leq 10,000$.

We minimize $C(x)$ by finding $C'(x)$, setting it equal to zero, and solving:

$$C'(x) = 0.10 - \frac{100,000}{x^2}$$

$$0.10 - \frac{100,000}{x^2} = 0$$

$$1 - \frac{1,000,000}{x^2} = 0$$

$$x^2 = 1,000,000$$

$$x = \pm 1000$$

We discard -1000 as meaningless. Since

$$C''(x) = \frac{2{,}000{,}000}{x^3} \quad \text{and} \quad C''(1000) > 0$$

we conclude that $x = 1000$ yields minimum cost. Based on our assumptions, then, the distributor should order 10,000/1,000 or 10 shipments, each of 1000 units, to be delivered at equal intervals throughout the year.

EXERCISES

1. In each of the following, the price in dollars per unit is given as a function of the number of units to be sold. Determine the revenue function, the marginal-revenue function, and the production level that maximizes revenue.

   (a) $p(x) = 5 - 0.025x$

   (b) $p(x) = 8 - 0.00002x$

   (c) $p(x) = 12 - 0.0006x$

   (d) $p(x) = \dfrac{10{,}000}{x}$

2. In each of the following, the total cost in dollars is given as a function of the number of the number of units produced. Determine the marginal-cost function, the average-cost function, and the production level that minimizes average cost.

   (a) $C(x) = 18 + 2x + 0.005x^2$

   (b) $C(x) = 6000 + 3x + 0.00006x^2$

   (c) $C(x) = 10{,}000 + 5x + 0.0001x^2$

   (d) $C(x) = 30{,}250 - 10x + 0.001x^2$

3. Pairing the corresponding parts in Exercises 1 and 2 above, determine the profit function, the marginal-profit function, the production level that maximizes profit, and the maximum profit.

4. For each of the following demand functions, find the expression for elasticity and determine when the demand is elastic, when inelastic.

   (a) $D(p) = 500(3 - p)$ $0 < p < 3$

   (b) $D(p) = 1000 - 250p$ $0 < p < 4$

   (c) $D(p) = 24 - 2p^2$ $0 < p < 3.46$

   (d) $D(p) = \dfrac{100}{p}$ $0 < p$

   (e) $D(p) = 3600 - 120p + p^2$ $0 < p < 60$

5. For each demand function in Exercise 4, use the expression for elasticity of demand to determine at which price the revenue will be maximized.

6. A candy manufacturer knows that the price in dollars at which x 2-lb

Goal 30 | **To Apply the Derivative to Problems in Economics** 293

boxes of chocolates can be sold is given by the function $p(x) = 8 - 0.0001x$. The cost in dollars of producing x boxes is $C(x) = 10{,}000 + 2x + 0.00015x^2$. What production level will maximize profit? What price corresponds to this?

7. A pizza maker knows that to sell x pizzas a week he can charge only $(5 - 0.003x)$ dollars for each. If his cost for producing x pizzas is $(250 + x + 0.002x^2)$ dollars, what is the maximum weekly profit he can obtain?

8. Suppose that Lectrue charges our regional distributor $100 (instead of $5) for each electronic unit, but that all other costs given on page 291 are unchanged. How will this affect the distributor's decision about the number of units to order per shipment, if her goal is still minimum total annual cost?

9. A musical instrument company manufactures plastic clarinets that it ships to Woody Wind, a distributor. Wind knows he can sell 500 clarinets a year at a constant rate. If he pays the manufacturer $40 per instrument and if he has storage costs of $1.50 per instrument per year and shipping costs of $15 per order, what is the optimal order size (the size leading to minimum overall cost)? How frequently should the clarinets be ordered? What is the total minimum annual cost?

10. The Goodwear Tire Company can sell 20,000 truck tires a year at a constant rate. It costs the company $100 to get its equipment ready for each production run, regardless of the size of the run. If it costs the firm $15 to produce a tire and $4 to store it for a year, how many tires should it manufacture during each run so as to minimize its total yearly cost? How many production runs a year should it plan for?

TEST A

1. Suppose that x units can be sold at a price of $(7 - 0.001x)$ dollars per unit and that the total cost of producing x units is $(1000 + 2x + 0.00025x^2)$ dollars. What production level maximizes revenue? What level minimizes average cost? What level maximizes profit? What is the maximum profit and what is the corresponding unit price?

2. Suppose that a demand function is given by $D(p) = 1500(5 - p)$. What is the elasticity-of-demand function, and how will the revenue be affected by an increase in price if the current price is $3?

3. A distributor can sell 1000 CB radios a year at a constant rate. He pays the manufacturer $8 for each radio and has storage charges of $1 per radio per yr and shipping costs of $20 per shipment. How many radios should he order per shipment to minimize his total yearly cost?

TEST B

1. Suppose that x units can be sold at a price of $(5 - 0.0002x)$ dollars per unit and the total cost of producing x units is $(1800 + x + 0.0002x^2)$ dollars. What production level maximizes revenue, what level minimizes average cost, and what level maximizes profit? What is the maximum profit and the corresponding unit price?
2. Suppose that a demand function is given by $D(p) = 3000(8 - p)$. What is the elasticity-of-demand function, and how will the revenue be affected by a small increase in price if the current price is $3?
3. A wholesaler can distribute 4000 sets of stainless steel knives a year in his area. If the manufacturer charges him $14 a set and if he has storage costs of 80¢ per set per yr and shipping costs of $16 per order, how many sets should he order at a time to minimize his total annual cost?

STUDY QUESTIONS

1. If a firm is small the number of units sold has little effect, if any, on the market price. In this situation the price is constant. What can be said about the marginal revenue?
2. The average revenue, defined to be the total revenue divided by the number of units, is the same as what other term defined in this goal? Note that we have

$$A(x) = \frac{R(x)}{x}; \qquad A'(x) = \frac{xR'(x) - R(x)}{x^2}$$

Using this, give a justification for the statement "the average revenue is maximized when the average revenue equals the marginal revenue."
3. Suppose that the cost function is linear; that is, $C(x) = ax + b$ ($a > 0$, $b > 0$). Sketch the graphs of the average-cost function, the marginal-cost function, and the marginal-average-cost function.
4. Suppose that the cost function is quadratic, of the type $C(x) = ax^2 + bx + c$ with $a, b, c > 0$. Sketch the graphs of the marginal-cost and average-cost functions.
5. Give a justification for the statement "the average cost is minimized when the average cost equals the marginal cost." HINT: see Question 2 above and Study Question 4 of Goal 14.
6. Suppose a small change Δp in the price p leads to corresponding change ΔD in the demand D. The percentage changes in price and

Goal 30 | To Apply the Derivative to Problems in Economics

demand are proportional to $\dfrac{\Delta p}{p}$ and $\dfrac{\Delta D}{D}$, respectively. The ratio of these changes (in the other order) is

$$\frac{\dfrac{\Delta D}{D}}{\dfrac{\Delta p}{p}} = \frac{p}{D} \frac{\Delta D}{\Delta p}$$

Take the limit as $\Delta p \to 0$ and justify the statement that the elasticity of demand is a measure of the percentage change in demand with respect to a percentage change in price.

Summary of Goals 25–30

You should now be able to

(1) translate a word problem calling for a max or min into mathematical terms, that is, write an expression for the function to be maximized or minimized;

(2) use the techniques of the calculus to ascertain where the function in (1) is a max or min;

(3) solve a given max or min problem;

(4) solve a related-rate problem: one in which there is a relationship involving two or more variables each of which is a function of time;

(5) given a particular function on a specified interval, find any points promised by the Mean Value Theorem, or explain why the theorem does not apply;

(6) find $f(x)$ if, say, $f'(x) = 4x^3$ and $f(2) = 5$;

(7) use differentials to

  (7a) approximate numbers like $\sqrt{4.2}$, $\sqrt[3]{7.9}$, $1/\sqrt{26}$;

  (7b) approximate the value of a function (for example, cost, volume, demand) at a particular point, given the value of the function at a nearby point;

(8) given a function for the position of a particle moving along a line,

  (8a) find what its velocity and acceleration are, when and where it reverses direction, when the particle is speeding up or slowing down;

  (8b) exhibit its motion in an appropriate picture;

(9) given the height of an object t sec after it is thrown up (or down), find what its maximum height is, when the latter is attained, when the object hits the ground, and what its velocity is at impact;

(10) solve problems in economics involving maximum revenue or profit or minimum average cost;

(11) use a given demand function for a product to find the elasticity of demand and to determine from the latter how revenue is affected by a small increase in the price of the product;

(12) solve problems on inventory control: given appropriate information about, say, annual storage and shipping costs of a product, determine the number of shipments annually leading to minimum total cost.

Review Exercises

GOAL 25

1. If a rectangular garden is to be fenced off on three sides using 200 ft of fencing, what should the dimensions of the garden be for maximum area?
2. In Exercise 18 on page 68, how many passengers will maximize Dayline Delight's revenue?
3. In Exercise 19 on page 68, what should be the dimensions of the carport floor to minimize the total cost of the carport?

GOAL 26

4. Two planes are scheduled to fly to Chicago. At noon, one is 1300 miles due west of Chicago and is approaching at 500 mph; the other is 1000 miles due south and is approaching at 300 mph. At what rate is the distance between the planes decreasing at 2 P.M.?
5. A spherical balloon is losing air at the rate of 8 cu in. per min. At what rate is its radius decreasing when the radius is 3 in.? ($V = \frac{4}{3}\pi r^3$)

GOAL 27

6. For the function $f(x) = x^2 - x + 2$ on the interval $0 \leq x \leq 2$, find any points promised by the MVT or explain why there are none.
7. Find $f(x)$ if $f'(x) = 3x^2$ and $f(-1) = 1$.

GOAL 28

8. Use differentials to approximate $\sqrt{8.9}$.
9. How much (approximately) will it cost to manufacture 10 more units if the cost of making x units is $6000 + 4x + 0.001x^2$ dollars and the current production level is 500 units?
10. Find dy if
    (a) $y = x^4 - 1/x + 2$ (b) $y = \sqrt{1 - x^2}$

GOAL 29

11. For the position function $f(t) = t^3 - 3t^2 - 9t + 20$ find the initial position, when and where the particle changes direction, and when it is speeding up. Show these results in an appropriate picture.

12. A ball is thrown straight up from the top of a flagpole so that its height above the ground in feet t sec later is $f(t) = -16t^2 + 64t + 192$. How high does the ball go, and how long does it take it to hit the ground?

GOAL 30

13. Suppose that x units of a product can be sold at a price, in dollars, of $6 - 0.002x$, and that the total cost of producing x units is, in dollars, $2000 + x + 0.0005x^2$. What production level maximizes revenue? What level maximizes profit? What level minimizes average cost?

14. Find the elasticity-of-demand function if the demand function is $D(p) = 1000(6 - p)$. If the current price is $4, how will the revenue be affected by a small increase in price?

15. A distributor of Dye-Vann furniture can sell 180 couches a year at a constant rate. If he pays the manufacturer $80 per couch, has storage costs of $8 per couch per yr, and shipping costs of $20 per order, how many shipments a year will minimize total yearly cost?

Chapter Test A

1. To make a box of volume 12 cu in. with square base, a manufacturer uses material costing 2¢ per sq in. for the bottom and 1¢ per sq in. for the sides and top. If other costs come to 5¢ a box, what dimensions will minimize his total cost?

2. An author knows he can sell 500 copies of his book *Get Rich Quick* at $15 each. He also knows he can sell 100 additional copies for each $1 reduction in the price per copy. What price per copy will maximize his income?

3. If a balloon is being inflated with gas at the rate of 2 cu ft per min, at what rate is the radius increasing when it is 4 ft? ($V = \frac{4}{3}\pi r^3$)

4. If a factory makes a profit, in dollars, of $P(x) = 0.01x^2 + 25x - 1000$ on x gadgets and its output after t h is $x(t) = 10t - 5$, at what rate is its profit changing after 4 h?

5. In each of the following, find any points promised by the MVT or explain why there are none.

(a) $f(x) = \dfrac{1}{x}$ on $1 \leqq x \leqq 3$

(b) $f(x) = \sqrt{x}$ on $0 \leqq x \leqq 4$

6. By about how much does the function $f(x) = x + 1/x$ change if x changes from 2 to 2.1?
7. Find dy if $y = \sqrt{x^2 + 5}$.
8. The velocity of an object moving along a line is given by

$$v(t) = (t - 1)^2(t - 2)$$

(a) For what values of t, if any, does the object reverse direction?
(b) Find the object's acceleration when $t = 2$.

9. If it costs, in dollars, $1800 + 2x + 0.0002x^2$ to produce x units, what is the minimum average cost?
10. A wholesaler can sell 2000 calculators a year in his region at a constant rate. If he pays the manufacturer $9 per calculator, if his storage costs are 25¢ per calculator per yr, and if each shipment costs him $10, how many calculators should he order in each shipment to minimize his yearly inventory costs?

Chapter 6

Exponential and Logarithmic Functions

So far we have restricted our attention to algebraic functions, especially polynomials and other rational functions, and to applications involving them. In this chapter we consider two different functions that are very closely related: the exponential function and the logarithmic function. These important functions are needed to solve problems of growth and decay that arise both in nature and in business and social science. An example of growth is the increase in a bank balance as interest is paid; an example of decay is the decrease in the amount of a radioactive substance as time passes.

We begin by reviewing the basic properties of exponents and logarithms and of functions involving them. Next we introduce the remarkable number e, the exponential function e^x, and its inverse $\ln x$, the so-called natural logarithm. After finding the derivatives of these and related functions, we turn to problems of growth and decay whose solutions depend on the calculus of these functions.

Goal 31. To understand exponents and logarithms and functions using them
Goal 32. To understand the number e and the exponential function e^x
Goal 33. To understand the natural logarithmic function
Goal 34. To differentiate $\ln x$ and related functions
Goal 35. To differentiate e^x and related functions
Goal 36. To solve problems involving exponential growth or decay
Summary
Review Exercises
Chapter Test A

GOAL 31

To understand exponents and logarithms and functions using them

We here review the basic facts and laws of exponents and logarithms and show how the two are related. Of special interest are the exponential and logarithmic functions introduced here.

EXPONENTS

In the definitions to follow we assume that b is a positive number and that n is a positive integer.

Definitions	Examples
$b^n = b \cdot b \cdot b \cdots b$, where the product on the right has n factors. b is called the *base*, n the *exponent*.	$4^3 = 4 \cdot 4 \cdot 4 = 64$; the base is 4; the exponent is 3; $(\frac{1}{2})^5 = \frac{1}{2} \cdot \frac{1}{2} \cdot \frac{1}{2} \cdot \frac{1}{2} \cdot \frac{1}{2} = \frac{1}{32}$; the base is $\frac{1}{2}$, the exponent is 5;
$b^0 = 1$	$5^0 = 1$; $(\frac{1}{2})^0 = 1$; $1397^0 = 1$;
$b^{-n} = \dfrac{1}{b^n}$	$2^{-4} = \dfrac{1}{2^4} = \dfrac{1}{16}$; $5^{-3} = \dfrac{1}{5^3} = \dfrac{1}{125}$; $7^{-1} = \dfrac{1}{7^1} = \dfrac{1}{7}$.

LAWS OF EXPONENTS

The definitions given above lead easily to the following laws of exponents, for positive bases and positive integral exponents:

Laws	Examples
(1) $b^m \cdot b^n = b^{m+n}$	$3^2 \cdot 3^1 = 3^3 = 27$; $5^2 \times 5^4 = 5^6$
(2) $b^m \div b^n = b^{m-n}$	$9^7 \div 9^4 = 9^3$;
	$2^3 \div 2^5 = 2^{-2} = \dfrac{1}{2^2} = \dfrac{1}{4}$
(3) $(b^m)^n = b^{mn}$	$(10^2)^3 = 10^6$; $(2^4)^3 = 2^{12}$
(4) $(ab)^m = a^m b^m$	$(3 \cdot 4)^2 = 3^2 \cdot 4^2 = 9 \cdot 16 = 144$;
	$(7 \cdot 9)^3 = 7^3 \cdot 9^3$

RATIONAL EXPONENTS

If exponential law (1) is to hold for fractional exponents, then we want

$$b^{1/2} \cdot b^{1/2} \text{ to equal } b^{1/2 + 1/2} \quad \text{or } b$$
$$b^{1/3} \cdot b^{1/3} \cdot b^{1/3} \text{ to equal } b^{1/3 + 1/3 + 1/3} \quad \text{or } b$$

$$\underbrace{b^{1/n} \cdot b^{1/n} \cdots b^{1/n}}_{n \text{ factors}} \text{ to equal } \underbrace{b^{1/n + 1/n + \cdots + 1/n}}_{\substack{n \text{ addends} \\ \text{in exponent}}} \quad \text{or } b$$

We therefore define

$$b^{1/2} \text{ equal to } \sqrt{b}, \; b^{1/3} \text{ equal to } \sqrt[3]{b}, \; \cdots, \; b^{1/n} \text{ equal to } \sqrt[n]{b}$$

We call $\sqrt{b}$ the *square root of* b, $\sqrt[3]{b}$ the *cube root of* b, $\sqrt[n]{b}$ the *nth root of* b. For example,

$$27^{1/3} = \sqrt[3]{27} = 3 \text{ because } 3^3 = 27$$
$$32^{1/5} = \sqrt[5]{32} = 2 \text{ because } 2^5 = 32$$

We now assert that the four laws of exponents above, plus the following two, hold for all positive bases and rational exponents:

Laws	Examples
(5) $b^{1/n} = \sqrt[n]{b}$	$9^{1/2} = \sqrt{9} = 3$;
	$27^{-1/3} = \dfrac{1}{27^{1/3}} = \dfrac{1}{\sqrt[3]{27}} = \dfrac{1}{3}$
(6) $b^{m/n} = \sqrt[n]{b^m} = (\sqrt[n]{b})^m, \; n > 0$	$8^{2/3} = \sqrt[3]{8^2} = (\sqrt[3]{8})^2 = 2^2 = 4$;
	$25^{-3/2} = \dfrac{1}{25^{3/2}} = \dfrac{1}{(\sqrt{25})^3}$
	$= \dfrac{1}{5^3} = \dfrac{1}{125}$

Goal 31 | To Understand Exponents and Logarithms and Functions Using Them

Note that in using law (6) it is almost always computationally easier to take the root first, then apply the power.

Exponential Functions

We have now defined b^x for all rational numbers (provided $b > 0$). Mathematicians have defined b^x even when x is an irrational number (see Study Question 4) and have shown that the six laws of exponents even hold for all real exponents. This means that the function

$$y = b^x$$

is now defined for all real x. It is called an *exponential function*. Note that the *exponent* varies in an exponential function, whereas the *base* varies in a power function:

2^x, 5^x, 7^x are all exponential functions; so are 3^{-x} and $(\frac{1}{2})^{x^2}$;
x^2, x^5, x^7 are power functions.

The base b of an exponential function is constant.

The shape of the graph of $y = b^x$ depends on whether $b > 1$ or $b < 1$ (the case of $b = 1$ is trivial since $1^x = 1$ for all x).

Example 31-1

Sketch the graph of $y = 2^x$.
We prepare a table of values, plot the points, then draw a smooth curve

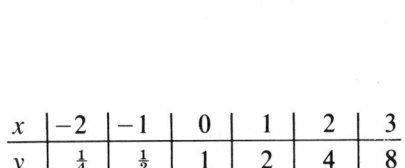

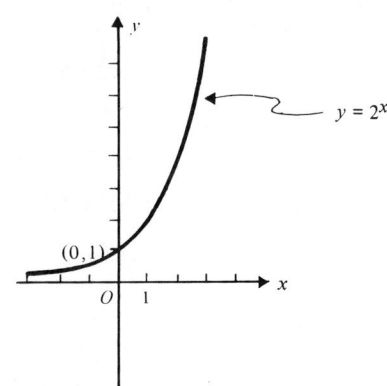

through them. Note that if we let $f(x) = 2^x$,

(a) $f(0) = 1$ (c) $f(x) \to \infty$ as $x \to \infty$
(b) $f(x) > 0$ for all (real) x (d) $f(x) \to 0$ as $x \to -\infty$

These four properties of the graph of $y = b^x$ hold whenever $b > 1$.

Example 31-2

Sketch the graph of $y = (\frac{1}{2})^x$.

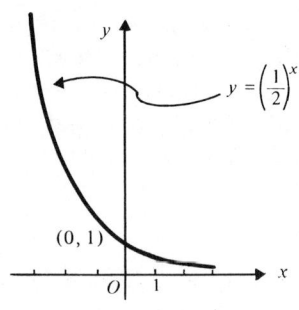

x	-2	-1	0	1	2	3
y	4	2	1	$\frac{1}{2}$	$\frac{1}{4}$	$\frac{1}{8}$

If $g(x) = (\frac{1}{2})^x$, then

(a) $g(0) = 1$

(b) $g(x) > 0$ for all (real) x

(c) $g(x) \to 0$ as $x \to \infty$

(d) $g(x) \to \infty$ as $x \to -\infty$

Every graph of $y = b^x$ with $b < 1$ has these four properties.

The domain of an exponential function, as noted above, is the set of all real numbers. Also, regardless of the magnitude of b, the function b^x is everywhere continuous. If $b > 1$ it is an increasing function, whereas if $b < 1$ it decreases.

LOGARITHMS

If you have used common logarithms (logarithms to the base 10) before, then you can recognize the equivalence of the following pairs of statements:

$$10^2 = 100 \quad \text{and} \quad \log_{10} 100 = 2$$
$$10^3 = 1000 \quad \text{and} \quad \log_{10} 1000 = 3$$
$$10^{-2} = \frac{1}{100} = 0.01 \quad \text{and} \quad \log_{10} 0.01 = -2$$

The *logarithm of a number N to the base* 10, denoted by $\log_{10} N$ and read "log of N to the base 10," is the power to which 10 is raised to yield N. We have, therefore, that

$$\log_b N = x \quad \text{is equivalent to} \quad b^x = N$$

The equation on the left here is said to be *logarithmic*; the one on the right is *exponential*.

Goal 31 | To Understand Exponents and Logarithms and Functions Using Them

Examples 31-3

Write as logarithmic equations:

(a) $10^4 = 10{,}000$
(b) $2^3 = 8$
(c) $5^{-2} = \frac{1}{25}$
(d) $2^{-3} = \frac{1}{8}$
(e) $3^{-1} = \frac{1}{3}$
(f) $10^1 = 10$
(g) $25^{1/2} = 5$
(h) $8^{1/3} = 2$
(i) $7^0 = 1$

Solutions:

(a) $\log_{10} 10{,}000 = 4$
(b) $\log_2 8 = 3$
(c) $\log_5 \frac{1}{25} = -2$
(d) $\log_2 \frac{1}{8} = -3$
(e) $\log_3 \frac{1}{3} = -1$
(f) $\log_{10} 10 = 1$
(g) $\log_{25} 5 = \frac{1}{2}$
(h) $\log_8 2 = \frac{1}{3}$
(i) $\log_7 1 = 0$

Examples 31-4

Write exponentially:

(a) $\log_{10} 1{,}000{,}000 = 6$
(b) $\log_2 2 = 1$
(c) $\log_9 27 = \frac{3}{2}$
(d) $\log_{16} \frac{1}{4} = -\frac{1}{2}$
(e) $\log_4 \frac{1}{16} = -2$
(f) $\log_{10} 1 = 0$

Solutions:

(a) $10^6 = 1{,}000{,}000$
(b) $2^1 = 2$
(c) $9^{3/2} = 27$
(d) $16^{-1/2} = \frac{1}{4}$
(e) $4^{-2} = \frac{1}{16}$
(f) $10^0 = 1$

LAWS OF LOGARITHMS

Since logarithms are exponents, the following properties of logarithms can be derived from the properties and laws of exponents. If $b > 0$ and $b \neq 1$, then

(1) $\log_b 1 = 0$ (since $b^0 = 1$)
(2) $\log_b b = 1$ (since $b^1 = b$)
(3) $\log_b pq = \log_b p + \log_b q$ The log of a product equals the sum of the logs.
(4) $\log_b \dfrac{p}{q} = \log_b p - \log_b q$ The log of a quotient equals the difference of the logs.
(5) $\log_b p^q = q \log_b p$ The log of a number to a power equals the power times the log.

To prove (3), for example, we let $m = \log_b p$ and $n = \log_b q$. Now

$m = \log_b p$ is equivalent to $p = b^m$
$n = \log_b q$ is equivalent to $q = b^n$

Then

$$p \cdot q = b^m \cdot b^n = b^{m+n} \qquad \text{(by exponent law 1)}$$

But

$$pq = b^{m+n} \text{ is equivalent to } \log_b pq = m + n$$

If we now replace m and n from above, we have

$$\log_b pq = \log_b p + \log_b q$$

These laws are the very ones that simplify numerical computations when logs to the base 10 are used, since they "reduce" products to sums (law 3), quotients to differences (law 4), and powers to products (law 5).

Examples 31-5

Suppose $\log_b 2 = r$ and $\log_b 3 = s$. Express in terms of r and s:

(a) $\log_b 6$ (e) $\log_b \frac{27}{16}$
(b) $\log_b 9$ (f) $\log_b \frac{1}{8}$
(c) $\log_b \frac{3}{2}$ (g) $\log_b \sqrt{3}$
(d) $\log_b 32$

Solutions:

(a) $\log_b 6 = \log_b (2 \cdot 3) = \log_b 2 + \log_b 3 = r + s$
(b) $\log_b 9 = \log_b 3^2 = 2 \log_b 3 = 2s$
(c) $\log_b \frac{3}{2} = \log_b 3 - \log_b 2 = s - r$
(d) $\log_b 32 = \log_b 2^5 = 5 \log_b 2 = 5r$
(e) $\log_b \frac{27}{16} = \log_b 27 - \log_b 16 = \log_b 3^3 - \log_b 2^4 = 3\log_b 3 - 4\log_b 2 = 3s - 4r$
(f) $\log_b \frac{1}{8} = \log_b 1 - \log_b 8 = 0 - \log_b 2^3 = -3 \log_b 2 = -3r$
(g) $\log_b \sqrt{3} = \log_b 3^{1/2} = \frac{1}{2} \log_b 3 = \frac{1}{2} s$

Example 31-6

Solve for t: $5^{3t+1} = 25$.

Method 1

$5^{3t+1} = 5^2$

$3t + 1 = 2$

$t = \frac{1}{3}$

Method 2

We take the log of each side to base 5:

$\log_5 (5^{3t+1}) = \log_5 25$

$3t + 1 = 2$ and $t = \frac{1}{3}$

Example 31-7

Solve for x: $\frac{1}{10} = 10^{5-3x}$.

Method 1

$$10^{-1} = 10^{5-3x}$$
$$-1 = 5 - 3x$$
$$x = 2$$

Method 2

Take logs to the base 10:

$$\log_{10} \tfrac{1}{10} = \log_{10} 10^{5-3x}$$
$$-1 = 5 - 3x \text{ and } x = 2$$

LOGARITHMIC FUNCTIONS

The function $y = \log_b x$ ($x > 0$) is called a *logarithmic function*, with b as the *base*. We shall consider only functions for which $b > 1$. Since

$$y = \log_b x \quad \text{is equivalent to} \quad x = b^y$$

we see that $f(x) = \log_b x$ and $g(x) = b^x$ are inverse functions (see Goal 5, pp. 47 and 48).

Example 31-8

Sketch the graph of $y = \log_2 x$.

We construct a table for $y = \log_2 x$ for convenient values of x. Remember

x	$\frac{1}{8}$	$\frac{1}{4}$	$\frac{1}{2}$	1	2	4	8
y	-3	-2	-1	0	1	2	3

that y is the *exponent* to which 2 is raised to yield x. We note that the graph of $f(x) = \log_2 x$ has the following properties:

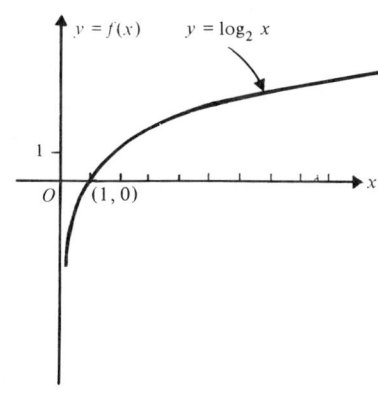

(1) $f(1) = 0$.
(2) The domain is $x > 0$.
(3) As $x \to \infty$, $f(x) \to \infty$.
(4) As $x \to 0^+$, $f(x) \to -\infty$.

Every function of the form $\log_b x$ has these properties provided $b > 1$.

We note, further, that $\log_b x$ is continuous on its domain and that it is increasing if $b > 1$. On p. 310 we sketch the graphs of the pair of inverse functions $f(x) = \log_2 x$ and $g(x) = 2^x$ on the same set of axes to show their symmetry with respect to the line $y = x$.

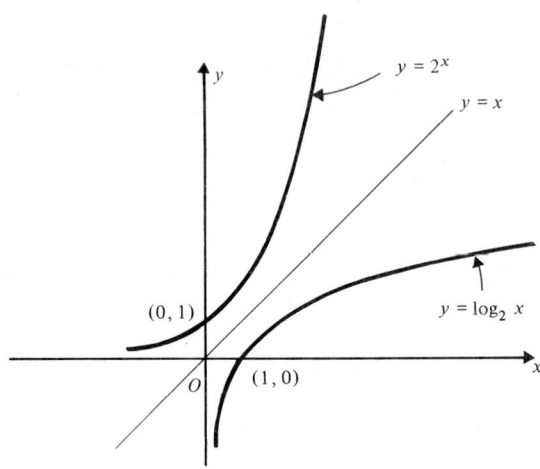

Example 31-9

Acousticians have determined that the sensation of loudness of a sound is a logarithmic function of intensity. Intensity is measured in watts per square meter. The softest audible sound (just audible) has intensity $I_0 = 10^{-12}$ watt/m². The loudness of a sound is measured in decibels, where it is assumed that the loudness of a sound of intensity I_0 is 0 decibels. If a sound has intensity I, then its loudness is given by

$$L = 10 \log_{10} \frac{I}{I_0}$$

Find the loudness, in decibels, of the sound of a rock band 100 ft from the band if the sound there has intensity 10^{-5} watt/m².

Setting $I = 10^{-5}$ we get

$$L = 10 \log_{10} \frac{10^{-5}}{10^{-12}} = 10 \log_{10} 10^7$$
$$= 10 \cdot 7 \log_{10} 10 = 70 \text{ decibels}$$

EXERCISES

1. Simplify each of the following:

   (a) 13_0 (b) $\dfrac{2}{8^0}$

(c) 6^{-2} (g) $49^{1/2}$
(d) 4^{-3} (h) $64^{2/3}$
(e) $(\frac{1}{2})^{-2}$ (i) $8^{-4/3}$
(f) $(\frac{5}{8})^{-1}$ (j) $25^{-3/2}$

2. Write the equivalent exponential equation for each of the following:
   (a) $\log_7 49 = 2$ (d) $\log_{10} \frac{1}{100} = -2$
   (b) $\log_8 2 = \frac{1}{3}$ (e) $\log_5 5 = 1$
   (c) $\log_9 1 = 0$

3. Write the equivalent logarithmic equation for each of the following:
   (a) $5^4 = 625$ (d) $9^{5/2} = 243$
   (b) $3^0 = 1$ (e) $7^1 = 7$
   (c) $2^{-5} = \frac{1}{32}$

4. Calculate each of the following:
   (a) $\log_{10} 1000$ (f) $\log_{10} 1$
   (b) $\log_{10} (10^8)$ (g) $10^{\log_{10} 25}$
   (c) $\log_5 \frac{1}{5}$ (h) $7^{\log_7 100}$
   (d) $\log_{36} 6$ (i) $e^{\log_e 3}$ ($e > 0, e \neq 1$)
   (e) $\log_4 \frac{1}{2}$

5. Simplify each of the following:
   (a) $x^4 x^5$ (d) $(t^3 t^4)^5$
   (b) $(t^4)^5$ (e) $\left(\dfrac{x^2}{x^3}\right)^{-2}$
   (c) $\dfrac{y^{10}}{y^5}$

6. Simplify each of the following:
   (a) $\log_b 10 + \log_b 5$ (d) $\log_b x^5 - \log_b x^3$
   (b) $\log_b x^3$ (e) $\log_b b^2$
   (c) $\log_b 8 - \log_b 4$ (f) $\log_e e^{19}$ ($e > 0, e \neq 1$)

7. Solve each of the following for t:
   (a) $10^{t+4} = 100$ (c) $5 = 10 - 5^{-2t}$
   (b) $\dfrac{1}{2} = 2^{3-4t}$ (d) $3 = \dfrac{5}{1 + 2(3^{4-t})}$

8. If $\log_b 3 = 0.4771$ and $\log_b 4 = 0.6021$, calculate the following:
   (a) $\log_b 12$ (d) $\log_b 6$
   (b) $\log_b 27$ (e) $\log_b \frac{1}{3}$
   (c) $\log_b 2$ (f) $\log_b 36$

9. Sketch each pair on a single set of axes:
   (a) $y = 2^x$ and $y = 3^x$
   (b) $y = 3^x$ and $y = (\frac{1}{3})^x$
   (c) $y = 3^x$ and $y = \log_3 x$

10. Does the graph of $y = b^x$ ($b > 0$, $b \neq 1$) have any asymptotes? How about the graph of $y = \log_b x$?

11. Prove:
   (a) $\log_b b^x = x$
   (b) $b^{\log_b q} = q$

12. True or false?
   (a) $\log_b \dfrac{M}{N} = \log_b M - \log_b N$
   (b) $\dfrac{\log_b M}{\log_b N} = \log_b M - \log_b N$
   (c) $\log_b MN = \log_b M + \log_b N$
   (d) $\log_b (M + N) = \log_b M + \log_b N$
   (e) $\log_b M^N = N \log_b M$
   (f) $(\log_b M)^N = N \log_b M$

13. Use the formula for loudness given in Example 31-9 to evaluate the intensity of a sound whose loudness is 60 decibels.

14. On the Richter scale, the magnitude M of an earthquake is measured logarithmically in terms of its intensity I according to the formula

$$M = \log_{10} \frac{I}{I_0}$$

where I_0 is the intensity of the smallest earthquake that registers on seismographic equipment.
   (a) Find the intensity, in terms of I_0, of an earthquake of magnitude 5 on the Richter scale.
   (b) How much more intense than an earthquake of magnitude 5 is one of magnitude 7?

TEST A

1. Simplify each of the following:
   (a) 3^0
   (b) $\left(\dfrac{1}{2}\right)^{-3}$
   (c) $27^{2/3}$
   (d) $\left(\dfrac{t^9}{t^3}\right)^2$
   (e) $\log_b 12 - \log_b 6$

Goal 31 | To Understand Exponents and Logarithms and Functions Using Them

2. Write the equivalent exponential or logarithmic statement:
   (a) $\log_2 64 = 6$
   (b) $3^{-3} = \frac{1}{27}$
   (c) $\log_9 27 = \frac{3}{2}$
   (d) $10^0 = 1$

3. Calculate each of the following:
   (a) $\log_{10}(10^{13})$
   (b) $\log_{10} \frac{1}{1000}$
   (c) $10^{\log_{10}(50)}$

4. Solve each of the following for t:
   (a) $6^{2t-4} = 36$
   (b) $2 = 8^{t-4}$

5. Sketch the graphs of $y = 2^x$ and $y = 2^{-x}$ on the same set of axes.

TEST B

1. Simplify each of the following:
   (a) $(\frac{1}{2})^0$
   (b) 5^{-3}
   (c) $(\frac{1}{4})^{3/2}$
   (d) $(x^3 x^5)^2$
   (e) $\log_b x^{10}$

2. Write the equivalent exponential or logarithmic statement:
   (a) $5^3 = 125$
   (b) $\log_{10}(10^6) = 6$
   (c) $9^{-3/2} = \frac{1}{27}$
   (d) $\log_2 1 = 0$

3. Calculate each of the following:
   (a) $\log_{10}(10^{-25})$
   (b) $\log_8(\frac{1}{2})$
   (c) $5^{\log_5 10}$

4. Solve each of the following for t:
   (a) $3^{5+t} = 27$
   (b) $\frac{1}{2} = 4^{2-3t}$

5. Sketch the graphs of $y = \log_2 x$ and $y = \log_3 x$ on the same set of axes.

STUDY QUESTIONS

1. Prove logarithmic law 4: $\log_b \frac{p}{q} = \log_b p - \log_b q$.
   HINT: let $m = \log_b p$ and $n = \log_b q$.

2. Prove logarithmic law 5: $\log_b p^q = q \log_b p$.

3. Prove: $(\log_{10} 2) \cdot (\log_2 10) = 1$. HINT: Let $m = \log_{10} 2$. Then $2 = 10^m$. Now take logs to the base 2. [Note that this proof can be generalized to show that $(\log_b a)(\log_a b) = 1$.]

4. To define b^x when x is irrational, consider the example $5^{\sqrt{3}}$. Since $\sqrt{3} = 1.7320508 \ldots$, where the infinite decimal is nonrepeating, we can define $5^{\sqrt{3}}$ as the limit of the sequence of numbers 5^1, $5^{1.7}$, $5^{1.73}$, $5^{1.732}$, ..., as the number of digits in the exponent becomes infinite. Each term of this sequence has a rational exponent.

Define: (a) $3^{\sqrt{2}}$; (b) 10^π.

GOAL 32

To understand the number e and the exponential function e^x

We now introduce the extraordinary number e and the associated exponential function e^x. The following three unrelated examples give some indication of the broad range of problems in which e appears.

Example 32-1

Little Bobby Jones invests his life savings of $1 in the Unbelievable Bank of New York, which pays 100% interest per yr. Find Bobby's balance at the end of 1 yr if interest is compounded (a) annually; (b) twice a year; (c) four times a year; (d) monthly; (e) weekly; (f) daily; (g) hourly; (h) n times a year.

When interest is compounded, the interest itself earns interest. (a) Annual compounding is just simple interest for 1 yr. (b) If compounded twice a year, the interest after the first 6 months is $1 \cdot \frac{1}{2}$ dollars and the balance is $(1 + \frac{1}{2})$ dollars; at the end of the second six months, the interest is $(1 + \frac{1}{2}) \cdot \frac{1}{2}$ dollars and the balance, in dollars, is

$$\left(1 + \frac{1}{2}\right) + \left(1 + \frac{1}{2}\right) \cdot \frac{1}{2} = \left(1 + \frac{1}{2}\right)\left(1 + \frac{1}{2}\right)$$

(c) Compounded four times a year, the interest at the end of the first quarter is $1 \cdot \frac{1}{4}$ dollars and the balance is $(1 + \frac{1}{4})$ dollars; at the end of the second quarter the interest is $(1 + \frac{1}{4}) \cdot \frac{1}{4}$ dollars and the balance, in dollars, is

$$\left(1 + \frac{1}{4}\right) + \left(1 + \frac{1}{4}\right) \cdot \frac{1}{4} = \left(1 + \frac{1}{4}\right)\left(1 + \frac{1}{4}\right) = \left(1 + \frac{1}{4}\right)^2$$

Continuing in this fashion, we see that the balance at the end of 1 yr (after four quarters) is $(1 + \frac{1}{4})^4$ dollars. (d) through (h) are done similarly.

With the aid of a hand calculator, we can produce the following table:

Frequency of Compounding	Balance ($) at the End of the Year	Amount ($) at End of Year
annually	$1 + 1$	2
twice a year	$\left(1 + \dfrac{1}{2}\right)^2$	2.25
four times a year	$\left(1 + \dfrac{1}{4}\right)^4$	2.44
monthly	$\left(1 + \dfrac{1}{12}\right)^{12}$	2.61
daily	$\left(1 + \dfrac{1}{365}\right)^{365}$	2.71
hourly	$\left(1 + \dfrac{1}{8760}\right)^{8760}$	2.718
n times a year	$\left(1 + \dfrac{1}{n}\right)^n$	

The missing answer in the rightmost column depends, of course, on n. The table suggests that while the balance increases with more frequent compounding, there is an upper bound to the amount that can be in Bobby's account at the end of the year. This is, in fact, the case. The upper bound is an irrational number, denoted by e:

$$e = \lim_{n \to \infty} \left(1 + \frac{1}{n}\right)^n$$

To twelve decimal places,

$$e \approx 2.718281828459$$

The decimal expansion of the constant e is both infinite and nonrepeating.

When we let $n \to \infty$, we say that the interest is *compounded continuously*. In Study Question 1 you will be asked to evaluate $[1 + 1/n]^n$ for some large values of n.

It can be shown that if A dollars is invested for t yr at 6% interest a year compounded continuously, it will grow to $Ae^{0.06t}$ dollars. It is worth noting that $1000 invested at 6% per yr compounded daily yields $1061.83 after one year, whereas compounded continuously it amounts to $1061.84, a difference of only one cent!

Goal 32 | To Understand the Number e and the Exponential Function e^x

Example 32-2

Since Hilda, the hatcheck girl, forgets to hand out check stubs, at the end of the evening she simply randomly returns hats to the customers and hopes for the best. What is the probability that, just by chance, at least one person will actually get back his own hat?

If you guess that the chances are small and become even smaller as the number of people (and hats) increases, then the following table will surprise you:

Number of People (and Hats)	Probability of at Least One Match
2	0.5
3	0.66667
5	0.63333
10	0.63212
25	0.63212
100	0.63212

The amazing conclusion is that, regardless of the number, the probability that there will be at least one match is substantial. Furthermore, this probability has a nonzero limiting value for large numbers of people and hats. This limiting value, to 10 decimal places, is 0.6321205588, and it is exactly equal to $1 - 1/e$, where e is the number defined in Example 32-1.

Example 32-3

Find any points of intersection of the line $y = x + 1$ and the exponential functions (a) $y = 2^x$; (b) $y = 4^x$; (c) $y = b^x$ $(2 < b < 4)$.

(a) The graphs of $y = 2^x$ and $y = x + 1$ intersect at the two points (0, 1) and (1, 2).

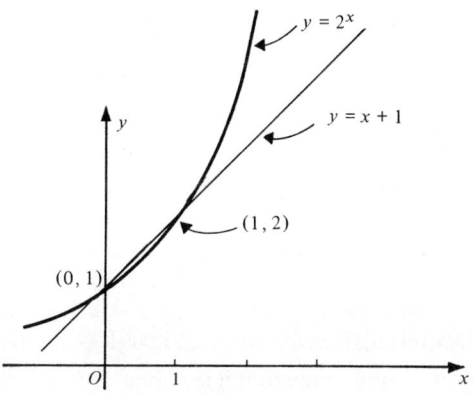

(b) The graphs of $y = 4^x$ and $y = x + 1$ intersect at the two points $(0, 1)$ and $[-\frac{1}{2}, \frac{1}{2}]$.

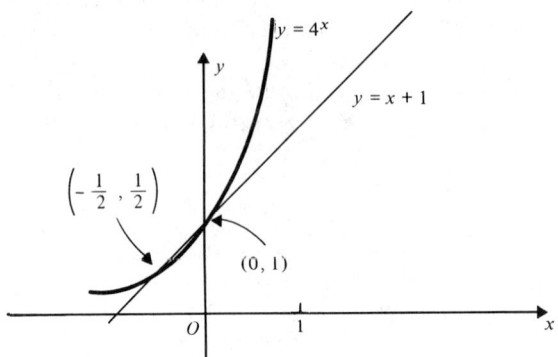

(c) It is clear that the graph of $y = b^x$ intersects the line $y = x + 1$ at $(0, 1)$ for every b. Is there some number b between 2 and 4 such that $(0, 1)$ is the *only* point of intersection? The answer is yes. It is precisely the number e. Below we see a sketch of the graph of $y = e^x$. Note its position relative to the graphs of $y = 2^x$ and $y = 4^x$.

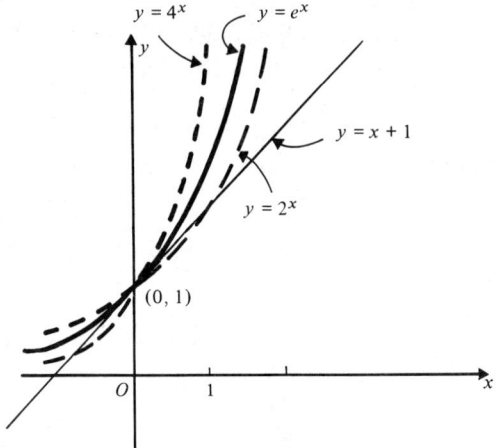

There are many more situations in which the number e arises in a natural way. Its importance in the calculus will be shown in the next few goals. It occurs in other branches of mathematics, too, in business and economics, and in the physical, biological, and social sciences.

THE EXPONENTIAL FUNCTION

The function $f(x) = e^x$ has such significance that it is called *the* exponential function. It arises repeatedly in applications ranging from population growth and radioactive decay to learning curves and probability distributions. These will be discussed in detail in later Goals.

PROPERTIES OF THE EXPONENTIAL FUNCTION

The function $f(x) = e^x$ enjoys all the properties of exponential functions listed in Goal 31 (p. 303). In particular, for all real m and n

$$e^0 = 1 \qquad e^{-n} = \frac{1}{e^n} \qquad e^m \cdot e^n = e^{m+n}$$

$$\frac{e^m}{e^n} = e^{m-n} \qquad (e^m)^n = e^{mn}$$

Example 32-4

Simplify: $e^3 e^4$, $(e^3)^4$, $(e^0)^2$, $e^3 \cdot e^{-4}$, $e^{-1/2}$.

$$e^3 \cdot e^4 = e^{3+4} = e^7 \qquad (e^3)^4 = e^{12}$$

$$(e^0)^2 = 1^2 = 1 \qquad e^3 \cdot e^{-4} = e^{3+(-4)} = e^{-1} = \frac{1}{e}$$

$$e^{-1/2} = \frac{1}{e^{1/2}} = \frac{1}{\sqrt{e}}$$

GRAPH OF THE EXPONENTIAL FUNCTION

Since $e \approx 2.7 > 1$, the graph of $f(x) = e^x$ has the shape of that of $y = b^x$ where $b > 1$ (see p. 305). In particular, as we see from the graph of $y = e^x$ on page 318,

(1) the domain is all real x;
(2) $e^x > 0$ for all x (the range of the function is all positive numbers);
(3) $e^x \to \infty$ as $x \to \infty$; $e^x \to 0$ as $x \to -\infty$;
(4) the graph is always rising and is always concave up.

In the next Goal we will prove statement (4).

GRAPHING FUNCTIONS INVOLVING EXPONENTIALS

Some examples of common functions that involve exponentials are (with t or x as the variable)

$P_0 e^{kt}$ for the growth of world population

$A_0 e^{-kt}$ for the decay of a radioactive isotope or for life expectancy

$a(1 - e^{-kt})$ to measure the amount of a particular new skill acquired or for the percentage of a population buying an advertised product

$Ce^{-x^2/2}$ for the normal distribution of grades or of some other random variable

To graph one of these functions we often need to know the value of e^x or e^{-x} for a specified x. Table II has approximations for both of these.* Thus, we can read off the following:

$$e^1 = 2.7183 \quad e^{3.5} = 33.115 \quad e^{-1} = 0.36788 \quad e^{-0.5} = 0.60653 \quad e^0 = 1$$

The methods of Goals 3 and 4 involving translation and reflection are especially useful in graphing functions involving exponentials. We will be interested here primarily in obtaining the general shape of the curve rather than in plotting many points. When the shape is determined, plotting a couple of points usually assures a reasonable sketch.

Example 32-5

Graph on the same set of axes: (a) $y = e^x$ and $y_1 = e^{-x}$; (b) $y = e^x$ and $y_2 = \frac{1}{2} e^x$.

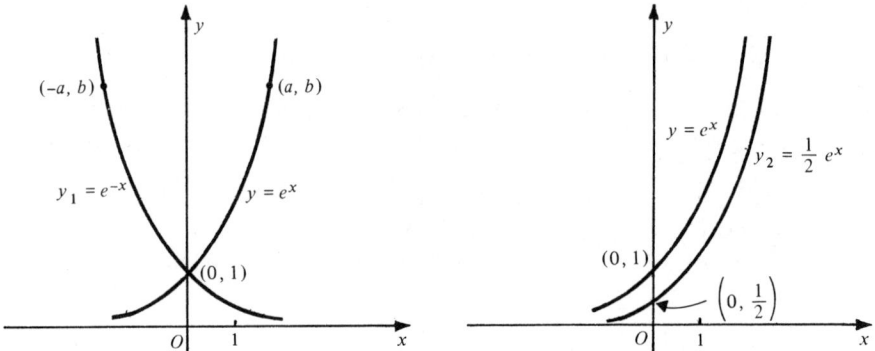

The graph of $y_1 = e^{-x}$ is the reflection of the graph of $y = e^x$ in the y-axis, since if (a, b) is on the graph of y, then $(-a, b)$ is on the graph of y_1.

The graph of $y_2 = \frac{1}{2} e^x$ has the same shape as that of $y = e^x$, but the ordinate for any x is half that of the corresponding ordinate of $y = e^x$.

Example 32-6

Sketch the graph of $y = 3e^{2x}$.

Note first that when $x = 0$, $y = 3$. The shape of the graph is similar to that of $y = e^x$ but is *steeper* if $x > 0$, less steep if $x < 0$. Since $e^{2x} = (e^x)^2$, the ordinates of this graph are the squares of the corresponding ones of $y = e^x$. Moreover, the coefficient 3 of $3e^{2x}$ triples the ordinates of e^{2x}. Note that we have used a much smaller unit here than for Example 32-5.

---

*Many hand calculators have a special function circuit to evaluate e^x. Books of mathematical tables also contain quite complete tables of exponential functions.

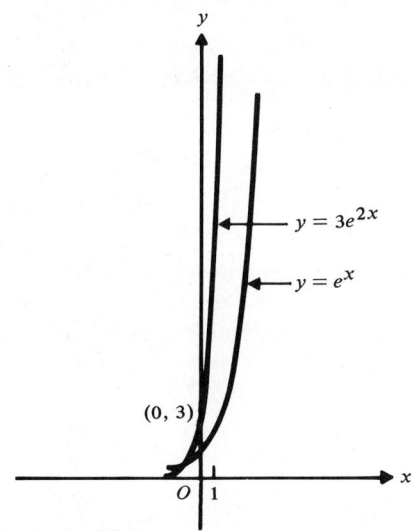

Example 32-7

Sketch the graph of $y = 4 - 2e^{-x/2}$.

We can get quite a good sketch in several steps:

(1) $y = e^{x/2}$ has the same general shape as $y = e^x$, but the ordinates of the former are the *square roots* of the corresponding ones of $y = e^x$.

(2) $y = e^{-x/2}$ is the reflection of $y = e^{x/2}$ in the y-axis.

(3) $y = 2e^{-x/2}$ has ordinates twice those of $y = e^{-x/2}$.

(4) $y = -2e^{-x/2}$ is the reflection of $y = 2e^{-x/2}$ in the x-axis.

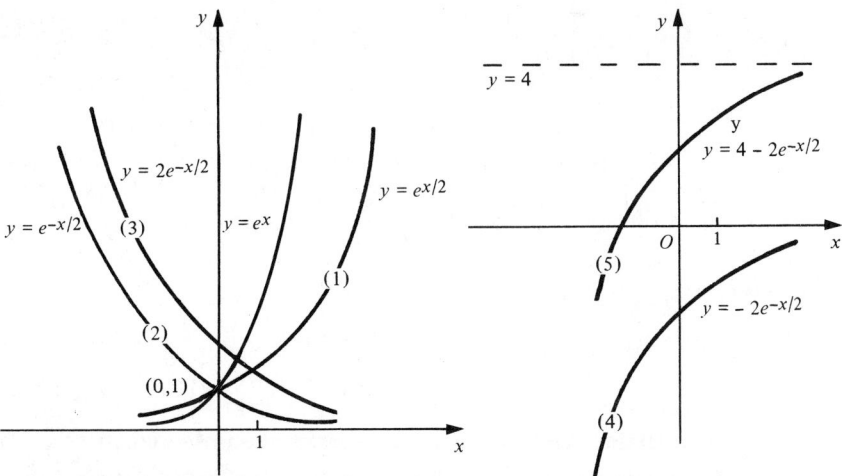

(5) Finally, to sketch $y = 4 - 2e^{-x/2}$ we simply translate the graph of $y = -2e^{-x/2}$ up 4 units.

Note, since $2e^{-x/2} \to 0$ as $x \to \infty$, that $y = 4$ is a horizontal asymptote of the curve.

EXERCISES

1. Simplify:
   - (a) $e^2 e^3$
   - (b) $(e^2)^3$
   - (c) $e^{-3} e^4$
   - (d) $(e^{-2})^4$
   - (e) e^0
   - (f) $(e^x)^2$
   - (g) $(t + e^t)^2$

2. Using Table II determine each of the following to four decimal places:
   - (a) $e^{1.8}$
   - (b) $e^{0.6}$
   - (c) e^3
   - (d) $e^{-1.9}$
   - (e) $e^{-0.3}$
   - (f) e^0

3. Make rough sketches of the graphs of
   - (a) $x = 2e^t$
   - (b) $x = -e^{2t}$
   - (c) $x = 1 - e^{-t/2}$
   - (d) $y = 2 - e^{-x}$

TEST A

1. Simplify: $(e^5)^2$; $e^t e^t$; $\dfrac{e^5}{e^2}$
2. Using Table II, determine $e^{-2.5}$; $e^{0.2}$; $e^{1.1}$.
3. Roughly sketch the graph of $y = 2e^{-x}$.

TEST B

1. Simplify: $e^8 e^2$; $(e^t)^3$; $\dfrac{e^8}{e^2}$.
2. Using Table II, determine: $e^{3.2}$; $e^{0.5}$; $e^{-1.6}$.
3. Roughly sketch the graph of $x = \frac{1}{2} e^t$.

STUDY QUESTIONS

1. Using a hand calculator with an X^Y key, find the value of $\left(1 + \dfrac{1}{n}\right)^n$ for $n = 10, 100, 1000, 10{,}000, 100{,}000$. If an X^Y key is not available, you will have to resort to common logarithms.

2. An amount A invested for t yr at an interest rate of 6% a year compounded m times a year will grow to $A\left(1 + \dfrac{0.06}{m}\right)^{mt}$. Use the definition

Goal 32 | **To Understand the Number e and the Exponential Function e^x**

$$\lim_{n \to \infty} \left(1 + \frac{1}{n}\right)^n = e$$

to show that more and more frequent compounding yields a balance that gets closer and closer to $Ae^{0.06t}$. HINT: Show that

$$A\left(1 + \frac{0.06}{m}\right)^{mt} = A\left[\left(1 + \frac{1}{\frac{m}{0.06}}\right)^{m/0.06}\right]^{0.06t}$$

Then let $n = \dfrac{m}{0.06}$; finally, let $n \to \infty$.

GOAL 33

To understand the natural logarithmic function

In Goal 31 we considered functions of the type $\log_b x$ ($b > 1$). When the base b is e, then $\log_e x$ is called the *natural logarithmic function* (or just the *natural log function*), and is symbolized by

$$\ln x$$

(read "ell en of eks"). Like the function e^x, $\ln x$ arises in a great many practical problems. These will be considered in the next few Goals. Our primary concern here is to state and use the properties of $\ln x$ and to show how e^x and $\ln x$ are related.

PROPERTIES OF $\ln x$

The properties of $\log_b x$, given on page 307, hold, of course, when $b = e$. Thus

(1) $\ln 1 = 0$
(2) $\ln e = 1$
(3) $\ln pq = \ln p + \ln q$
(4) $\ln p/q = \ln p - \ln q$
(5) $\ln p^q = q \ln p$

Example 33-1

If $\ln x = a$ and $\ln y = b$, express each of the following in terms of a and b, if possible:

(a) $\ln xy$; (b) $\ln x^2$; (c) $\ln (1/y)$; (d) $\ln \sqrt{y}$;
(e) $\ln (x^3/y^2)$; (f) $\ln (x + y)$; (g) $(\ln x)/(\ln y)$

ANSWERS:

(a) $\ln xy = \ln x + \ln y = a + b$ (b) $\ln x^2 = 2 \ln x = 2a$

Goal 33 | To Understand the Natural Logarithmic Function

(c) $\ln (1/y) = \ln 1 - \ln y = 0 - b = -b$
(d) $\ln \sqrt{y} = \frac{1}{2} \ln y = b/2$
(e) $\ln (x^3/y^2) = \ln x^3 - \ln y^2 = 3 \ln x - 2 \ln y = 3a - 2b$
(f) (cannot be done)
(g) $(\ln x)/(\ln y) = a/b$

APPROXIMATE VALUES OF $\ln x$

In many applications we need to know the value of $\ln x$ for a specified x. Table I has approximations for some numbers.* From Table I and properties of the ln function, we can read off

$\ln 1 = 0 \quad \ln 3.9 = 1.3610 \quad \ln 10 = 2.3026 \quad \ln 0.1 = -2.3026$

$\ln 2 = 0.6931 \quad \ln \frac{1}{2} = -0.6932$

$\ln 10^6 = 6 \ln 10 = 6(2.3026) = 13.8156$

$\ln 29{,}000 = \ln (2.9 \times 10^4) = \ln 2.9 + \ln 10^4 = \ln 2.9 + 4 \ln 10 = 1.0647 + 9.2104 = 10.2751$

GRAPH OF $y = \ln x$

We sketch the graph by using the fact that $\ln 2 \approx 0.693$ and by then evaluating $\ln x$ for values of x that are powers of 2. Note that $\ln 2^n = n \ln 2$.

x	$\frac{1}{8} = 2^{-3}$	$\frac{1}{4} = 2^{-2}$	$\frac{1}{2} = 2^{-1}$	$1 = 2^0$	$2 = 2^1$	$4 = 2^2$	$8 = 2^3$
$\ln x$	-2.079	-1.386	-0.693	0	0.693	1.386	2.079

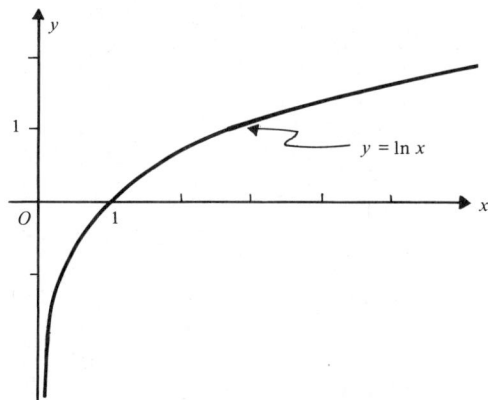

---

*Books of mathematical tables give more complete tables of values of the natural log function. Also, many hand calculators have a special LN X key.

We note that

(1) $\ln x$ is defined only for $x > 0$;
(2) $\ln x < 0$ for $x < 1$; $\ln x = 0$ if $x = 1$; $\ln x > 0$ for $x > 1$;
(3) as $x \to \infty$, $\ln x \to \infty$; as $x \to 0^+$, $\ln x \to -\infty$.

The natural log function and its graph thus exhibit the same characteristics as does $\log_b x$ where $b > 1$ (see p. 309). We note further that $\ln x$ is continuous for all $x > 0$, and that its graph is both rising and concave down.

RELATION OF THE FUNCTIONS $\ln x$ AND e^x

In Goal 31 we noted that $\log_b x$ and b^x are inverse functions. It follows that the natural log function $\ln x$ and the exponential function e^x are inverse functions. So

$$y = e^x \quad \text{if and only if} \quad x = \ln y$$

Example 33-2

Prove the identities: (1) $\ln e^x = x$ for all x; (2) $e^{\ln x} = x$ if $x > 0$.

(1) $\ln e^x = x \ln e = x \cdot 1 = x$.
(2) Let $q = e^{\ln x}$ and take the logarithms of both sides: $\ln q = \ln x \ln e = \ln x \cdot 1 = \ln x$. But $\ln q = \ln x$ implies that $q = x$, so $x = e^{\ln x}$.

Note that (1) and (2) illustrate the following identities that hold for *every* pair of inverse functions $f(x)$ and $g(x)$:

$$f(g(x)) = x \quad \text{and} \quad g(f(x)) = x$$

In particular, if $f(x) = \ln x$ and $g(x) = e^x$, then

$$f(g(x)) = \ln(e^x) = x \quad \text{and} \quad g(f(x)) = e^{\ln x} = x$$

Identity (1) reaffirms that $\ln e\ [\ = \ln e^1] = 1$.

Example 33-3

Simplify (1) $\ln e^{3.9}$; (2) $e^{\ln 11.3}$; (3) $e^{4 \ln 2}$; (4) $\ln (\ln e)$; (5) $\ln (xe^x)$; (6) $e^{x + \ln x}$.

We use the identities of Example 33-2.

(1) $\ln e^{3.9} = 3.9$.
(2) $e^{\ln 11.3} = 11.3$.
(3) $e^{4 \ln 2} = (e^{\ln 2})^4 = 2^4 = 16$.

Goal 33 | To Understand the Natural Logarithmic Function

(4) $\ln (\ln e) = \ln 1 = 0$

(5) $\ln (xe^x) = \ln x + \ln e^x = \ln x + x$

(6) $e^{x + \ln x} = e^x \cdot e^{\ln x} = e^x \cdot x = xe^x$

Example 33-4

The radioactive isotope carbon-14 decays at such a rate that, if Q_0 is the original amount, after t yr the quantity remaining is $Q = Q_0 e^{-0.00012t}$. Find the half-life of carbon-14.

The *half-life* of a radioactive substance is the length of time it takes for half of the original amount to decay. We therefore seek the value of t here for which $Q = \frac{1}{2} Q_0$. We have

$$\frac{1}{2} Q_0 = Q_0 e^{-0.00012t}$$

$$\frac{1}{2} = e^{-0.00012t}$$

Now take the ln of each side:

$$\ln \frac{1}{2} = \ln e^{-0.00012t}$$

$$\ln 1 - \ln 2 = -0.00012t$$

Since $\ln 1 = 0$, we have

$$t = \frac{\ln 2}{0.00012} = \frac{0.693}{0.00012} \approx 5775 \text{ yr}$$

Therefore, using the three-place approximation for ln 2 we get from Table I, our estimate of the half-life of carbon-14 is 5775 yr.

Example 33-5

Solve for t: $e^{3.5t} = 56$.

We take the ln of each side:

$$\ln e^{3.5t} = \ln 56 \qquad 3.5t = \ln 56$$

Since (using Table I)

$$\ln 56 = \ln 5.6 + \ln 10 = 1.7228 + 2.3026 = 4.0254$$

we have

$$t = \frac{4.025}{3.5} \approx 1.15$$

Example 33-6

Approximate e^{99} by an integral power of 10.
We seek an integer k such that $e^{99} \approx 10^k$. Taking the ln of each side yields

$$\ln e^{99} = \ln 10^k$$
$$99 = k \ln 10 \approx k \cdot (2.3)$$
$$k \approx \frac{99}{2.3} \approx 43 \text{ to the nearest integer}$$

So $e^{99} \approx 10^{43}$.

EXERCISES

1. Given that $\ln 2 = 0.693$ and $\ln 3 = 1.099$, use properties of logarithms to determine
   - (a) $\ln 6$
   - (b) $\ln 16$
   - (c) $\ln \frac{1}{3}$
   - (d) $\ln \sqrt{2}$
   - (e) $\ln 36$

2. Simplify:
   - (a) $e^{\ln 5}$
   - (b) $\ln e^{10}$
   - (c) $e^{6 \ln 2}$
   - (d) $e^{-\ln (1/x)}$
   - (e) $\ln (xe^{-x})$
   - (f) $\ln (e^{-x^2})$
   - (g) $e^{x - \ln x}$
   - (h) $\ln \left(-\ln \frac{1}{e}\right)$

3. Use Table I to determine each of the following:
   - (a) $\ln 3.2$
   - (b) $\ln 7.9$
   - (c) $\ln 0.2$
   - (d) $\ln 0.8$
   - (e) $\ln 75$
   - (f) $\ln 38$
   - (g) $\ln 120$
   - (h) $\ln 570$
   - (i) $\ln 10^8$
   - (j) $\ln 190{,}000$

4. Graph each pair of functions on a single set of axes:
   - (a) $y = \ln x$ and $y = \ln \frac{1}{x}$
   - (b) $y = \ln x$ and $y = \ln 2x$

5. Solve for t:
   - (a) $e^{3.1t} = 8.3$
   - (b) $e^{0.001t} = 95$
   - (c) $e^{-0.0025t} = 0.5$
   - (d) $e^{5t} = 5300$
   - (e) $e^{-0.00001t} = \frac{1}{2}$
   - (f) $e^{-0.03t} = 0.3$

Goal 33 — To Understand the Natural Logarithmic Function

6. One isotope of uranium (U_{232}) has a half-life of 73.6 yr. If the amount Q of uranium-232 present after t yr is given by $Q = Q_0 e^{-kt}$, where the initial amount was Q_0, find the constant k. HINT: Solve $\frac{1}{2} = e^{-73.6k}$ for k.

7. For what integral k does each of the following equal approximately 10^k?
   (a) e^{25} (b) e^{39}

TEST A

1. Simplify:
   (a) $\ln \dfrac{1}{e}$ (b) $e^{\ln 2}$ (c) $\ln \sqrt{e}$ (d) $e^{-5 \ln 3}$

2. Solve for t:
   (a) $e^{3t} = 51$ (b) $e^{-0.2t} = 0.1$ (c) $e^{-0.0005t} = \frac{1}{2}$

3. Graph on a single set of axes: $y = e^x$ and $y = \ln x$.

4. For what integer k does $e^{36} \approx 10^k$?

TEST B

1. Simplify:
   (a) $e^{-2 \ln 2}$ (b) $\ln \sqrt[3]{e}$ (c) $e^{\ln 1/x}$ (d) $\ln e^{-x}$

2. Solve for t:
   (a) $e^{-0.005t} = 0.7$ (b) $e^{2t} = 33$ (c) $e^{-0.0011t} = \frac{1}{2}$

3. Graph on a single set of axes: $y = \ln x$ and $y = \log_4 x$.

4. For what integer k does $e^{50} \approx 10^k$?

STUDY QUESTIONS

1. If $Q = Q_0 e^{-kt}$, solve for t in terms of k when $Q = \frac{1}{2} Q_0$.

2. Prove:
$$\log_{10} x = \frac{\ln x}{\ln 10}$$

HINT: Let $q = \log_{10} x$; write this equation exponentially, then take the ln of each side.

GOAL 34

To differentiate ln x and related functions

The definition of derivative can be used to obtain $\dfrac{d}{dx}\log_b x$. When b is replaced by e, the derivative of ln x is obtained immediately. With the chain rule we can then find $\dfrac{d}{dx}\ln u$, where u is a differentiable function of x.

DERIVATIVE OF $\log_b x$, OF ln x, AND OF ln u

The derivative of $\log_b x$ is $\dfrac{1}{x}\log_b e$. The proof of this rather surprising fact is omitted because it is complicated. It is usually given in more advanced texts.

Since

$$\frac{d}{dx}\log_b x = \frac{1}{x}\log_b e$$

it follows, if we replace b by e, that

$$\frac{d}{dx}\ln x = \frac{d}{dx}\log_e x = \frac{1}{x}\log_e e = \frac{1}{x}$$

The simplicity of this derivative explains why e is the logarithmic base most commonly used in the calculus.

If u is a differentiable function of x, then by the chain rule we have immediately

$$\frac{d}{dx}\ln u = \frac{1}{u}\frac{du}{dx}$$

Goal 34 | To Differentiate ln x and Related Functions

Examples 34-1

In the following we assume that x takes on only permissible values (remember that $\ln u$ is defined only if u is greater than 0).

(a) $\dfrac{d}{dx}(3 \ln x) = 3 \cdot \dfrac{1}{x} = \dfrac{3}{x}$

(b) $\dfrac{d}{dx}(x^2 + \ln x) = 2x + \dfrac{1}{x}$

(c) $\dfrac{d}{dx}(x^3 + 2 \ln x + 1) = 3x^2 + \dfrac{2}{x}$

In the next three parts, we use the formula given just above for $\dfrac{d}{dx}\ln u$. What does u equal in each part?

(d) $\dfrac{d}{dx}\ln(x^2 + x) = \dfrac{1}{x^2 + x}(2x + 1) = \dfrac{2x + 1}{x^2 + x}$

(e) $\dfrac{d}{dx}\ln(x^3) = \dfrac{1}{x^3}(3x^2) = \dfrac{3}{x}$

What is the answer in (e) if you rewrite $\ln(x^3)$ as $3 \ln x$, then differentiate?

(f) $\dfrac{d}{dx}\ln(\ln x) = \dfrac{1}{\ln x} \cdot \dfrac{1}{x} = \dfrac{1}{x \ln x}$

In part (g) we use the product rule:

(g) $\dfrac{d}{dx} x \ln x = x \cdot \dfrac{1}{x} + \ln x \cdot 1 = 1 + \ln x$

Now we invoke the quotient rule:

(h) $\dfrac{d}{dx}\dfrac{\ln x}{x} = \dfrac{x \cdot \dfrac{1}{x} - \ln x \cdot 1}{x^2} = \dfrac{1 - \ln x}{x^2}$

In part (i) we regard $(\ln x)^3$ as u^3 and use the extended power rule, $\dfrac{d}{dx}u^3 = 3u^2 \dfrac{du}{dx}$:

(i) $\dfrac{d}{dx}(\ln x)^3 = 3(\ln x)^2 \cdot \dfrac{1}{x} = \dfrac{3(\ln x)^2}{x}$

NOTE: Pay special attention to the difference between $(\ln x)^3$ and $\ln(x^3)$; the first is the cube of the ln function, the second the ln of x cubed. Thus

$$(\ln x)^3 = (\ln x)(\ln x)(\ln x) = \ln^3 x$$

whereas

$$\ln(x^3) = \ln(x \cdot x \cdot x) = \ln x^3$$

(j) $\dfrac{d}{dx} \ln \left(\dfrac{x^2 + 5}{2x} \right) = \dfrac{d}{dx} \left[\ln (x^2 + 5) - \ln (2x) \right]$

$= \dfrac{2x}{x^2 + 5} - \dfrac{2}{2x} = \dfrac{2x}{x^2 + 5} - \dfrac{1}{x}$

Here, in the first step, we used the fact that $\ln (m/n) = \ln m - \ln n$ to avoid having to apply the quotient rule to $\dfrac{x^2 + 5}{2x}$.

What happens in (j) if you rewrite $\ln (2x)$ as $\ln 2 + \ln x$, then take the derivative?

Example 34-2

Find the equation of the tangent to the graph of $f(x) = \ln x$ at the point $(e, 1)$.

Since $f'(x) = 1/x$, $f'(e) = 1/e$ and the equation of the tangent is

$$\dfrac{y - 1}{x - e} = \dfrac{1}{e} \to y - 1 = \dfrac{1}{e}(x - e) \to y - 1 = \dfrac{1}{e}x - 1 \to y = \dfrac{1}{e}x$$

Practical problems involving the natural logarithmic function will be taken up in Goal 36.

EXERCISES

Assume that x takes on only permissible values.

1. Differentiate:
   (a) $8 \ln x$
   (b) $\ln 5x$
   (c) $\ln (x^2)$
   (d) $(\ln x)^2$
   (e) $\ln (x^3 - 5x + 2)$
   (f) $x^2 \ln x$
   (g) $\ln [(x^2 + 1)(3x^2 - 2)]$ HINT: Use $\ln mn = \ln m + \ln n$
   (h) $\ln \dfrac{2x^3 - x + 4}{3 - 2x}$
   (i) $\dfrac{\ln x}{x^2}$
   (j) $\sqrt{1 + \ln x}$ HINT: This is $u^{1/2}$ with $u = 1 + \ln x$
   (k) $\ln (\ln 2x)$
   (l) $\ln e$

2. Find the equations of the tangents to the graphs of the following functions at the designated points:
   (a) $f(x) = \ln x$ at $(1, 0)$
   (b) $g(x) = x \ln x$ at (e, e)
   (c) $h(x) = \ln (x^2 + 1)$ at $(0, 0)$

3. Find $f'(x)$ if
   (a) $f(x) = \log_{10} x$
   (b) $f(x) = \log_{10} (x^2 + 1)$

   HINT: $\dfrac{d}{dx} \log_b u = \dfrac{1}{u} \dfrac{du}{dx} \log_b e = \dfrac{1}{(\ln b) \cdot u} \dfrac{du}{dx}$

4. In Example 31-9 (p. 310), the loudness of a sound of intensity I is given by $L = 10 \log_{10} \dfrac{I}{I_0}$, where I_0 is the intensity of a sound that is just barely audible. Find $\dfrac{dL}{dI}$, the rate of change of loudness with respect to intensity.

5. In Exercise 14 (p. 312), the magnitude M of an earthquake in terms of its intensity I is $M = \log_{10} \dfrac{I}{I_0}$, where I_0 is the intensity of the smallest earthquake that registers on seismographic equipment. Find $\dfrac{dM}{dI}$.

6. Use the first and second derivatives of $f(x) = \ln x$ to show that the graph is always rising and that it is everywhere concave down.

7. Show that the tangent to the graph of $f(x) = \ln 2x$ at the point where $x = a$ is parallel to the tangent to the graph of $g(x) = \ln x$ at $x = a$.

8. Find any max, min, or inflection points of the graph of $f(x) = x \ln x$ and sketch it.

TEST A

1. Differentiate:
   (a) $4 \ln 3x$
   (b) $\ln (x^3 + x^2 - 1)$
   (c) $\ln \dfrac{1 - x}{x}$
   (d) $x \ln (1 - x)$
   (e) $[\ln (2x + 1)]^2$
   (f) $\ln (\ln x)$

2. Find the equation of the tangent to the graph of $y = \ln (x + 1)$ at the point $(0, 0)$.

3. Show that, if $x > 0$, the graph of $f(x) = \dfrac{\ln x}{x}$ has an absolute max at $\left(e, \dfrac{1}{e}\right)$.

TEST B

1. Differentiate:
   (a) $3 \ln x^2$
   (b) $\ln (x^5 + x - 2)$
   (c) $(x^2 + 1) \ln x$
   (d) $\dfrac{x}{\ln x}$
   (e) $\ln \dfrac{x - 1}{x^2 + 1}$
   (f) $\ln (\ln 3x)$

2. Find the equation of the tangent to the graph of $f(x) = \ln (x^2)$ at the point $(1, 0)$.

3. Prove that the graph of $y = x \ln x$ is everywhere concave up.

STUDY QUESTIONS

1. In some applications one is interested in $\dfrac{f'(x)}{f(x)}$, called the *relative rate of change*, instead of in $f'(x)$. Show that the derivative of $\ln f(x)$ is the relative rate of change of f, and use this technique to find the relative rate of change of $f(x) = x^2 + 1$.

2. Consider the following technique, here applied to finding the derivative of $y = x^x$. First take the logarithm of both sides:

$$\ln y = \ln x^x = x \ln x$$

Now differentiate implicitly (see Goal 16):

$$\frac{1}{y} \frac{dy}{dx} = x\left(\frac{1}{x}\right) + (1) \cdot \ln x = 1 + \ln x$$

$$\frac{dy}{dx} = y(1 + \ln x) = x^x(1 + \ln x)$$

This technique is called *logarithmic differentiation*. Use it to show that

(a) $\dfrac{d}{dx} 2^x = 2^x \ln 2$

(b) $\dfrac{d}{dx} e^x = e^x$

GOAL 35

To differentiate e^x and related functions

The derivative of the exponential function e^x is e^x itself! This fact is the basis for many applications. In Goal 36 we will consider numerous problems whose solutions involve exponential functions. Here we prepare for those applications by acquiring the necessary techniques for differentiating such functions.

DERIVATIVES OF THE FUNCTIONS e^x AND e^u

From Goal 34 we know that

$$\frac{d}{dx} \ln u = \frac{1}{u} \frac{du}{dx}$$

To find the derivative of e^x, we use the fact that

$$y = e^x \quad \text{if and only if} \quad x = \ln y$$

Now to obtain dy/dx we differentiate the equation on the right implicitly (see p. 158) with respect to x:

$$\frac{d}{dx}(x) = \frac{d}{dx}(\ln y)$$

$$1 = \frac{1}{y} \frac{dy}{dx}$$

$$\frac{dy}{dx} = y = e^x$$

And there it is! Only a function of the form ce^x, where c is a constant, has a derivative equal to itself. No other function has this property. Moreover, if $y = e^x$, then

$$y' = y'' = y''' = \ldots = e^x$$

Every derivative of e^x equals e^x.

If u is a differentiable function of x, then the chain rule yields immediately

$$\frac{d}{dx} e^u = e^u \frac{du}{dx}$$

In particular, if k is a constant then we have

$$\frac{d}{dx} e^{kx} = k e^{kx}$$

Examples 35-1

(a) $\dfrac{d}{dx} 3e^x = 3e^x$

(b) $\dfrac{d}{dx} e^{2x} = 2e^{2x}$

(c) $\dfrac{d}{dx} e^{-x} = -e^{-x}$

(d) $\dfrac{d}{dt}\left(\dfrac{1}{7} e^t + 5t^2 + 3\right) = \dfrac{1}{7} e^t + 10t$

(e) $\dfrac{d}{dx} e^{x^2+x} = e^{x^2+x}(2x + 1)$

(f) $\dfrac{d}{dx}\left(1 - e^{-(1/2)x}\right) = \dfrac{1}{2} e^{-x/2}$

In parts (g), (h), and (i) we use the product rule:

(g) $\dfrac{d}{dt}(te^t) = t \cdot e^t + e^t \cdot 1 = te^t + e^t$

(h) $\dfrac{d}{dx}(x^2 e^{3x}) = x^2 \cdot e^{3x} \cdot 3 + e^{3x} \cdot 2x = 3x^2 e^{3x} + 2x e^{3x}$

(i) $\dfrac{d}{dt}(e^t \ln t) = e^t \cdot \dfrac{1}{t} + (\ln t)e^t = \dfrac{e^t}{t} + e^t \ln t$

In part (j) we use the quotient role:

(j) $\dfrac{d}{dx} \dfrac{e^x}{x} = \dfrac{xe^x - e^x}{x^2}$

In part (k) we use the formula for $\dfrac{d}{dx} \ln y$:

(k) $\dfrac{d}{dx} \ln(e^x + 2) = \dfrac{1}{e^x + 2} \cdot e^x = \dfrac{e^x}{e^x + 2}$

Goal 35 | To Differentiate e^x and Related Functions

In part (1) we use the extended power rule:

(1) $\dfrac{d}{dx}(e^x - e^{-x})^2 = 2(e^x - e^{-x})(e^x + e^{-x})$ or $2(e^{2x} - e^{-2x})$

SLOPE OF THE GRAPH OF $f(x) = e^x$

Since $\dfrac{d}{dx} e^x = e^x$, the slope of the graph of the exponential function at any point is precisely equal to the value of the function at that point.

Example 35-2

Find the equation of the tangent to the curve $y = e^x$ at the point $(1, e)$. The slope of the tangent to $y = e^x$ at $x = 1$ is e^1, or e. Thus we have

$$\frac{y - e}{x - 1} = e$$

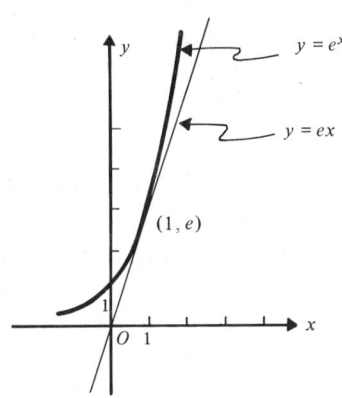

which yields

$$y - e = ex - e$$

or

$$y = ex$$

Example 35-3

Find any max, min, or inflection points and sketch the graph of $f(x) = xe^x$. We have

$$f'(x) = x \cdot e^x + e^x \cdot 1 = e^x(x + 1)$$
$$f''(x) = e^x \cdot 1 + (x + 1) \cdot e^x = e^x(x + 2)$$

Since $e^x > 0$ for all x, $f'(x) = 0$ only if $x = -1$. Since $f''(-1) > 0$, the point $\left(-1, -\dfrac{1}{e}\right)$ is a minimum. Since $f''(x) = 0$ only if $x = -2$ and $f''(x)$ changes

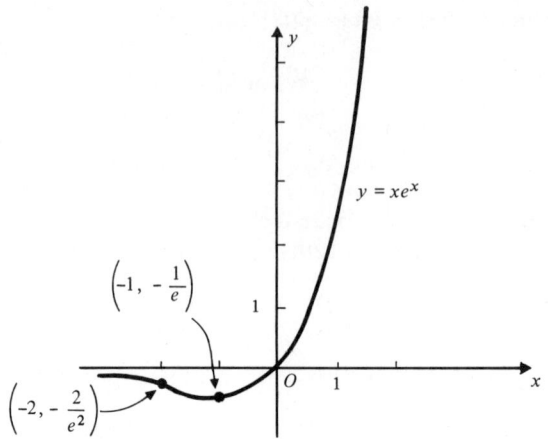

from negative to positive there, $x = -2$, $y = -2/e^2$ is the only point of inflection. Note also that (1) $f(0) = 0$; (2) the curve is concave down if $x < -2$, up if $x > -2$; (3) as $x \to \infty$, $f(x) \to \infty$; (4) as $x \to -\infty$, $f(x) \to 0$.

RATE OF CHANGE INVOLVING EXPONENTIALS

In the next Goal we will take up practical applications of exponential functions that are special cases of those to be considered in the following three examples. In each example we will be especially interested in translating an equation for the derivative of a function into a statement relating the rate of change of the function to the function itself.

Example 35-4

Find the derivative of $f(t) = ce^{kt}$, where c and k are constants.

$$f'(t) = c\frac{d}{dt}e^{kt} = c \cdot ke^{kt}$$

Now note that

$$f'(t) = k(ce^{kt}) = kf(t)$$

This equation says that the rate of change of the given function at any time is proportional to the function itself at that time. (Recall that if u is *proportional to* v then there is some constant k such that $u = kv$; k may be either positive or negative.)

Example 35-5

Find $f'(t)$ if $f(t) = A - ce^{-kt}$, where a, c, and k are constants.

$$f'(t) = 0 - c\left(\frac{d}{dt}e^{-kt}\right) = cke^{-kt}$$

Goal 35 | To Differentiate e^x and Related Functions

If we add and subtract the constant kA to $f'(t)$, we get

$$f'(t) = kA - kA + cke^{-kt} = kA - k(A - ce^{-kt})$$
$$= kA - kf(t) = k[A - f(t)]$$

The algebraic manipulation just performed shows that the rate of change of the given function is at any time proportional to the difference between the fixed constant A and the (value of) the function itself at that time.

Example 35-6

If $f(t) = \dfrac{A}{1 + ce^{-Akt}}$, where A, c, and k are constants, show that $f'(t) = kf(t)[A - f(t)]$.

This equation says that the rate of change of the given function is proportional both to the function and to the difference between the constant A and the function. Although the algebraic manipulation that follows is messy, it is included here for the sake of completeness. After verifying that the derivative of $f(t)$ is as advertised above, we will comment on the importance of functions of the types given in this and the preceding two examples.

To find $f'(t)$, we first rewrite $f(t)$ as follows:

$$f(t) = A(1 + ce^{-Akt})^{-1}$$

Then we use the formula for $\dfrac{d}{dx} Au^{-1}$, where $u = 1 + ce^{-Akt}$:

$$f'(t) = -A(1 + ce^{-Akt})^{-2} \cdot (ce^{-Akt} \cdot -Ak)$$
$$= \frac{kA^2 ce^{-Akt}}{(1 + ce^{-Akt})^2} = \frac{kA}{1 + ce^{-Akt}} \cdot \frac{Ace^{-Akt}}{1 + ce^{-Akt}}$$

where the form at the right reveals that

$$f'(t) = kf(t) \cdot \frac{Ace^{-Akt}}{1 + ce^{-Akt}}$$

To get the required answer, it would have to be the case that the fraction in the last equation is equal to $A - f(t)$. Is it? Let's see:

$$A - f(t) = A - \frac{A}{1 + ce^{-Akt}} = \frac{A + Ace^{-Akt} - A}{1 + ce^{-Akt}} = \frac{Ace^{-Akt}}{1 + ce^{-Akt}}$$

It is! Therefore we have shown that

$$f'(t) = kf(t)[A - f(t)]$$

SIGNIFICANCE OF EXAMPLES 35-4, 35-5, AND 35-6

In the last three examples we have verified that

$$\text{if } f(t) = ce^{kt} \quad \text{then } f'(t) = kf(t)$$
$$\text{if } f(t) = A - ce^{-kt} \quad \text{then } f'(t) = k[A - f(t)]$$
$$\text{if } f(t) = \frac{A}{1 + ce^{-Akt}} \quad \text{then } f'(t) = kf(t)[A - f(t)]$$

The reverse implications are also true! That is: if we know that an equation for $f'(t)$ on the right holds, then the corresponding one for $f(t)$ on the left also holds. We will prove these reverse implications in the chapter on differential equations.

As we will see in the next Goal, there are numerous applications in biology, physics, chemistry, business, economics, psychology, and sociology in which the rate of change of a quantity satisfies one of the three equations for $f'(t)$ given above on the right. We can then always conclude that $f(t)$ satisfies the corresponding equation given just to its left.

DETERMINING THE CONSTANTS

We now show how given data may be used to evaluate the constants in functions like those of the last three examples.

Example 35-7

If $f'(t) = f(t)$ and $f(0) = 6$, find $f(t)$.

From our preceding discussion we know that $f(t)$ is of the form ce^{kt} and that here $k = 1$. So $f(t) = ce^t$. Since $f(0) = 6$, we have $6 = ce^0 = c$ and

$$f(t) = 6e^t$$

Example 35-8

If $f'(t) = 2[A - f(t)]$ and if $f(0) = 4$ and $f(2) = 5$, find $f(t)$.

Since $f'(t)$ is of the form $k[A - f(t)]$, $f(t)$ is of the form $A - ce^{-kt}$. We are given that $k = 2$. Now we use the other data given to determine A and c.

$$f(0) = 4 = A - ce^{-2 \cdot 0} = A - ce^0 = A - c \cdot 1 = A - c$$
$$f(2) = 5 = A - ce^{-2 \cdot 2} = A - ce^{-4}$$

From the first of these equations we get that $A = 4 + c$. We replace A in the second equation:

$$5 = 4 + c - ce^{-4} \qquad 1 = c(1 - e^{-4}) \qquad c = \frac{1}{1 - e^{-4}}$$

$$c \approx 1.02 \qquad A \approx 4 + 1.02 = 5.02$$

Goal 35 | To Differentiate e^x and Related Functions

So
$$f(t) = 5.02 - 1.02e^{-2t}$$

EXERCISES

1. Differentiate:
   (a) $5e^x$
   (b) $\frac{11}{7}e^t$
   (c) e^{x^2}
   (d) e^{-8t}
   (e) $e^{x^3-x^2+2x+1}$
   (f) $e^{1/t}$
   (g) $e^{\sqrt{x}}$
   (h) $x^2 e^x$
   (i) $(t^2 - t + 1)e^{2t}$
   (j) e^{t^2}/t^2
   (k) $\dfrac{1}{1 + e^x}$
   (l) $\dfrac{2}{3 + e^{-4x}}$
   (m) $e^x \ln x$
   (n) e^3
   (o) $\ln e^x$
   (p) $e^{\ln x}$

2. Find the equations of the tangents to the given curves at the designated points.
   (a) $y = e^x$ at $(0, 1)$
   (b) $y = xe^x$ at $(1, e)$
   (c) $y = e^{x^2}$ at $(0, 1)$

3. For each of the following, show that $f'(t) = kf(t)$. What is k?
   (a) $f(t) = e^{10t}$
   (b) $f(t) = 5e^{t/2}$
   (c) $f(t) = \frac{1}{6}e^{-3t}$

4. For each of the following show that $f'(t) = k[A - f(t)]$. What are k and A?
   (a) $f(t) = 7 - e^{-t}$
   (b) $f(t) = 5 - e^{-2t}$
   (c) $f(t) = 3 - 2e^{-4t}$

5. If $f(t) = 1/(1 + e^{-t})$, show that $f'(t)$ can be written in the form $f'(t) = kf(t)[A - f(t)]$. What are k and A?

   In Exercises 6, 7, and 8 use the following facts:

   if $f'(t) = kf(t)$ then $f(t) = ce^{kt}$
   if $f'(t) = k[A - f(t)]$ then $f(t) = A - ce^{-kt}$
   if $f'(t) = kf(t)[A - f(t)]$ then $f(t) = \dfrac{A}{1 + ce^{-Akt}}$

6. Find $f(t)$ if
   (a) $f'(t) = 4f(t)$, $f(0) = 3$
   (b) $f'(t) = -3f(t)$, $f(0) = 8$
   (c) $f'(t) = kf(t)$, $f(0) = 3$, $f(1) = 9$

7. Find $f(t)$ if
   (a) $f'(t) = 3[8 - f(t)]$, $f(0) = 6$
   (b) $f'(t) = A - f(t)$, $f(0) = 3$, $f(1) = 4$

8. Find $f(t)$ if $f'(t) = 3f(t)[5 - f(t)]$ and $f(0) = 1$.

9. Find any maximum, minimum, or inflection points of the graph of
$$y = e^{-x^2/2}$$
and sketch it.

TEST A

1. Differentiate:

   (a) $f(x) = e^{x^2 - 5x + 2}$ (c) $g(x) = \dfrac{x}{e^x}$

   (b) $f(t) = te^{3t}$ (d) $y = \dfrac{3}{1 + e^{-t}}$

2. Find the equation of the tangent to the curve $y = 2e^{2x}$ at $(0, 2)$.
3. If $f(t) = -5e^{-2t}$, show that $f'(t) = kf(t)$ for some constant k. What is k?
4. If $f'(t) = 2f(t)$ and $f(0) = 8$, find $f(t)$. HINT: If $f'(t) = kf(t)$, then $f(t) = ce^{kt}$.

TEST B

1. Differentiate:

   (a) $f(x) = e^{x^3 + 3x - 1}$ (c) $y = \dfrac{e^x}{x^2}$

   (b) $g(t) = te^{-t}$ (d) $f(t) = \dfrac{4}{1 + e^{-2t}}$

2. Find the equation of the tangent to $y = e^{-x}$ at $(0, 1)$.
3. If $f(t) = 3e^{4t}$, show that $f'(t) = kf(t)$ for some k. What is k?
4. Find $f(t)$ if $f'(t) = kf(t)$ and if $f(0) = 4$ and $f(2) = 1$. HINT: $f(t) = ce^{kt}$.

STUDY QUESTIONS

1. Use a hand calculator to determine $(e^h - 1)/h$ for values of h closer and closer to zero. [It is true that as $h \to 0$, $\lim (e^h - 1)/h = 1$.]
2. Using Study Question 1, give an argument that $\dfrac{d}{dx} e^x = e^x$. HINT: Go back to the definition of the derivative as a limit.
3. Show that the tangent to the curve $y = e^x$, at the point where $x = a$, intersects the x-axis at $(a - 1, 0)$.
4. Use the technique of logarithmic differentiation (see Study Question 2, p. 334) to show that if $y = a^u$, where u is a differentiable function of x, then

Goal 35 | To Differentiate e^x and Related Functions

$$\frac{dy}{dx} = a^u \frac{du}{dx} \ln a$$

HINT: $y = a^u$ is equivalent to $\ln y = \ln a^u = u \ln a$. Now differentiate $\ln y = u \ln a$ implicitly with respect to x. What does this formula yield if $a = e$?

5. Given $f'(t) = kf(t)$. Consider the function $f(t)/e^{kt}$; show, using the quotient rule, that the derivative of this function is zero. Then conclude, by Corollary I of the MVT (p. 264), that $f(t)/e^{kt}$ equals a constant, say c, thus proving that $f(t) = ce^{kt}$.

GOAL 36

To solve problems involving exponential growth or decay

When a quantity increases, its rate of change is called its *rate of growth*; when it decreases, its rate of change is called its rate of *decay*.

In the last Goal we showed that the rates of change of three classes of functions satisfied certain conditions. In this Goal we analyze, in turn, the particular class of functions associated with each of three derivatives, suggest some applications involving each, and then solve practical problems on growth and decay.

On page 340 of Goal 35 we noted that, if we know that $f'(t)$ satisfied a certain given equation, then we can always conclude that $f(t)$ is of a specified form. Although we have not yet proved this, we will use these implications throughout this Goal.

CASE I

The rate of change of a quantity is proportional to the amount or magnitude of the quantity present (see p. 338).

If $f(t)$ is the amount at time t and $k > 0$, then

If f is increasing or growing	If f is decreasing or decaying
$f'(t) = kf(t)$	$f'(t) = -kf(t)$

which implies that

$f(t) = ce^{kt}$	$f(t) = ce^{-kt}$

for some positive constant c

Goal 36 | To Solve Problems Involving Exponential Growth or Decay

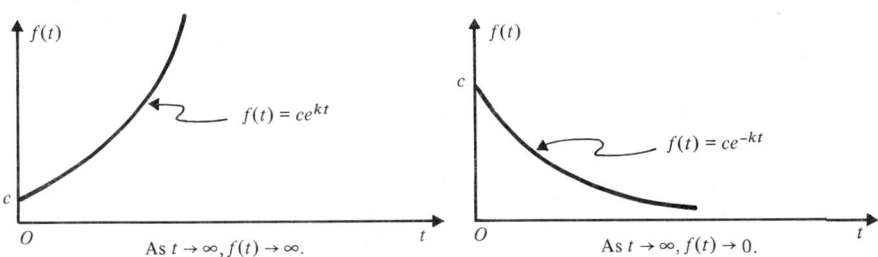

In both cases, note that $f(0) = c \cdot e^0 = c \cdot 1 = c$. So c is the *initial amount present*, the amount at time $t = 0$.

SOME APPLICATIONS OF EXPONENTIAL GROWTH

(1) A colony of bacteria may grow at a rate proportional to its size. (2) Other populations, such as of humans, rodents, or fruit flies, whose supply of food is unlimited, may also grow at a rate proportional to the size of the population. (3) Money invested at interest which is compounded continuously accumulates at a rate proportional to the amount present. The constant of proportionality here is the interest rate. (4) The demand for certain precious commodities (gas, oil, electricity, valuable metals) is growing at a rate proportional to the existing demand.

Each of the above quantities (population, amount, demand) is a function of the form ce^{kt} ($k > 0$).

SOME APPLICATIONS OF EXPONENTIAL DECAY

(1) Radioactive isotopes, such as uranium-235, strontium 90, iodine-131, and carbon-14, decay at a rate proportional to the amount still present. (2) If P is the *present value* of a fixed sum of money A due t yr from now, where the interest is compounded continuously, then P decreases at a rate proportional to the value of the investment. (3) The concentration of a drug in the bloodstream often drops at a rate proportional to the existing concentration. (4) As a beam of light passes through murky water (or air), its intensity at any depth (or distance) decreases at a rate proportional to the intensity at that depth.

Each of the above quantities (amount, present value, concentration, intensity) is a function of the form ce^{-kt} ($k > 0$).

Example 36-1

In 1970 the world population was $3\frac{1}{2}$ billion. Since then it has been growing at a rate proportional to the population and the factor of proportionality is 1.8% per yr. At that rate, how many years would it take for there to be one person per square foot of land? (The land area of Earth is approximately 200,000,000 sq miles, or about 5.5×10^{15} sq ft.)

If $P(t)$ is the population at time t, the problem tells us that P satisfies the equation $dP/dt = 0.018P$. Its solution is the exponential growth equation

$$P(t) = P_0 e^{0.018t}$$

where P_0 is the initial population. Letting $t = 0$ correspond to 1970, we have

$$3.5 \times 10^9 = P(0) = P_0 e^0 = P_0$$

So

$$P(t) = (3.5 \times 10^9) e^{0.018t}$$

The question is: for what t does $P(t) = 5.5 \times 10^{15}$? We solve:

$$(3.5)(10^9) e^{0.018t} = (5.5) 10^{15}$$

$$e^{0.018t} \approx (1.6) 10^6$$

Taking the logarithm of each side yields

$$0.018t \approx \ln 1.6 + 6 \ln 10 \approx 14.3$$

$$t \approx 800 \text{ yr}*$$

So, if the human population continued to grow at the present rate, there would be one person for every square foot of land in just 800 yr.

Example 36-2

If 1¢ had been invested 2000 yr ago at 6% yearly interest compounded continuously, what would it be worth now?

From Goal 32 (p. 316) we know that the amount $A(t)$, at time t, satisfies the exponential growth equation $A(t) = A_0 e^{0.06t}$. Since 1¢ = \$0.01 = the initial amount A_0, we have

$$A(t) = 0.01^{0.06t}$$

For $t = 2000$ this gives

$$A(2000) = 0.01 e^{(0.06)(2000)} = 0.01 e^{120}$$

Taking the logarithm of each side yields

$$\ln [A(2000)] = \ln 0.01 + 120 = -04.6 + 120 = 115.4$$

---

*Results in this Goal are rounded off whenever the resulting approximation seems reasonable.

Goal 36 | To Solve Problems Involving Exponential Growth or Decay

So when $t = 2000$, $A = e^{115.4}$. Let's express this answer as an integral power of 10, say 10^k:

$$e^{115.4} = 10^k$$
$$115.4 = k \ln 10 \approx k(2.3)$$
$$k = 50$$

Thus the penny would by now have grown to $\$10^{50}$; that is, to $\$100,000,000,000,000,000,000,000,000,000,000,000,000,000,000,000,000$.

Example 36-3

The radioactive isotope iodine-131 is injected into the body to test how fast the thyroid gland absorbs iodine. It is important to note how fast this isotope decays. It is known that I_{131} decays at a rate proportional to the amount present and that the proportionality factor is 8.6% per day. How long will it be before half of a given quantity has decayed?

If $Q(t)$ is the amount of I_{131} present at time t, then the problem tells us that Q satisfies the equation $dQ/dt = -0.086Q$. Its solution is the exponential decay equation

$$Q(t) = Q_0 e^{-0.086t}$$

where Q_0 is the initial amount present. The question is: for what t does $Q(t) = \frac{1}{2} Q_0$? We have

$$\frac{1}{2} Q_0 = Q_0 e^{-0.086t} \qquad \frac{1}{2} = e^{-0.086t}$$

$$\ln \frac{1}{2} = -0.086t \qquad -\ln 2 = -0.086t$$

$$0.6931 = 0.086t \qquad t \approx 8$$

Thus the half-life (see p. 327) of iodine-131 is about 8 days.

Example 36-4

One important method of dating fossil remains is to determine what portion of the carbon content of a fossil is the radioactive isotope carbon-14. During life, any organism exchanges carbon with its environment. Upon death this circulation ceases, and the C_{14} in the organism then decays at a rate proportional to the amount present. The proportionality factor is 0.012% per yr. When did an animal die if an archaeologist determines that only 25% of the original amount of C_{14} is still present in its fossil remains?

The quantity Q of C_{14} present at time t satisfies the equation $dQ/dt = -0.00012Q$ with solution

$$Q(t) = Q_0 e^{-0.00012t}$$

We are asked to find t when $Q(t) = 0.25Q_0$.

$$0.25Q_0 = Q_0 e^{-0.00012t} \qquad 0.25 = e^{-0.00012t}$$
$$\ln 0.25 = -0.00012t \qquad -1.386 \approx -0.00012t$$
$$t \approx 11{,}550$$

Rounding to the nearest 500 yr, we see that the animal died approximately 11,500 yr ago.

NOTE. The answers obtained in Examples 36-3 and 36-4 did *not* depend on Q_0. This is always true in problems in which we seek the value of t for which $Q(t)$ is a given part or percentage of Q_0.

CASE II

The rate of change of a quantity is proportional to a difference. There are two situations. The rate of change is proportional

to a fixed constant A minus the amount (or magnitude) of the quantity present:

$$f'(t) = k[A - f(t)]$$

or

to the amount (or magnitude) of the quantity present minus a fixed constant A:

$$f'(t) = -k[f(t) - A]$$

where $f(t)$ is the amount at time t and k and A are both positive.

We may conclude that

$f(t)$ is increasing or growing,
$$f(t) = A - ce^{-kt}$$

$f(t)$ is decreasing or diminishing,
$$f(t) = A + ce^{-kt}$$

for some positive constant c.

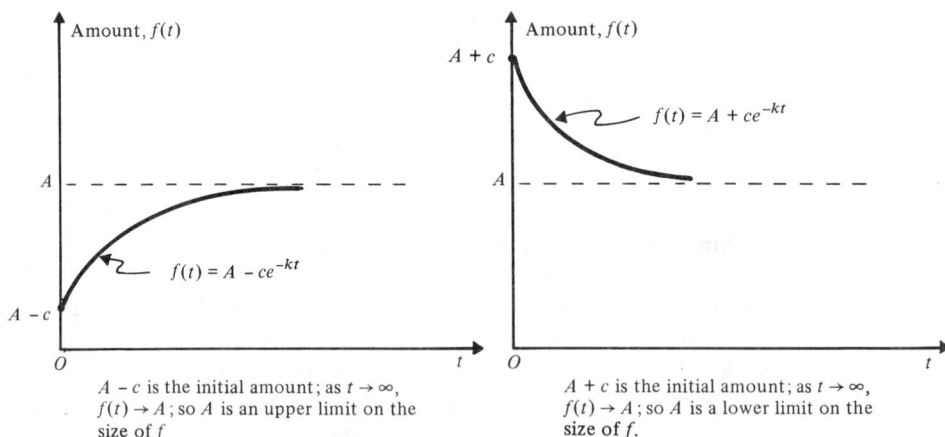

$A - c$ is the initial amount; as $t \to \infty$, $f(t) \to A$; so A is an upper limit on the size of f

$A + c$ is the initial amount; as $t \to \infty$, $f(t) \to A$; so A is a lower limit on the size of f.

SOME APPLICATIONS OF CASE II

(1) The rate of change of the temperature of a roast put into an oven is proportional to the difference between the oven temperature T and its

Goal 36 | To Solve Problems Involving Exponential Growth or Decay

own temperature. If $R(t)$ is the roast's temperature at time t, then $R(t)$ is a function of the form $R(t) = T - ce^{-kt}$.

(2) If the roast is removed from the oven, then it cools at a rate proportional to the difference between its own temperature and the room temperature F. Now $R(t)$ is of the form $R(t) = F + ce^{-kt}$.

(3) In advertising, it is generally assumed that the rate at which people hear about a product is proportional to the number of people who have not yet heard about it. If initially ($t = 0$) in a population P_0 no one knows of the product, then the number $N(t)$ of people who are aware of the product at time t is of the form $N(t) = P_0(1 - e^{-kt})$. [Note that this is a special case of $f(t) = A - ce^{-kt}$ with $A = c = P_0$.]

(4) Because of air friction, the velocity of a falling object approaches a limiting value L (rather than increasing without bound). The acceleration (rate of change of velocity) is proportional to the difference between the limiting velocity and the object's velocity. If initial velocity is zero, then at time t the object's velocity $V(t) = L(1 - e^{-kt})$.

(5) Some psychologists propose that there is a limit D to the number of words a person will learn, no matter how long the lifetime, and that the rate at which new words are learned is proportional to the number not yet learned. If $W(t)$ is the vocabulary size at time t, then, since $W(0) = 0$, $W(t) = D(1 - e^{-kt})$.

(6) If a tire has a small leak, then the air pressure inside drops at a rate proportional to the difference between the inside pressure and the fixed outside pressure O. At time t the inside pressure $P(t) = O + ce^{-kt}$.

Example 36-5

A roast at 68° F is put into an oven with temperature 350°. After 1 h the internal temperature of the roast is 110°. What is its temperature after 2 h? Assume that the rate of change of the roast's temperature $R(t)$ is proportional to the difference between the oven temperature and $R(t)$.

We know that $dR/dt = k(350 - R)$ and that $R(t) = 350 - ce^{-kt}$. Since $R(0) = 68$, we have

$$68 = 350 - c \qquad c = 282 \qquad R(t) = 350 - 282e^{-kt}$$

Also, $R(1) = 110$ yields

$$110 = 350 - 282e^{-k} \qquad 282e^{-k} = 240 \qquad e^{-k} \approx 0.85$$

Thus

$$R(t) = 350 - 282(e^{-k})^t = 350 - 282(0.85)^t$$

Next we find $R(2)$:

$$R(2) = 350 - 282(0.85)^2 \approx 350 - 204 \approx 146° \text{ F}$$

NOTE: Using natural logarithms and two-place decimals, the answer obtained is 144° F. The method shown above makes it unnecessary to evaluate k.

Example 36-6

Meat at 72° F is put into a freezer whose temperature is 10°. If the temperature of the meat is 55° after 1 h, what is its temperature after 3 h?

If $M(t)$ is the temperature of the meat at time t, then $dM/dt = -k(M - 10)$ and $M(t) = 10 + ce^{-kt}$. Since $M(0) = 72$, we have

$$72 = 10 + c \qquad c = 62 \qquad M(t) = 10 + 62e^{-kt}$$

From $M(1) = 55$ we get

$$55 = 10 + 62e^{-k} \qquad e^{-k} = \frac{45}{62} \approx 0.73$$

Now we compute $M(3)$:

$$M(3) = 10 + 62(e^{-k})^3 = 10 + 62(0.73)^3 \approx 10 + 24 \approx 34$$

So after 3 h the temperature of the meat has dropped to approximately 34°F.

CASE III

The rate of change of a quantity is proportional both to the amount (or magnitude) of the quantity and to the difference between a fixed constant A and the magnitude of the quantity.

If $f(t)$ is the magnitude at time t, then

$$f'(t) = kf(t)[A - f(t)]$$

where k and A are both positive. In this case we may always conclude that

$$f(t) = \frac{A}{1 + ce^{-Akt}}$$

for some positive constant c.

In most applications, $c > 1$. In these cases, the initial amount $A/(1 + c)$ is *less* than $A/2$. In all applications, since the exponent of e in the expression for $f(t)$ is negative for all positive t, therefore, as $t \to \infty$, (1) $ce^{-Akt} \to 0$; (2) the denominator of $f(t) \to 1$; and (3) $f(t) \to A$. Thus A is an upper limit of f in this growth model.

UNRESTRICTED VERSUS RESTRICTED GROWTH

On p. 351 we show the graphs of the growth functions of Cases I and III.

The growth function of Case I is known as the *unrestricted* (or *uninhibited or unchecked*) model. It is not a very realistic one for most populations. It is clear that human populations cannot continue endlessly to grow expo-

Goal 36 | To Solve Problems Involving Exponential Growth or Decay

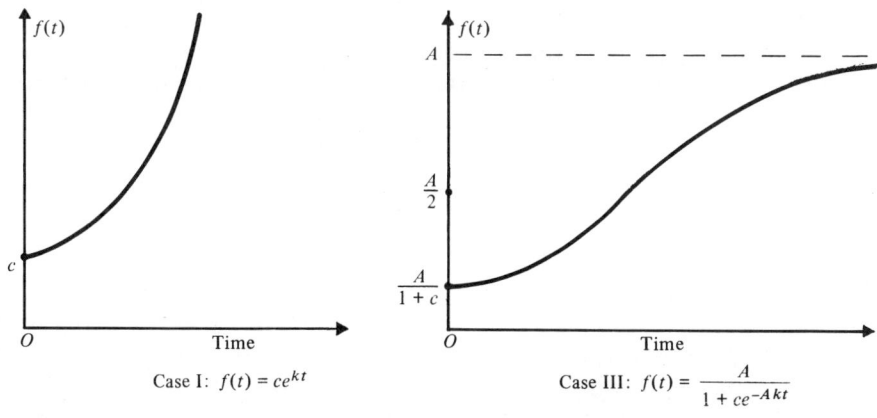

Case I: $f(t) = ce^{kt}$

Case III: $f(t) = \dfrac{A}{1 + ce^{-Akt}}$

nentially. Not only is Earth's land area fixed, but also there are limited supplies of food, energy, and other natural resources. The growth function in Case III allows for these factors which serve to check growth. It is therefore referred to as the *restricted* (or *inhibited*) model.

The S-shaped graph in Case III is often called a *logistic curve*. It shows that

(1) The rate of growth, $f'(t)$, increases for a while, i.e., $f''(t) > 0$.
(2) The growth rate attains a maximum when $f(t) = A/2$, at the point of inflection.
(3) The growth rate then decreases ($f''(t) < 0$), approaching zero as $f(t)$ approaches its upper limit.

It is not difficult to verify these statements.

When applied to populations, the fixed constant A in the restricted growth model (Case III) is often called the *maximum sustainable population*. Since

$$f'(t) = kf(t)[A - f(t)]$$

we see that the rate of growth is proportional to both the population attained by time t and the additional population still attainable.

APPLICATIONS OF CASE III

(1) Some diseases spread through a (finite) population P at a rate proportional to the number of people $N(t)$ infected by time t and the number, $P - N(t)$, not yet infected. Thus $N'(t) = kN(P - N)$ and, for some positive c and k,

$$N(t) = \dfrac{P}{1 + ce^{-Pkt}}$$

(2) A rumor (or fad, or new religious cult) often spreads through a population P according to the formula in (1), where $N(t)$ is the number of people who have heard the rumor (acquired the fad, converted to the cult) and $P - N(t)$ is the number who have not.

(3) Bacteria in a culture on a Petri dish grow at a rate proportional to the product of the existing population and the difference between the maximum sustainable population and the existing population. (Replace bacteria on a Petri dish by fish in a small lake, ants confined to a small receptacle, fruit flies supplied with only a limited amount of food, yeast cells, and so on.)

(4) Advertisers sometimes assume that sales of a particular product depend on the number of TV commercials for the product and that the rate of increase in sales is proportional both to the existing sales and the additional sales they conjecture as possible.

(5) In an autocatalytic reaction a substance changes into a new one at a rate proportional to the product of the amount of the new substance present and the amount of the original substance still unchanged.

Example 36-7

In a town of 10,000, two people start a rumor that spreads to 25 people by the end of 1 day. How many know the rumor by the end of 3 days? [Assume the rumor spreads as in application (1) above.]

Since the rumor can reach at most $P = 10,000$ people, $N'(t) = kN(t)[10,000 - N(t)]$, and

$$N(t) = \frac{10,000}{1 + ce^{-10,000kt}}$$

From $N(0) = 2$ we get

$$2 = \frac{10,000}{1 + c} \qquad 1 + c = 5000 \qquad c = 4999$$

From $N(1) = 25$ we obtain

$$25 = \frac{10,000}{1 + 4999e^{-10,000k \cdot 1}} \qquad 1 + 4999e^{-10,000k} = 400$$

$$4999e^{-10,000k} = 399 \qquad e^{-10,000k} = \frac{399}{4999} \approx 0.08$$

We seek $N(3)$:

$$N(3) = \frac{10,000}{1 + 4999(e^{-10,000k})^3} \approx \frac{10,000}{1 + 4999(0.08)^3} \approx 2800$$

By the end of 3 days, approximately 2800 people will know the rumor.

Goal 36 | **To Solve Problems Involving Exponential Growth or Decay**

SUMMARY OF CASES

For convenience in doing the exercises, we summarize here the cases of growth or decay considered in this Goal.

I. The rate of change of a quantity is proportional to the amount present. Then, if $k > 0$,

growth: $f'(t) = kf(t) \quad \rightarrow \quad f(t) = ce^{kt}$
decay: $f'(t) = -kf(t) \quad \rightarrow \quad f(t) = ce^{-kt}$

II. The rate of change of a quantity is proportional to the difference between a fixed constant A and the amount of the quantity present. Then, if $k > 0$,

growth: $f'(t) = k[A - f(t)] \quad \rightarrow \quad f(t) = A - ce^{-kt}$
decay: $f'(t) = -k[f(t) - A] \quad \rightarrow \quad f(t) = A + ce^{-kt}$

III. The rate of change of a quantity is proportional both to the amount of the quantity and to the difference between a fixed constant A and the magnitude of the quantity. Then, if $k > 0$, $A > 0$,

growth: $f'(t) = kf(t)[A - f(t)] \quad \rightarrow \quad f(t) = \dfrac{A}{1 + ce^{-Akt}}$

Most applications are in terms of the rate of change (growth or decay) of some quantity, say $f(t)$. This should be translated into an equation relating $f'(t)$ to $f(t)$. In the exercises it will be one of the five equations to the left of the arrows above. You may then immediately write down the corresponding equation for $f(t)$ from the right, immediately replacing k or A, if possible. Any still undetermined constants may usually be found from other data given in the problem.

EXERCISES

1. Suppose that a bacterial culture is growing at a rate proportional to its size. If the initial size is 1000 and the growth rate is 3% per h, write down an expression for the number of bacteria as a function of time.

2. Suppose that a colony of fruit flies is growing at a rate proportional to its size. If the initial size is 500 and the size after 3 days is 1600, derive an expression for the size of the colony as a function of time.

3. Suppose that the population of a country is growing at a rate proportional to its size. If the growth rate is 4% per yr, how long will it take the population to double?

4. One isotope of plutonium, an important by-product in nuclear reactors, decays at a rate proportional to the quantity present, with a proportionality factor of 0.003% per yr. How long before a given quantity will decay to one half its original amount?

5. Radium-226 decays with a proportionality factor of 0.043% per yr. What is its half-life?

6. Carbon-14 decays with a proportionality factor of 0.012% per yr. When did an animal die if a technician determines that only 10% of the original C_{14} is still present in its remains?

7. With continuous compounding the rate of growth of money is proportional to the amount of money. At a yearly interest rate of 5% compounded continuously, how long does it take an investment to triple?

8. If a bank claims that it will double your money in 9.9 yr, what yearly interest rate, compounded continuously, is it offering?

9. If with continuous compounding a $500 investment grows to $1000 in 7.7 yr, how much will it be worth in 10 yr?

10. How much was invested 10 yr ago at a yearly interest rate of 6% compounded continuously if the balance is now $2733?

11. On the birth of their first child a couple decided to invest a sum of money that would accumulate at 8% a yr compounded continuously to $10,000 on the child's twentieth birthday. How much did they have to invest?

12. Suppose that the volume of an object grows at a rate proportional to its existing volume. If it takes 10 days for the volume to triple, how long will it take it to quadruple?

13. If the half-life of uranium-235 is 7.1×10^8 yr, express the quantity Q present in t yr if the initial amount is Q_0.

14. When did an animal die if 80% of the original carbon-14 is still present in its remains? ($k = 0.012\%$ per yr.)

15. Potassium-40 (K_{40}) is a radioactive isotope used in dating fossil remains. If its half-life is 1.4×10^9, when was a tree cut down if a log cut from it retains 90% of the original K_{40}?

16. After an intravenous injection of 100 mg of demerol into a 150-lb man, the concentration of the drug in the bloodstream is 0.02 mg per milliliter (ml) of blood. Assuming that the concentration drops at a rate proportional to the existing concentration with a proportionality factor of 35% per h, what will the drug concentration be after 4 h?

17. Suppose the rate of growth of an animal population, whose food and space are limited, is proportional to the difference between a maximum population of 5000 and the existing population. If the population was 2000 five yr ago and is now 3000, express the population as a function of time using 5 yr ago as $t = 0$.

18. Suppose a frozen pie is taken from a 20° F freezer and placed in a 400° oven. If its temperature after 1 h is 210°, derive an expression for the temperature of the pie t h after it is put into the oven.

19. Meat put into a freezer cools at a rate proportional to the difference between its own temperature and that of the freezer. If a roast at room

temperature of 68°F is put into a 20° freezer, and if the temperature of the roast after 2 h is 40°, what is its temperature after 5 h?

20. On a cold arctic day with temperature $-100°$ F (100 degrees below zero), the temperature of a polar bear drops from 98° to 90° in just 5 min. What will its temperature be after 1 h? Assume that the bear's temperature changes as in Case II. (HINT: $e^{-0.48} \approx 0.62$.)

21. Suppose the rate at which one learns new words is proportional to the number of words still to be learned. Suppose further that the limiting value is 50,000 words, that there is an initial vocabulary of 0 words, and that after 10 yr the vocabulary is 27,500 words. Find the vocabulary after 25 yr.

22. As the result of a certain kind of advertising, the rate at which more people hear about a product is proportional to the number of people who have not yet heard about it. Suppose that the size of a community is 15,000, that at time $t = 0$ no one has heard about the product, and that after a week 1500 people know of it. How long will it take for 2150 people to have heard of the product?

23. Because of air friction, the acceleration of an object is proportional to the difference between the limiting velocity L and its own velocity. Suppose that an object falling from rest achieves a velocity of 50 ft/sec after 2 sec and a velocity of 80 ft/sec after 4 sec. What is the limiting velocity?

24. If the concentration of an injected medication in the bloodstream drops at a rate proportional to the existing concentration, with a factor of 30% per h, after how long will the concentration be one-tenth that of the initial concentration?

25. Suppose a flu-like virus has spread through a town of population 50,000 at a rate proportional both to the number of people already infected and to the number still uninfected. If 100 people were infected initially and 500 had been after 10 days, derive an expression for the number of sick people as a function of time.

26. Suppose the rate of growth of a population is proportional both to the existing population and to the difference between a limiting size of 25,000 and the existing size. If the population was 10,000 four yr ago and is 15,000 at present, what will it be 4 yr from now?

27. A rumor spreads in a town of 1000 at a rate proportional to both the number who know it and the number who do not. If one person starts the rumor and 10 people know it by the end of 1 day, how many know it by the end of 3 days?

28. In an autocatalytic reaction (see p. 352) there are initially 150 units of the original substance and 15 units of the new one. If the factor of proportionality is 0.06% per sec, express the amount of the new substance as a function of time.

29. As a beam of light passes through a medium, its intensity at any depth changes at a rate proportional to the existing intensity at that depth. Suppose the absorption factor for salt water is 1.5% per cm. After how many meters will the intensity be 1% of the original strength?
30. If the intensity of light has an absorption factor, for a particular smog condition, of 1% per ft, after how many feet will the intensity of an entering beam of light be cut in half?

TEST A

1. Suppose that a population is growing at a rate proportional to its size. If the growth rate is 3% per yr, how long will it take for the population to triple?
2. Carbon-14 decays with a proportionality factor of 0.012% per yr. When did an animal die if an archeologist finds that only 20% of the original C_{14} is still present in the animal's remains?
3. If under continuous compounding a $1000 investment grows to $1500 in 10 yr, how much will it be worth after 15 yr?
4. If water at 70° F is placed in a 20° freezer and the water temperature is 50° after 1 h, when will there be ice (32°)? Assume that the water cools at a rate proportional to the difference between its own temperature and that of the freezer, and neglect the time required for water at 32° to change to ice at 32°.
5. An epidemic is spreading through a city of 15,000 at a rate proportional both to the number who have already caught the disease and to the number still uninfected. If 100 people had been infected as of 1 week ago and 400 more people came down with it this past week, how many in all will have had it 1 week from now?

TEST B

1. A culture of bacteria is growing at a rate proportional to its size. If the culture increases from 1000 to 1500 in 5 h, how long does it take to double in size?
2. Cobalt-60 is a radioactive isotope used in cancer treatment. If its half-life is 5.25 yr, determine the function for the quantity present in terms of the number t of years and the original quantity Q_0.
3. How much will an investment of $1000 be worth in 4 yr at a yearly interest rate of 10% compounded continuously?
4. A roast at 75° F is put into an oven with temperature 375°. After 1 h the internal temperature of the roast is 125°. What is its temperature after 2 h? Assume that the rate of change of the roast's temperature

Goal 36 | **To Solve Problems Involving Exponential Growth or Decay**

is proportional to the difference between the oven temperature and its own.

5. Because of limited food and space, a squirrel population cannot exceed 1000. It grows at a rate proportional both to the existing population and to the attainable additional population. If there were 100 squirrels 2 yr ago and the population was 400 one yr ago, about how many squirrels are there now?

STUDY QUESTIONS

1. An important characteristic of the logistic curve (Case III) is its inflection point, since this is the point at which the rate of growth is a maximum. Suppose a quantity q satisfies the equation

$$q(t) = \frac{1}{1 + 2e^{-t}}$$

Show that the point of inflection is at $(\ln 2, \frac{1}{2})$.

HINT: Note that

$$q''(t) = \frac{-2e^{-t}(1 - 2e^{-t})}{(1 + 2e^{-t})^3}$$

2. If q changes at a rate proportional both to the existing amount of q and to the difference $A - q$, where A is a fixed constant, then

$$q'(t) = kq(A - q)$$

Show that when the rate of growth—$q'(t)$—is a maximum, q equals $\frac{A}{2}$. HINT: Find $q''(t)$ by differentiating $q'(t)$ with respect to t; then solve $q''(t) = 0$.

Summary of Goals 31–36

You should now be able to

(1) use the laws of exponents and logarithms to simplify expressions involving them;

(2) sketch the graph of simple exponential or logarithmic functions;

(3) use the fact that the functions b^x and $\log_b x$ are inverses;

(4) solve simple exponential equations;

(5) use the properties of e^x and $\ln x$ to simplify expressions involving them;

(6) sketch the graphs of $f(x) = e^x$ and $g(x) = \ln x$ and related functions;

(7) use the fact that e^x and $\ln x$ are inverses; for example, be able to find an integer k such that $e^{25} \approx 10^k$.

(8) differentiate exponential and logarithmic functions;

(9) find the equation of the tangent to an exponential or logarithmic curve at a specified point.

(10) solve problems involving exponential growth or decay: given information about the rate of change of a function and other appropriate data, determine the function in terms of time or evaluate it at a specific time.

Review Exercises

GOAL 31

1. Simplify
   (a) $16^{3/2}$
   (b) $\dfrac{2}{4^0}$
   (c) $\dfrac{x^2 \cdot x^4}{(x^3)^2}$
   (d) $(t^2 t^3)^4$

2. Simplify
   (a) $\log_2 12 - \log_2 6$
   (b) $\log_b b^3$
   (c) $10^{\log_{10} 4}$

3. Solve for t: $5^{3t-1} = \frac{1}{25}$.

4. Sketch on a single set of axes: $y = 2^x$ and $y = \log_2 x$.

GOAL 32

5. Simplify

(a) $e^4 e$ (b) $\dfrac{e^2}{(e^2)^3}$ (c) $e^x \cdot e^{-x}$

6. Sketch the graph of $y = 2e^{-x}$.

GOAL 33

7. Simplify

(a) $e^{\ln 3}$ (b) $e^{\ln (x/2)}$ (c) $\ln \dfrac{1}{e^2}$

8. Find x correct to the nearest hundredth if $e^{3x} = 2900$.

9. Graph on a single set of axes: $y = \log_2 x$ and $y = \ln x$.

10. Find an integer k for which $e^{39} = 10^k$.

11. If Q_0 is the amount of radioactive isotope uranium-235 present in a certain specimen now, then the quantity that will be left t yr from now is given by

$$Q = Q_0 e^{-(9.76 \times 10^{-10} t)}$$

Find the half-life of uranium-235.

GOAL 34

12. Differentiate

(a) $3 \ln 5x$ (d) $[\ln (3x + 2)]^2$

(b) $\dfrac{1}{3} \ln \dfrac{4}{x^3 + 3x + 1}$ (e) $\ln (\ln 2x)$

(c) $(x^2 - 1) \ln x$ (f) $\ln x^4$

GOAL 35

13. Differentiate

(a) e^{4x} (c) xe^x (e) $(\ln x)e^x$

(b) e^{-x^2} (d) $\dfrac{e^{-x}}{x}$ (f) $(e^{2x} + 1)^3$

14. Find the equation of the tangent to the graph of $y = e^{-2x}$ at the point where $x = 0$.

15. If $f(t) = 2e^{3t}$ find k such that $f'(t) = kf(t)$.

16. Find $f(t)$ if $f'(t) = -f(t)$ and if $f(0) = 4$. HINT: Use the fact that if $f'(t) = kf(t)$ then $f(t) = ce^{kt}$.

GOAL 36

(You may refer to the summary of cases on p. 353.)

17. A population is growing at a rate proportional to its size. If its growth rate is 4% per yr, how long will it take the population to double?
18. If $500 is deposited in an account that earns 6% compounded continuously, how much will it accumulate to in 3 yr?
19. Uranium-232 has a half-life of 74 yr. Find the amount present t yr from now if the current amount is Q_0.
20. A population of animals grows at a rate proportional to the difference between a limiting value of 2500 and the existing population. If the population was 1000 three yr ago and is 1500 at present, what will it be 3 yr from now?
21. A disease spreads in a town of 20,000 at a rate proportional to both the number who have caught it and the number who have not. If 200 people had been infected by noon yesterday and 50 more people became infected by noon today, how many in all will have had it by noon tomorrow?

Chapter Test A

1. Simplify:
   (a) $27^{1/3}$
   (b) $\log_3 54 - \log_3 6$
   (c) $2^{\log_2(1/8)}$
   (d) $\left(\dfrac{x^3 \cdot x}{x^2}\right)^3$
   (e) $\ln e^{\sqrt{x}}$

2. Solve for t: $3^{1-2t} = \sqrt{27}$

3. Sketch on a single set of axes:
   (a) $y = e^x$
   (b) $y = \ln x$
   (c) $y = \ln \dfrac{1}{x}$

4. Find x to the nearest hundredth if $e^{-2x} = 0.4$.

5. Find an integer k such that $e^{60} \approx 10^k$.

6. Differentiate:
   (a) $y = e^{2x} - e^{-x}$
   (b) $y = x^2 \ln x^2$
   (c) $y = x^2 e^x$
   (d) $y = \sqrt{1 + \ln 2x}$
   (e) $y = \ln \sqrt{4 - x^2}$
   (f) $y = 1 - e^{-(x/2)}$

7. Find the equation of the tangent to the graph of $y = x \ln x$ at the point where $x = 1$.

8. If $f(t) = ce^{kt}$ prove that $f'(t) = kf(t)$.

9. If there are Q_0 units of a substance now and there will be $Q_0 e^{-0.1t}$ units left in t yr, when will half of the present quantity have decayed?
10. If $3000 accumulates to $3700 in 3 yr under continuous compounding, what is the yearly interest rate?
11. Ice cream whose temperature is 40° F is placed in a 20° freezer and drops to 30° in 1 h. What is its temperature after 2 h? (Assume the temperature decreases as in Case II on page 353.)
12. A rumor spreads in a town of 600 people at a rate proportional to both the number who know it and the number who do not. If 2 people start the rumor and 100 people know it by the end of 1 week, how many will know it by the end of 2 weeks?

Chapter 7
The Definite Integral

The calculus has two major branches. One focuses on the derivative; the other deals with the definite integral.

We know that the derivative of a function is defined as the limit of a certain quotient involving the function. In this chapter we will see that the definite integral is also defined in terms of limits, but now involving an "infinite sum"; that is, the limit of a sum as the number of terms in the sum becomes infinite. We will see, too, that the definite integral is intimately linked to the derivative.

We begin by finding approximate answers to four problems whose solutions can be expressed as infinite sums. One of these problems is finding the area under a curve. To find exact answers to the problems, we introduce the definite integral. Its power and usefulness derive from the Fundamental Theorem of the Calculus, which enables us to evaluate definite integrals without resorting to infinite sums. An important part of this chapter is concerned with techniques for finding an antiderivative (obtaining a function from its derivative). We will see that this is crucial to the evaluation of definite integrals.

Goal 37. To approximate limits of sums
Goal 38. To find the area under a curve
Goal 39. To understand and use the definite integral
Goal 40. To use the Fundamental Theorem of the Calculus
Goal 41. To find antiderivatives
Goal 42. To integrate by substitution
Goal 43. To use integration by parts
Summary
Review Exercises
Chapter Test A

GOAL 37

To approximate limits of sums

Here, as in Goal 7, we consider four seemingly different problems. However, each can be viewed as involving an infinite sum, that is, the limit of the sum of n terms as n gets larger and larger. The four examples, together with many others, are actually disguises for a single basic problem that can easily be solved using still another important tool of the calculus. We will not solve any of these problems in this Goal, but will instead obtain approximations to their solutions. Once again here, as in Goal 7, we call attention to our method of attack, which is the same in all four examples. This approach will help us develop the powerful tool needed to find the exact solutions.

THE EXAMPLES

Example 37-1

An object is dropped so that its velocity is $32t$ ft/sec after t sec. Find the distance it travels during the first 3 sec.

From physics we know that if the velocity is constant then the distance traveled equals the product of this velocity and the time elapsed. But how can we find the distance if the velocity is changing? Our approach to this problem is to approximate the distance traveled over small time intervals, then to add these distances together to approximate the total distance. To approximate the distance over a small interval, we multiply the velocity at some point in the interval by the duration of the interval. The following table and graph summarize this computation when we choose intervals of 1 sec and take the velocity at the *end* of each interval:

Time Interval	Velocity at End of Time Interval	Approximated Distance Traveled during Interval
$t = 0$ to $t = 1$	$32(1) = 32$ ft/sec	32 ft/sec × 1 sec = 32 ft
$t = 1$ to $t = 2$	$32(2) = 64$ ft/sec	64 ft/sec × 1 sec = 64 ft
$t = 2$ to $t = 3$	$32(3) = 96$ ft/sec	96 ft/sec × 1 sec = 96 ft

Approximate total distance = (32 + 64 + 96) = 192 ft

We now divide the total time of 3 sec into smaller and smaller intervals (usually called *subintervals*). In the cases that follow, we approximate the distance over each small interval by the product of the velocity at the *end* of the interval and the duration of the interval.

(1) Subintervals of length 0.5 sec (there are six such subintervals):

Total distance $\approx 16(\frac{1}{2}) + 32(\frac{1}{2}) + 48(\frac{1}{2}) + 64(\frac{1}{2}) + 80(\frac{1}{2}) + 96(\frac{1}{2}) = 168$ ft

(2) Subintervals of length 0.1 sec (there are 30 such subintervals):

Total distance $\approx (3.2)(0.1) + (6.4)(0.1) + \ldots + (96)(0.1) = 148.8$ ft

(3) Subintervals of length 0.01 sec (there are 300 such subintervals): Total distance $\approx (0.32)(0.01) + (0.64)(0.01) + \ldots + (96)(0.01) = 144.48$ ft.

So 192, 168, 148.8, and 144.48 ft are all approximations to the distance traveled by the object during the first 3 sec. We will see shortly that the exact distance is 144 ft.

Example 37-2

A colony of bacteria grows so that at the end of t h its rate of growth is $1000(1 + t^2)$ bacteria per h. By how much does the colony increase during the first 2 h?

From biology we know that if the rate of growth is constant then the increase in the colony over some period of time equals the product of the rate of growth and the time elapsed. But how do we find the increase if the rate of growth changes? We approximate the increases over small time intervals, then add these increases together to approximate the total increase. In each case that follows, we approximate the colony's increase over a small interval by the product of the growth rate at the *midpoint* of the interval and the length of the interval (midpoints are as useful as endpoints).

(1) Subintervals of length 1 h (there are two such subintervals):

Goal 37 | **To Approximate Limits of Sums**

$$\text{Growth rate} = 1000(1 + t^2) \text{ bac/hr}$$

Total increase $\approx$
$$1000(1 + \tfrac{1}{2}^2) + 1000(1 + \tfrac{3}{2}^2) = 1250 + 3250 = 4500 \text{ bacteria}$$

(2) Subintervals of length 0.4 h (there are five such subintervals):

$1000(1 + 0.2^2)$ bac/h	$1000(1 + 0.6^2)$ bac/h	$1000(1 + 1^2)$ bac/h	$1000(1 + 1.4^2)$ bac/h	$1000(1 + 1.8^2)$ bac/h	*growth rate*
0 h 0.2 0.4 h	0.6 0.8 h	1 1.2 h	1.4 1.6 h	1.8 2 h	*time*

Total increase $\approx$
$$(0.4)(1000)[(1 + 0.2^2) + (1 + 0.6^2) + (1 + 1^2) + (1 + 1.4^2) + (1 + 1.8^2)]$$
$$\approx 4640 \text{ bacteria}$$

where we factored out the fixed subinterval 0.4 and the constant 1000 to simplify the computation.

(3) Subintervals of length 0.1 h (there are 20 such subintervals):

$1000(1 + 0.05^2)$ bac/h			$1000(1 + 1.95^2)$ bac/h	*growth rate*
0 0.1 0.2	...	1.8 1.9 2		*time (h)*

Total increase $\approx$
$$(0.1)(1000)[(1 + 0.05^2) + (1 + 0.15^2) + \ldots + (1 + 1.85^2) + (1 + 1.95^2)]$$
$$\approx 4665 \text{ bacteria}$$

So 4500, 4640, and 4665 bacteria are approximations to the increase in the colony during the first 2 h. We shall soon see that the actual increase is 4667 bacteria (rounded off to a whole number).

Example 37-3

Suppose a company determines that its marginal cost of production is, in dollars, $(0.04x - 5)$ per gadget when the production level is x gadgets. Find the additional cost when the production level is raised from 500 to 600 gadgets.

From economics we know that if the marginal cost is constant then the additional cost equals the marginal cost times the number of additional gadgets. But how do we find the additional cost if the marginal cost is changing? We approximate the additional costs over small changes in production level, then add these together to approximate the total additional cost. This time we shall approximate the added cost over a small interval by the product of the marginal cost at the *beginning* of the interval and the length of the interval (left endpoints work as well as right endpoints or midpoints!).

(1) Subintervals of length 25 units (there are four such subintervals). To

evaluate the marginal cost, we replace x by the production level at the beginning of the interval in the expression $0.04x - 5$.

Production level (gadgets): 500   525   550   575
Marginal cost/gadget ($): 15   16   17   18
Total added cost $\approx (15)(25) + (16)(25) + (17)(25) + (18)(25) = \$1650$.

(2) Subintervals (10 of them) of length 10 units. It can be verified that
Total added cost $\approx (15)(10) + (15.40)(10) + (15.80)(10) + \ldots + (18.60)(10) \approx \$1680$

(3) Subintervals (25 of them) of length four units:
Total added cost $\approx (15)(4) + (15.16)(4) + \ldots + (18.84)(4) \approx \$1692$

So $1650, $1680, and $1692 are approximations to the additional cost involved in raising the production level from 500 to 600 gadgets. The exact added cost, we will see shortly, is $1700.

Example 37-4

Find the shaded area: under the curve $y = x^2$, above the x-axis, and between the lines $x = 1$ and $x = 3$.

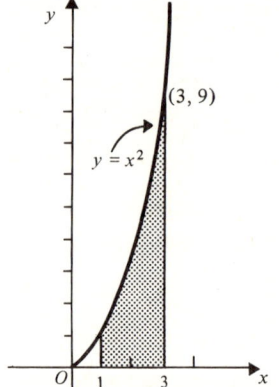

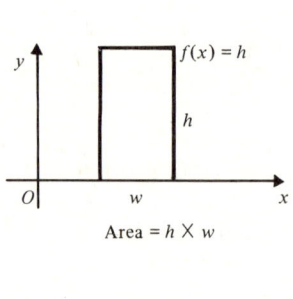

Area $= h \times w$

From geometry we know that if a function is constant, then the area is the product of height and width. But how do we find the area if the function is not constant? In the manner established in the preceding examples, we approximate the areas over small intervals, then add these together to approximate the total area. On p. 369, we show the interval from $x = 1$ to $x = 3$ divided into two one-unit subintervals. We choose to approximate the first small area under the curve, the region $ADEB$, by the rectangle $ADGB$, and to approximate the second small area, the region $BEFC$, by the rectangle $BEKC$. Thus

Total area $\approx (AD \times 1) + (BE \times 1) = (1 \times 1) + (4 \times 1) = 5$ sq units

Note that this approximation chooses the functional values at the *left* endpoints as the heights of the rectangles.

Goal 37 | To Approximate Limits of Sums

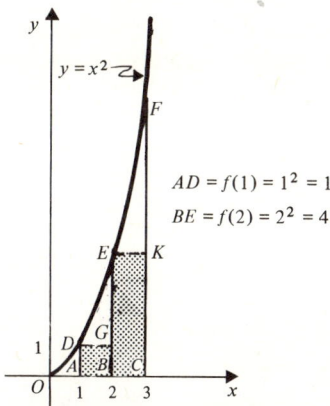

$AD = f(1) = 1^2 = 1$
$BE = f(2) = 2^2 = 4$

This does not appear to be a very good approximation to the desired area, does it? So, proceeding as before, we divide the interval from $x = 1$ to $x = 3$ into smaller and smaller subintervals, continuing to use rectangles for approximations whose heights are the left–endpoint ordinates.

(1) Subintervals of length 0.2 (10 of them). For convenience the scales on the x- and y-axes are different.

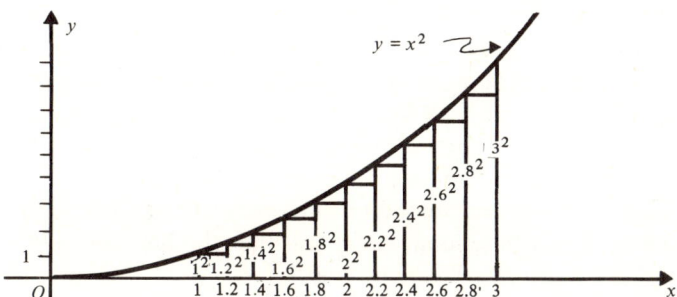

Total area ≈ 1^2 (0.2) + 1.2^2 (0.2) + 1.4^2 (0.2) + 1.6^2 (0.2)
+ 1.8^2 (0.2) + 2^2 (0.2) + 2.2^2 (0.2) + 2.4^2 (0.2) + 2.6^2 (0.2)
+ 2.8^2 (0.2) ≈ 7.88 sq units

(2) Subintervals (200 of them) of length 0.01.

Total area ≈ 1^2 (0.01) + $(1.01)^2$ (0.01) + $(1.02)^2$ (0.01) + ...
+ $(2.98)^2$ (0.01) + $(2.99)^2$ (0.01) ≈ 8.6267 sq units

So 5, 7.88, and 8.6267 sq units are approximations to the area under the curve $y = x^2$, above the x-axis, between the lines $x = 1$ and $x = 3$. The exact area will be shown to be $8\frac{2}{3}$ sq units.

HOW GOOD ARE THESE APPROXIMATIONS?

It seems clear that the smaller the subintervals (or, equivalently, the greater their number) the better is the approximation to the total area. This is the central idea of the next Goal.

In Example 37-4 we choose the rectangle height at the left endpoint of each subinterval. We might have chosen right endpoints, midpoints, or, indeed, a point at random differing from subinterval to subinterval. This, too, will be discussed in Goal 38.

FOUR SIMILAR PROBLEMS

Example 37-5

How are the following problems related?

(a) An object is dropped so that its velocity after t sec is $(32t + 10)$ ft/sec. Find the distance it travels between $t = 20$ and $t = 25$.

(b) A colony of bacteria grows so that at the end of t h the rate of growth is $32t + 10$ bac/h. By how much does the colony increase between $t = 20$ and $t = 25$?

(c) Suppose a company determines that the marginal cost of production is $(32x + 10)$ cents per gadget when the production level is x gadgets. Find the additional cost when the production level is raised from 20 to 25 gadgets.

(d) Find the area under the curve $y = 32x + 10$, above the x-axis, and between the lines $x = 20$ and $x = 25$.

The answer to the question about the relationship of the four problems is: We find approximations for all four different problems at the same time.

One approximation for all the problems is obtained as follows: divide the interval from $x = 20$ to $x = 25$ into five equal subintervals each of length 1, choose the value of the given function at the midpoint of each subinterval, form the product of this functional value and the length of the subinterval, and then add these products. We note that

$f(20.5) = 666$; $f(21.5) = 698$; $f(22.5) = 730$; $f(23.5) = 763$; $f(24.5) = 794$

Since $666(1) + 698(1) + 730(1) + 762(1) + 794(1) = 3650$, we see that 3650 is an approximation to

(a) the number of feet the object travels between $t = 20$ and $t = 25$;

(b) the increase in the size of the colony between $t = 20$ and $t = 25$;

(c) the additional cost (in cents) in raising the production level of the company from 20 to 25 gadgets;

(d) the area under the graph of the function and above the x-axis, between the lines $x = 20$ and $x = 25$.

Increasingly better approximations are obtainable by taking smaller and smaller subintervals. But the crucial conclusion here is that the technique developed in this Goal is applicable to all four problems. In the next few Goals we will see that when we find the exact solution for one of these four problems we have it immediately for all of them.

Goal 37 | To Approximate Limits of Sums

EXERCISES

(A hand calculator will be helpful in doing the following exercises.)

1. Suppose that the speed of a rocket t sec after blasting off is $\frac{1}{100}t^2$ miles per sec. Find two approximations to the distance traveled during the first 10 sec. For the first approximation divide the interval into two equal subintervals; for the second, into five. Choose the midpoints of the subintervals for calculating the relevant velocities.

2. Suppose that a certain animal population is growing so that after w weeks the rate of growth is $500(1 + 4w)$ animals per week. Find two approximations to the increase in the population during the fifth week, that is, from $w = 4$ to $w = 5$. Divide the total time into four equal subintervals, using (a) the left-hand endpoints of the subintervals for one approximation and (b) the right-hand endpoints for the other.

3. If the marginal cost at a production level of x units is $(0.002x - 0.3)$ dollars/unit, find an approximation to the additional cost in raising the production level from 1000 to 1050 units. Use five equal subintervals and choose the midpoint of each to calculate a relevant marginal cost.

4. Find three approximations to the area under the graph of $f(x) = x^2 + x - 1$ and above the x-axis, from $x = 1$ to $x = 6$, using (a) two, (b) five, and (c) ten equal subintervals. Always use the left-hand endpoint of each subinterval to obtain the approximations.

5. Find four approximations to the area under the graph of $f(x) = x^3$ from $x = 0$ to $x = 2$. Divide the interval into (a) two and (b) four equal subintervals; in each case find one approximation using left-hand endpoints and one using right-hand endpoints.

TEST A

1. Find an approximation to the area under the graph of $f(x) = x^2 + 2x - 3$ from $x = 2$ to $x = 4$. Divide the interval into four subintervals and use the midpoints of the latter.

2. Suppose a colony of bacteria is growing so that after t h the growth rate is $100(0.1 + 2t)$ bac/h. Find an approximation to the growth during the fourth hour, that is, from $t = 3$ to $t = 4$. Divide the time into five subintervals and use the left-hand endpoints.

TEST B

1. Find an approximation to the area under the graph of $f(x) = x^3 + x$ from $x = 1$ to $x = 4$. Divide the interval into six equal subintervals and use the right-hand endpoints.

2. If the marginal cost at a production level of x units is $(0.05x - 0.2)$ dollars/unit, find an approximation to the additional cost when the

production level is raised from 100 to 120 units. Use five equal subintervals and choose the midpoints for the calculations.

STUDY QUESTION

1. Suppose the curve $y = f(x)$ is both above the x-axis and increasing on the interval from $x = a$ to $x = b$. If the area under the curve between $x = a$ and $x = b$ is approximated by using rectangles whose heights are those at right-hand endpoints, how does any approximation compare with the actual area? What if left-hand endpoints are used? Answer both questions if the conditions are the same except that the curve is decreasing.

GOAL 38

To find the area under a curve

As we showed in Goal 37, there are many problems, from a variety of fields, whose solutions can all be approximated using the same mathematical approach. Here we develop a somewhat painful technique for finding the exact solution to one of these problems: that of finding the area under a curve. We can then use this procedure to solve various disguised versions of the same problem. In Goal 40 we will present a much more powerful and efficient technique for solving this problem along with many others.

EXAMPLES

Example 38-1

Find the area under the curve $y = x^2$, above the x-axis, and between $x = 0$ and $x = 1$.

In Goal 37 we found approximations to the area under a curve by dividing the interval into equal subintervals, then summing the areas of rectangles. In each of the following we choose the height of each rectangle to be the functional value at the *right* endpoint.

(a) Subintervals of length $\frac{1}{2}$.

Area $\approx (\frac{1}{2})^2(\frac{1}{2}) + (1)^2(\frac{1}{2}) = \frac{5}{8} = 0.625$ sq units

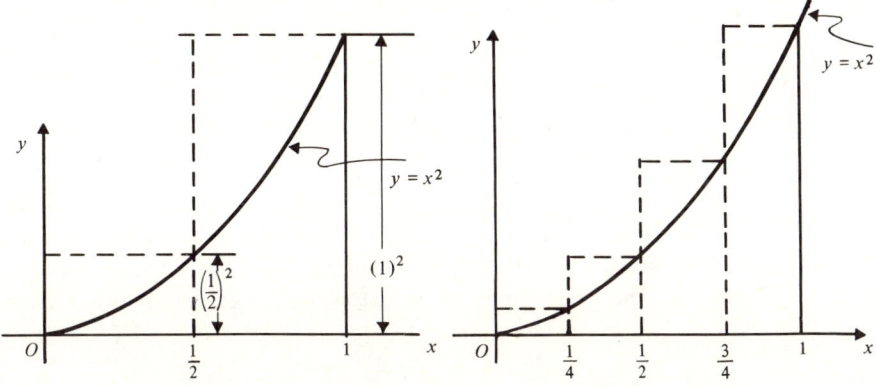

(b) Subintervals of length $\frac{1}{4}$.

Area $\approx (\frac{1}{4})^2(\frac{1}{4}) + (\frac{2}{4})^2(\frac{1}{4}) + (\frac{3}{4})^2(\frac{1}{4}) + (1)^2(\frac{1}{4}) = \frac{15}{32} = 0.46875$ sq units

(c) Subintervals of length $\frac{1}{10}$.

Area $\approx (\frac{1}{10})^2(\frac{1}{10}) + (\frac{2}{10})^2(\frac{1}{10}) + (\frac{3}{10})^2(\frac{1}{10}) + (\frac{4}{10})^2(\frac{1}{10}) + (\frac{5}{10})^2(\frac{1}{10}) +$
$(\frac{6}{10})^2(\frac{1}{10}) + (\frac{7}{10})^2(\frac{1}{10}) + (\frac{8}{10})^2(\frac{1}{10}) + (\frac{9}{10})^2(\frac{1}{10}) + (\frac{10}{10})^2(\frac{1}{10}) =$
$\frac{77}{200} = 0.385$ sq units

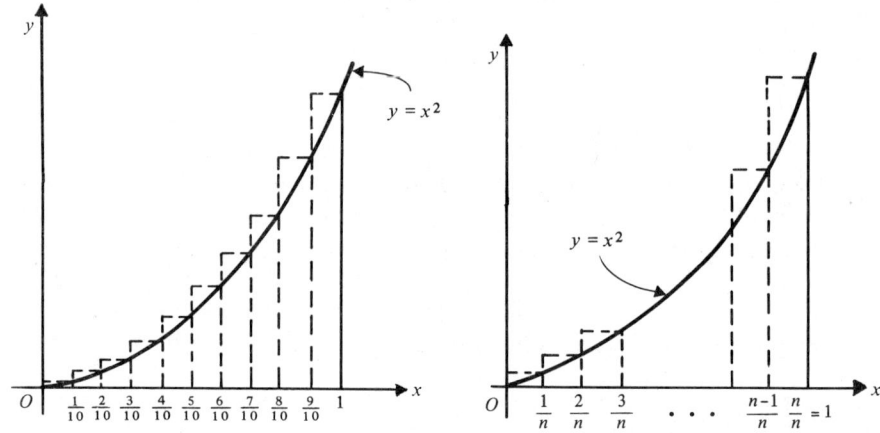

So 0.625, 0.46875, and 0.385 sq units are three approximations to the specified area under the curve. As we divide the interval from $x = 0$ to $x = 1$ into smaller and smaller subintervals the resulting approximations become closer and closer to the exact area. If we let $1/n$ represent the length of each subinterval (above we used, in turn, $n = 2$, 4, and 10), then we can generalize our approximation as follows:

(d) Subintervals of length $1/n$.

$$\text{Area} \approx \left(\frac{1}{n}\right)^2\left(\frac{1}{n}\right) + \left(\frac{2}{n}\right)^2\left(\frac{1}{n}\right) + \left(\frac{3}{n}\right)^2\left(\frac{1}{n}\right) + \dots$$
$$+ \left(\frac{n-1}{n}\right)^2\left(\frac{1}{n}\right) + \left(\frac{n}{n}\right)^2\left(\frac{1}{n}\right)$$
$$= \frac{1}{n^3} + \frac{2^2}{n^3} + \frac{3^2}{n^3} + \dots + \frac{(n-1)^2}{n^3} + \frac{n^2}{n^3}$$

As n becomes larger, the length of each subinterval becomes smaller and the approximation to the exact area gets increasingly better. If we simplify the approximation in (d) above by factoring out $1/n^3$, we get

$$\text{Area} \approx \frac{1}{n^3}[1^2 + 2^2 + 3^2 + \dots + (n-1)^2 + n^2]$$

Goal 38 | To Find the Area Under a Curve

But the sum within parentheses here is just the sum of the first n squares, for which there is a well-known formula:*

$$1^2 + 2^2 + 3^2 + \ldots + n^2 = \frac{n^3}{3} + \frac{n^2}{2} + \frac{n}{6}$$

The above approximation can therefore be rewritten:

$$\text{Area} \approx \frac{1}{n^3}\left(\frac{n^3}{3} + \frac{n^2}{2} + \frac{n}{6}\right) = \frac{1}{3} + \frac{1}{2n} + \frac{1}{6n^2}$$

What happens to this sum if we now let n get bigger and bigger (that is, as $n \to \infty$)?

$$\lim_{n\to\infty}\left(\frac{1}{3} + \frac{1}{2n} + \frac{1}{6n^2}\right) = \lim_{n\to\infty}\frac{1}{3} + \lim_{n\to\infty}\frac{1}{2n} + \lim_{n\to\infty}\frac{1}{6n^2}$$

$$= \frac{1}{3} + 0 + 0 = \frac{1}{3}$$

We define the exact area we are seeking as equal precisely to this limit. Thus the area under $y = x^2$, above the x-axis, and between $x = 0$ and $x = 1$ is equal exactly to $\frac{1}{3}$ sq unit.

Example 38-2

Find the area under the curve $y = 3x^2 + 1$, above the x-axis, and between $x = 0$ and $x = 1$.

For illustration we find one approximation, using subintervals of length $\frac{1}{5}$ (there are five) and right endpoints. See the figure on p. 376.

$$\text{Area} \approx \left[3\left(\frac{1}{5}\right)^2 + 1\right]\left(\frac{1}{5}\right) + \left[3\left(\frac{2}{5}\right)^2 + 1\right]\left(\frac{1}{5}\right) +$$

$$\left[3\left(\frac{3}{5}\right)^2 + 1\right]\left(\frac{1}{5}\right) + \left[3\left(\frac{4}{5}\right)^2 + 1\right]\left(\frac{1}{5}\right) + \left[3\left(\frac{5}{5}\right)^2 + 1\right]\left(\frac{1}{5}\right)$$

$$= \left(\frac{1}{5}\right)\left[3\left(\frac{1}{5}\right)^2 (1^2 + 2^2 + 3^2 + 4^2 + 5^2) + (1 + 1 + 1 + 1 + 1)\right]$$

$$= 3\left(\frac{1}{5^3}\right)(55) + \frac{1}{5}(5) = 2.32$$

---

*This formula and similar ones involving integers can easily be proved using a method known as *mathematical induction*. See, for example, H. Flanders and J. J. Price, *Algebra and Trigonometry* (New York: Academic Press, 1975), pp. 364 ff.

So one approximation for the specified area is 2.32 sq units.

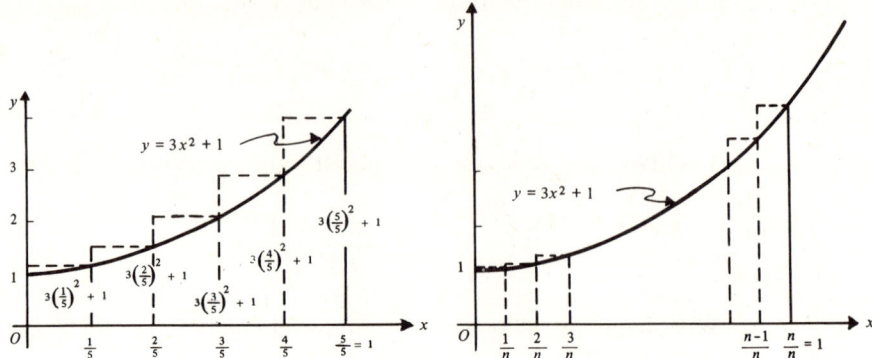

(Note that for convenience different scales are used on the axes)

Now we use n subintervals, each of length $1/n$:

$$\text{Area} = \lim_{n \to \infty} \left\{ \left[3\left(\frac{1}{n}\right)^2 + 1 \right]\left(\frac{1}{n}\right) + \left[3\left(\frac{2}{n}\right)^2 + 1 \right]\left(\frac{1}{n}\right) + \left[3\left(\frac{3}{n}\right)^2 + 1 \right]\left(\frac{1}{n}\right) + \ldots \right.$$

$$\left. + \left[3\left(\frac{n-1}{n}\right)^2 + 1 \right]\left(\frac{1}{n}\right) + \left[3\left(\frac{n}{n}\right)^2 + 1 \right]\left(\frac{1}{n}\right) \right\}$$

We now factor out $1/n$ and simplify further:

$$\text{Area} = \lim_{n \to \infty} \frac{1}{n} \left[3\left(\frac{1}{n}\right)^2 (1^2 + 2^2 + 3^2 + \ldots + n^2) + \underbrace{(1 + 1 + 1 + \ldots + 1)}_{n \text{ of these}} \right]$$

$$= \lim_{n \to \infty} \left[3\left(\frac{1}{n^3}\right)\left(\frac{n^3}{3} + \frac{n^2}{2} + \frac{n}{6}\right) + \left(\frac{1}{n}\right)n \right]$$

$$= \lim_{n \to \infty} \left(\frac{3}{3} + \frac{3}{2n} + \frac{1}{2n^2} + 1 \right)$$

$$= 1 + 0 + 0 + 1 = 2$$

The exact area is thus 2 sq units.

Example 38-3

Use the method of the preceding examples to find exactly the area under the line $y = 2x + 3$, above the x-axis, and between the lines $x = 1$ and $x = 5$.

Goal 38 | To Find the Area Under a Curve

If we divide the interval from $x = 1$ to $x = 5$ into n subintervals, then each is of length $(5 - 1)/n$ or $4/n$ units. Here, too, we choose the rectangle heights at right endpoints.

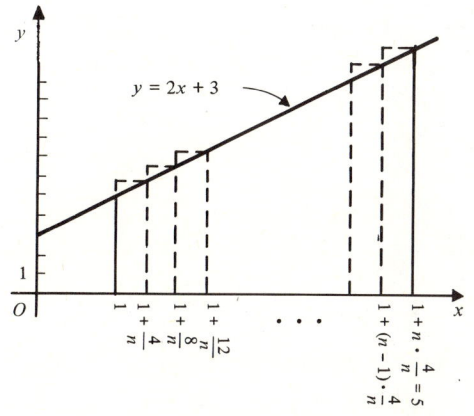

(Again, for convenience, different scales are used on the two axes)

$$\text{Area} = \lim_{n \to \infty} \left\{ \left[2\left(1 + \frac{4}{n}\right) + 3\right]\left(\frac{4}{n}\right) + \left[2\left(1 + \frac{8}{n}\right) + 3\right]\left(\frac{4}{n}\right) + \right.$$
$$\left. \left[2\left(1 + \frac{12}{n}\right) + 3\right]\left(\frac{4}{n}\right) + \ldots + \left[2\left(1 + \frac{4n}{n}\right) + 3\right]\left(\frac{4}{n}\right) \right\}$$

$$= \lim_{n \to \infty} \frac{4}{n} [2\underbrace{(1 + 1 + 1 + \ldots + 1)}_{n \text{ ones}}$$
$$+ \frac{8}{n}(1 + 2 + 3 + \ldots + n) + \underbrace{(3 + 3 + 3 + \ldots + 3)}_{n \text{ threes}}]$$

Once again we invoke a formula, this one for the sum of the first n integers:

$$1 + 2 + 3 + \ldots + n = \frac{n(n + 1)}{2} = \frac{n^2}{2} + \frac{n}{2}$$

So for the exact area we have

$$\lim_{n \to \infty} \frac{4}{n} \left[2 \cdot n + \frac{8}{n}\left(\frac{n^2}{2} + \frac{n}{2}\right) + 3 \cdot n \right]$$
$$= \lim_{n \to \infty} \frac{4}{n} [5n + 4n + 4]$$
$$= \lim_{n \to \infty} \frac{4}{n} (9n + 4)$$

$$= \lim_{n \to \infty} \left(36 + \frac{16}{n} \right) = 36 + 0 = 36$$

Note that the region whose area we have just found is the interior of a trapezoid whose height $h = 4$ units and whose parallel bases are $b_1 = 2(1) + 3 = 5$, $b_2 = 2(5) + 3 = 13$. The area of this trapezoid is $\frac{1}{2}h(b_1 + b_2) = \frac{1}{2}(4)(5 + 13) = 36$ sq units, precisely the answer obtained by our more cumbersome approach.

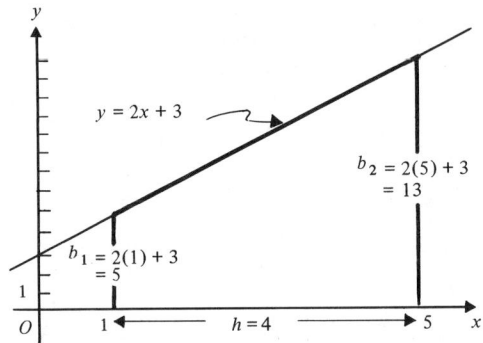

REMARKS ON THE METHOD

(1) Despite the extensive manipulations in the preceding examples, the main purpose of this Goal is *not* to put you through your algebraic paces. It is to motivate, and to help explain, the much more efficient technique for finding areas (and other similar quantities) that will be presented in the next few Goals.

(2) In the three examples we always chose right-hand endpoints of the subintervals. We could have used left-hand endpoints, or midpoints, or, indeed, we could have chosen points at random, one in each subinterval (but then it would not have been easy to find infinite sums!).

(3) In all our approximations we divided the given interval into equal subintervals. That is usually a great convenience, but it is not theoretically necessary. As long as both the number of subintervals becomes infinite and the length of each subinterval approaches zero, we can define the area of any of the regions so far considered as the limit, as $n \to \infty$, of a sum of the areas of n rectangles. Having said this, we shall restrict ourselves in further discussions to subintervals of equal length.

FURTHER EXAMPLES

Example 38-4

Find (again!) the area under $y = x^2$, above the x-axis, and between $x = 0$ and $x = 1$, choosing left-hand endpoints of the subintervals.

Goal 38 | To Find the Area Under a Curve

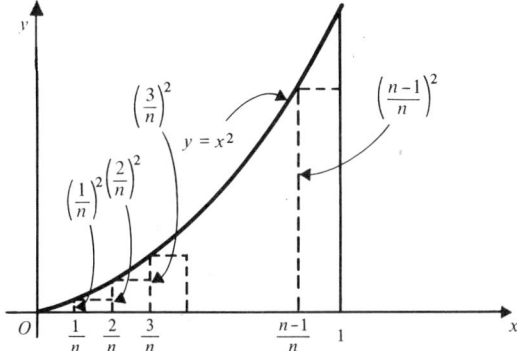

The base of each rectangle has length $1/n$. Using heights at left-hand endpoints, note that the first rectangle has height $0(!)$; the last has height $(n-1)/n$.

$$\text{Area} = \lim_{n\to\infty}\left[0\cdot\frac{1}{n} + \left(\frac{1}{n}\right)^2\left(\frac{1}{n}\right) + \left(\frac{2}{n}\right)^2\left(\frac{1}{n}\right) + \left(\frac{3}{n}\right)^2\left(\frac{1}{n}\right) + \ldots\right.$$
$$\left. + \left(\frac{n-1}{n}\right)^2\left(\frac{1}{n}\right)\right] = \lim_{n\to\infty}\frac{1}{n^3}[1^2 + 2^2 + 3^2 + \ldots + (n-1)^2]$$

The last term in this sum of squares is the square of $(n-1)$; so the sum, from the formula on page 375, is

$$\frac{(n-1)^3}{3} + \frac{(n-1)^2}{2} + \frac{n-1}{6}$$

Therefore

$$\text{Area} = \lim_{n\to\infty}\left(\frac{(n-1)^3}{3n^3} + \frac{(n-1)^2}{2n^3} + \frac{n-1}{6n^3}\right) = \frac{1}{3} + 0 + 0 = \frac{1}{3}$$

where we used the rules on page 222 for finding the limit, as $n \to \infty$, of a rational function $P(n)/Q(n)$.

Note that the area obtained using left-hand endpoints is just what we got in Example 38-1 using right-hand endpoints.

Example 38-5

A culture of bacteria is growing so that its rate of increase after t min is $2t + 3$ bac/min. By how much does the culture increase during the four min from $t = 1$ to $t = 5$?

From Goal 37 we know that the solution and numerical answer to this example are identical to those of Example 38-3. Thus, without further fuss or bother we conclude immediately that the increase is 36 bacteria.

EXERCISES

1. Verify each of the following formulas for the values of n specified:

    (a) $1 + 2 + 3 + \ldots + n = \dfrac{n^2}{2} + \dfrac{n}{2}$ for $n = 3, 7, 9$

    (b) $1^2 + 2^2 + 3^2 + \ldots + n^2 = \dfrac{n^3}{3} + \dfrac{n^2}{2} + \dfrac{n}{6}$ for $n = 2, 8$

    (c) $1^3 + 2^3 + 3^3 + \ldots + n^3 = \dfrac{n^4}{4} + \dfrac{n^3}{2} + \dfrac{n^2}{4}$ for $n = 1, 6$

    In Exercises 2 through 10 find the area under the given curve, above the x-axis, and between the two specified x-coordinates. After sketching the region, write down the generalized approximation using n subintervals and the designated points. Simplify algebraically using the above identities. Then find the exact area by letting n become infinite.

2. $y = 3x^2$ between $x = 0$ and $x = 1$, using (a) right-hand endpoints, and (b) left-hand endpoints.
3. $y = x^3$ between $x = 0$ and $x = 1$ using right-hand endpoints.
4. $y = 5x$ between $x = 0$ and $x = 1$ using right-hand endpoints. Check by using the formula for the area of a triangle.
5. $y = 5x$ between $x = 0$ and $x = 1$ using (a) left-hand endpoints, and (b) midpoints. HINT: $1 + 3 + 5 + \ldots + (2n - 1) = n^2$.
6. $y = 3x + 1$ between $x = 2$ and $x = 3$ using right-hand endpoints. Check by using the formula for the area of a trapezoid.
7. $y = x^2 + x$ between $x = 0$ and $x = 1$, using right-hand endpoints.
8. $y = x^2 + x$ between $x = 0$ and $x = 5$, using right-hand endpoints.
9. $y = x^2 - x + 2$ between $x = 3$ and $x = 5$, using right-hand endpoints.
10. $y = 2x^2 - 1$ between $x = 1$ and $x = 4$, using right-hand endpoints.

TEST A

Using the technique developed in this Goal, find the area under the curve $y = 2x^2 + 1$, above the x-axis, and between $x = 0$ and $x = 1$. HINT: Call as necessary on the formulas listed in Exercise 1 above.

TEST B

Using the technique developed in this Goal, find the area under the curve $y = x^3 + 4$ between $x = 0$ and $x = 1$, above the x-axis. HINT: Call as necessary on the formulas listed in Exercise 1 above.

GOAL **39**

To understand and use the definite integral

In Goal 37 we considered four different problems each of whose solutions we approximated with similar summations. In Goal 38 we isolated one of these problems, the area under a curve, and presented some examples in which we could obtain the exact area by evaluating an infinite sum. The technique used, while tedious and inefficient, can also be applied to a wide variety of other problems. In this Goal we define the extremely useful definite integral as an infinite sum. In the next two Goals we will see how to evaluate definite integrals (and therefore infinite sums) efficiently and neatly.

DEFINITION OF THE DEFINITE INTEGRAL

Generally, all the problems being considered involve a function $f(x)$, an interval $a \leq x \leq b$, and approximating sums obtained as follows:

(1) Divide the interval into n subintervals of equal length. The length is usually denoted by Δx:

$$\Delta x = \frac{b - a}{n}$$

(2) Choose a point in each subinterval:

(3) Evaluate the function at each chosen point; multiply this value by the length of the subinterval, Δx; then sum together all these products:

$$f(x_1)\Delta x + f(x_2)\Delta x + \ldots + f(x_{n-1})\Delta x + f(x_n)\Delta x$$

If these sums have a limiting value as $n \to \infty$ (or as $\Delta x \to 0$), then this limit is called the *definite integral* of f over the interval from a to b, and is denoted by

$$\int_a^b f(x) \, dx$$

which is read "the definite integral (or just the integral) from a to b of f of x dx." Thus we have

$$\int_a^b f(x) \, dx = \lim_{n \to \infty} [\, f(x_1) \Delta x + f(x_2) \Delta x + \ldots + f(x_n) \Delta x \,]$$

provided the limit on the right exists.

NOTES ON THE DEFINITE INTEGRAL

(1) The definite integral is defined only if the limiting value does not depend on the choice of points within the subintervals. In Goal 38 we sometimes chose left endpoints, sometimes right endpoints, sometimes midpoints.

(2) $\int_a^b f(x) \, dx$ is a *number*. This is illustrated below in the examples.

(3) It can be shown that, if $f(x)$ is continuous on $a \leq x \leq b$, then $\int_a^b f(x) \, dx$ exists. Most of the functions in this book are continuous everywhere.

(4) The infinite sum at the right above is called a *Riemann sum*. (We evaluated numerous Riemann sums in Goal 38.) The definite integral as defined above is called the *Riemann integral*.

(5) If f is given as a function of t, or of u, or of some other variable, then we write $\int_a^b f(t) \, dt$ or $\int_a^b f(u) \, du$ or . . . . The variable x, t, u, . . . is called a *dummy variable*.

(6) The function $f(x)$ in $\int_a^b f(x) \, dx$ is called the *integrand*; a and b are called the *limits of integration* or just the *limits*. We call a the *lower limit*, b the *upper limit*.

Example 39-1

Express each of the following areas (which were found exactly in Examples 38-1, 38-2, and 38-3) as a definite integral: (a) the area under $y = x^2$, above the x-axis, and between $x = 0$ and $x = 1$; (b) the area under $y = 3x^2 + 1$, above the x-axis, and between $x = 0$ and $x = 1$; (c) the area under $y = 2x + 3$, above the x-axis, and between $x = 1$ and $x = 5$.

(a) $\int_0^1 x^2 \, dx$. From Example 38-1 we now know that this definite integral equals $\frac{1}{3}$.
(b) $\int_0^1 (3x^2 + 1) \, dx = 2$. (From Example 38-2.)
(c) $\int_1^5 (2x + 3) \, dx = 36$. (From Example 38-3.)

WHAT GOOD IS THE DEFINITE INTEGRAL?

We can't blame you if you wonder whether we've really made much progress in finding areas under curves efficiently (or in solving the other problems posed in Goal 37). At this stage, to find an area under a curve exactly we must still evaluate an infinite sum—right? Although this is true, we are really much better off now because of the following two points:

(1) If f is a continuous function, *every* infinite sum of the type we have been considering can be expressed as a definite integral.
(2) Very soon we will be able to write down the answers to many definite integrals after just a few seconds of computation. The three definite integrals in Example 39-1 above are typical ones.

In the remainder of this Goal we will present a variety of problems whose solutions are definite integrals. In addition to their use in finding the area under a curve, definite integrals can also be used, given appropriate information, to find distance traveled along a line, additional cost (or revenue or profit) resulting from increased production, number of new marriages (or divorces), growth of an animal (or of money, of information, of an epidemic), and more. The examples and exercises below illustrate many of these applications.

In Examples 39-2 through 39-12 assume that the given function is continuous and positive over the specified interval; express the answers as definite integrals.

Example 39-2

Find the distance traveled between times t_1 and t_2 by an object moving along a line if its velocity at time t is $v(t)$.

The distance traveled over a small subinterval of time is equal approximately to the product of its velocity at some point in the subinterval and the duration of the subinterval. The sum of all these products in an approximation to the total distance. The exact distance is the limit of these sums as the number of subintervals becomes infinite. Thus the total distance D traveled between t_1 and t_2 is given by

$$D = \int_{t_1}^{t_2} v(t) \, dt$$

Example 39-3

Find the additional cost resulting from a rise in production level from L_1 to L_2 units if $M(x)$ is the marginal cost when the production level is x units.

The added cost resulting from a small rise in production level (a subinterval) is equal approximatly to the product of the marginal cost at some level in the subinterval and the length of the subinterval. The total additional cost is the limit of the sum of these products over n subintervals as $n \to \infty$. The exact additional cost C, therefore, resulting from a rise in production from L_1 to L_2 units is given by

$$C = \int_{L_1}^{L_2} M(x)\, dx$$

Example 39-4

Find the change R in revenue resulting from a change in production level from 400 to 450 units if $f(x)$ is the marginal revenue when the production level is x units.

It is easy to conclude that

$$R = \int_{400}^{450} f(x)\, dx$$

Example 39-5

Find the change P in profit resulting from a change in production level from 750 to 830 units if the marginal profit, in dollars, is $(610 - 0.02x)$ per unit when x units are produced.

$$P = \int_{750}^{830} (610 - 0.02x)\, dx$$

(Why?)

Example 39-6

Find the total number of new marriages between times $t = a$ and $t = b$ if $M(t)$ is the number of marriages per day t days after some fixed point in time.

The number of marriages taking place over a small subinterval of time is equal approximately to the product of the marriage rate at some point (of time) in the subinterval and the length (in days) of the subinterval. The total number T of new marriages occurring between $t = a$ and $t = b$ is the limit of the sum of these products over n subintervals as $n \to \infty$. Therefore

Goal 39 | To Understand and Use the Definite Integral

$$T = \int_a^b M(t)\, dt$$

Example 39-7

Find the total number N of divorces during February 1978 if $D(t)$ is the number of divorces per day at the end of the tth day of 1978.

$$T = \int_{31}^{59} D(t)\, dt$$

(Why?)

Example 39-8

Find the change P in a population between times t_1 and t_2 if $G(t)$ is its growth rate at time t.

Explain why

$$P = \int_{t_1}^{t_2} G(t)\, dt$$

Example 39-9

Find the change C in weight of an animal during its first 5 yr of life if $A(t)$ is its rate of growth after t yr, in pounds per year.

Explain why

$$C = \int_0^5 A(t)\, dt$$

Example 39-10

Find the amount A of money earned during the third year if the rate of growth in dollars per year, after t yr, of a $100 investment is $6e^{0.06t}$.

$$A = \int_2^3 6e^{0.06t}\, dt$$

(Why?)

Example 39-11

Find the number N of people who hear a rumor during the first 24 h if $R(t)$ is the rate at which it spreads, in people per hour, t h after it began.

The number who hear the rumor over a small subinterval of time is equal approximately to its rate of spread (in people per hour) at some point of time in the subinterval and the duration of the subinterval. The total number of people is the limit of the sum of these products over n subintervals. The number of people who hear the rumor during the first 24 h is therefore given by

$$N = \int_0^{24} R(t)\, dt$$

Example 39-12

Find the number Q of people who contact a disease during the second week of an epidemic if $D(t)$ is the rate, in people per day, at which it spreads after t days.

$$Q = \int_7^{14} D(t)\, dt$$

(Why?)

EXERCISES

Express the answer to each of the following in terms of a definite integral. Assume each function given is continuous and positive over the interval specified.

1. What is the area under the curve $y = 5x^3 + x^2 + 2$, above the x-axis, and between $x = 1$ and $x = 3$?
2. What is the area under the curve $y = \dfrac{1}{x} + e^x + 2$, above the x-axis, and between $x = 4$ and $x = 8$?
3. What is the area under the graph of $f(x)$, above the x-axis, and between $x = a$ and $x = b$?
4. If $M(x) = 5 + 0.001x$ is the marginal cost, in dollars per unit, at a production level of x units, what change in cost is brought about by an increase in production from 200 to 225 units?
5. If $(3 + 0.01x - 0.0003x^2)$ dollars per unit is the marginal profit at a production level of x units, what is the change in profit when production increases from 75 to 100 units?
6. If after t days a colony of fruit flies is increasing at the rate of $15(2^t)$ flies per day, by how many does the colony increase during the third week?

Goal 39 | To Understand and Use the Definite Integral

7. If $W(t)$ gives the rate, in ounces per week, at which a tumor is growing after t weeks, by how much does the tumor increase in weight during the first week?
8. If after t days a rumor is spreading at the rate of $100e^{-2t}$ new people per day, how many people hear of the rumor during the fifth and sixth days; that is, from $t = 4$ to $t = 6$?
9. If, t h into a 24-h workday, $(25 + 2t)$ tons of waste per h are being dumped into a river, how many tons are dumped during the whole day?
10. If at time t an invalid's temperature is increasing at the rate of $R(t)$ degrees per h, what is the net change in temperature during the second 8-hour shift?

TEST A

For each of the following express the answer in terms of a definite integral.

1. What is the area under the graph of $f(x) = \dfrac{1}{x} + 2$, above the x-axis, and between $x = 1$ and $x = 3$?
2. If after t min a chemical is decomposing at the rate of $10e^{-t}$ g per min, how much of the chemical decomposes during the first 3 min?

TEST B

For each of the following express the answer in terms of a definite integral.

1. What is the area under the curve $y = e^x$, above the x-axis, and between $x = -4$ and $x = 2$?
2. If $(5 - 0.001x)$ is the marginal revenue, in dollars per unit, at a production level of x units, what is the change in revenue when production increases from 1500 to 2000 units?

STUDY QUESTION

1. Laminar flow of blood through an artery is greater at the center of the artery than it is close to the arterial wall. The flow through a small patch, of cross-sectional area A, equals vA, where v is the velocity of the blood through that patch. The velocity is a function of the distance r from the center and is given by $v(r) = C(R^2 - r^2)$, where R is the radius of the artery and C is a positive constant which depends upon other characteristics of the particular artery.

To find the flow through an artery, divide it into thin concentric rings of equal width Δr, as shown, approximate the flow through each ring, then add

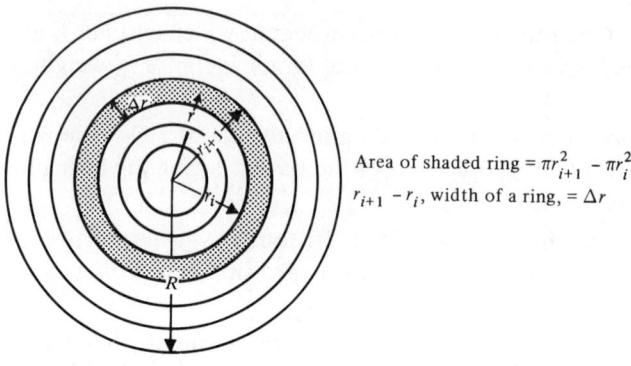

Area of shaded ring = $\pi r_{i+1}^2 - \pi r_i^2$
$r_{i+1} - r_i$, width of a ring, = Δr

up these approximations. If the number of rings becomes infinite, then the exact total flow is equal to

$$\int_0^R 2\pi C(R^2 r - r^3)\, dr$$

Show that this is true.

HINT: The ring between circles of radii r_i and r_{i+1} has area

$$\pi r_{i+1}^2 - \pi r_i^2 = \pi(r_{i+1} + r_i)(r_{i+1} - r_i)$$

Note that

$$\frac{r_{i+1} + r_i}{2} = r$$

where r is the distance from the center of the artery to the midpoint of the ring. And $r_{i+1} - r_i$ serves as Δr, the width of a ring.

GOAL **40**

To use the Fundamental Theorem of the Calculus

We now know (1) that the answers to a great many problems can be expressed as infinite sums of a particular form, and (2) that if the limit of the infinite sum exists, then it can be expressed as a definite integral. In this Goal we find out how to evaluate definite integrals without resorting to the awkward definition as a limit. In the process we discover the interrelationship between the definite integral and the derivative.

In preparation, we reconsider four examples from the preceding Goal. In each one we assume, as previously, that the given function is continuous and positive over the specified interval.

Example 40-1

If the velocity of an object moving along a line at time t is $v(t)$, then the distance traveled between times $t = a$ and $t = b$ is given by $\int_a^b v(t)\,dt$. If the distance traveled by time t is explicitly given by the function $f(t)$, then the distance traveled between times $t = a$ and $t = b$ can also be expressed as $f(b) - f(a)$. We know that the velocity function is the derivative of the distance function; thus we conclude that

$$\int_a^b v(t)\,dt = f(b) - f(a) \qquad \text{where } v'(t) = f(t)$$

Example 40-2

If $M(x)$ is the marginal cost of production at a production level of x units, then $\int_a^b M(x)\,dx$ gives the additional cost in raising production from a to b units. If $C(x)$ is the total cost of producing x units, then the above change in cost is also given by $C(b) - C(a)$. We know that the marginal cost function is the derivative of the cost function; thus we conclude that

$$\int_a^b M(x)\,dx = C(b) - C(a) \qquad \text{where } C'(x) = M(x)$$

Example 40-3

If $G(t)$ is the growth rate of some population at time t, then the increase in population between times $t = a$ and $t = b$ is given by $\int_a^b G(t)\, dt$. If $P(t)$ is the population size at time t, then the above change is also given by $P(b) - P(a)$. We know that the growth rate function is the derivative of the population function; thus we conclude that

$$\int_a^b G(t)\, dt = P(b) - P(a) \qquad \text{where } P'(t) = G(t)$$

Example 40-4

The area under the curve $y = f(x)$, above the x-axis, and between $x = a$ and $x = b$ is equal to $\int_a^b f(x)\, dx$. To parallel the conclusions of the preceding three examples, we now define an area function. Let $A(x)$ be the area under the graph of $f(x)$, above the x-axis, and between some fixed line (say $x = x_0$)

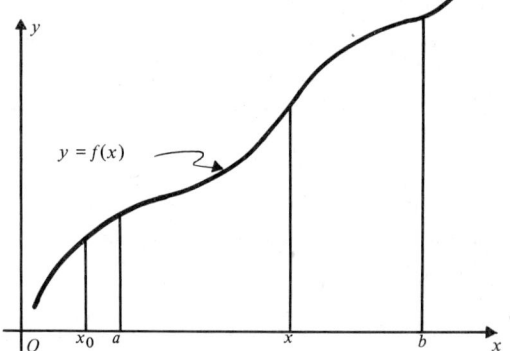

and the variable line at x. (Think of the area, if you like, as that swept out by a vertical line as it moves from $x = x_0$ toward the right to the line at x.) Then $A(b)$ is the area under the curve from $x = x_0$ to $x = b$, $A(a)$ is the area from $x = x_0$ to $x = a$, and $A(b) - A(a)$ is the area under the curve from $x = a$ to $x = b$. So we have

$$\int_a^b f(x)\, dx = A(b) - A(a)$$

We now show that $A'(x) = f(x)$. In the figure on p. 391 the shaded region has area $A(x + h) - A(x)$. This is equal approximately to the area of rectangle $EFLH$. So

$$A(x + h) - A(x) \approx hf(x)$$
$$\frac{A(x + h) - A(x)}{h} \approx f(x)$$

Goal 40 | To Use the Fundamental Theorem of the Calculus

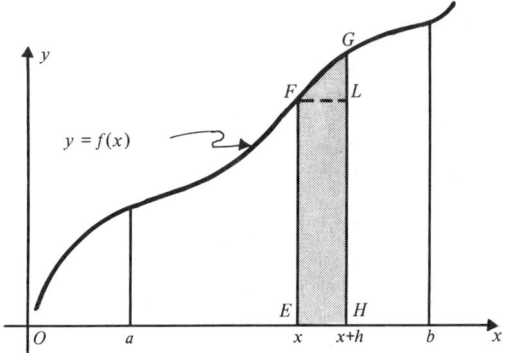

$A(x + h) - A(x)$ = area of region $EFGH$

As $h \to 0$, the left side of the last equation on p. 390 approaches $A'(x)$ (from the definition of the derivative); the right side approaches $f(x)$. So $A'(x) = f(x)$, and we have that the area specified is

$$\int_a^b f(x)\, dx = A(b) - A(a) \qquad \text{where } A'(x) = f(x)$$

THE FUNDAMENTAL THEOREM OF THE CALCULUS

The four examples above illustrate this extremely important and useful theorem:

If $\int_a^b f(x)\, dx$ exists and $F'(x) = f(x)$, then $\int_a^b f(x)\, dx = F(b) - F(a)$.

This theorem enables us to evaluate definite integrals without resorting to the definition as the limit of an (often formidable) infinite sum. What do we now do instead? To evaluate $\int_a^b f(x)\, dx$, we need only find a function F whose derivative is f, evaluate F at b, then subtract from this the value of F at a.

The quantity $F(b) - F(a)$ is usually denoted by $[F(x)]_a^b$.

Example 40-5

Evaluate $\int_1^5 3x^2\, dx$.

We want a function whose derivative is $3x^2$. But that's easy: the derivative of x^3 is $3x^2$. So

$$\int_1^5 3x^2\, dx = [x^3]_1^5 = (5)^3 - (1)^3 = 124$$

Example 40-6

Evaluate $\int_0^1 e^x \, dx$.

Can we find a function whose derivative is e^x? Of course! One such function is e^x. So

$$\int_0^1 e^x \, dx = [e^x]_0^1 = e^1 - e^0 = e - 1$$

Example 40-7

Evaluate $\int_1^3 \left(\frac{1}{x} - 2x\right) dx$.

By inspection or trial and error or working backwards, we observe that $\ln x - x^2$ is a function whose derivative is $(1/x) - 2x$. Therefore

$$\int_1^3 \left(\frac{1}{x} - 2x\right) dx = [\ln x - x^2]_1^3$$
$$= (\ln 3 - 3^2) - (\ln 1 - 1^2) = \ln 3 - 9 - 0 + 1 = \ln 3 - 8$$

Example 40-8

Use the Fundamental Theorem of the Calculus to find the three areas of Goal 38:

(a) $\int_0^1 x^2 \, dx$ (b) $\int_0^1 (3x^2 + 1) \, dx$ (c) $\int_1^5 (2x + 3) \, dx$

Solutions:

(a) $\int_0^1 x^2 \, dx = \left[\frac{x^3}{3}\right]_0^1 = \frac{1^3}{3} - \frac{0^3}{3} = \frac{1}{3}$ sq unit

(b) $\int_0^1 (3x^2 + 1) \, dx = [x^3 + x]_0^1 = (1 + 1) - (0 = 0) = 2$ sq units

(c) $\int_1^5 (2x + 3) \, dx = [x^2 + 3x]_1^5 = (5^2 + 3 \cdot 5) - (1^2 + 3 \cdot 1) = 40 - 4 = 36$ sq units

These are, of course, precisely the answers we got in Goal 38 by evaluating formidable sums!

Goal 40 | To Use the Fundamental Theorem of the Calculus

THE ANTIDERIVATIVE

If the derivative of $F(x)$ is $f(x)$, then $F(x)$ is called an *antiderivative* of $f(x)$. To evaluate integrals we need to be able to find antiderivatives. We will attend to this in the remaining Goals of this chapter.

EXERCISES

1. Compute the following definite integrals:

    (a) $\int_3^5 2x \, dx$

    (b) $\int_0^2 6x^2 \, dx$

    (c) $\int_1^2 (4x^3 - 3x^2) \, dx$

    (d) $\int_1^2 \frac{1}{x} \, dx$

    (e) $\int_0^1 e^x \, dx$

2. If the velocity of an object moving along a line at time t is $5t^4$ ft/sec, how far does it travel in the first 2 sec?

3. If the marginal cost of production is $1/x$ dollars per unit when the production level is x units, what is the change in cost when the production level increases from 50 to 75 units? (HINT: Use Table I in the Appendix.)

4. If the growth rate of a bacterial colony is $7t^6 + 3t^2$ bac/h after t h, what is the change in population during the third hour—that is, between times $t = 2$ and $t = 3$?

5. Find the area under the graph of $f(x) = 3x^2$ between $x = 1$ and $x = 3$.

TEST A

1. Compute the following definite integrals:

    (a) $\int_0^1 (5x^4 - 3x^2 + 1) \, dx$

    (b) $\int_1^2 \left(2x + \frac{1}{x}\right) dx$

2. Find the area under the graph of $f(x) = 4x^3$ between $x = 0$ and $x = 2$.

TEST B

1. Compute the following definite integrals:

    (a) $\int_1^3 (4x^3 - 2x) \, dx$

    (b) $\int_0^1 (e^x - 1) \, dx$

2. If the velocity of an object at time t is $3t^2$ ft/sec, how far does it travel during the first 3 sec?

STUDY QUESTIONS

1. Another version of the Fundamental Theorem of the Calculus states that if f is continuous and if $F(x) = \int_c^x f(t)\, dt$ for some fixed constant c, then $F'(x) = f(x)$.

   (a) If $F(x) = \int_1^x (t^{10} - t^2)\, dt$, then $F'(x) = x^{10} - x^2$. What is $F'(x)$ if $F(x) = \int_3^x (\sqrt{t} - \ln t)\, dt$?

   (b) If $F(x) = \int_1^{x^2} (t^3 - 1)\, dt$ then using the chain rule gives $F'(x) = [(x^2)^3 - 1](2x) = 2x^7 - 2x$. What is $F'(x)$ if $F(x) = \int_2^{x^3} (t^2 - \sqrt{t})\, dt$?

2. The following justification of the Fundamental Theorem uses the Mean Value Theorem (p. 261). To prove that $\int_a^b f(x)\, dx = F(b) - F(a)$ where $F'(x) = f(x)$ and where the definite integral exists, first divide the interval from a to b into n equal subintervals with endpoints $a, x_1, x_2, x_3, \ldots, x_{n-1}, b$. Each subinterval has length Δx.

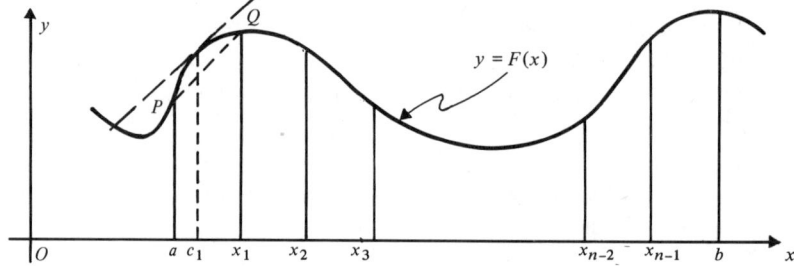

Now justify each equals sign below:

$$F(b) - F(a) = F(b) - F(x_{n-1}) + F(x_{n-1}) - F(x_{n-2}) + \ldots - F(x_1) + F(x_1) - F(a)$$
$$= [F(b) - F(x_{n-1})] + [F(x_{n-1}) - F(x_{n-2})] + \ldots + [F(x_1) - F(a)]$$
$$= F'(c_1)\,\Delta x + F'(c_2)\,\Delta x + \ldots + F'(c_n)\,\Delta x$$

where c_1 is in the first subinterval, c_2 in the second, ..., c_n in the nth;

$$F(b) - F(a) = f(c_1)\,\Delta x + f(c_2)\,\Delta x + \ldots + f(c_n)\,\Delta x$$

where $F'(x) = f(x)$. Therefore

$$\lim_{n \to \infty} [F(b) - F(a)] = \lim_{n \to \infty} [f(c_1)\,\Delta x + f(c_2)\,\Delta x + \ldots + f(c_n)\,\Delta x]$$

$$F(b) - F(a) = \int_a^b f(x)\, dx$$

GOAL 41

To find antiderivatives

As noted in Goal 40, to apply the powerful Fundamental Theorem of the Calculus we need to find a function, given its derivative. This process is called *finding antiderivatives*, *finding indefinite integrals*, or just *integration*.

ANTIDERIVATIVES OR ANTIDIFFERENTIALS; INDEFINITE INTEGRALS

We know that one antiderivative of $3x^2$ is x^3. But there are others: $x^3 + 23$, $x^3 - 10$, $x^3 + 1.27$ for example. Any two antiderivatives of a given function will differ by a constant (see Corollary 2 of the Mean Value Theorem, p. 265). We say that the *indefinite integral* of $3x^2$ is $x^3 + C$, where C is an arbitrary constant; and for this we write

$$\int 3x^2 \, dx = x^3 + C$$

In Goal 39 we pointed out that the *definite* integral $\int_a^b f(x) \, dx$, where a and b are numbers, is a *number*. When the limits of integration are omitted, so that we have an *indefinite* integral, the answer is not only not a number, it is not even a unique function. This is the sense in which it is "indefinite"— since any of an infinite number of functions will do.

Of course, x^3 is an antiderivative of $3x^2$. We may also say that x^3 is an *antidifferential* of $3x^2 \, dx$. This term, which arises from our definition of differential on page 273, will prove especially useful in later Goals. As further examples,

u^3 is an antidifferential of $3u^2 \, du$;
t^5 is an antidifferential of $5t^4 \, dt$;
$4x$ is an antidifferential of $4 \, dx$;
$4u$ is an antidifferential of $4 \, du$; etc.

RULES OF INTEGRATION

Most of the rules set forth below are obtainable by inspection (or by working backwards). If we want to check that $\int f(x)\, dx = F(x) + C$, we need only verify that the derivative of $F(x)$ is $f(x)$—or that the differential of $F(x)$ is $f(x)\, dx$.

INTEGRATION OF MONOMIALS

We can use the power rule from the study of differentiation to find certain indefinite integrals immediately.

Examples 41-1

(a) $\int 1\, dx = x + C$ \qquad (d) $\int 7x^6\, dx = x^7 + C$

(b) $\int 2x\, dx = x^2 + C$ \qquad (e) $\int 10x^9\, dx = x^{10} + C$

(c) $\int 3x^2\, dx = x^3 + C$ \qquad (f) $\int 33x^{32}\, dx = x^{33} + C$

Examples 41-2

(a) $\int 8\, dx = 8x + C$ \qquad (b) $\int x\, dx = \dfrac{1}{2}x^2 + C$

(c) $\int x^2\, dx = \dfrac{1}{3}x^3 + C$

(d) $\int 18x^2\, dx = 18\left(\dfrac{1}{3}x^3\right) + C = 6x^3 + C$

(e) $\int 4x^{11}\, dx = 4\left(\dfrac{1}{12}x^{12}\right) + C = \dfrac{1}{3}x^{12} + C$

In each of these we also used the fact that the derivative of a constant times a function equals the constant times the derivative of the function. To check Example 41-2 part (e), we note that

$$\frac{d}{dx}\left(\frac{1}{3}x^{12} + C\right) = \frac{1}{3}(12x^{11}) = 4x^{11}$$

The techniques we are using lead to the *power rule for integration*:

$$\int x^n\, dx = \frac{x^{n+1}}{n+1} + C \qquad (n \text{ is a nonnegative integer})$$

TWO GENERAL RULES

From corresponding rules for differentiation we obtain the following:

Goal 41 | To Find Antiderivatives

I. $\int kf(x)\, dx = k \int f(x)\, dx$ (k any constant)

The integral of a constant times a function equals the constant times the integral of the function.

II. $\int [f(x) + g(x)]\, dx = \int f(x)\, dx + \int g(x)\, dx$

The integral of a sum is the sum of the integrals.

Examples 41-3

(a) $\int [2x + 3x^2]\, dx = \int 2x\, dx + \int 3x^2\, dx = x^2 + C_1 + x^3 + C_2 = x^2 + x^3 + C$

where we replaced the sum of the arbitrary constants C_1 and C_2 by the single arbitrary constant C.

(b) $\int (x^3 - 2x^2 + 1)\, dx = x^4/4 - 2x^2/3 + x + C$

(c) $\int \dfrac{x^5 - 3x^4 + 4x^2}{x^2}\, dx = \int (x^3 - 3x^2 + 4)\, dx = \dfrac{x^4}{4} - x^3 + 4x + C$

where we *needed* to divide through by x^2 *before* integrating. In general the antiderivative of a quotient is *not* the quotient of the antiderivatives.

THE GENERAL POWER RULE

As in differentiation, the power rule actually applies to all numbers n different from -1:

$$\int x^n\, dx = \dfrac{x^{n+1}}{n+1} \qquad (n \neq -1)$$

Examples 41-4

(a) $\int x^4\, dx = x^5/5 + C$

(b) $\int x\, dx = \int x^1\, dx = x^2/2 + C$

(c) $\int \dfrac{1}{x^2}\, dx = \int x^{-2} = \dfrac{x^{-1}}{-1} + C = -\dfrac{1}{x} + C$

(d) $\int \dfrac{5}{x^9}\, dx = \int 5x^{-9}\, dx = 5\left(\dfrac{x^{-8}}{-8}\right) + C = -\dfrac{5}{8x^8} + C$

(e) $\int \sqrt{x}\, dx = \int x^{1/2}\, dx = \dfrac{x^{3/2}}{3/2} + C = \dfrac{2}{3} x^{3/2} + C$

(f) $\int \dfrac{1}{\sqrt[3]{x}}\, dx = \int x^{-1/3}\, dx = \dfrac{x^{2/3}}{2/3} = \dfrac{3}{2} x^{2/3} + C$

(g) $\int \dfrac{x^3 + 2x^2 - 5}{x^2}\, dx = \int \left(x + 2 - \dfrac{5}{x^2}\right) dx = \int (x + 2 - 5x^{-2})\, dx = \dfrac{x^2}{2} + 2x - \dfrac{5x^{-1}}{-1} + C = \dfrac{x^2}{2} + 2x + \dfrac{5}{x} + C$

INTEGRATION OF e^x AND $1/x$

Using the facts that $\dfrac{d}{dx} e^x = e^x$ and that $\dfrac{d}{dx} \ln x = \dfrac{1}{x}$, we conclude that

$$\int e^x\, dx = e^x + C$$

$$\int \dfrac{1}{x}\, dx = \ln x + C \qquad (x > 0)$$

Note especially that the second of these rules applies to $\int x^n\, dx$ only when $n = -1$. Since the general power rule gives us an antiderivative of x^n only when n is *not* equal to -1, we therefore now know antiderivatives of x^n for all real n.

Examples 41-5

(a) $\int -8e^x\, dx = -8e^x + C$ \qquad (b) $\int \dfrac{5}{x}\, dx = 5 \ln x + C$

APPLICATIONS TO DEFINITE INTEGRALS

We can now evaluate a great many definite integrals, since the rules just stated enable us to find antiderivatives of numerous functions. The Fundamental Theorem of the Calculus, it will be remembered, says

$$\int_a^b f(x)\, dx = F(b) - F(a) \qquad \text{where} \quad F'(x) = f(x)$$

We usually choose the simplest antiderivative of f—that is, the one with $C = 0$—although, of course, *any* antiderivative would do.

Examples 41-6

(a) $\int_1^3 x^2\, dx = \left[\dfrac{x^3}{3}\right]_1^3 = \dfrac{3^3}{3} - \dfrac{1^3}{3} = 9 - \dfrac{1}{3} = 8\dfrac{2}{3}$

(b) $\int_4^9 \sqrt{x}\, dx = \int_4^9 x^{1/2}\, dx = \left[\dfrac{2}{3} x^{3/2}\right]_4^9 = \dfrac{2}{3}(27 - 8) = \dfrac{2}{3}(19) = \dfrac{38}{3}$

Note that it is usually computationally easier in examples of this sort to factor out any constant multiplier before evaluating $F(b) - F(a)$.

(c) $\displaystyle\int_0^1 (x^3 - 2x^2 + 3x - 4)\, dx = \left[\dfrac{x^4}{4} - \dfrac{2x^3}{3} + \dfrac{3x^2}{2} - 4x\right]_0^1$

$= \left(\dfrac{1}{4} - \dfrac{2}{3} + \dfrac{3}{2} - 4\right) - 0 = -\dfrac{35}{12}$

Example 41-7

The marriage rate in a large city during the month of June after t days is given by $M(t) = -t^2 + 16t + 100$. Find the total number of new marriages during the month.

In Example 39-6 we expressed the answer as a definite integral, which we can now evaluate. The total number of new marriages is

$\displaystyle\int_0^{30} (-t^2 + 16t + 100)\, dt$

$= \left[-\dfrac{t^3}{3} + 8t^2 + 100t\right]_0^{30} = -9000 + 7200 + 3000 = 1200$

Example 41-8

The growth of a herd of deer after t yr is $G(t) = 100e^t$ deer per yr. Find the change in population during the third year. From Example 39-8 we know that the change in population is

$\displaystyle\int_2^3 100e^t\, dt = 100[e^t]_2^3 = 100(e^3 - e^2)$

$= 100(20.086 - 7.389) = 100(12.697) \approx 1270$ deer

Example 41-9

The rate of spread of a rumor after t days is $R(t) = t^2 + 10t$ new people per day. How many people hear the rumor during the second week?

From Example 39-11 we know that the number is

$\displaystyle\int_7^{14} (t^2 + 10t)\, dt = \left[\dfrac{t^3}{3} + 5t^2\right]_7^{14} = \dfrac{14^3}{3} + 5(14^2)$

$- \left[\dfrac{7^3}{3} + 5(7^2)\right] \approx 1535$ people

IMPROPER INTEGRALS

In some applications we will be confronted by an integral of this form:

$\displaystyle\int_a^\infty f(x)\, dx$

Such an integral, with an infinite limit of integration, is called an *improper integral*. We define an improper integral in terms of a definite integral, as follows:

$$\int_a^\infty f(x)\, dx = \lim_{n \to \infty} \int_a^n f(x)\, dx \qquad \text{if the limit on the right exists.}$$

Examples 41-10

(a) $\int_1^\infty \dfrac{1}{x^2}\, dx = \lim_{n \to \infty} \int_1^n x^{-2}\, dx$

$= \lim_{n \to \infty} \left[-\dfrac{1}{x} \right]_1^n = \lim_{n \to \infty} -\left[\dfrac{1}{n} - \dfrac{1}{1} \right] = -(0 - 1) = 1$

(b) $\int_1^\infty \dfrac{1}{\sqrt{x}}\, dx = \lim_{n \to \infty} \int_1^n x^{-1/2}\, dx = \lim_{n \to \infty} [2\sqrt{x}]_1^n = \lim_{n \to \infty} 2[\sqrt{n} - 1]$

—but as $n \to \infty$, the quantity $(\sqrt{n} - 1)$ becomes infinitely large. Therefore there is no limit and there is no answer for the given improper integral. We say that the improper integral in (a) *converges* or that it *converges to* 1, whereas the improper integral in (b) is said to *diverge*, or to *diverge to infinity*.

Example 41-11

Find the area, if it exists, under the graph of $y = 1/x^2$, above the x-axis, and bounded on the left by the line $x = 1$.

The specified area is given by

$\int_1^\infty \dfrac{1}{x^2}\, dx = \lim_{n \to \infty} \int_1^n \dfrac{1}{x^2}\, dx$

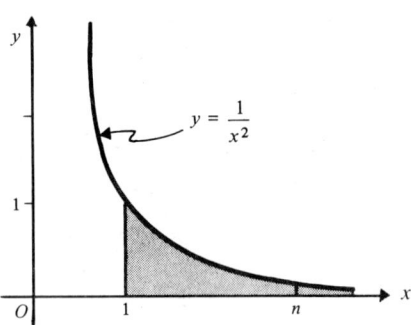

From part (a) of Example 41-10 we see that the area is equal to 1. This means that the larger n is, the closer is the area under the curve between $x = 1$ and $x = n$ to 1 sq unit. For example, if $n = 100$, the latter area equals 0.99 sq units; if $n = 10,000$, it equals 0.9999 sq units; and so on. (See also Study Question 2 below.)

Further applications of improper integrals will appear in later Goals.

Goal 41 | To Find Antiderivatives

EXERCISES

1. Find the following indefinite integrals.
   (a) $\int 7x^6 \, dx$
   (b) $\int -3 \frac{1}{x^4} \, dx$
   (c) $\int 6x^3 \, dx$
   (d) $\int (4x^3 - 2x^2 + x - 6) \, dx$
   (e) $\int (x^{10} - x^9 + 7x^4 + 2) \, dx$
   (f) $\int (x^{100} + 32x^{31} - 25x^4 - 18) \, dx$
   (g) $\int \left(x^5 + x - \frac{1}{x^2} + \frac{1}{x^3} \right) dx$
   (h) $\int \left(x + 1 + \frac{3}{x^2} - \frac{1}{x^5} \right) dx$
   (i) $\int \left(x^2 + 1 + \frac{1}{x} - \frac{2}{x^2} \right) dx$
   (j) $\int \frac{x^2 - x + 2}{x} \, dx$
   (k) $\int \frac{x^{10} - x^4 + x - 1}{2x^3} \, dx$
   (l) $\int \frac{1}{\sqrt{x}} \, dx$
   (m) $\int x^{7/2} \, dx$
   (n) $\int \sqrt[4]{x} \, dx$
   (o) $\int \frac{x^2 + x + 1}{\sqrt{x}} \, dx$
   (p) $\int 18e^x \, dx$
   (q) $\int \left(3e^x - \frac{1}{x} \right) dx$
   (r) $\int \left(x^2 + 2e^x + \frac{1}{x} - 5 \right) dx$

2. Calculate the following definite integrals.
   (a) $\int_0^1 (x^3 + x^2 - 2) \, dx$
   (b) $\int_1^3 (u^2 + 3u - 1) \, du$
   (c) $\int_{-1}^1 (x^2 - 1) \, dx$
   (d) $\int_1^2 \left(t + \frac{1}{t^2} \right) dt$
   (e) $\int_1^e \frac{3}{y} \, dy$
   (f) $\int_0^4 3\sqrt{u} \, du$

(g) $\int_0^1 (x^{99} - x^{49})\, dx$ (i) $\int_0^1 e^t\, dt$

(h) $\int_1^8 \sqrt[3]{x}\, dx$

3. Find actual solutions to the following Exercises of Goal 39:
   (a) #1   (b) #2   (c) #4   (d) #5   (e) #9

4. Evaluate if possible

   (a) $\int_1^\infty \dfrac{1}{x^3}\, dx$ (b) $\int_1^\infty \dfrac{1}{\sqrt[3]{x}}\, dx$ (c) $\int_1^\infty \dfrac{1}{t}\, dt$

5. Find the area under the graph of $f(x)$, above the x-axis, and bounded at the left by $x = 2$, if it exists. If not, explain why not.

   (a) $f(x) = \dfrac{1}{x^2}$ (b) $f(x) = \dfrac{1}{x}$

TEST A

1. Find the following indefinite integrals:

   (a) $\int (x^8 - 7x^3 + 3)\, dx$ (c) $\int (2e^x - 3x)\, dx$

   (b) $\int \dfrac{3 + x^2 - x^3}{x^4}\, dx$

2. Calculate the following definite integrals:

   (a) $\int_0^1 (x^9 + x^4)\, dx$ (b) $\int_0^4 (3\sqrt{x} - x^{3/2})\, dx$

3. Evaluate if possible:

   $$\int_3^\infty \dfrac{1}{x^2}\, dx$$

TEST B

1. Find the following indefinite integrals:

   (a) $\int (3x^4 - 2x^5 - 1)\, dx$ (c) $\int \left(e^x + \dfrac{3}{x} - 1\right) dx$

   (b) $\int \left(\dfrac{1}{x^2} + x^{1/4}\right) dx$

2. Calculate the following definite integrals:

   (a) $\int_1^2 \dfrac{x^4 - x^2}{2x}\, dx$ (b) $\int_e^{e^2} \dfrac{1}{x}\, dx$

Goal 41 | To Find Antiderivatives

3. Evaluate if possible:

$$\int_2^\infty \frac{1}{\sqrt{x}}\, dx$$

STUDY QUESTIONS

1. We know that $\int_a^b f(x)\, dx = F(b) - F(a)$, where $F'(x) = f(x)$. Show why it is unnecessary to use $F(x) + C$ as the antiderivative of $f(x)$ when evaluating a definite integral.

2. Example 41-11 shows that it makes sense to speak of the area under the curve $y = 1/x^2$, above the x-axis, and bounded on the left by $x = 1$.
   (a) Show that the area under $y = 1/\sqrt{x}$, above the x-axis, and to the right of $x = 1$ is infinite.
   (b) Draw the graphs of $y = 1/x^2$ and $y = 1/\sqrt{x}$ on the same axes, then tell how these graphs differ.

3. (a) Find three antiderivatives of $f(x) = 3x^2$ and draw their graphs on the same set of axes.
   (b) Suppose a graph of each antiderivative of $3x^2$ were sketched. Describe the resulting figure.
   (c) Which antiderivative of $3x^2$ goes through the point $(2, -5)$?

4. In Study Question 39-1 (p. 387), show that the total flow through the artery is equal to $\dfrac{\pi C}{2} R^4$. Thus the flow will go up (or down!) as does the fourth power of R. This is a mathematical justification for medication that dilates arteries.

GOAL 42

To integrate by substitution

Finding derivatives is usually straightforward; finding antiderivatives (or indefinite integrals) usually involves trial and error. Often we guess what the answer might be, then check it out. Sometimes it takes several guesses before the correct one is found. Skill and insight are, naturally, helpful, and these normally increase with practice. Tables of integral formulas are available, but to use them efficiently you need to have done many examples and to be familiar with a variety of techniques. The most basic of these is *substitution*.

USING SUBSTITUTION IN INTEGRATION

Every table of integrals has this formula:

$$\text{I.} \quad \int u^n \, du = \frac{u^{n+1}}{n+1} + C \qquad (n \neq -1)$$

However, you would undoubtedly not find any of the following in most tables of integrals:

(a) $\int t^9 \, dt$ (b) $\int x^{1/2} \, dx$ (c) $\int \frac{1}{y^2} \, dy$ (d) $\int u^3 \, du$

To obtain answers for these, of course, we use the formula just given:

In (a) we let $u = t$ and $n = 9$, note that $du = dt$, and get the answer $\frac{1}{10} t^{10} + C$.

In (b) we let $u = x$ and $n = \frac{1}{2}$, note that $du = dx$, and get the answer $\frac{2}{3} x^{3/2} + C$.

In (c) we let $u = y$ and $n = -2$, note that $du = dy$, and get the answer $-1/y + C$.

In (d) the answer is $u^4/4 + C$.

For (a), (b), and (c) we used substitution, or change of variable, in the simplest sense to obtain indefinite integrals.

THE CHAIN RULE

The power of the technique of substitution is far greater than suggested just above. Using it together with the chain rule, we can write down almost immediately, with a bit of practice, answers for the following three integrals:

(e) $\int (x^2 + 1)^6 2x \, dx$

(f) $\int \sqrt{3x^3 + 5} \, 9x^2 \, dx$

(g) $\int \dfrac{1}{(e^x + 2)^3} e^x \, dx$

For each of these we use the same formula used for (a) through (d) above—Formula I. Our approach in these examples is to seek antidifferentials. Remember that the final answer must be expressed in terms of the given variable of integration.

(e) In this one we seek an antidifferential of $(x^2 + 1)^6 \, 2x \, dx$. We let

$$u = x^2 + 1 \quad \text{and} \quad n = 6$$

and note that

$$du = 2x \, dx$$

Then

$$\int (x^2 + 1)^6 \, 2x \, dx = \int u^6 \, du = \frac{1}{7} u^7 + C = \frac{1}{7}(x^2 + 1)^7 + C$$

(f) To integrate $\int \sqrt{3x^3 + 5} \, 9x^2 \, dx$, we let

$$u = 3x^3 + 5 \quad \text{and} \quad n = \frac{1}{2}$$

and note that

$$du = 9x^2 \, dx$$

So

$$\int \sqrt{3x^3 + 5} \, 9x^2 \, dx = \int (3x^3 + 5)^{1/2} \, 9x^2 \, dx = \int u^{1/2} \, du$$

$$= \frac{2}{3} u^{3/2} + C = \frac{2}{3}(3x^3 + 5)^{3/2} + C$$

(g) Since

$$\int \frac{1}{(e^x + 2)^3} e^x \, dx = \int (e^x + 2)^{-3} e^x \, dx$$

we let

$$u = e^x + 2 \quad \text{and} \quad n = -3$$

and observe that $du = e^x \, dx$. Then

$$\int (e^x + 2)^{-3} e^x \, dx = \int u^{-3} \, du = -\frac{1}{2} u^{-2} + C$$

$$= -\frac{1}{2u^2} + C = -\frac{1}{2(e^x + 2)^2} + C$$

The clue to integrating (e), (f), and (g) is the recognition that each integral is of the form $\int u^n \, du$. To verify this we applied the chain rule. Note, in the following examples, how we use the technique of substitution not only with Formula I but also with these two formulas:

$$\text{II.} \quad \int \frac{1}{u} \, du = \ln u + C \qquad (u > 0)$$

$$\text{III.} \quad \int e^u \, du = e^u + C$$

Whenever applying Formula II in examples or exercises in this Goal, assume that u is positive.

Example 42-1

Find $\int \dfrac{1}{4x + 3} \, 4 \, dx$.

We use Formula II, letting $u = 4x + 3$ and noting that $du = 4 \, dx$. Then we have $\int \dfrac{1}{u} \, du = \ln u + C$ and the answer to the given integral is $\ln(4x + 3) + C$.

Example 42-2

Find $\int e^{-x^2}(-2x \, dx)$.

Let u in Formula III be $-x^2$. Then $du = -2x \, dx$ (parenthesized above in the statement of the integral) and our integral becomes $\int e^u \, du = e^u + C = e^{-x^2} + C$.

Example 42-3

Find $\int \dfrac{e^t}{e^t + 4} \, dt$.

Let $u = e^t + 4$. Then $du = e^t \, dt$ and the given integral is of the form $\int \dfrac{1}{u} \, du$ (Formula II), with answer $\ln u + C$ or $\ln(e^t + 4) + C$.

NOTE: Each answer obtained can be checked quickly by differentiating. You should verify the correctness of the three integrals done above.

MORE COMPLICATED EXAMPLES. Each example given so far in this Goal was "rigged," in the following sense: for each choice of u, du turned out to be just what we already had. Thus in 42-1, du was $4 \, dx$; in 42-2, du was $-2x \, dx$; and in 42-3, du was $e^t \, dt$. In each case the given integral was in "perfect" form. Not so in the following examples; therefore more manipulation is called for.

Example 42-4

Find $\int (x^2 + 7)^3 \, x \, dx$.

Hoping to apply Formula I, we let $u = x^2 + 7$ and $n = 3$. Then $du = 2x \, dx$.

Goal 42 | To Integrate by Substitution

But we have only $x\, dx$, not $2x\, dx$! No problem: multiply by 1 in the form $\frac{2}{2}$ or $(\frac{1}{2})(2)$; pull the factor $\frac{1}{2}$ out in front of the integral sign; and put the factor 2 where we need it, with $x\, dx$. Thus

$$\int (x^2 + 7)^3 x\, dx = \frac{1}{2}\int (x^2 + 7)^3 (2x\, dx) = \frac{1}{2}\int u^3\, du$$

$$= \frac{1}{2}\left(\frac{u^4}{4}\right) + C = \frac{1}{8}(x^2 + 7)^4 + C$$

Check:

$$\frac{d}{dx}\frac{1}{8}(x^2 + 7)^4 = \frac{1}{8}\cdot 4(x^2 + 7)^3 \cdot 2x = (x^2 + 7)^3 x$$

or, with differentials,

$$d\left[\frac{1}{8}(x^2 + 7)^4\right] = (x^2 + 7)^3 x\, dx.$$

Example 42-5

Find $\int e^{1-x^3} x^2\, dx$.

Formula III looks hopeful. Let $u = 1 - x^3$; then $du = -3x^2\, dx$. We multiply by 1 in the form $(-\frac{1}{3})(-3)$; then

$$\int e^{1-x^3} x^2\, dx = -\frac{1}{3}\int e^{1-x^3}(-3x^2\, dx) = -\frac{1}{3}\int e^u\, du$$

$$= -\frac{1}{3}e^u + C = -\frac{1}{3}e^{1-x^3} + C$$

Example 42-6

Find $\int (x^4 - 4x)^{80}(x^3 - 1)\, dx$.

Try Formula I with $u = x^4 - 4x$ and $n = 80$. Then $du = (4x^3 - 4)\, dx$. Since $du = 4(x^3 - 1)\, dx$, and we have $(x^3 - 1)\, dx$, we multiply by 1, this time in the form $(\frac{1}{4})(4)$, yielding

$$\int (x^4 - 4x)^{80}(x^3 - 1)\, dx = \frac{1}{4}\int (x^4 - 4x)^{80} 4(x^3 - 1)\, dx$$

$$= \frac{1}{4}\int (x^4 - 4x)^{80}[(4x^3 - 4)\, dx]$$

$$= \frac{1}{4}\int u^{80}\, du = \frac{1}{4}\cdot\frac{u^{81}}{81} + C$$

$$= \frac{1}{324}(x^4 - 4x)^{81} + C$$

Example 42-7

Find $\int (x^3 + 1)^2 \, x \, dx$.

This looks like $\int u^n \, du$ (Formula I) with $u = x^3 + 1$ and $n = 2$. Note that in this case $du = 3x^2 \, dx$. We have $x \, dx$. Can we adjust as in the preceding three examples? No! Only a constant factor can "move across" the integral sign (see Study Question 1 below). The given integral is *not*, as it stands, in the form $\int u^n \, du$. However, it can be integrated by expanding as follows:

$$\int (x^3 + 1)^2 \, x \, dx = \int (x^6 + 2x^3 + 1) \, x \, dx$$

$$= \int (x^7 + 2x^4 + x) \, dx = \frac{x^8}{8} + \frac{2x^5}{5} + \frac{x^2}{2} + C$$

where we applied Formula I to each term in the sum.

SUMMARY OF METHODS USED SO FAR

As of the moment, three formulas are available (I on p. 404, II and III on p. 406): for $\int u^n \, du$ (given $n \neq -1$), $\int (du/u)$, and $\int e^u \, du$. Given an integral, proceed as follows:

(1) Decide which of the three forms the integral fits best. Choose u and, if Formula I is to be applied, n.

(2) Find du for your choice of u.

(3) If a constant factor, say k, is needed to transform the given integral into proper form, multiply by $(1/k)(k)$ (which equals $k/k = 1$), using the factor $1/k$ before the integral sign and the factor k where needed in the integrand.

(4) Keep in mind the possibility of expanding if necessary (as in Example 42-7).

(5) When using substitution on indefinite integrals, be sure to express the final answer in terms of the original variable.

Example 42-8

(a) $\int \sqrt{5x^2 + 2} \, x \, dx = \int (5x^2 + 2)^{1/2} x \, dx = \frac{1}{10} \int (5x^2 + 2)^{1/2} (10x \, dx)$

$= \frac{1}{10} \cdot \frac{2}{3} (5x^2 + 2)^{3/2} + C = \frac{1}{15} (5x^2 + 2)^{3/2} + C$

Here we used $\int u^n \, du$ with $u = 5x^2 + 2$ and $n = \frac{1}{2}$. In this case, $du = 10x \, dx$.

(b) $\int (\ln x)^{13} \left(\frac{1}{x} \right) dx = \frac{(\ln x)^{14}}{14} + C$

Goal 42 | To Integrate by Substitution

We used $\int u^n \, du$ with $u = \ln x$ and $n = 13$. Note that $du = \dfrac{1}{x} \, dx$.

(c) $\displaystyle\int \dfrac{1}{x^3 - x + 2}(6x^2 - 2) \, dx = 2\int \dfrac{1}{x^3 - x + 2}(3x^2 - 1) \, dx$
$= 2 \ln(x^3 - x + 2) + C$

Formula II was used. In $\displaystyle\int \dfrac{du}{u}$, we chose u equal to $(x^3 - x + 2)$. Since $du = (3x - 1) \, dx$, we "pulled" the constant 2 across the integral sign.

(d) $\displaystyle\int \left(\dfrac{e^{\sqrt{x}}}{\sqrt{x}}\right) dx = \int e^{x^{1/2}}(x^{-1/2} \, dx) = 2\int e^{x^{1/2}}\left(\dfrac{1}{2}x^{-1/2} \, dx\right) = 2e^{x^{1/2}} + C$

We used Formula III with $u = x^{1/2}$. Since $du = \tfrac{1}{2}x^{-1/2} \, dx$, we multiplied by 1 in the form $(2)(\tfrac{1}{2})$.

EVALUATING DEFINITE INTEGRALS

The technique of substitution, when applicable, provides an antiderivative for use in evaluating a definite integral, as illustrated in the following examples.

Example 42-9

Evaluate $\displaystyle\int_0^4 \sqrt{x^2 + 9} \, x \, dx$.

To find an antiderivative we note that this fits $\int u^n \, du$ with $u = x^2 + 9$, $n = \tfrac{1}{2}$, $du = 2x \, dx$. So

$\displaystyle\int_0^4 (x^2 + 9)^{1/2} x \, dx = \dfrac{1}{2}\int_0^4 (x^2 + 9)^{1/2} \, 2x \, dx = \dfrac{1}{2}\left[\dfrac{2}{3}(x^2 + 9)^{3/2}\right]_0^4$

$= \dfrac{1}{3}[(4^2 + 9)^{3/2} - (0 + 9)^{3/2}] = \dfrac{1}{3}(25^{3/2} - 9^{3/2})$

$= \dfrac{1}{3}(125 - 27) = \dfrac{98}{3}$

Example 42-10

Evaluate $\displaystyle\int_1^e (4 + 2 \ln x)^2 \dfrac{1}{x} \, dx$.

Formula I applies here too, this time with $u = 4 + 2 \ln x$ and $n = 2$; and $du = (2/x) \, dx$. Therefore

$$\int_1^e (4 + 2\ln x)^2 \frac{1}{x}\, dx = \frac{1}{2} \int_1^e (4 + 2\ln x)^2 \frac{2}{x}\, dx$$

$$= \frac{1}{2}\left[\frac{(4 + 2\ln x)^3}{3} \right]_1^e = \frac{1}{6}[(4 + 2)^3 - (4 + 0)^3]$$

$$= \frac{1}{6}(216 - 64) = \frac{152}{6} = \frac{76}{3}$$

EXERCISES

Find the following indefinite integrals (1 through 29):

1. $\int (x^4 - x^2)^4 (4x^3 - 2x)\, dx$
2. $\int (x^3 + 2x^2 - 3x + 4)^9 (3x^2 + 4x - 3)\, dx$
3. $\int (x^5 - 2x + 1)^6 (5x^4 - 2)\, dx$
4. $\int (\ln x)^7 (1/x)\, dx$
5. $\int 25(x^2 - 5x + 1)^{24} (2x - 5)\, dx$
6. $\int 18(x^3 + 2x^2 - 3)^{17} (3x^2 + 4x)\, dx$
7. $\int e^{(5x^2 - x + 1)}(10x - 1)\, dx$
8. $\int e^{(x^3 + 8x)}(3x^2 + 8)\, dx$
9. $\int \frac{1}{x^3 + x^2 - x} (3x^2 + 2x - 1)\, dx$
10. $\int \frac{1}{4x^2 - 5} (8x)\, dx$
11. $\int (x^4 + 2x^2)^{10} (x^3 + x)\, dx$
12. $\int (x^3 + 3x^2 - 3x)^5 (x^2 + 2x - 1)\, dx$
13. $\int e^{(x^2 - 2x + 1)}(x - 1)\, dx$
14. $\int e^{(x^5 - 5x)}(10x^4 - 10)\, dx$
15. $\int \frac{1}{3x^4 + x^3 + 1} (4x^3 + x^2)\, dx$
16. $\int \frac{1}{2x^2 + 3x - 1} (8x + 6)\, dx$
17. $\int \sqrt{5x^2 + 3x - 2}\, (10x + 3)\, dx$
18. $\int \sqrt[3]{4x^3 - 1}\, (12x^2)\, dx$
19. $\int \sqrt{x^7 - 7x^2}\, (x^6 - 2x)\, dx$
20. $\int (x^5 + 3x^2 + 1)^{3/4} (10x^4 + 12x)\, dx$
21. $\int \frac{1}{(x^2 + 2x - 1)^5} (2x + 2)\, dx$

Goal 42 | To Integrate by Substitution

22. $\int \dfrac{1}{(x^3 - 3x - 2)^3} (x^2 - 1)\, dx$

23. $\int \dfrac{1}{(x^2 + x)^2} (10x + 5)\, dx$

24. $\int \dfrac{1}{\sqrt{5x - 3}}\, dx$

25. $\int e^{\sqrt{x+1}} \dfrac{1}{\sqrt{x+1}}\, dx$

26. $\int (3x + 4)^5\, dx$

27. $\int \dfrac{dx}{5x + 2}$

28. $\int (5x + 3)^2\, x\, dx$

29. $\int \dfrac{dx}{(3 - 4x)^5}$

30. Explain why we are unable at present to integrate the following:

   (a) $\int \sqrt{x^2 + 1}\, x^2\, dx$

   (b) $\int \dfrac{x}{x^3 + 4}\, dx$

   (c) $\int \dfrac{x}{(2x + 1)^2}\, dx$

   (d) $\int \dfrac{x^2}{\sqrt{x^2 + 1}}\, dx$

Evaluate the definite integrals in 31 through 36:

31. $\int_1^e \dfrac{\ln x}{x}\, dx$

32. $\int_0^1 (x^2 + 3)^4\, x\, dx$

33. $\int_0^2 \sqrt{4 - x^2}\, x\, dx$

34. $\int_1^2 \dfrac{x\, dx}{x^2 + 1}$

35. $\int_0^2 e^{-(x^2/2)} x\, dx$

36. $\int_0^3 (x - 1)^2\, x\, dx$

37. We know that

$$\int (x^2 + 1)^5\, 2x\, dx = \dfrac{(x^2 + 1)^6}{6} + C$$

Students often wonder what "happened" to the factor $2x$ when integrating. Can you explain this?

TEST A

In 1 through 5 find the indefinite integrals.

1. $\int (x^3 + 3x^2 - 1)^{10} (3x^2 + 6x)\, dx$
2. $\int e^{(5x^2 + 2)} (10x)\, dx$

3. $\int (x^6 + 2x^3 + 3x^2)^4 (x^5 + x^2 + x)\, dx$

4. $\int (8x - 5)^{20}\, dx$

5. $\int \dfrac{1}{x^2 + 2}\, x\, dx$

6. Evaluate $\displaystyle\int_0^1 (x^2 + 1)^3\, x\, dx$.

TEST B

In 1 through 5 find the indefinite integrals.

1. $\int (x^2 - x - 1)^8 (2x - 1)\, dx$

2. $\int (5x + 1)^9\, dx$

3. $\int \dfrac{1}{(3x^2 - 1)^3}\, x\, dx$

4. $\int e^{(5x^2 - 5x)}(2x - 1)\, dx$

5. $\int \dfrac{1}{x^3 - 1}\, x^2\, dx$

6. Evaluate $\displaystyle\int_0^2 \dfrac{x^2\, dx}{\sqrt{x^3 + 1}}$.

STUDY QUESTIONS

1. Show that $\int x\, dx \neq (1/x) \int x^2\, dx$. This verifies that the variable factor x cannot be moved across the integral sign.

2. Integrate $\int (x^2 + 1)^2\, x\, dx$ in two ways: (a) by substitution, using the formula for $\int u^n\, du$; (b) by expanding $(x^2 + 1)^2$, multiplying by x, and integrating the resulting sum term by term. Explain any discrepancy in the answers you obtain in (a) and (b).

3. How, if at all, are $\int_a^b f(x)\, dx$ and $\int f(x)\, dx$ related?

4. (a) Justify the following statement:

$$\int f'(g(x))\, g'(x)\, dx = f(g(x)) + C$$

HINT: Use the chain rule to obtain $\dfrac{d}{dx} f(g(x))$.

(b) Use part (a) to show that

$$\int 100(x^2 + 1)^{99} (2x\, dx) = (x^2 + 1)^{100} + C$$

(c) Show how

$$\int [g(x)]^n [g'(x)\, dx] = \dfrac{[g(x)]^{n+1}}{n + 1} + C$$

is a special case of the statement in part (a).

GOAL **43**

To use integration by parts

Integration by parts is a powerful technique because its use often enables us to replace a complicated integral by a simpler one. In this Goal we state the so-called Parts Formula, apply it to several integrals, and then justify the formula.

THE PARTS FORMULA

If u and v are differentiable functions then

$$\int u \, dv = uv - \int v \, du$$

To apply this formula to a given integral, say $\int f(x) \, dx$, we break $f(x) \, dx$ (into parts!) so that one part is u and the other is dv. Successful use of the formula, if it leads to a solution at all, almost always depends on the choices made in this division.

Example 43-1

Integrate $\int xe^x \, dx$.

Let $u = x$ and $dv = e^x \, dx$; then $du = dx$ and $v = e^x$. (It is not necessary to write "$v = e^x + C$"; in Study Question 1 of this Goal you are asked to show why.) By the Parts Formula we have

$$\int u \, dv = uv - \int v \, du$$
$$\int x \, e^x \, dx = x \, e^x - \int e^x \, dx$$
$$= xe^x - e^x + C$$

Check: $\dfrac{d}{dx}(xe^x - e^x + C) = x \cdot e^x + e^x \cdot 1 - e^x = xe^x$.

The choices for u and dv made above led to an easy integral, $\int e^x \, dx$. Let us see what happens if we start over again with the same integral $\int xe^x \, dx$

but choose u equal to e^x and dv equal to $x\,dx$. Then $du = e^x\,dx$, $v = \tfrac{1}{2}x^2$, and the Parts Formula yields

$$\int u\,dv = u\,v - \int v\,du$$

$$\int xe^x\,dx = \int e^x x\,dx = e^x \cdot \frac{1}{2}x^2 - \int \frac{1}{2}x^2 \cdot e^x\,dx$$

$$= \frac{1}{2}[x^2 e^x - \int x^2 e^x\,dx]$$

But $\int x^2 e^x\,dx$ is more difficult to integrate than the expression we started with! So the second choice led to a more complicated integral. Of course, it is most efficient to make the correct choice on the first trial; but if you do not succeed then (at worst) you simply try again. As with other methods of integration, practice helps.

Example 43-2

Find $\int x^2 \ln x\,dx$.

Suppose we let $u = x^2$ and $dv = \ln x\,dx$. Then $du = 2x\,dx$, and v is what? At present we do not yet have an antiderivative for $\ln x$. So let's try again. This time we let $u = \ln x$ and $dv = x^2\,dx$. Then $du = dx/x$ and $v = x^3/3$. Now

$$\int u\,dv = u\,v - \int v\,du$$

$$\int x^2 \ln x\,dx = \int \ln x \cdot x^2\,dx = \ln x \cdot \frac{x^3}{3} - \int \frac{x^3}{3} \cdot \frac{1}{x}\,dx$$

$$= \frac{1}{3}x^3 \ln x - \frac{1}{3}\int x^2\,dx$$

$$= \frac{1}{3}x^3 \ln x - \frac{1}{3}\cdot\frac{x^3}{3} + C = \frac{1}{3}x^3\left(\ln x - \frac{1}{3}\right) + C$$

This time it worked. We check the answer by differentiating it, using the product rule:

$$\frac{d}{dx}\left[\frac{1}{3}x^3\left(\ln x - \frac{1}{3}\right) + C\right] = \frac{1}{3}\left[x^3 \cdot \frac{1}{x} + \left(\ln x - \frac{1}{3}\right)\cdot 3x^2\right]$$

$$= \frac{1}{3}(x^2 + 3x^2 \ln x - x^2) = x^2 \ln x$$

NOTES ON THE METHOD

(1) The Parts Formula is often useful in integrating a product of functions, such as xe^x, $x^3 \ln x$, $x\sqrt{x+1}$, $(2+x)\sqrt{4-x}$, and so on.

Goal 43 | To Use Integration by Parts

(2) Given $\int f(x)\, dx$, whatever choice is made for u and dv, dv must include dx.

(3) The division of $f(x)\, dx$ should be made in such a way that $\int v\, du$ can be integrated. If $\int v\, du$ is simpler than the original integral, optimism is in order. If not, another choice for u and dv should be tried.

(4) It often works well to choose as u a function that is simplified by differentiation: say $\ln x$, or x^n where n is a positive integer.

(5) If a function changes only slightly under integration (for example, e^x, e^{-x}, e^{2x}) it frequently works well to include it within dv.

These generalizations are like others: they have many exceptions. In fact, integration by parts often reduces to trial and error, and even then it sometimes happens that no solution is forthcoming.

Example 43-3

Use the Parts Formula to integrate $\int x^2 e^x\, dx$.

Try letting $u = x^2$ and $dv = e^x\, dx$. Then $du = 2x\, dx$ and $v = e^x$; and

$$\int u\, dv = u\, v - \int v\, du$$

$$\int x^2 e^x\, dx = x^2 e^x - \int e^x \cdot 2x\, dx$$
$$= x^2 e^x - 2\int x e^x\, dx$$

You may think we made a bad choice because we have no "instant" antiderivative for xe^x. But we do! In Example 43-1 we integrated $\int xe^x\, dx$ by parts and got the answer $xe^x - e^x + C$. Therefore

$$\int x^2 e^x\, dx = x^2 e^x - 2(xe^x - e^x) + C = x^2 e^x - 2xe^x + 2e^x + C$$

Check this answer by differentiating it.

As mentioned above in the Notes on the Method, whenever $\int v\, du$ is simpler than the original integral (as in the last example) we should be encouraged. Sometimes an integral requires the parts treatment two or more times. It is instructive to try this example by parts again, now letting $u = e^x$ and $dv = x^2\, dx$.

Example 43-4

Evaluate the definite integral $\int_1^e \ln x\, dx$.

To find an antiderivative of $\ln x$, we use parts. The choice of u and dv is here clear: $u = \ln x$ and $dv = dx$. Then $du = dx/x$ and $v = x$; and

$$\int u\, dv = u\, v - \int v\, du$$

$$\int \ln x\, dx = \ln x \cdot x - \int x \cdot (1/x)\, dx = x \ln x - x + C$$

Therefore for the definite integral we have

$$\int_1^e \ln x \, dx = [x \ln x - x]_1^e = (e \ln e - e) - (1 \ln 1 - 1) = 1$$

Example 43-5

Evaluate $\int_0^3 x\sqrt{x+1} \, dx$.

Let's try to integrate by parts. Noting that $(x + 1)^{1/2}$ does not simplify under differentiation but that the function x does, we let $u = x$ and $dv = (x + 1)^{1/2} \, dx$; then $du = dx$ and $v = \frac{2}{3}(x + 1)^{3/2}$. So

$$\int x\sqrt{x+1} \, dx = \frac{2}{3} x(x+1)^{3/2} - \frac{2}{3} \int (x+1)^{3/2} \, dx$$

and

$$\int_0^3 x\sqrt{x+1} \, dx = \left[\frac{2}{3} x(x+1)^{3/2} - \frac{2}{3} \cdot \frac{2}{5}(x+1)^{5/2}\right]_0^3$$

In this step we recognized $\int (x + 1)^{3/2} \, dx$ as of the form $\int U^n \, dU$, with $U = x + 1$, $n = \frac{3}{2}$, $dU = dx$. Proceeding, the last term above gives us

$$\left[\frac{2}{3}(3)(3+1)^{3/2} - \frac{4}{15}(3+1)^{5/2}\right] - \left[\frac{2}{3}(0)(1)^{3/2} - \frac{4}{15}(1)^{5/2}\right]$$

$$= 16 - \frac{128}{15} + \frac{4}{15} = \frac{116}{15}$$

THE BASIS OF THE PARTS FORMULA

If u and v are differentiable functions of, say, x, then by the product rule

$$\frac{d}{dx}(uv) = u\frac{dv}{dx} + v\frac{du}{dx}$$

or in terms of differentials

$$d(uv) = u \, dv + v \, du$$

Rearranging terms yields

$$u \, dv = d(uv) - v \, du$$

It then follows that

$$\int u \, dv = \int d(uv) - \int v \, du$$

Goal 43 | To Use Integration by Parts

But $\int d(uv)$ is just an antidifferential of the differential of uv, namely uv. Therefore

$$\int u\, dv = uv - \int v\, du$$

which is the Parts Formula.

[Why, here, can we write uv for $\int d(uv)$, rather than $uv + C$?]

EXERCISES

Find each indefinite integral in 1 through 7.

1. $\int xe^{3x}\, dx$
2. $\int xe^{-x}\, dx$
3. $\int x^2 e^{-x}\, dx$
4. $\int x^5 \ln x\, dx$
5. $\int x\sqrt{x+2}\, dx$
6. $\int x(x+3)^{10}\, dx$
7. $\int \dfrac{\ln x}{\sqrt{x}}\, dx$

Evaluate each definite integral in 8 through 10. (You may want to make use of some of the answers to the preceding exercises or to examples in this Goal.)

8. $\int_0^1 xe^x\, dx$
9. $\int_0^2 x^2 e^x\, dx$
10. $\int_1^2 x^2 \ln x\, dx$

TEST A

Integrate in 1 through 4.

1. $\int xe^{2x}\, dx$
2. $\int x(x+9)^5\, dx$
3. $\int x \ln x\, dx$
4. $\int \dfrac{x+1}{e^x}\, dx$

5. Using the answer you obtained for exercise 3, evaluate the definite integral

$$\int_1^3 x \ln x\, dx$$

TEST B

Integrate in 1 through 4.

1. $\int x(x-3)^7\, dx$
2. $\int \dfrac{\ln x}{x^2}\, dx$
3. $\int xe^{-5x}\, dx$
4. $\int x\sqrt{2x+1}\, dx$

5. Using your answer to Exercise 4, evaluate the definite integral

$$\int_0^4 x\sqrt{2x+1}\,dx$$

STUDY QUESTIONS

1. In Example 43-1 we integrated $\int xe^x\,dx$ by parts, letting $u = x$ and $dv = e^x\,dx$. We noted that $du = dx$ and $v = e^x$. What difference does it make if instead we use v equal to $e^x + k$? Show, in general, that it is not necessary to replace v by $v + k$ in the Parts Formula.

2. Consider $\int \ln(x+1)\,dx$. It is clear that we should try $u = \ln(x+1)$ and $dv = dx$. There are actually an infinite variety of possibilities for v. (Why?) Discuss the two choices $v = x$ and $v = x + 1$ in this integration.

3. Discuss the following:

$$\int u\,dv = uv - \int v\,du$$

$$\int \frac{1}{x}\,dx = \int x \cdot \frac{1}{x^2}\,dx = x\left(-\frac{1}{x}\right) - \int -\frac{1}{x}\,dx$$

$$\int \frac{1}{x}\,dx = -1 + \int \frac{1}{x}\,dx$$

$$0 = -1(!)$$

Summary of Goals 37–43

You should now be able to

(1) obtain approximations to the distance traveled by an object during an interval (or to the increase in a population, to the additional cost, to the area under a curve, and so on) by adding up the distances (or other appropriate quantities) over small equal subintervals;

(2) find the exact area under a curve by summing the areas of n rectangles, then evaluating this sum as n becomes infinite;

(3) given $y = f((x)$, where f is continuous and positive, express the area under the curve between $x = a$ and $x = b$ as a definite integral;

(4) given marginal cost (or revenue or profit), or the rate at which a population (or investment, or epidemic, or the like) grows, express the change in cost (or other appropriate quantity) over a specified interval as a definite integral;

(5) evaluate $\int_a^b f(x)\, dx$ if you know that $F'(x) = f(x)$, by finding $F(b) - F(a)$;

(6) find antiderivatives of x^n ($n \neq 1$), e^x, $1/x$, and polynomials;

(7) evaluate definite integrals of the functions listed in (6);

(8) evaluate an improper integral of the form $\int_a^\infty f(x)\, dx$ if it exists and if you know an antiderivative of $f(x)$;

(9) integrate (find antiderivatives) using the technique of substitution and the formulas for

$$\int u^n\, du \quad (n \neq -1), \qquad \int \frac{1}{u}\, du, \qquad \int e^u\, du$$

(10) integrate by using the Parts Formula: $\int u\, dv = uv - \int v\, du$.

Review Exercises

GOAL 37

1. Find an approximation to the distance traveled in 4 sec by an object moving along a line, if its speed is $(t^2 + 1)$ ft/sec after t sec. Divide the interval into four equal subintervals and use left-hand endpoints.

2. Find an approximation to the area above the x-axis and under the graph of $f(x) = 2x^2 - 1$ from $x = 1$ to $x = 4$ using six equal subintervals and right-hand endpoints.

3. If a company's marginal cost function is, in dollars per unit, $0.01x - 0.4$ when the production level is x units, find an approximation to the increase in cost when the production level is raised from 450 to 850 units. Use four equal subintervals and midpoints of the subintervals.

GOAL 38

In Exercises 4 and 5 use the definition of area as the limit of an infinite sum of rectangles to find the specified area. You may use a formula from Exercise 1 on page 380.

4. The area above the x-axis, under $y = 2x^2$, and between $x = 0$ and $x = 1$, using right-hand endpoints.

5. The area under $y = 4x + 3$, above the x-axis, and between $x = 1$ and $x = 2$, using left-hand endpoints.

GOAL 39

Express each quantity in Exercises 6, 7, and 8 in terms of a definite integral.

6. The area under $y = 4x^3 + x - 3$, above the x-axis, and between $x = 0$ and $x = 2$.

7. The additional profit resulting from a change in production from 400 to 500 units if the marginal profit is $-1.4x + 800$ dollars per unit.

8. The amount of a radioactive substance that decays in 4 days if it is decomposing at the rate of $5e^{-0.1t}$ units per day after t days.

GOAL 40

9. Evaluate the definite integrals

   (a) $\displaystyle\int_1^3 (2x + 1)\, dx$ (b) $\displaystyle\int_0^2 e^x\, dx$

10. If the velocity of an object moving along a line at time t equals $1/t$ ft/sec, how far does the object travel during the second and third seconds?

11. Find the area under the graph of $f(x) = 3x^2$ between $x = 1$ and $x = 3$.

GOAL 41

12. Find each indefinite integral:

(a) $\int (3x^2 + 2x + 4)\, dx$

(b) $\int \left(x^2 + \dfrac{1}{x}\right) dx$

(c) $\int \left(x^6 - \dfrac{4}{x^2} + 2\right) dx$

(d) $\int (2e^x - 3)\, dx$

(e) $\int \left(\sqrt{x} - \dfrac{1}{\sqrt{x}}\right) dx$

(f) $\int 5\sqrt[3]{x^2}\, dx$

13. Evaluate

(a) $\displaystyle\int_1^2 \left(x - \dfrac{1}{x}\right) dx$

(b) $\displaystyle\int_0^1 e^x\, dx$

(c) $\displaystyle\int_1^4 (4\sqrt{x} - 1)\, dx$

14. Evaluate if possible

(a) $\displaystyle\int_4^\infty \dfrac{1}{x^2}\, dx$

(b) $\displaystyle\int_4^\infty \dfrac{dx}{\sqrt{x}}$

GOAL 42

15. Integrate

(a) $\int (x^2 + 1)^6 (2x\, dx)$

(b) $\int e^{-x^2} x\, dx$

(c) $\int \dfrac{3x^2 + 2x}{2x^3 + 2x^2 + 5}\, dx$

(d) $\int \dfrac{x^2}{\sqrt{x^3 - 1}}\, dx$

(e) $\int \sqrt{3x - 1}\, dx$

16. Evaluate

(a) $\displaystyle\int_0^1 (x^3 + 4)^3 x^2\, dx$

(b) $\displaystyle\int_2^{10} \sqrt{2x + 5}\, dx$

(c) $\displaystyle\int_0^2 \dfrac{x}{x^2 + 1}\, dx$

GOAL 43

17. Integrate

(a) $\int xe^{-x}\, dx$

(b) $\int x^4 \ln x\, dx$

18. Evaluate

$$\int_1^5 x\sqrt{x - 1}\, dx$$

HINT: Use the Parts Formula.

Chapter Test A

1. Find an approximation to the area under the graph of $f(x) = 2x^2 + 1$ from $x = 0$ to $x = 5$ using five equal subintervals and left-hand endpoints.

2. Use the definition of area as the limit of an infinite sum of rectangles to find the area under $y = 4x^2$, above the x-axis, and between $x = 0$ and $x = 1$. (Choose right-hand endpoints.)

3. Express the area in Exercise two as a definite integral and evaluate it.

4. If the rate at which a population is growing at time t is equal to $g(t)$, tell what the following represents:

$$\int_{t_1}^{t_2} g(t)\, dt$$

5. Express as a definite integral: the change in profit when the production level is increased from 100 to 120 units, if the marginal profit is $(5 + 0.04x - 0.003x^2)$ dollars per unit when x units are produced.

6. Evaluate the definite integral in Exercise 5.

7. Integrate

   (a) $\int \left(2x^3 - x + \dfrac{5}{x}\right) dx$

   (b) $\int \left(x - \dfrac{1}{x}\right)^2 dx$

   (c) $\int x\sqrt{1 - x^2}\, dx$

   (d) $\int x \ln x\, dx$

   (e) $\int \dfrac{dx}{4x + 1}$

   (f) $\int \dfrac{dx}{\sqrt{4x + 1}}$

8. Evaluate

   (a) $\int_0^1 (2x + 1)^4\, dx$

   (b) $\int_0^1 xe^x\, dx$

   (c) $\int_2^3 \dfrac{x}{x^2 - 3}\, dx$

Chapter 8

Applications of Integration; Differential Equations

In the last chapter we used the definite integral to find area under a curve. In this chapter we use it to find area between curves, and also to solve numerous problems of many different kinds in physics, biology, social science, and economics.

So far most of our applications have involved finding a *particular value* of a given function, such as its maximum or minimum, its rate of change at a designated time, the value of its definite integral as noted above over constant limits, and so on. A fundamentally different problem is one calling for a *function* that satisfies specific conditions; for example, finding a distance function given its acceleration, or inferring a profit function given the marginal profit. In such problems the unknown we seek is not a number but a function! Differential equations constitute the most important class of such problems. They are used extensively in the study of natural phenomena, from astronomy and mechanics to growth and decay. In the second part of this chapter we show how to solve some simple differential equations, then consider applications.

Goal 44. To compute areas between curves
Goal 45. To analyze motion along a line and related problems
Goal 46. To understand and apply the average value of a function
Goal 47. To use integration in problems in economics
Goal 48. To use integration on probability problems
Goal 49. To understand and solve simple differential equations
Goal 50. To apply separation of variables
Summary
Review Exercises
Chapter Test A

GOAL 44

To compute areas between curves

In Chapter 7 we calculated the area between a curve and the x-axis, first as a limit of sums of approximating rectangles, then as a definite integral. Specifically, we noted that if f is continuous and nonnegative on $[a, b]$ then the area between the graph of f, the x-axis, and the lines $x = a$ and $x = b$ is equal to $\int_a^b f(x)\, dx$. This definite integral is actually the area of the shaded

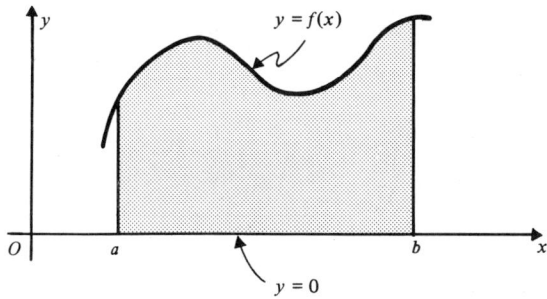

region *between the two curves* $y = f(x)$ and $y = 0$. In this Goal we generalize this concept of area between curves; we do not require that one of them be the x-axis.

Suppose f and g are continuous on $[a, b]$ and that $f(x) \geq g(x)$ for all x in $[a, b]$. To compute the area between the curves from $x = a$ to $x = b$, we look at a typical approximating rectangle whose height is $[f(x_i) - g(x_i)]$* and whose base is Δx (see figure on p. 426). Its area is $[f(x_i) - g(x_i)]\, \Delta x$. The area between the curves is the limit, as $n \to \infty$, of the sum

$$[f(x_1) - g(x_1)]\, \Delta x + [f(x_2) - g(x_2)]\, \Delta x + \ldots + [f(x_n) - g(x_n)]\, \Delta x$$

---

*Convince yourself that this is the height of any rectangle, whether (1) f and g are both positive, (2) $f > 0$ and $g < 0$, or (3) f and g are both negative. Draw a picture for each case.

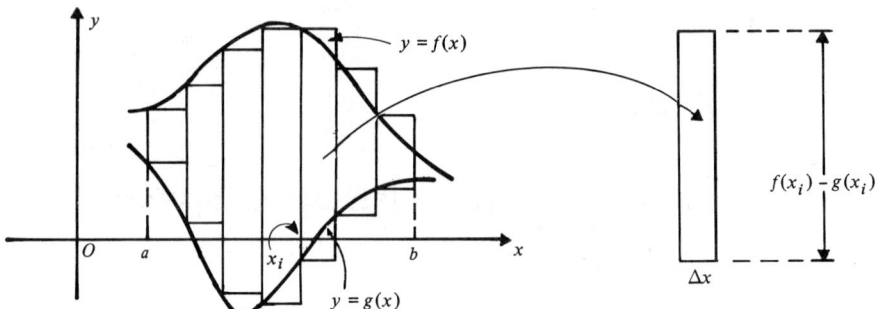

where x_i is a point in the ith subinterval, whose width is Δx. (In the figure, we chose these points to be the left endpoints.) By the definition of the definite integral (p. 382), this limit is equal to

$$\int_a^b [f(x) - g(x)]\, dx$$

Note that the area under $y = f(x)$ and above the x-axis from $x = a$ to $x = b$ is a special case of this, with $g(x) = 0$.

Example 44-1

Find the area between the curves $y = x + 5$ and $y = x^2 - 1$ over the interval $[-1, 2]$.

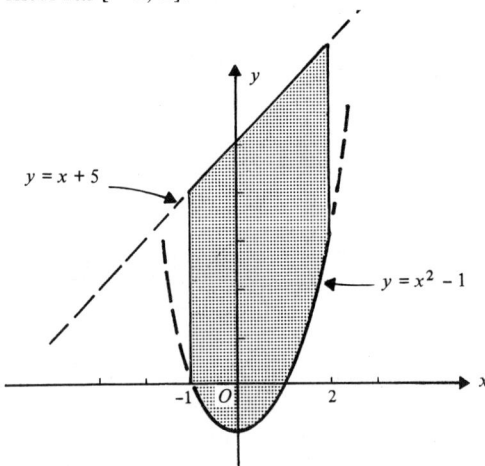

We draw a rough sketch of the region to aid in determining which is the upper curve. The area of the shaded region is equal to

$$\int_{-1}^{2} [(x + 5) - (x^2 - 1)]\, dx = \int_{-1}^{2} (-x^2 + x + 6)\, dx$$

Goal 44 | To Compute Areas Between Curves

$$= \left[-\frac{x^3}{3} + \frac{x^2}{2} + 6x \right]_{-1}^{2} = \frac{34}{3} - \left(-\frac{31}{6} \right) = \frac{33}{2}$$

Example 44-2

Find the area between the curves $y = x^2$ and $y = -x + 6$.

In this case we first determine the limits of integration by finding where the curves intersect. We equate the two expressions for y:

$x^2 = -x + 6$

$x^2 + x - 6 = 0$

Factoring yields

$(x + 3)(x - 2) = 0$

with solutions

$x = -3$ or 2

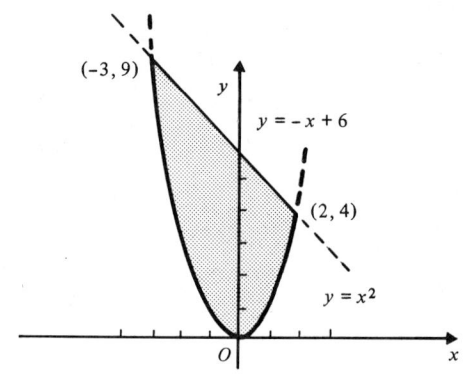

The specified area is thus

$$\int_{-3}^{2} [(-x + 6) - x^2] \, dx = \int_{-3}^{2} (-x^2 - x + 6) \, dx$$

$$= \left[-\frac{x^3}{3} - \frac{x^2}{2} + 6x \right]_{-3}^{2}$$

$$= \frac{22}{3} - \left(-\frac{27}{2} \right) = \frac{125}{6}$$

Example 44-3

Find the area between the curve $y = x^3 - 3x^2 - x$ and the x-axis over the interval $[0, 3]$.

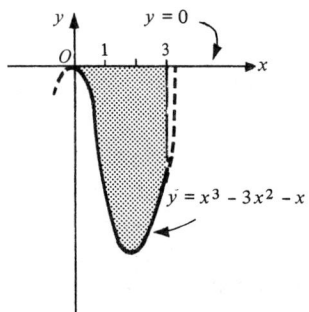

In this example the upper curve is $y = 0$, the x-axis. The area of the shaded region equals

$$\int_0^3 [0 - (x^3 - 3x^2 - x)]\, dx = \int_0^3 (-x^3 + 3x^2 + x)\, dx$$

$$= \left[-\frac{x^4}{4} + x^3 + \frac{x^2}{2} \right]_0^3 = 11\frac{1}{4}$$

Example 44-4

Find the area between the curves $y = x^2 - 2x + 2$ and $y = -x^2 + 6x - 4$. We first find the limits of integration by determining where the curves intersect:

$$x^2 - 2x + 2 = -x^2 + 6x - 4$$
$$2x^2 - 8x + 6 = 0$$
$$2(x - 1)(x - 3) = 0$$
$$x = 1 \text{ or } 3$$

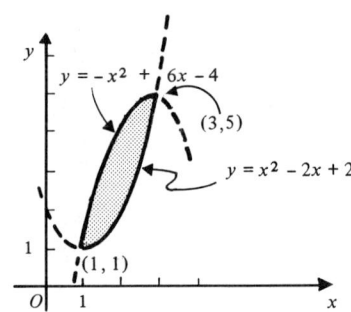

The shaded region thus has area

$$\int_1^3 [(-x^2 + 6x - 4) - (x^2 - 2x + 2)]\, dx = \int_1^3 (-2x^2 + 8x - 6)\, dx$$

$$= \left[-\frac{2x^3}{3} + 4x^2 - 6x \right]_1^3 = 2\frac{2}{3}$$

REMARK: Drawing a rough sketch is often a valuable aid not only in deciding which is the upper curve but also in finding the limits of integration.

EXERCISES

1. For each of the following, draw a rough sketch of the region between the given curves over the indicated interval; express the area of the region as a definite integral; then evaluate the integral.
    (a) $y = x^2 - 4x$ and $y = 2x$ over $[1, 4]$
    (b) $y = x^2 + 2x$ and $y = x - 1$ over $[0, 2]$
    (c) $y = 4 - x^2$ and $y = x$ over $[-1, 0]$
    (d) $y = x^2 - 6x + 9$ and $y = -x^2 + 6x - 2$ over $[2, 4]$

(e) $y = -x^2 + 4x + 5$ and the x-axis over $[1, 4]$
(f) $y = -x^3 + x^2 + 6x$ and $y = 2$ over $[-1, 0]$
(g) $y = -x^3 + x^2 + 6x$ and $y = 2$ over $[1, 2]$
(h) $y = \sqrt{x} + 2$ and $y = x$ over $[0, 4]$
(i) $y = \dfrac{1}{x}$ and $y = x^2$ over $[1, 2]$
(j) $y = e^x$ and $y = x$ over $[0, 1]$

2. Draw a rough sketch, then find the area between the given curves.
   (a) $y = x^2 - 3x + 4$ and $y = x + 1$
   (b) $y = 5 - x^2$ and $y = x^2 - 3$
   (c) $y = -x^2 + 5x - 6$ and the x-axis
   (d) $y = x^2$ and $y = \sqrt{x}$
   (e) $y = \dfrac{1}{x}$ and $y = \dfrac{5}{2} - x$

3. Suppose $g(x) < 0$ over the interval $[a, b]$. Tell why the area bounded by the graph of g, the x-axis, and the lines $x = a$ and $x = b$ is equal to $-\int_a^b g(x)\, dx$.

TEST A

1. Find the area between the curves $y = x^2 - 3x$ and $y = x^3$ over the interval $[0, 2]$.
2. Find the area between the curves $y = 1 - x^2$ and $y = -2$ from $x = -1$ to $x = 0$.
3. Find the area between the curves $y = x^2 - 3x$ and $y = x$.

TEST B

1. Find the area between the curves $y = 3 - x^2$ and $y = x^2 - 3$ over the interval $[-1, 1]$.
2. Find the area between the curve $y = x^3 - 4x$ and the x-axis from $x = 0$ to $x = 1$.
3. Find the area between the curves $y = 4x - x^2$ and $y = -x$.

STUDY QUESTIONS

1. Consider the problem of finding the area between the curve $x = 1 - y^2$ and the y-axis, as shown in the figure at the top of p. 430.

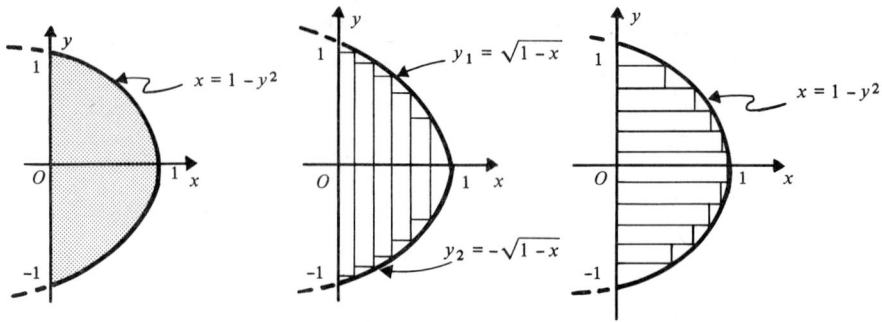

While it is possible to find curves y_1 and y_2 such that the area lies between the two curves, it is simpler to integrate with respect to the variable y:

$$\int_{-1}^{1} (1 - y^2)\, dy = \left[y - \frac{y^3}{3} \right]_{-1}^{1} = \frac{4}{3}$$

(a) Justify the definite integral above as the limit of an infinite sum.

(b) Using this technique, find the area between the curves $x = y^2 - 6y + 5$ and $x = y - 1$.

2. The area of any region must be a nonnegative number. How does our treatment of area between curves assure this?

3. (a) Write down in terms of definite integrals the total area of the shaded region (shown below) between $y = f(x)$ and the x-axis. (b) Why is $\int_a^d f(x)\, dx$ an incorrect answer in (a)? (c) Sketch the graph of $y = x^3 - 3x^2 + 2x$, then find the total area between the curve and the x-axis.

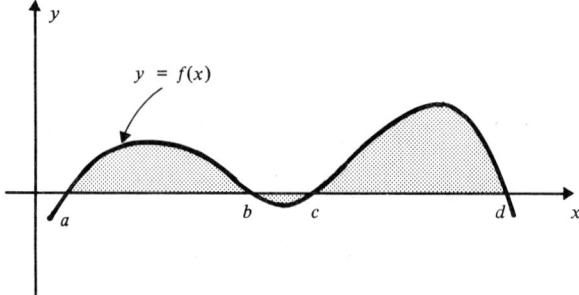

4. (a) Sketch the region bounded by the curves $y = x + 10$ and $y = x^2 + 4$ over the interval $[-1, 2]$ and find its area. (b) Compare your sketch and answer with those of Example 44-1. (c) What does this problem illustrate?

GOAL 45

To analyze motion along a line and related problems

In the first part of this Goal we use integration to obtain information about the motion of an object, given its velocity or its acceleration along with other useful data. The technique developed will enable us to find a function in numerous other applications in which we are given the derivative of a function along with some additional information. These related problems are discussed in the second part of this Goal.

MOTION ALONG A LINE

As noted in Goals 16 and 17, if $f(t)$, $v(t)$, and $a(t)$ are the position, velocity, and acceleration functions, respectively, of a particle P (say, on a coordinate axis) at time t, then

$$v(t) = f'(t) \quad \text{and} \quad a(t) = v'(t)$$

In Goal 29 we started with the position function $f(t)$, derived the velocity and acceleration functions through differentiation, and then analyzed in depth the motion of the particle. In this Goal we reverse that process.

GIVEN THE VELOCITY FUNCTION

If we know $v(t)$ we can obtain the acceleration function $a(t)$ immediately by differentiation. But how do we obtain the position function $f(t)$? Since $v(t)$ is the derivative of $f(t)$, we are looking for an antiderivative of $v(t)$. However, as we already know, there is an infinite set of antiderivatives of $v(t)$, each differing from the others by a constant. To find a particular one from this set, or, equivalently, to determine the specific constant of integration, we must have additional information.

Example 45-1

A particle moves along the x-axis so that its velocity at time t is given by $v(t) = 3t^2 - 4t$. If initially it is at $x = 5$, find (a) $a(t)$, and (b) $f(t)$.

(a) $a(t) = v'(t) = 6t - 4$.

(b) $f(t) = \int v(t)\, dt = \int (3t^2 - 4t)\, dt = t^3 - 2t^2 + C$. To find C we use the additional hypothesis that $f(0) = 5$. Thus

$$5 = f(0) = 0^3 - 2(0)^2 + C \qquad C = 5$$

So

$$f(t) = t^3 - 2t^2 + 5$$

Example 45-2

A bullet is fired at a target with such force that its velocity after t sec is $(100 - 20t)$ ft/sec. If the bullet hits the target in 1 sec, how far is the target from the gun?

If $f(t)$ is the distance the bullet travels in t sec, then $f(t) = \int v(t)\, dt = \int (100 - 20t)\, dt = 100t - 10t^2 + C$. The initial position of the bullet is $x = 0$, so

$$f(0) = 0 = 100(0) - 10(0)^2 + C \qquad C = 0$$

Therefore

$$f(t) = 100t - 10t^2$$

The distance between gun and target is equal to $f(1)$ or $100(1) - 10(1)^2$, or 90 ft.

GIVEN THE ACCELERATION FUNCTION

If we are given the acceleration $a(t)$, then, since $a(t) = v'(t)$, we know that $v(t)$ is a function whose derivative is $a(t)$. When we find the indefinite integral $\int a(t)\, dt$, we obtain an infinite set of functions differing from one another by constants. With additional information, we are often able to solve for the particular constant of integration so as to determine a unique $v(t)$. We can then find $f(t)$ by determining the indefinite integral $\int v(t)\, dt$. The process yields a second constant of integration that must also be determined.

Example 45-3

Suppose $a(t) = 6t - 2$, $v(1) = 4$, and $f(2) = 3$. Find $v(t)$ and $f(t)$.

First we have
$$v(t) = \int a(t)\, dt = \int (6t - 2)\, dt = 3t^2 - 2t + C_1$$
Since $v(1) = 4$, this gives

$$4 = v(1) = 3(1)^2 - 2(1) + C_1 \qquad C_1 = 3$$

Goal 45 | To Analyze Motion Along a Line and Related Problems

Now we integrate again to find $f(t)$:

$$f(t) = \int v(t)\, dt = \int (3t^2 - 2t + 3)\, dt = t^3 - t^2 + 3t + C_2$$

Since $f(2) = 3$, we have

$$3 = f(2) = 2^3 - 2^2 + 3(2) + C_2 = 10 + C_2 \qquad C_2 = -7$$

Thus $v(t) = 3t^2 - 2t + 3$ and $f(t) = t^3 - t^2 + 3t - 7$.

Example 45-4

Suppose $a(t) = 20t^3 - 12t$, $f(1) = -1$, and $f(-1) = -5$. Find $v(t)$ and $f(t)$.
First we integrate twice:

$$v(t) = \int a(t)\, dt = \int (20t^3 - 12t)\, dt = 5t^4 - 6t^2 + C_1$$
$$f(t) = \int v(t)\, dt = \int (5t^4 - 6t^2 + C_1)\, dt = t^5 - 2t^3 + C_1 t + C_2$$

We now use the additional information to determine C_1 and C_2:

$$-1 = f(1) = 1^5 - 2(1)^3 + C_1(1) + C_2 = -1 + C_1 + C_2$$
$$-5 = f(-1) = (-1)^5 - 2(-1)^3 + C_1(-1) + C_2 = 1 - C_1 + C_2$$

These equations reduce to

$$C_1 + C_2 = 0$$
$$-C_1 + C_2 = -6$$

and the simultaneous solution is $C_1 = 3$ and $C_2 = -3$. Thus

$$v(t) = 5t^4 - 6t^2 + 3 \qquad \text{and} \qquad f(t) = t^5 - 2t^3 + 3t - 3$$

The reader should check these results.

Example 45-5

A ball is thrown directly upward from the top of a flagpole with an initial velocity of 50 ft/sec. If it hits the ground 5 sec later, how tall is the flagpole?
HINT: The acceleration $a(t)$ due to gravity is constant and equals -32 ft/sec².

To start, we have

$$a(t) = -32 \qquad \text{and} \qquad v = \int a(t)\, dt = -32t + C_1$$

Since the ball's initial velocity is 50 ft/sec,

$$50 = v(0) = -32(0) + C_1 \qquad \text{and} \qquad C_1 = 50$$

Since $v(t) = -32t + 50$,

$$f(t) = \int(-32t + 50)\, dt = -16t^2 + 50t + C_2$$

The ball hits the ground after 5 sec, and $f(t)$ is the height of the ball after t sec, so

$$0 = f(5) = -16(5)^2 + 50(5) + C_2 \quad \text{and} \quad C_2 = 150$$

Therefore the height function is $f(t) = -16t^2 + 50t + 150$. To find the height of the flagpole, since the ball is at the top of the flagpole at time $t = 0$, we evaluate $f(0)$:

$$f(0) = -16(0)^2 + 50(0) + 150 = 150 \text{ ft}$$

OTHER APPLICATIONS

We now consider several fields in which problems can be solved by the methods illustrated above.

BUSINESS PROBLEMS

Example 45-6

Suppose a company's marginal cost for x units is given by $M(x) = (6 + 0.002x)$ dollars per unit. Find its cost function $C(x)$, if its overhead is $5000.

Since $M(x) = C'(x)$,

$$C(x) = \int M(x)\, dx = \int(6 + 0.002x)\, dx = 6x + 0.001x^2 + k$$

The overhead cost is $C(0)$. Therefore

$$5000 = 6(0) + 0.001(0)^2 + k \text{ and } k = 5000$$

So

$$C(x) = 6x + 0.001x^2 + 5000$$

Example 45-7

How much does it cost the company in the preceding example to manufacture the first 200 items?

$$C(200) = 6(200) + 0.001(200)^2 + 5000 = \$6240$$

Goal 45 | To Analyze Motion Along a Line and Related Problems

Example 45-8

If $50e^{0.05t}$ is the rate of growth, in dollars per year, after t yr, of an initial investment of $1000, find the worth of the investment after 6 yr.

If $W(t)$ is the value of the investment after t yr, then $W'(t) = 50e^{0.05t}$ and

$$W(t) = \int 50e^{0.05t}\,dt = 50\left(\frac{1}{0.05}\right)\int e^{0.05t}\,(0.05\,dt) = 1000e^{0.05t} + C$$

Since $W(0) = 1000$, we have

$$1000 = 1000e^{0.05(0)} + C = 1000 + C \text{ and } C = 0$$

Therefore $W(t) = 1000e^{0.05t}$, and

$$W(6) = 1000e^{0.05(6)} = 1000e^{0.3} \approx \$1350$$

GROWTH OR DECAY

Example 45-9

If after t yr an animal grows at the rate of $[10 - \tfrac{1}{2}t]$ lb per yr and if the animal weighs 10 lb at birth, find its weight $W(t)$ as a function of age in years.

Since $W'(t) = 10 - \tfrac{1}{2}t$,

$$W(t) = \int\left(10 - \frac{1}{2}t\right)dt = 10t - \frac{1}{4}t^2 + C$$

Now we find C:

$$10 = W(0) = 10(0) - \frac{1}{4}(0)^2 + C \text{ and } C = 10$$

Therefore

$$W(t) = 10t - \frac{1}{4}t^2 + 10$$

Example 45-10

After t yr, a substance is decaying at the rate of $0.01e^{-0.001t}$ oz per yr. If there were 10 oz of the substance originally, find the amount $Q(t)$ of the substance left after t yr.

Since the amount is diminishing, $Q'(t)$ is negative. Thus $Q'(t) = -0.01e^{-0.001t}$ and

$$Q(t) = \int -0.01e^{-0.001t}\,dt = 0.01\left(\frac{1}{0.001}\right)\int e^{-0.001t}\,(-0.001\,dt)$$
$$= 10e^{-0.001t} + C$$

It is given that $Q(0) = 10$. Therefore

$$10 = 10e^{-0.001(0)} + C \quad \text{and} \quad C = 0$$

So $Q(t) = 10e^{-0.001t}$.

Example 45-11

What is the half-life of the substance in the preceding example?
We seek t for which $Q(t) = 5$ (see p. 327).

$$5 = 10e^{-0.001t} \qquad \frac{1}{2} = e^{-0.001t} \qquad -\ln 2 = -0.001t$$

$$t = \frac{0.693}{0.001} \approx 693 \text{ yr}$$

SPREAD OF A DISEASE

Example 45-12

After t days a viral infection spreads through a population at the rate of $(6t + 15)$ people per day. If 235 people have become infected after 5 days, how many will have contracted the disease after 10 days?

If $N(t)$ is the number who have become infected after t days, then $N'(t) = 6t + 15$ and

$$N(t) = \int (6t + 15)\, dt = 3t^2 + 15t + C$$

Since $N(5) = 235$,

$$235 = 3(5)^2 + 15(5) + C \text{ and } C = 85$$

So $N(t) = 3t^2 + 15t + 85$ and

$$N(10) = 3(10)^2 + 15(10) + 85 = 535$$

EXERCISES

1. For each of the following, determine the acceleration function $a(t)$ and the position function $f(t)$.
    (a) $v(t) = 4t^3 - 2t$
    $f(2) = 13$
    (b) $v(t) = -\dfrac{1}{(t+1)^2}$
    $f(0) = 1$
    (c) $v(t) = 6e^{2t} + 5$
    $f(0) = 10$

Goal 45 | **To Analyze Motion Along a Line and Related Problems**

2. For each of the following, determine the velocity and position functions.

   (a) $a(t) = 6t$
   $v(1) = 5$
   $f(1) = 0$

   (b) $a(t) = 12t^2 + 6t$
   $f(0) = 2$
   $f(1) = -1$

   (c) $a(t) = 2$
   $f(1) = 8$
   $f(2) = 14$

   (d) $a(t) = \dfrac{4}{(t+1)^3}$
   $v(0) = -2$
   $f(0) = 3$

   (e) $a(t) = 9e^{3t}$
   $f(0) = 2$
   $f(1) = e^3 + 3$

   (f) $a(t) = e^{-t}$
   $f(1) = e^{-1} - 1$
   $f(2) = e^{-2}$

   (g) $a(t) = -\dfrac{1}{(t+1)^2}$
   $v(0) = 1$
   $f(0) = 0$

3. A ball is thrown directly upward from the top of a flagpole. In each of the following parts we give both the initial velocity and the time it takes the ball to hit the ground; determine the height of the flagpole in each part. [HINT: $a(t) = -32$.]

   (a) 25 ft/sec, 3 sec

   (b) 100 ft/sec, 10 sec

   (c) -10 ft/sec, 2 sec (the ball is thrown directly *down*)

4. To figure out how deep a well is, a woman drops a stone into it and then listens for the sound of the splash. If she hears the splash $2\frac{1}{2}$ sec after the stone is dropped, how deep is the well? (Assume that the initial velocity is zero, that the acceleration due to gravity is 32 ft/sec², and that the time it takes the sound of the splash to reach her ears can be neglected.)

5. If a car can accelerate from 0 to 60 mph in 10 sec, how far will it travel during these 10 sec? (Assume that the acceleration is constant and note that 60 mph = 88 ft/sec.)

6. In order to hurl an object into the air so that it will reach a height of 100 ft, what initial velocity is necessary? (HINT: At the 100-ft height the velocity of the object will be zero.)

7. If a manufacturer determines that his marginal revenue is $M(x) = 10 - 0.001x$ dollars per unit when the production level is x units, find the manufacturer's revenue when 500 units are produced [assume that $M(0) = 0$].

8. If $90e^{0.06t}$ is the rate of growth in dollars per year, after t yr, of an initial investment of $1500, determine the worth of the investment after 5 yr.

9. A rumor is spreading through a population at the rate of $26t + 100$ people per day. If 2500 people have heard the rumor after 10 days, find the number of people who will have heard the rumor after 20 days.
10. A colony of fruit flies is increasing, after t days, at the rate of $10{,}000e^{0.2t}$ flies per day. If the initial population was 10,000, how long does it take for the colony to double?

TEST A

In questions 1 and 2, $f(t)$, $v(t)$, and $a(t)$ are the position, velocity, and acceleration functions, respectively, of a particle moving along a line.

1. If $v(t) = 2t + 3$ and $f(0) = 4$, find $a(t)$ and $f(t)$.
2. If $a(t) = 12t - 8$, $f(1) = -2$, and $f(2) = 0$, determine $v(t)$ and $f(t)$.
3. If a substance is decaying at the rate of $0.375e^{-0.025t}$ g per yr, determine the amount left after 100 yr if the initial quantity is 15 g.

TEST B

In questions 1 and 2, $f(t)$, $v(t)$, and $a(t)$ are the position, velocity, and acceleration functions, respectively, of a particle moving along a line.

1. If $v(t) = 6t^2 - 4$ and $f(0) = 1$, find $a(t)$ and $f(t)$.
2. If $a(t) = 25e^{5t}$, $v(0) = 7$, and $f(0) = 4$, determine $v(t)$ and $f(t)$.
3. An animal exhibits a growth rate after t yr of $5 - 0.4t$ lb per yr. If its weight at birth is 8 lb, determine its weight after 2 yr.

GOAL 46

To understand and apply the average value of a function

To find maxima or minima of a function we use the derivative of the function. To find the average (or mean) value of a function over some interval, we use the definite integral of the function over the interval. In this Goal we define the average value of a function and interpret it geometrically; we also consider several of its applications.

BACKGROUND: AVERAGE VELOCITY

Before we define the average value of a function $f(x)$ over an interval $[a, b]$, let us consider the average velocity (or speed), say of a car, from time $t = a$ to $t = b$, if its velocity at time t is given by the continuous function $v(t)$. We can approach this problem in two ways:

WAY I

We can approximate the average speed by adding the car's speeds at, say, 10 different times, equally spaced, between $t = a$ and $t = b$, then dividing the sum by 10. Thus

$$\text{Av speed} \approx \frac{v(t_1) + v(t_2) + \ldots + v(t_{10})}{10}$$

It is not surprising that this approximation improves as the number of points of time used increases. In fact, it is to be expected that the average speed is equal exactly to

$$\lim_{n \to \infty} \frac{v(t_1) + v(t_2) + \ldots + v(t_n)}{n}$$

But how do we get a definite integral from this limit?

If we let $\Delta t = (b - a)/n$, then we can replace $1/n$ by $\Delta t/(b - a)$, and rewrite the infinite sum this way:

$$\text{Av speed} = \lim_{n \to \infty} \frac{1}{b - a} [v(t_1) \Delta t + v(t_2) \Delta t + \ldots + v(t_n) \Delta t]$$

If we now transpose the limit symbol and the constant $1/(b - a)$, we see that the right-hand side is equal precisely to the constant $1/(b - a)$ times the definite integral from a to b of $v(t)$. Thus the average speed on the interval $a \leq t \leq b$ is

$$\frac{1}{b - a} \int_a^b v(t) \, dt$$

Way II

If $v(t)$ is the car's speed on the interval $[a, b]$, then its average speed equals the distance traveled between $t = a$ and $t = b$ divided by the time of travel, $b - a$. If $f(t)$ is the position of the car at time t, then

$$\text{Av speed} = \frac{f(b) - f(a)}{b - a}$$

But (see p. 389)

$$f(b) - f(a) = \int_a^b v(t) \, dt$$

So

$$\text{Av speed} = \frac{1}{b - a} \int_a^b v(t) \, dt$$

as before.

DEFINITION OF AVERAGE (OR MEAN) VALUE

If a function $f(x)$ is continuous on $a \leq x \leq b$, then the *average value* f_{av} *of the function over* $[a, b]$ is

$$f_{av} = \frac{1}{b - a} \int_a^b f(x) \, dx$$

From this definition we see that the average speed considered above is just a special case.

GEOMETRIC INTERPRETATION
Suppose f is a continuous nonnegative function over the interval $[a, b]$.

Goal 46 | To Understand and Apply the Average Value of a Function

From the defining equation just above, by multiplying both sides by $(b - a)$, we get

$$f_{av} \cdot (b - a) = \int_a^b f(x)\, dx$$

The left side is the area of a rectangle of height f_{av} and base $b - a$. The right side is the area under the graph of f from $x = a$ to $x = b$. f_{av} is the height of rectangle $RSTQ$, of base $b - a$, whose area equals the shaded area under the curve.

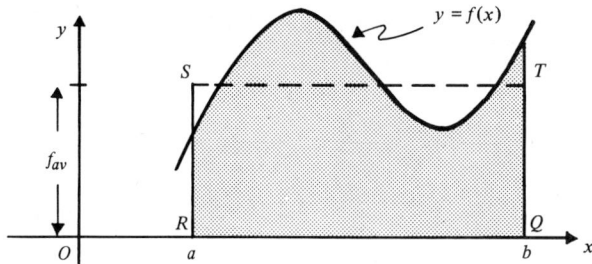

In Study Question 1, below, we state and use the Mean Value Theorem for Integrals, which assures the existence, if f is continuous, of a number c between a and b such that $f(c)$ is the average value of f on $[a, b]$.

EXAMPLES

Example 46-1

Sketch the graph of $f(x) = x^2 - 2x + 1$ and find the maximum, minimum, and average values of the function over $[0, 3]$.

Since $f'(x) = 2x - 2$, which equals zero only at $x = 1$, we evaluate f at 1 and at the endpoints 0 and 3 to find the max and min. $f(0) = 1$, $f(1) = 0$, $f(3) = 4$; so the max value of f on $[0, 3]$ is 4 and the min is 0.

$$f_{av} = \frac{1}{3 - 0} \int_0^3 (x^2 - 2x + 1)\, dx$$

$$= \frac{1}{3}\left[\frac{x^3}{3} - x^2 + x\right]_0^3 = 1$$

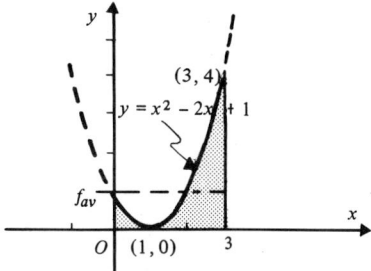

Example 46-2

Find the average value of $f(x) = x^3$ over $[-2, 1]$.

$$f_{av} = \frac{1}{1-(-2)} \int_{-2}^{1} x^3 \, dx = \frac{1}{3} \int_{-2}^{1} x^3 \, dx$$
$$= \frac{1}{3} \left[\frac{x^4}{4} \right]_{-2}^{1} = \frac{1}{3} \left(-\frac{15}{4} \right) = -\frac{5}{4}$$

Example 46-3

If the temperature recorded on a thermometer at the top of a bank building is calculated to be, in degrees Fahrenheit,

$$T(t) = \left(\frac{t^3}{24} - t^2 + 6t + 50 \right)$$

after t h, what is the average temperature during a 24-h period starting with $t = 0$?

$$T_{av} = \frac{1}{24 - 0} \int_0^{24} \left(\frac{t^3}{24} - t^2 + 6t + 50 \right) dt = 74°$$

Example 46-4

During a 10-yr period it is noted that the population of a certain animal colony is $P(t) = 1000 + 3000t - 300t^2$. What are the maximum, minimum, and average values of the population during this period?

$$P_{av} = \frac{1}{10 - 0} \int_0^{10} (1000 + 3000t - 300t^2) \, dt = 6000$$

Since $P'(t) = 3000 - 600t$, which is zero when $t = 5$, and $P(0) = 1000$, $P(5) = 8500$, $P(10) = 1000$, the minimum value is 1000 and the maximum value is 8500.

Example 46-5

Suppose that the net value of an investment is $5000e^{0.05t}$ dollars after t yr. What is the average net value of the investment during the fourth year, that is, between $t = 3$ and $t = 4$?

$$\text{Av net value} = \frac{1}{4 - 3} \int_3^4 5000 e^{0.05t} \, dt = \left[100{,}000 e^{0.05t} \right]_3^4 \approx \$5960$$

Example 46-6

If velocity as a function of time is given by $v(t) = 1/(1 + t)^2$, $0 \leq t \leq 5$, what is the average velocity during the time period $1 \leq t \leq 3$?

$$v_{av} = \frac{1}{3 - 1} \int_1^3 \frac{1}{(1+t)^2} \, dt = \frac{1}{2} \left[-\frac{1}{1+t} \right]_1^3 = \frac{1}{8}$$

Goal 46 — To Understand and Apply the Average Value of a Function

Example 46-7

The depth throughout a cross-section of a lake situated between two oceans is given by

$$D(x) = \frac{30{,}000}{x + 100} \qquad 0 \leq x \leq 100$$

When locks are closed simultaneously at both ends, to what even depth will the water level off?

The even depth is given by D_{av}.

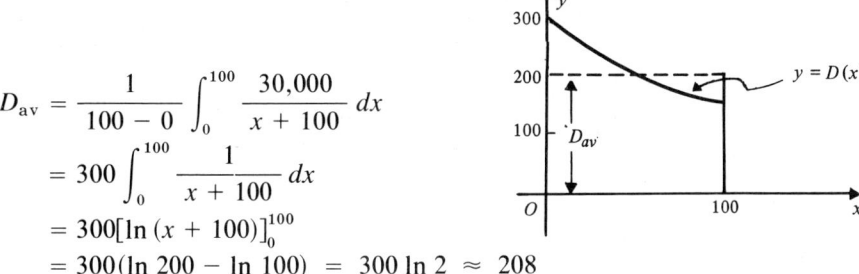

$$D_{av} = \frac{1}{100 - 0} \int_0^{100} \frac{30{,}000}{x + 100} \, dx$$

$$= 300 \int_0^{100} \frac{1}{x + 100} \, dx$$

$$= 300[\ln (x + 100)]_0^{100}$$

$$= 300(\ln 200 - \ln 100) = 300 \ln 2 \approx 208$$

Example 46-8

Suppose that a piece of machinery, because of wear and tear, is worth in dollars $7800 - 555x - 12.5x^2$ after x yr. What is its average worth during its second three years of use?

The average value V_{av} is given by

$$V_{av} = \frac{1}{6 - 3} \int_3^6 (7800 - 555x - 12.5x^2) \, dx$$

$$= \frac{1}{3} \left[7800x - \frac{555}{2} x^2 - \frac{12.5}{3} x^3 \right]_3^6 = \$5040$$

EXERCISES

1. Find the average value of each of the following functions over the indicated intervals.
   (a) $f(x) = x^2 + 5,\ 1 \leq x \leq 3$
   (b) $C(x) = 5000 + 0.01x + 0.0001x^2,\ 0 \leq x \leq 10{,}000$
   (c) $T(t) = \dfrac{t^3}{12} - 3t^2 + 2t + 70,\ 0 \leq t \leq 6$
   (d) $v(t) = 144 - 16t,\ 5 \leq t \leq 9$
   (e) $V(t) = 1500e^{0.06t},\ 0 \leq t \leq 5$

(f) $R(t) = 1000e^{-0.01t}$, $100 \leq t \leq 150$

(g) $f(x) = \dfrac{1}{x}$, $1 \leq x \leq 2$

(h) $g(x) = \sqrt{x}$, $1 \leq x \leq 4$

2. If the temperature after t h, in degrees Fahrenheit, is calculated to be $T(t) = 60 + 5t - \dfrac{t^2}{2}$, $0 \leq t \leq 10$, find the minimum, maximum, and average temperatures during the first 10-h period.

3. If $50e^{-0.015t}$ lb of a radioactive substance are left after t yr, find the average amount left during the second hundred years.

4. A stone is thrown upward from the ground with an initial velocity of 96 ft/sec. What is its average velocity during the third second? HINT: $a(t) = -32$.

5. If a piece of machinery is valued at $(6500 - 120x - 51x^2)$ dollars after x yr, find its average value during (a) the first year of use, (b) the third year, (c) the first 10 yr.

TEST A

1. Find the average value of each of the following functions over the indicated intervals.

    (a) $f(x) = x^2 + x + 1$, $-1 \leq x \leq 2$
    (b) $f(x) = \dfrac{1}{x^2}$, $1 \leq x \leq 4$

2. If the net value of an investment is $500e^{0.1t}$ dollars after t yr, what is the average net value during the first year?

TEST B

1. Find the average value of each of the following functions over the indicated intervals.

    (a) $f(x) = x^3 - x + 2$, $-2 \leq x \leq 1$
    (b) $f(x) = 3e^x$, $0 \leq x \leq 1$

2. During the worst 4 h of a hurricane the wind velocity in miles per hour is given by $V(t) = 5t - t^2 + 100$, $0 \leq t \leq 4$. What is the average wind velocity during this period?

STUDY QUESTIONS

1. The average value of a function is the subject of the Mean Value Theorem for Integrals, which is of use mainly in proving other theorems. It says: If f is continuous on $[a, b]$ then there exists a c between a and b such that

Goal 46 | To Understand and Apply the Average Value of a Function

$$f(c) = \frac{1}{b-a} \int_a^b f(x)\,dx$$

(a) Draw a sketch, where $f \geq 0$ on $[a, b]$, and interpret the theorem geometrically.

(b) Find the number c promised by the theorem for the function $f(x) = \frac{1}{x}$, on $[1, e]$.

2. Suppose f is continuous on $[a, b]$ and we define $F(x) = \int_a^x f(t)\,dt$. Justify each equals sign in the following proof that $F'(x) = f(x)$.

$$F'(x) = \lim_{h \to 0} \frac{F(x+h) - F(x)}{h} = \lim_{h \to 0} \frac{\int_a^{x+h} f(t)\,dt - \int_a^x f(t)\,dt}{h}$$

$$= \lim_{h \to 0} \frac{\int_x^{x+h} f(t)\,dt}{h} = \lim_{h \to 0} \frac{hf(c)}{h} = \lim_{h \to 0} f(c) = f(x)$$

where, in the last three steps, c is between x and $x + h$.

3. Show that if f is a linear function of the form $f(x) = mx + c$, then the average value of f over $[a, b]$ is equal to $f\left(\dfrac{a+b}{2}\right)$.

4. Suppose a car is driven at 10 mph for 100 miles and then at 20 mph for the next 100 miles. Find the average velocity, if velocity is expressed as a function of distance. Then find the average velocity if velocity is expressed as a function of time. (HINT: In each case graph the function and calculate the necessary integral by noting the area under the graph.)

GOAL **47**

To use integration in problems in economics

Some applications of the definite integral to problems in business and economics were considered in Goal 45 (see Examples 45-6 through 45-8 and Exercises 7 and 8). From these we know that given the marginal cost (or revenue or profit) and other data we can find the cost (or revenue or profit) by integrating. In Goal 46 we applied the concept of average value of a function to business problems calling for the average net value of an investment and for the average worth of machinery (Examples 46-5 through 46-8 and Exercise 5).

In this Goal we look at some other applications of the integral calculus to problems in economics. We shall discuss the problems of the R. U. Reading Publishing Company, which is planning to publish a new mathematics magazine. We will consider questions this publisher might face regarding market conditions, maintenance costs, the number of potential subscribers, and the present value of its venture. As in Goal 30, the questions will be simplified versions of those typically of concern to many businessmen and economists.

SUPPLY AND DEMAND

The demand function was defined on page 19 as the price per unit at which x units can be sold. Current market conditions are such that R. U. Reading figures that the price in dollars per magazine at which x magazines can be sold is

$$D(x) = 5 - 0.00005x - 0.00000003x^2$$

As expected, more customers are willing to buy the magazine if the price is lowered. However, manufacturing costs are such that the price in dollars per magazine at which the company is willing to sell x magazines is given by the supply function $S(x)$ on p. 447.

Goal 47 | To Use Integration in Problems in Economics

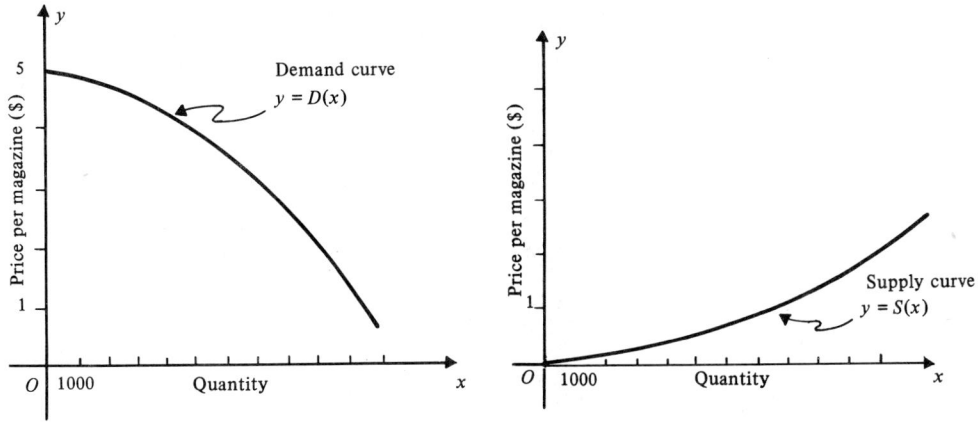

$$S(x) = 0.00005x + 0.00000001x^2$$

Again, as expected, since higher production involves more equipment, maintenance, and labor, it is worthwhile only if the price per magazine is raised.

EQUILIBRIUM POINT

This is the point at which the demand curve and the supply curve intersect. Below we find the equilibrium point for R. U. Reading, both algebraically and graphically.

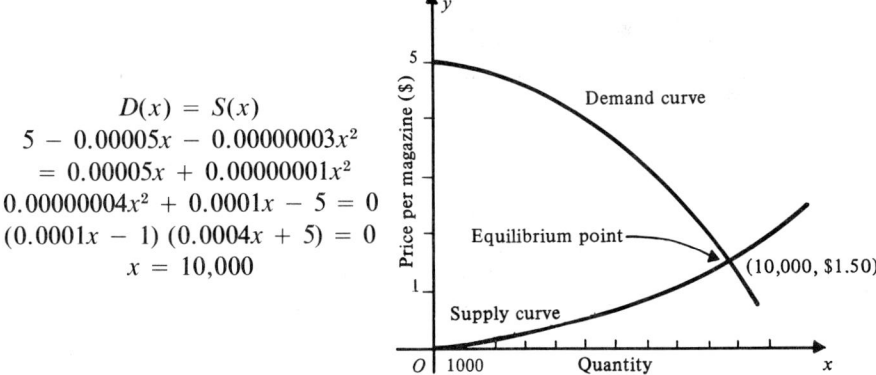

$$D(x) = S(x)$$
$$5 - 0.00005x - 0.00000003x^2$$
$$= 0.00005x + 0.00000001x^2$$
$$0.00000004x^2 + 0.0001x - 5 = 0$$
$$(0.0001x - 1)(0.0004x + 5) = 0$$
$$x = 10{,}000$$

As usual, we discard the negative solution.

According to the law of supply and demand, in a free competitive market, the actual selling price of a commodity will be the equilibrium price, in this case $1.50; the corresponding production level here is 10,000 magazines.

CONSUMERS' SURPLUS

Inspection of the demand curve for the R. U. Reading magazine shows that some consumers are willing to pay more than the equilibrium price

($1.50) for the magazine; indeed, even if the magazine were to sell for close to $5 some consumers would buy it. The amount saved by these consumers when they pay only the equilibrium price is known as the *consumers' surplus*.

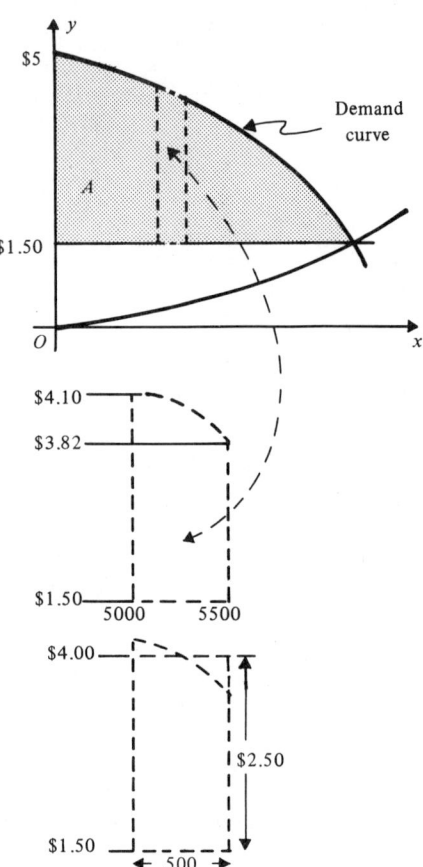

Before we define this surplus carefully, consider the shaded region A, and, in particular, one small part of it whose boundary we have dotted. An enlargement of this small region is shown below it. Note that, as the price decreases from $4.10 to $3.82, a total of 500 new people will buy the magazine. How much will these 500 consumers save, from what they would have been willing to pay, if the magazine now sells for $1.50? Since each person saves approximately $2.50 (the savings actually range from $2.32 to $2.60), this particular group of 500 customers saves approximately (500 × $2.50), or about $1250. At the bottom we view this product 500 × 2.50 as the area of a rectangle that approximates the area of our small region.

A familiar pattern is emerging. The sum of n such areas, as $n \to \infty$, will give the area of region A, and the latter is (with a dollar sign!) precisely the consumers' surplus. For our mathematics magazine this is given by the definite integral

$$\int_0^{10,000} [D(x) - 1.50] \, dx$$

$$= \int_0^{10,000} (5 - 0.00005x - 0.00000003x^2 - 1.50) \, dx = \$22,500$$

Mathematically the consumers' surplus is given by the area between two curves; less abstractly, as noted above, it represents the savings to consumers who would have paid more than the equilibrium price.

PRODUCER'S SURPLUS

Similarly, inspection of the supply curve for R. U. Reading shows that it would have been willing to furnish magazines at a price lower than $1.50 per magazine, the equilibrium price. The publisher's additional income, gained by selling at the equilibrium price rather than at a lower one, is known as the *producer's surplus*.

This time we consider the shaded region B, in particular the small portion of B whose enlargement appears below it. Note now that as the price increases from $1.04 to $1.08, the publisher is willing to print an additional 200 copies of the magazine. How much additional income will R. U. Reading receive on these 200 copies if the magazine sells for $1.50?

Since each copy will sell for approximately $0.45 more (the increase actually ranges from $0.42 to $0.46), this particular set of 200 copies will yield approximately (200 × $0.45), or an additional $90. At the right we show this product (200 × 0.45) as the area of a rectangle which approximates that of our small region. As expected, we define the producer's surplus to be the area of region B (also with a dollar sign!). It is equal to

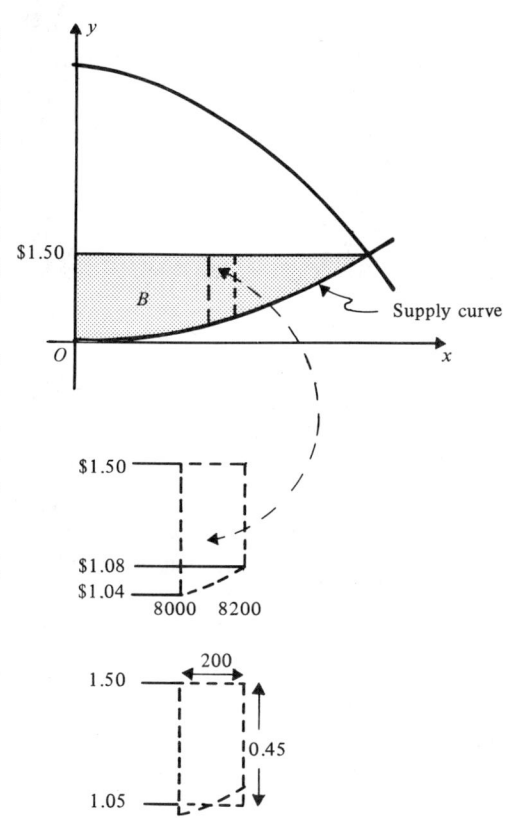

$$\int_0^{10,000} [1.5 - S(x)]\, dx$$
$$= \int_0^{10,000} (1.5 - 0.00005x - 0.00000001x^2)\, dx = \$9166.67$$

We think of $9166.67 as the additional income R. U. Reading receives over what the company would have gotten if the price (and production level) had been lower.

It is not hard to generalize from the above:

consumers' surplus

$$= \int_0^q [D(x) - p]\, dx$$

producer's surplus

$$= \int_0^q [p - S(x)]\, dx$$

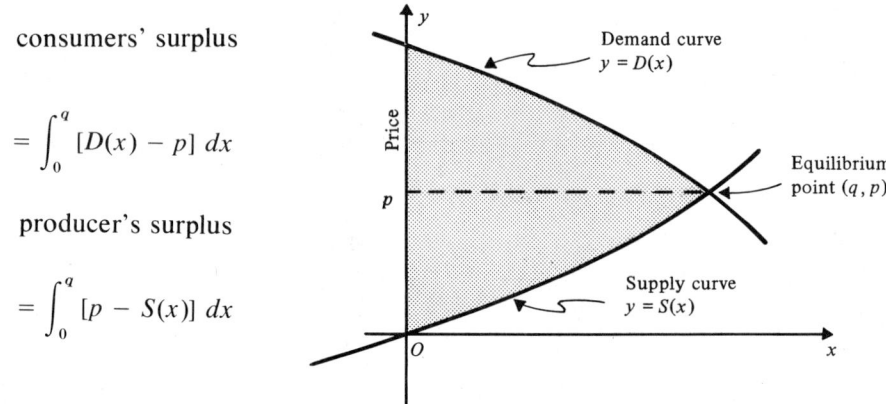

TOTAL MAINTENANCE CHARGES

Because of inflation and projected expansions, the maintenance cost for equipment and buildings associated with the printing of the new magazine will constantly increase. Management expects the variety of expenditures that make up the maintenance charges to change daily, and estimates that they will rise, after t months, to the rate of $(1200 + 50t)$ dollars per month. What will be the total maintenance bill for the first 5 yr?

Since the total maintenance cost over an interval of time is equal to the sum of the costs over smaller intervals of time, we should immediately suspect that a definite integral might be an appropriate tool here. With this in mind we therefore consider the region under the graph of $f(t) = 1200 + 50t$, in particular the smaller typical region indicated, enlarged at the right.

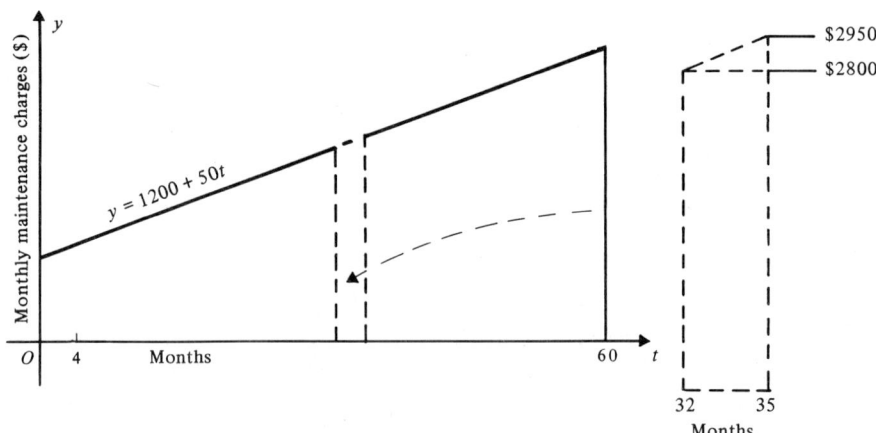

What are the charges for the indicated 3-month period from 32 to 35 months? Since the cost for each of these three months is approximately $2900 (they actually range from $2800 to $2950), the maintenance cost for the entire 3-

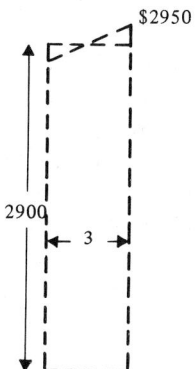

month period is approximately (3 × 2900) dollars, or $8700. Note that 8700 is the area of a rectangle that approximates the area of our small region.

As expected, we conclude that the total maintenance cost over the first 5 yr (60 months) is equal to the area under the graph of $f(t) = 1200 + 50t$ on the interval [0, 60]. It is given by the definite integral

$$\int_0^{60} (1200 + 50t)\, dt = [1200t + 25t^2]_0^{60} = \$162{,}000$$

(The answer obtained by using the formula for the area of a trapezoid, p. 378, is of course the same.)

In general, if the maintenance charges of some company are changing, after t months, at a rate given by the function $f(t)$ (in dollars per month), then the company's total maintenance cost for the first k yr is given by the definite integral

$$\int_0^{12k} f(t)\, dt$$

PREDICTING GROUP SIZE FROM DROPOUT EXPECTANCY

R. U. Reading plans to sell some of its mathematics magazines through subscriptions. The company estimates that it will receive 100 new subscriptions per month for many years and that the fraction of subscribers who will retain their subscriptions for at least m months is $R(m) = e^{-m/50}$.

(a) How many subscriptions will Reading have 12 months after starting?

(b) How many q months after starting?

(c) Is there a limiting value to the number of subscriptions?

(a) Consider a small interval of time, say from $8\frac{1}{2}$ to 9 months. During this $\frac{1}{2}$-month period there will be approximately $100 \times \frac{1}{2}$, or 50, new subscriptions. How many of these will still be active at the end of the first 12 months? To be active then, a subscription must last approximately 3 months (actually from 3 to $3\frac{1}{2}$ months, depending on the subscription). The fraction of subscriptions lasting at least 3 months equals $R(3)$ or $e^{-3/50}$, and of our particular group of 50 subscriptions, the number still active at the 12-month point is therefore approximately $50e^{-3/50}$ or 47.

In general, if we divide the interval from 0 to 12 months into n equal subintervals, each of length Δt months, and choose a time in each subinterval, say t_k in the kth, then the number of new subscriptions in the kth subinterval of time is $100\Delta t$ and the fraction of these still active at the 12-month point is approximately

$$R(12 - t_k) = e^{-(12 - t_k)/50}$$

So the number of subscriptions new in the kth subinterval of time *and* still active at the 12-month point is approximately $100\Delta t$ times this, or

$$100 e^{-(12 - t_k)/50} \Delta t$$

As the number of subintervals becomes infinite, the sum of these approximations becomes

$$\int_0^{12} 100 e^{-(12 - t)/50} \, dt = 100(50) \int_0^{12} e^{-(12 - t)/50} \frac{1}{50} \, dt$$

$$= 5000 \left[e^{-(12 - t)/50} \right]_0^{12} = 5000(1 - e^{-0.24}) \approx 1065*$$

So R. U. Reading will have about 1065 subscriptions at the end of its first year of operation.

(b) To find the number of subscribers after q months we simply replace 12 in (a) by q, getting

$$\int_0^q 100 e^{-(q - t)/50} \, dt = 5000 \left[e^{-(q - t)/50} \right]_0^q = 5000(1 - e^{-q/50})$$

---

*Many answers in this Goal are rounded off (e.g., above, to the nearest 5 subscriptions). We do this whenever the resulting approximation seems reasonable.

(c) In this part we are interested in what happens to the number of subscriptions with increasing time. The answer is given by

$$\lim_{q \to \infty} 5000(1 - e^{-q/50}) = 5000(1 - 0) = 5000$$

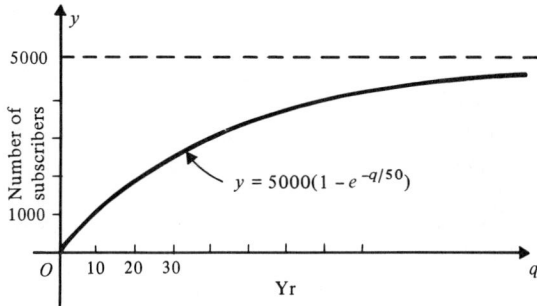

The figure shows that the graph of the subscribers function approaches the horizontal line at 5000 as an asymptote.

PRESENT VALUE—

In order to sell investment shares in its new magazine, R. U. Reading must determine the present value of profits to be earned in the future.

—OF MONEY WORTH A DOLLARS IN t YEARS

In Goal 36 we defined the present value of money worth A dollars in t yr to be the amount (in a lump sum) we must invest now so as to get a return of A dollars t yr from now. We know, for example, that an investment of $1000 now at 6% interest compounded continuously will, in 5 yr, be worth $1000e^{(0.06)(5)}$, or about $1350. Thus the present value of $1350 due in 5 yr (at 6% compounded continuously) is $1000. In general, the present value I of money worth A dollars in t yr, if the interest rate is 6% compounded continuously, is given by

$$I = Ae^{-0.06t}$$

—OF FUTURE PERIODIC INCOME

Suppose our publisher decides, for philanthropic (and other) reasons, to award an R. U. Reading $2000 scholarship at the end of each of the next 4 yr to the most promising freshman in its town's college. How much must it invest now, at 7% compounded continuously, to generate this income?

The total present value of this income equals

$$2000e^{-0.07} + 2000e^{(-0.07)(2)} + 2000e^{(-0.07)(3)} + 2000e^{(-0.07)(4)}$$

(Why?) With our tables, this equals approximately

$$2000\,(0.932 + 0.869 + 0.811 + 0.756) \approx \$6735$$

—OF FUTURE CONTINUOUS INCOME

Our optimistic publisher estimates that its mathematics magazine will earn profits of $10,000 per yr, to be received continuously and at a constant rate, over the next 10 yr. What is the present value of its future earnings, assuming an interest rate of 5% compounded continuously?

As in previous applications, we consider a small time interval, say from the next $3\frac{3}{4}$ to $4\frac{1}{4}$ yr. The profits during this $\frac{1}{2}$-yr period equal $(10,000 \times \frac{1}{2})$ dollars, or $5000. The present value of this $5000 is approximately $5000e^{-0.05(4)}$ or about $4095. (This is an approximation because the profits are not all to be received at the 4-yr point, but rather are to be received evenly throughout the $\frac{1}{2}$-yr period.)

Now we generalize. If we divide the 10-yr period into n subintervals of equal duration, Δt yr, and let t_k be a point of time in the kth subinterval, then the profits from this subinterval are $10,000\Delta t$ and the present value of these profits is approximately

$$10,000\,\Delta t e^{-0.05 t_k} = 10,000 e^{-0.05 t_k}\,\Delta t$$

The total present value is equal to the sum of approximations of present value over all n subintervals, as $n \to \infty$. As expected, this infinite sum is equal exactly to

$$\int_0^{10} 10,000 e^{-0.05t}\,dt = \frac{10,000}{-0.05}\left[e^{-0.05t}\right]_0^{10} = -200,000(e^{-0.5} - 1) \approx \$78,700$$

NOTE: Some authors use the term "capital value" instead of "present value" when income is to be received periodically or continuously.

EXERCISES

1. In each of the following, the marginal revenue $M(x)$ is given as a function of the production level x in dollars per unit. Assuming that the revenue is 0 when $x = 0$, determine each of the revenue functions.
   (a) $M(x) = 15 - 0.002x$
   (b) $M(x) = 10 - 0.001x$
   (c) $M(x) = 250 - 0.4x - 0.006x^2$

Goal 47 | To Use Integration in Problems in Economics

2. Determine the total cost function for each of the following marginal cost functions (in dollars per unit), given the associated fixed overhead costs.
   (a) $4 + 0.002x$; $6000
   (b) $9 + 0.06x$; $9500
   (c) $25 + 0.8x + 0.012x^2$; $25,000

3. For each of the following pairs of demand and supply functions determine the equilibrium point, the consumers' surplus, and the producer's surplus.
   (a) $D(x) = -0.0003x + 3$; $S(x) = 0.0003x$
   (b) $D(x) = 5.4 - 0.0002x^2$; $S(x) = 0.00004x^2$
   (c) $D(x) = 6 - 0.0002x - 0.00000006x^2$; $S(x) = 0.00162x$

4. Suppose it is estimated by a manufacturer that the monthly maintenance expenditures after n months will be $(800 + 60n)$ dollars, incurred at a constant rate. Determine the total maintenance bill for (a) the first half year; (b) the second year; (c) the first 5 yr.

5. In this exercise we consider two subsidiaries of R. U. Reading, each one publishing a magazine. For each magazine we give first the number of new subscriptions per month, then the fraction of subscribers who will retain their subscriptions for at least m months. For each magazine, find the number of subscriptions it has, first after 12 months, then after q months; and finally find the limiting value, if it exists.
   (a) 150; $e^{-m/25}$ (b) 500; $e^{-m/100}$

6. How much money must you invest now at 6% compounded continuously for it to accumulate to $10,000 in 10 yr?

7. How much must a college invest now at 7% compounded continuously if it wants to distribute scholarships of $1000 one year from now, $2000 two years from now, $4000 three years from now, and $5000 four years from now?

8. For each of the following projected yearly profit functions, determine the present value of the expected profits over the next 5 yr. Assume that profits are received continuously and at a constant rate throughout the year and that the interest rate is 5% compounded continuously.
   (a) $7500
   (b) $3000
   (c) $5000t$ dollars after t yr   (HINT: Integrate by parts)

9. An orthodontist is setting up a new practice. He estimates that he will attract 25 new patients per month at a continuous and constant rate throughout the month. If the fraction of patients who will still be seeing him after m months is given by $e^{-m/20}$, determine the number of patients

he will have (a) after 5 months, and (b) after q months. (c) Is there a limiting size to his practice?

10. A college professor is setting up a fund that will pay her heirs $1000 per yr. How much must she put into the fund if she wishes it to last for the next 10 yr? Assume that money is to be paid out continuously and at a constant rate and that the interest rate is 5% compounded continuously.

TEST A

1. If the marginal profit is the function $-0.01x + 6.5$ (in dollars per unit), find the profit function, given that the profit is $200 when the production level is 100.
2. Determine the equilibrium point, the consumers' surplus, and the producer's surplus for the demand and supply functions $D(x) = -0.004x + 3$ and $S(x) = 0.000004x^2$.
3. Suppose that there are 150 new subscriptions per month and that the fraction of subscribers who retain their subscriptions for at least 30 months is given by $e^{-m/15}$. Determine the number of subscriptions after $2\frac{1}{2}$ yr.
4. If the projected yearly profits are $3600, to be received continuously and at a constant rate, what is the present value of the first 5 yr profit, assuming an interest rate of 6% compounded continuously?

TEST B

1. Determine the total cost function if the marginal cost is $5 + 0.01x$ dollars per unit and the fixed overhead is $750.
2. Determine the equilibrium point, the consumers' surplus, and the producer's surplus for the demand and supply functions $D(x) = -0.02x + 18$ and $S(x) = 0.004x$.
3. Suppose that there are 100 new subscriptions per month and that the fraction of subscribers who retain their subscriptions for at least m months is given by $e^{-m/12}$. Determine the number of subscriptions after 6 months.
4. If the expected yearly profits are $2000, to be received continuously and at a constant rate, what is the present value of the first 3 years' profits, assuming an interest rate of 5% compounded continuously?

STUDY QUESTIONS

1. Suppose that profits of $10,000 per yr are to be received continuously and at a constant rate for the next 10 yr. If all profits are invested at

Goal 47 | To Use Integration in Problems in Economics

6% compounded continuously, what will be the *future value* of these profits (that is, the amount accumulated at the end of 10 yr)?

2. Do Exercise 5 above for the following three additional subsidiaries of R. U. Reading, given both the number of new subscriptions per month and the fraction who retain their subscriptions for at least m months:
   (a) $25t$ after t months; $e^{-m/50}$ (HINT: Integrate by parts)
   (b) 100; $1/(1 + m)^2$
   (c) 50; 1 if $m \leq 5$ and 0 if $m > 5$

3. Suppose the orthodontist in Exercise 9 above has 250 patients when he goes into practice. If all the other conditions are as given in the exercise, find answers to (a), (b), and (c).

4. Suppose our college professor in Exercise 10 wants her heirs' income to continue forever. How much must she invest now? HINT: Evaluate the improper integral $\int_0^\infty f(t)\, dt$ for an appropriate function $f(t)$.

GOAL 48

To use integration on probability problems

The theory of probability has important applications throughout the social, biological, and physical sciences. In this Goal we will see that the definitions and methods we use to solve problems involving probability depend on the calculus and, in particular, on the definite integral.

AN EXAMPLE

A telephone company has gathered data on the duration of calls from a large set of calls lasting no more than 60 min. The table below at the left shows the fraction $F(q)$ of calls that lasted q min or less. These points are plotted at the right, and a smooth curve is drawn through them and labeled $F(x)$.

q (min)	$F(q)$
5	0.16
10	0.31
15	0.44
20	0.56
25	0.66
30	0.75
35	0.83
40	0.89
45	0.94
50	0.97
55	0.99

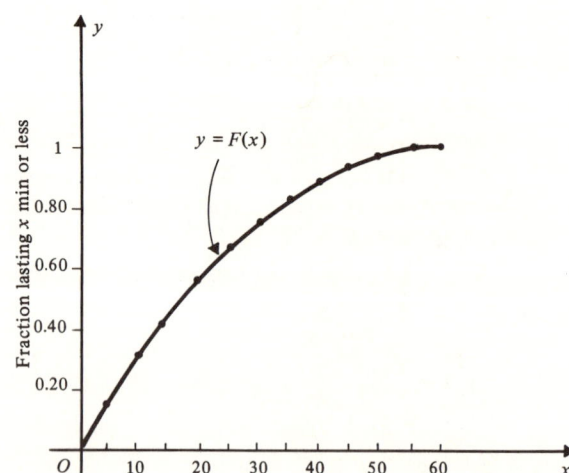

458

Note that $F(0) = 0$ (no calls lasted 0 min) and that $F(60) = 1$ (every call in the sample lasted 60 min or less).

We now use the data above as the basis for finding probabilities.

From the graph of $F(x)$ we can read off the probability that a call taken at random will last, say, 3 min or less. It's just $F(3)$, the height at $x = 3$, which is equal approximately to 0.10. About 1 in 10 calls will last up to 3 min.

What proportion of calls will last from 15 to 45 min? The answer is $F(45) - F(15)$ (why?), or about $(0.94 - 0.44)$, which equals 0.5.

What is the probability that a call will last more than 45 min? It's $F(60) - F(45)$ (why?), or about 0.06.

In general, the probability that a call will last $\leq q$ min is $F(q)$.

The probability that a call will last between a and b mins equals $F(b) - F(a)$.

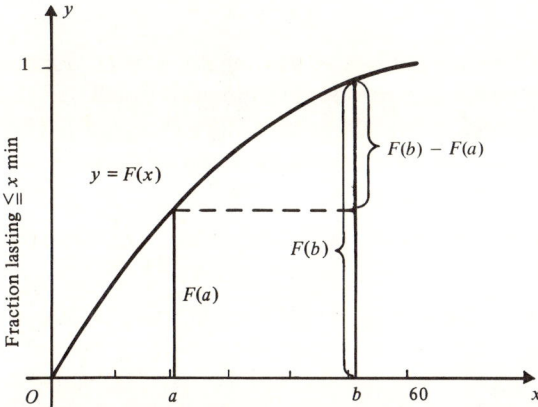

PROBABILITY DENSITY FUNCTIONS

In most practical problems we do not have a function such as $F(x)$ above. Instead we are given the derivative of $F(x)$, denoted by $f(x)$. We then obtain $F(b) - F(a)$ by evaluating the definite integral $\int_a^b f(x)\,dx$. In general, if a function $f(x)$ is nonnegative on its domain and if the definite integral of $f(x)$ over its domain is equal to 1, then $f(x)$ is called a *probability density* (or just a *density*) function. In our example on telephone calls, the domain of f is $0 \leq x \leq 60$ and

$$\int_0^{60} f(x)\,dx = \left[F(x)\right]_0^{60} = F(60) - F(0) = 1 - 0 = 1$$

Therefore, by our definition, the function $f(x)$ is a density function for the duration of a phone call. We also say that f is a density function for F, the proportion of calls that last up to q mins.

In general, if $f(x)$ is a density function with domain $c \leq x \leq d$, then

$$\int_c^d f(x)\, dx = 1$$

The function $F = \int f(x)\, dx$ is called a (*probability*) *distribution function*.

Example 48-1

Verify that each of the following is a density function over the domain indicated:

(a) $f(x) = \frac{3}{8}x^2$ over $0 \leq x \leq 2$ (c) $f(x) = 1/x^2$ over $1 \leq x < \infty$

(b) $f(x) = \frac{1}{4}$ on $1 \leq x \leq 5$ (d) $f(x) = 3e^{-3x}$ over $0 \leq x < \infty$

Note that each of the given functions is nonnegative on its domain. We need therefore only verify that the appropriate integrals equal 1.

(a) $\displaystyle\int_0^2 \frac{3}{8} x^2\, dx = \frac{3}{8}\left[\frac{x^3}{3}\right]_0^2 = 1$

(b) $\displaystyle\int_1^5 \frac{1}{4}\, dx = \frac{1}{4}[x]_1^5 = 1$

(c) $\displaystyle\int_1^\infty \frac{1}{x^2}\, dx = \lim_{b\to\infty} \int_1^b x^{-2}\, dx = \lim_{b\to\infty} -\left[\frac{1}{x}\right]_1^b = \lim_{b\to\infty}\left(1 - \frac{1}{b}\right) = 1$

(d) $\displaystyle\int_0^\infty 3e^{-3x}\, dx = \lim_{b\to\infty} \int_0^b 3e^{-3x}\, dx = \lim_{b\to\infty} -\int_0^b e^{-3x}(-3\, dx) =$
$\lim_{b\to\infty} -[e^{-3x}]_0^b = \lim_{b\to\infty}(1 - e^{-3b}) = 1$

A FAMOUS PROBABILITY DENSITY FUNCTION

Below we see a graph that shows how the weights of college women are distributed in the United States. In advanced courses it is shown that the function graphed here is a probability density function. The familiar bell-

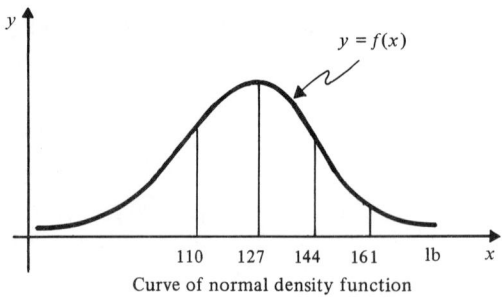

Curve of normal density function

Goal 48 | **To Use Integration on Probability Problems**

shaped curve depicts a so-called *normal* probability density function for the distribution of many biological traits, like height or weight, and of psychological measurements of intelligence, achievement, and so on.

The ordinate $f(q)$ is the "density" of q lb. It is a measure of the relative "popularity" of a weight of q lb among college women. This density graph tells us, for example, that there are more college women who weigh between 127 and 144 lb than there are between 144 and 161 lb.

FINDING PROBABILITIES FROM A GIVEN DENSITY FUNCTION

To find the probability that a college woman chosen at random weighs between 110 and 127 lb, we evaluate the definite integral

$$\int_{110}^{127} f(x)\, dx$$

where $f(x)$ is the density function for the weight of college women, graphed above.

In general, if $f(x)$ is a probability density function for a variable x, then the probability that x falls between two values a and b (in the domain of f) is

$$\int_a^b f(x)\, dx$$

This says that the probability that x falls between a and b is simply the area under the curve from $x = a$ to $x = b$. (Study Question 1 in this Goal suggests how this conclusion may be justified mathematically.) We can also view $\int_a^b f(x)\, dx$ as the proportion (of the population in question) for which x is between a and b.

NOTE: You may want to verify that each probability density function in the following examples satisfies the definition given above.

Example 48-2

The telephone company considered earlier decided to use $f(x) = (60 - x)/1800$ as the density function for the duration of a phone call, where $0 \leq x \leq 60$ (mins). (a) Find the probability that a randomly selected call lasts 3 min or less. (b) What proportion of calls last from 15 to 45 min? (c) What is the probability that a call will last more than 45 min?

(a) $\int_0^3 \dfrac{1}{1800} (60 - x)\, dx = \dfrac{1}{1800} \left[60x - \dfrac{x^2}{2} \right]_0^3 = 0.0975 \approx 0.10$

(b) $\int_{15}^{45} \dfrac{1}{1800} (60 - x)\, dx = \dfrac{1}{1800} \left[60x - \dfrac{x^2}{2} \right]_{15}^{45} = 0.5$

(c) $\int_{45}^{60} \frac{1}{1800}(60-x)\,dx = \frac{1}{1800}\left[60x - \frac{x^2}{2}\right]_{45}^{60} = 0.0625 \approx 0.06$

Now check the answers we obtained on page 459 to these very questions. They're exactly the same answers! Why? The density function $f(x)$ given here is just the derivative of $F(x)$ whose graph is shown on page 458. There we found the answer to (b) by evaluating $F(45) - F(15)$. Just above we evaluated the definite integral $\int_{15}^{45} f(x)\,dx$. Since $F'(x) = f(x)$, we know from the Fundamental Theorem that $F(45) - F(15)$ equals this definite integral.

Example 48-3

The Luminate Company manufactures light bulbs with a life expectancy of between 1 and 3 yr. The company determines experimentally that the density function for the life t in years of its light bulbs is $t^3/20$. (a) What percentage of bulbs are expected to last 2 yr or less? (b) What is the probability that a bulb chosen at random will last between 2 and 3 yr?

(a) $\int_1^2 \frac{t^3}{20}\,dt = \frac{1}{20}\left[\frac{t^4}{4}\right]_1^2 = \frac{3}{16} = 18\frac{3}{4}\%$

(b) $\int_2^3 \frac{t^3}{20}\,dt = \frac{1}{20}\left[\frac{t^4}{4}\right]_2^3 = \frac{13}{16} = 81\frac{1}{4}\%$

Note, however, that having found the answer to (a) it is not necessary to compute the answer to (b) directly from the definite integral. From the conditions given, we know that a bulb that lasts more than 2 yr does not last more than 3. Hence to get the answer to (b) we merely subtract the answer to (a) from unity: $1 - \frac{3}{16} = \frac{13}{16}$.

Example 48-4

A psychologist determines experimentally that the density function for the time t in minutes it takes to solve a puzzle is $f(t) = 2/(1+t)^3$, $t \geq 0$. (a) What is the probability that a person chosen at random will be able to solve the puzzle in 1 min or less? (b) What is the probability that it will take more than 2 min?

(a) $\int_0^1 \frac{2}{(1+t)^3}\,dt = 2\int_0^1 (1+t)^{-3}\,dt$

$= -\left[\frac{1}{(1+t)^2}\right]_0^1$

$= -\left(\frac{1}{4} - 1\right) = \frac{3}{4}$

Goal 48 | To Use Integration on Probability Problems

(b) Note that the domain of f is $0 \leq t < \infty$. So the probability the solution will take more than 2 min is

$$\int_2^\infty \frac{2}{(1+t)^3} \, dt = \lim_{b \to \infty} \int_2^b \frac{2}{(1+t)^3} \, dt$$

$$= \lim_{b \to \infty} -\left[\frac{1}{(1+t)^2}\right]_2^b$$

$$= \lim_{b \to \infty} \left(\frac{1}{9} - \frac{1}{(1+b)^2}\right) = \frac{1}{9}$$

Example 48-5

A radio manufacturer estimates that the proportion of his radios that should last up to x yr before breaking down has probability density $2/(x+2)^2$. (a) What proportion of the radios should last between 3 and 5 yr? (b) What percentage should last more than 3 yr?

(a) $\int_3^5 \frac{2}{(x+2)^2} \, dx = \left[-\frac{2}{x+2}\right]_3^5 = \frac{4}{35}$

(b) since the domain of the function is $0 \leq x < \infty$, we have

$$\int_3^\infty \frac{2}{(x+2)^2} \, dx = \lim_{n \to \infty} \int_3^n \frac{2}{(x+2)^2} \, dx$$

$$= \lim_{n \to \infty} \left(\frac{2}{5} - \frac{2}{n+2}\right) = \frac{2}{5} \text{ or } 40\%$$

Example 48-6

The highway department ascertains that the number of accidents occurring per year at major intersections in the county has a density function $f(x) = 0.15e^{-0.15x}$. If an intersection is chosen at random, what is the probability that there will be at least 20 accidents there in 1 yr?

$$\int_{20}^\infty 0.15e^{-0.15x} \, dx = \lim_{n \to \infty} \int_{20}^n 0.15e^{-0.15x} \, dx = \lim_{n \to \infty} \left[-e^{-0.15x}\right]_{20}^n$$

$$= \lim_{n \to \infty} (e^{-3} - e^{-0.15n}) = e^{-3} = 0.05$$

There is therefore only a 5% probability that there will be at least 20 accidents per yr at the randomly chosen intersection.

Example 48-7

A distributor determines experimentally that the demand for his commodities per month has a probability density function $f(x) = 0.01e^{-0.01x}$.

What is the probability that the demand for a randomly chosen commodity will be at least 45 but fewer than 100 units?

$$\int_{45}^{100} 0.01 e^{-0.01x}\, dx = \Big[-e^{-0.01x}\Big]_{45}^{100} = e^{-0.45} - e^{-1} \approx 0.27$$

Example 48-8

An engineer determines that the life expectancy, in years, of fuses has the density function $f(x) = \frac{3}{128}(16 - x^2)$, $0 \le x \le 4$. What percentage of the fuses are likely to last less than 2 yr.

$$\int_0^2 \frac{3}{128}(16 - x^2)\, dx = \left[\frac{3}{128}\left(16x - \frac{x^3}{3}\right)\right]_0^2 = \frac{11}{16} = 68\frac{3}{4}\,\%$$

EXERCISES

1. Which of the following are probability densities?
   (a) $f(x) = 3x^2$ over $0 \le x \le 1$
   (b) $f(x) = 1 - 2x$ over $-\frac{1}{2} \le x \le \frac{1}{2}$
   (c) $f(x) = \dfrac{1}{x}$ over $1 \le x \le e$
   (d) $f(x) = \frac{1}{2} e^{-x/2}$ over $0 \le x < \infty$
   (e) $f(x) = x - \frac{1}{2}$ over $0 \le x \le 2$
   (f) $f(x) = 3/(x + 1)^4$ over $0 \le x < \infty$
   (g) $f(x) = \frac{1}{10}$ over $5 \le x \le 15$

2. A certain brand of television tubes has a life expectancy of between 1 and 10 yr. The proportion of tubes that will fail within q yr, $1 \le q \le 10$, is equal to $\int_1^q f(x)\, dx$, where f is the probability density function $\frac{2}{81}(x - 1)$. What proportion of the tubes have a life expectancy of less than 4 yr? More than 7?

3. The Internal Revenue Service has determined that the annual income of families in a certain community has density function $f(x) = 5000/x^2$ ($x \ge 5000$). What proportion of families have an income between \$10,000 and \$20,000? More than \$5000?

4. A mattress manufacturer knows that the proportion of adults who weigh no more than w lb has probability density $85/w^2$, $w \ge 85$. What proportion of adults weigh less than 100 lb? Between 100 and 150? Over 150?

Goal 48 — To Use Integration on Probability Problems

5. A biologist determines that the number of offspring per year from a certain wildlife population has a density $0.02e^{-0.02x}$. What is the probability that the number of offspring next year will be between 100 and 200?

6. A manufacturer claims that the probability that the light intensity of one of its flashlights is between a and b units is given by $\int_a^b f(x)\,dx$, where f is the probability density $\frac{6}{125}(-x^2 + 15x - 50)$, $5 \le x \le 10$. What is the probability that the light intensity of a randomly selected flashlight is between 6 and 9 units? More than 8 units?

7. A highway engineer determines that the interval in minutes between cars arriving successively at a given intersection has the probability density function $f(x) = \frac{1}{8}(4 - x)$, $0 \le x \le 4$. What is the probability that the next car will arrive within 1 min? Within 2 min?

8. A supermarket analyst finds that the proportion of adults who spend q min or less in front of a cereal display before making a choice has the probability density $f(q) = \frac{3}{1000}(q - 10)^2$, $0 \le q \le 10$. What proportion of adults spend less than 1 min before the display? Over 5 min?

9. A psychologist determines that the proportion of rats who learn a route through a maze in q min or less has the density function $f(q) = \frac{1}{10}e^{-q/10}$, $q \ge 0$. What proportion of the rats learn the route in 10 min or less? In between 7 and 14 min? In over 23 min?

10. A consumer analyst finds that the amount x of money in dollars spent by shoppers per visit to the supermarket has a probability density $20/(x + 20)^2$. What is the probability that a shopper chosen at random spends between \$10 and \$20?

TEST A

1. Which of the following are probability density functions?
   (a) $f(x) = \frac{1}{12}(3x^2 + x + 1)$, $0 \le x \le 2$
   (b) $f(x) = \frac{2}{3}x(x - 1)$, $-1 \le x \le 2$
   (c) $f(x) = \dfrac{2}{x^3}$, $1 \le x < \infty$

2. An actuary has determined that the proportion of Eskimos who die before they reach the age n is given by $\int_0^n f(x)\,dx$, where $f(x)$ is the probability density $\frac{1}{20}e^{-x/20}$. What proportion of Eskimo children die before the age of 5?

3. A highway engineer determines that the distance x in feet between cars on a highway has the probability density function $f(x) = 1/100$, $0 \le x \le 100$. If a pair of adjacent cars is chosen at random, what is the probability that they are no more than 25 ft apart?

TEST B

1. Which of the following are probability densities?

   (a) $f(x) = \dfrac{x^2}{63}$, $3 \leq x \leq 6$

   (b) $f(x) = 5e^{-5x}$, $0 \leq x < \infty$

   (c) $f(x) = 1$, $0 \leq x \leq 2$

2. A sportswriter has found that the proportion of basketball players whose height is less than h in. above $5\frac{1}{2}$ ft is equal to $\int_0^h f(x)\,dx$, where f is the probability density $\frac{1}{2304}(24x - x^2)$, $0 \leq x \leq 24$. What proportion of the players are between 6 and $6\frac{1}{2}$ ft tall?

3. An educator notes that the probability density function for scores on an SAT examination is $f(x) = (-x^2 + 1000x - 160{,}000)/36{,}000{,}000$, $200 \leq x \leq 800$. What is the probability that a student chosen at random will score less than 400?

STUDY QUESTIONS

1. Suppose $F(x)$ is a probability distribution function over the interval $c \leq x \leq d$. Given that $F'(x) = f(x)$ on $[c, d]$ and that $c \leq a < b \leq d$, use the Mean Value Theorem and the definition of definite integral to prove that

$$F(b) - F(a) = \int_a^b f(x)\,dx$$

   HINT: See Study Question 2 in Goal 40.

2. (a) Find k such that kx is a probability density function over the interval $[1, 4]$. (b) Find k such that kx^2 is a probability density function on $[0, 3]$.

3. If f is nonnegative on $[a, b]$ and $\int_a^b f(x) = k > 0$, then show how to construct a probability density involving the function f.

4. If f is a probability density over the interval $[a, b]$, then we define the mean μ of f to be $\mu = \int_a^b x f(x)\,dx$. Determine the mean of each of the following functions:

   (a) $f(x) = 3x^2$ over $0 \leq x \leq 1$

   (b) $f(x) = \dfrac{1}{x}$ over $1 \leq x \leq e$

   (c) $f(x) = \frac{1}{10}$ over $5 \leq x \leq 15$

Goal 48 | **To Use Integration on Probability Problems**

5. If f is a probability density over the interval $[a, b]$ then we define the *median m* of f to be the number between a and b such that

$$\int_a^m f(x)\, dx = \int_m^b f(x)\, dx = \frac{1}{2}$$

Determine the median of each function in Study Question 4.

GOAL 49

To understand and solve simple differential equations

In this Goal we define and classify differential equations, distinguish between general and particular solutions, and solve simple examples.

DEFINITIONS

A *differential equation* is an equation that contains either a derivative or a differential. We have already seen (and solved) many of them:

(a) $dP/dt = 0.018P$ (d) $R'(t) = k(350 - R)$
(b) $y'' = -32$ (e) $dy = (x^2 + 1)^6 \, 2x \, dx$
(c) $M'(x) = 1/x$

How might we interpret these equations?

(a) That a population P is growing at a rate (1.8%) proportional to the existing population (p. 345).
(b) That the acceleration of a moving object is equal to the constant -32 (p. 433).
(c) That the marginal cost of production is $1/x$ dollars per gadget when x gadgets are produced (p. 393).
(d) That the temperature of a roast put into a 350°F oven rises at a rate (k) proportional to the difference between the oven temperature and its own temperature R (p. 349).
(e) That the differential of a function y equals $[(x^2 + 1)^6 2x] \, dx$ (p. 405). Other examples of differential equations are

$$y \, dx = (x^3 + 3) \, dy \qquad \frac{d^3y}{dx^3} = \frac{dy}{dx} + x \qquad (y'')^3 - (y')^2 = e^x$$

Goal 49 | To Understand and Solve Simple Differential Equations

CLASSIFYING DIFFERENTIAL EQUATIONS

The *order* of a differential equation is the order of the highest derivative that occurs. For example:

$$y' = 4y, \quad \frac{dy}{dx} = x^3 + 2, \quad 3xy + (y')^2 = 2 \quad \text{are } \textit{first order};$$

$$y'' = 3y, \quad \frac{d^2y}{dx^2} = \left(\frac{dy}{dx}\right)^3 - 3x, \quad (y'')^5 - 2y = x \quad \text{are } \textit{second order};$$

$$\frac{d^5y}{dx^5} = 5\left(\frac{d^3y}{dx^3}\right)^{10} + 3y \quad \text{is of the } \textit{fifth order}.$$

SOLUTIONS

A *solution* of a differential equation is any function that satisfies it.

Example 49-1

Show that $f(x) = x^2 + 2$ is a solution of the differential equation $y'' = y - x^2$.

We replace y by $f(x)$ and y'' by $f''(x)$. Since $f'(x) = 2x$ and $f''(x) = 2$, we get

$$2 = (x^2 + 2) - x^2$$
$$= x^2 + 2 - x^2$$
$$= 2$$

Therefore $f(x) = x^2 + 2$ is a solution.

Example 49-2

Verify that $f(x) = ce^{kx}$ is a solution of the differential equation $y' = ky$.
Here we replace y by $f(x)$ and y' by $f'(x)$. Since $f'(x) = ce^{kx} \cdot k = (ck)e^{kx}$, we get

$$cke^{kx} = k(ce^{kx})$$

The substitution yields an identity, so $f(x) = ce^{kx}$ is a solution.

GENERAL VERSUS PARTICULAR SOLUTIONS

The simplest differential equations to solve (and the only ones we've asked you to solve so far) are those of the form

$$y' = g(x) \quad \text{or equivalently} \quad dy = g(x) \, dx$$

which can be solved directly by integration. The general solution is $y = f(x) = \int g(x) \, dx$.

Examples 49-3

(a) If $y' = 2x$, then y or $f(x) = x^2 + C$.
(b) If $dy = e^x \, dx$, then $y = e^x + C$.
(c) If $dy/dx = (x^2 + 1)^6 \, 2x$, then $dy = (x^2 + 1)^6 \, 2x \, dx$, and

$$y = \int (x^2 + 1)^6 2x \, dx = \frac{(x^2 + 1)^7}{7} + C$$

Each *general* solution above involves an arbitrary constant, which can be replaced by a specific number to yield a *particular* solution. For example, in (a) we see that $x^2 + 7$, $x^2 - 9$, $x^2 + 3.4$ are all particular solutions. Each is obtainable from the general solution.

Example 49-4

Find the particular solution of $y' = 4x - 2$ if $f(1) = 3$.

$$f(x) = \int (4x - 2) \, dx = 2x^2 - 2x + C$$
$$f(1) = 3 = 2(1)^2 - 2(1) + C$$
$$C = 3$$

So $f(x) = 2x^2 - 2x + 3$ is the particular solution sought.

A condition such as $f(1) = 3$ is called a *boundary* or *side condition*. If the value of f (or of a derivative) is given at 0, the condition is usually called an *initial* condition. To obtain a particular solution from a general one, supplementary information of this sort is necessary. In applications we are most often interested in particular solutions.

Example 49-5

Find the particular solution of $y'' = e^x + 6x$ for which $f(0) = 1$ and $f'(0) = 2$.

To solve this second-order equation we integrate twice:

$$y' = \int (e^x + 6x) \, dx = e^x + 3x^2 + C_1$$
$$y = \int (e^x + 3x^2 + C_1) \, dx = e^x + x^3 + C_1 x + C_2$$

This is the general solution. To find the particular solution we use the given initial conditions:

$f(0) = e^0 + 0 + C_1(0) + C_2 = 1 \qquad 1 + C_2 = 1 \qquad C_2 = 0$
$f'(0) = e^0 + 3(0)^2 + C_1 \qquad\quad = 2 \qquad 1 + C_1 = 2 \qquad C_1 = 1$

The particular solution is therefore $f(x) = e^x + x^3 + x$.

Goal 49 | To Understand and Solve Simple Differential Equations

NOTES:

(1) It is often very difficult to find solutions even to first order differential equations. In fact, most differential equations do not have solutions that can be expressed explicitly as simple functions.
(2) When a general solution exists, there are usually an infinite number of particular solutions.
(3) As may be inferred from the examples given, the general solution of a differential equation of order n usually involves n arbitrary constants.

EXERCISES

1. Classify each differential equation as to order.

   (a) $\dfrac{dy}{dx} + 3x^2 y = 0$ (d) $\left(\dfrac{d^2 y}{dx^2}\right)^3 = \dfrac{dy}{dx} + y^5$

   (b) $y'' - 3y = (y')^5$ (e) $\dfrac{dy}{dx} + (\ln x)y = e^x$

   (c) $(y')^3 = 3xy + 2$ (f) $y''' = y' + x$

2. Verify that the following functions are solutions of the corresponding differential equations.

   (a) $f(x) = x^3 + x + C$ $y' = 3x^2 + 1$
   (b) $f(x) = e^x + x^2 + C$ $y' = e^x + 2x$
   (c) $f(x) = x^2 + Cx$ $y' = \dfrac{y}{x} + x$
   (d) $f(x) = ce^{5x}$ $y' = 5y$
   (e) $f(x) = \dfrac{e^{Cx}}{x}$ $y' = \dfrac{y}{x}(\ln xy - 1)$
   (f) $f(x) = x^4 - x^3 + Cx + k$ $y'' = 12x^2 - 6x$
   (g) $f(x) = Ce^{2x} + ke^{-2x}$ $y'' = 4y$
   (h) $f(x) = C \ln kx$ $y'' = -\dfrac{y'}{x}$

3. Find the particular solution of each of the differential equations in Exercise 2 that satisfies the following boundary conditions:

   (a) $f(1) = 5$ (e) $f(1) = e^2$
   (b) $f(0) = 3$ (f) $f(0) = 3$, $f(1) = 4$
   (c) $f(-1) = 0$ (g) $f(0) = 2$, $f(1) = 2e^2$
   (d) $f(0) = 8$ (h) $f(1) = 0$, $f(e) = 3$

4. Find the general solution to each of the following differential equations.
   (a) $y' = x^2 + 4x - 1$
   (b) $\dfrac{dy}{dx} = e^x + 3$
   (c) $y' = \dfrac{1}{x} - x$
   (d) $y'' = x + 2$
   (e) $\dfrac{d^2y}{dx^2} = e^{2x}$
   (f) $y''' = x^2 + \dfrac{1}{x^3}$

5. Find the particular solution to each of the following differential equations for the given side conditions.
   (a) $y' = 3x^2 + 6x - 1$, $f(1) = 2$
   (b) $\dfrac{dy}{dx} = e^{2x} + 5$, $f(0) = -1$
   (c) $y' = \dfrac{1}{x}$, $f(1) = 0$
   (d) $\dfrac{d^2y}{dx^2} = x^2 - 2x$, $f(0) = 2$, $f(1) = 0$
   (e) $y''' = 6x^2 + 12x$, $f(0) = 1$, $f(1) = 3\frac{3}{5}$, $f(-1) = -\frac{3}{5}$

TEST A

1. What is the order of the differential equation $y'' = (y''')^2 - x^4$?
2. Verify that $f(x) = x^3 + Cx$ is a solution of $y' = 2x^2 + \dfrac{y}{x}$.
3. Find the general solution to $y'' = 6x^5 - x^2$.
4. Find the particular solution to $\dfrac{dy}{dx} = e^{3x} - 6x + 1$ such that $f(0) = 2$.

TEST B

1. What is the order of the differential equation $\left(\dfrac{dy}{dx}\right)^3 + 3\dfrac{d^2y}{dx^2} = 5y$?
2. Verify that $f(x) = Ce^{3x}$ is a solution of $y'' = 9y$.
3. Find the general solution to $\dfrac{dy}{dx} = \dfrac{1}{x} + e^x$.
4. Find the particular solution to $y' = x^2 - 1$ such that $f(1) = -1$.

Goal 49 | To Understand and Solve Simple Differential Equations

STUDY QUESTIONS
1. A *second-order linear differential equation* is one of the form

$$y'' + a(x)y' + b(x)y = c(x)$$

where $a(x)$, $b(x)$, and $c(x)$ are functions of x. If $c(x) = 0$ the equation is said to be *homogeneous*. Show that if $f_1(x)$ and $f_2(x)$ are both solutions to the homogeneous equation, then

$$f(x) = c_1 f_1(x) + c_2 f_2(x)$$

where c_1 and c_2 are constants, is also a solution.

2. In the preceding question, if $a(x)$ and $b(x)$ are the *constants* a and b, respectively, then we have a *homogeneous equation with constant coefficients*. There is good reason to believe that the solution to such an equation will involve the exponential function e^{mx} because its derivatives are all multiples of itself. If $f(x) = e^{mx}$, then $f'(x) = me^{mx}$, $f''(x) = m^2 e^{mx}$, and substituting into the differential equation gives

$$m^2 e^{mx} + am e^{mx} + b e^{mx} = 0$$

or

$$m^2 + am + b = 0$$

If this has two real roots, m_1 and m_2, then $e^{m_1 x}$ and $e^{m_2 x}$, and thus $c_1 e^{m_1 x} + c_2 e^{m_2 x}$, are solutions to the differential equation.

Find the solution to

$$\frac{d^2 y}{dx^2} - 7\frac{dy}{dx} + 10y = 0$$

and the solution to

$$y'' + y' - 12y = 0$$

GOAL 50

To apply separation of variables

Many differentiable equations of the first order are *separable*; that is, the equation can be written so that all the terms involving one variable (say, y) are on one side and all the terms involving the other (say, x) are on the other. In this Goal we describe the technique known as *separation of variables*. We can then actually solve the three types of separable differential equations applied so extensively in Goal 36.

SEPARATING VARIABLES

We illustrate the technique with examples. Assume, as usual, that division by zero is excluded.

Example 50-1

Solve $\dfrac{dy}{dx} = \dfrac{x}{y}$.

The equation can be rewritten in terms of differentials:

$$y\, dy = x\, dx$$
$$\int y\, dy = \int x\, dx$$
$$\frac{1}{2} y^2 = \frac{1}{2} x^2 + C$$

(Why is a constant not necessary on the left side of the equation?) Note that the solution can also be written $y^2 = x^2 + k$ (why?) or $y = \pm\sqrt{x^2 + k}$.

Goal 50 | To Apply Separation of Variables

Example 50-2

Solve the equation $dy/dx = x^2y$ (where $y > 0$).*
We rewrite in differential form:

$$\frac{dy}{y} = x^2\, dx$$

$$\int \frac{dy}{y} = \int x^2\, dx$$

$$\ln y = \frac{x^3}{3} + c$$

$$y = e^{(x^3/3)+c} = e^{(x^3/3)} \cdot e^c = k e^{(x^3/3)}$$

where in the last step we replaced the arbitrary constant e^c by k.

Example 50-3

Solve $y' = xy + y$.

First we rewrite the equation: $\quad \dfrac{dy}{dx} = xy + y$ or $dy = (xy + y)\, dx$

Then we factor the right-hand side: $\quad dy = y(x + 1)\, dx$

and separate the variables: $\quad \dfrac{dy}{y} = (x + 1)\, dx$

Now we integrate both sides: $\quad \int \dfrac{dy}{y} = \int (x + 1)\, dx$

$$\ln y = x^2/2 + x + C$$

And finally we solve for y: $\quad y = k e^{(x^2/2)+x}$

Why is the last step valid?

Example 50-4

Find the particular solution of Example 50-1 for which $y(0) = 2$.
The general solution found above was

$$y = \pm\sqrt{x^2 + k}$$

Since we are given that $y(0) = 2$, we must choose the positive solution for y and then determine k for it. From $y(0) = \sqrt{0^2 + k} = 2$ we get $k = 4$. So the particular solution sought is $y = \sqrt{x^2 + 4}$.

---

*We will assume throughout this Goal, whenever we have $\int \dfrac{du}{u}$, that $u > 0$, so the antiderivative can be written $\ln u$ rather than $\ln |u|$.

PROCEDURE
To solve a separable differential equation:

(1) Write the equation in differential form.
(2) Separate the variables.
(3) Integrate both sides.

A separable differential equation can always be expressed in the form

$$h(y)\, dy = g(x)\, dx$$

THREE IMPORTANT SEPARABLE DIFFERENTIAL EQUATIONS

In Goal 36 we considered numerous applications of three differential equations. We supplied the general solution for each, having verified them by direct substitution into the differential equations in Goal 35.

We now solve these equations using the technique of separation of variables.

CASE I
The rate of change of a quantity is proportional to the amount or magnitude of the quantity present.

$$y' = ky \qquad \frac{dy}{dx} = ky \qquad \frac{dy}{y} = k\, dx$$

$$\int \frac{dy}{y} = \int k\, dx$$

$$\ln y = kx + C_1 \qquad y = e^{kx+C_1} = e^{C_1} \cdot e^{kx}$$

$$y = ce^{kx}$$

where $c = e^{C_1}$, a positive constant. So the general solution of $y' = ky$ is $f(x) = ce^{kx}$, where c is a positive constant. [On p. 344 we started with the differential equation $f'(t) = kf(t)$ and gave as its solution $f(t) = ce^{kt}$.]

CASE II
The rate of change of a quantity is proportional to the difference between a fixed constant A and the amount or magnitude of the quantity present. In the following, k denotes the positive constant of proportionality.

CASE IIa.
$A - y > 0$; that is, the quantity is increasing toward the limiting value A. We have

Goal 50 | To Apply Separation of Variables

$$y' = k(A - y) \qquad \frac{dy}{dx} = k(A - y) \qquad \frac{dy}{A - y} = k\,dx$$

$$\int \frac{dy}{A - y} = \int k\,dx$$

$$-\ln(A - y) = kx + C_1 \qquad \ln(A - y) = -kx - C_1$$
$$(A - y) = e^{-kx - C_1} \qquad y = A - e^{-kx - C_1} = A - e^{-kx} \cdot e^{-C_1}$$

Replacing the positive constant e^{-C_1} gives $y = A - ce^{-kx}$.

CASE IIb.

$y - A > 0$; that is, the quantity is decreasing toward the limiting value A. In this case,

$$y' = -k(y - A) \qquad \frac{dy}{y - A} = -k\,dx$$

Integration yields

$$\ln(y - A) = -kx + C \qquad y - A = e^{-kx + C}$$
$$y = A + e^{-kx} \cdot e^{C} = A + ce^{-kx}$$

where $c = e^{C} > 0$.

CASE III

The rate of change of a quantity is proportional to both the amount (or magnitude) of the quantity and the difference between a fixed constant A and the magnitude of the quantity.

$$y' = ky(A - y) \qquad \frac{dy}{dx} = ky(A - y)$$

$$\frac{dy}{y(A - y)} = k\,dx \qquad \int \frac{dy}{y(A - y)} = \int k\,dx$$

To integrate the left side, we use an algebraic trick: We note that

$$\frac{1}{y(A - y)} = \frac{1}{A}\left(\frac{1}{y} + \frac{1}{A - y}\right)$$

(Verify this!) So we have

$$\frac{1}{A}\int \left(\frac{1}{y} + \frac{1}{A - y}\right) dy = \int k\,dx$$

$$\frac{\ln y - \ln(A - y)}{A} = kx + C_1$$
$$\ln y - \ln(A - y) = Akx + C_2$$

where $C_2 = AC_1$;

$$\ln (A - y) - \ln y = - Akx - C_2$$
$$\ln \frac{A - y}{y} = - Akx - C_2$$

$$\frac{A - y}{y} = e^{-Akx-C_2} = e^{-C_2} \cdot e^{-Akx} = ce^{-Akx}$$

where $c = e^{-C_2} > 0$;

$$\frac{A}{y} - 1 = ce^{-kx} \qquad \frac{A}{y} = 1 + ce^{-kx}$$

$$\frac{y}{A} = \frac{1}{1 + ce^{-kx}} \qquad y = \frac{A}{1 + ce^{-kx}}$$

where c is a positive constant.

Although the solution of the differential equation in Case III is a bit more complicated than the first two, the final result is obtained essentially by using straightforward algebraic manipulations and basic laws of logs and exponents. Justifying each step above will help you review both of these.

NOTES:

(1) As noted earlier, numerous applications of these three differential equations were presented in Goal 36. It will be instructive at this point to reread the examples given there.

(2) In Case II, we have

$$y = A - ce^{-kx} \qquad \text{or} \qquad y = A + ce^{-kx}$$

depending on whether y increases or decreases. In either case, as $x \to \infty$, $y \to A$. Since the differential equation in Case II is

$$y' = k(A - y) \qquad \text{or} \qquad y' = -k(y - A)$$

we note that as $y \to A$, $y' = dy/dx \to 0$. Therefore the limiting value of the quantity y is obtained simply by solving $y' = 0$.

(3) In Case III, similarly,

$$y = \frac{A}{1 + ce^{-Akx}} \qquad \text{and} \qquad y' = ky(A - y)$$

As $x \to \infty$, $y \to A$ and $y' \to 0$. So again, the solution of the equation $y' = 0$ yields the limiting value for the quantity y.

Goal 50 | To Apply Separation of Variables

(4) We are now in a position to solve any differential equation for which Case I, II, or III is a mathematical model (if we want to!). The justification for solving the equation rather than using the known solution is to offer practice that will provide further skill and insight in solving differential equations.

Example 50-5

The rate at which a certain drug diffuses from the injection site into the bloodstream satisfies the differential equation $dy/dt = 10 - 2y$, where y is in milligrams and t in minutes. If at time $t = 0$ there is no drug in the bloodstream, how much is in the blood after 1 min?

We have $dy/dt = 10 - 2y$ (which case is this?). So

$$\frac{dy}{10 - 2y} = dt \qquad \int \frac{dy}{10 - 2y} = \int dt \qquad -\frac{1}{2} \ln(10 - 2y) = t + C_1$$

$$\ln(10 - 2y) = -2t + C_2 \qquad \text{(where } C_2 = -2C_1\text{)}$$

$$10 - 2y = e^{-2t + C_2} = C_3 e^{-2t} \qquad \text{(where } C_3 = e^{C_2}\text{)}$$

$$2y = 10 - C_3 e^{-2t}$$

$$y = 5 - c e^{-2t} \qquad \text{(where } c = \tfrac{1}{2} C_3, \text{ a positive constant)}$$

Note how we repeatedly simplify the arbitrary constant as the algebraic simplification of the solution proceeds. To determine c, we note that when $t = 0$, $y = 0$; so $0 = 5 - ce^0 = 5 - c$, and $c = 5$. Thus the solution of the given differential equation is

$$y = 5 - 5e^{-2t} = 5(1 - e^{-2t})$$

When $t = 1$, $y = 5(1 - e^{-2}) \approx 4.3$ mg.

Example 50-6

If a falling feather is subject only to gravity, then its velocity increases (downward) at a constant rate of 32 ft/sec². Suppose the feather is also subject to air resistance (acting in an upward direction), and this is equal at any time to 16 times the velocity at that time. What is the limiting velocity of the feather?

The feather's velocity satisfies the differential equation

$$\frac{dv}{dt} = 32 - 16v$$

This is an example of Case II, with v and t instead of y and x. Note (2) tells us that as $t \to \infty$, $dv/dt \to 0$. We set dv/dt equal to zero and solve for v:

$$32 - 16v = 0 \qquad v = 2$$

Therefore 2 ft/sec is the limiting velocity.

Example 50-7

Ecologists believe that the world's human population as a function of time is logistic (see p. 351) and that it satisfies the differential equation (for restricted or inhibited growth)

$$\frac{dP}{dt} = (0.029 - 3 \times 10^{-12} P) \, P$$

If this is so, what is the limiting value of the size of Earth's human population?

If we recognize that the differential equation is an example of Case III, then Note (3) tells us that we can answer the question quickly. As $t \to \infty$, $dP/dt \to 0$. We therefore set dP/dt equal to zero and solve for P:

$$P = \frac{0.029}{3(10^{-12})} \approx 9.67 \text{ billion}$$

Since the population was about 3.5 billion in 1970, we have growing pains ahead.

FURTHER APPLICATIONS

There are many applications of differential equations other than those found in the three cases above. The following two examples also use the technique of separation of variables.

Example 50-8

The differential equation

$$\frac{dy}{dx} = \frac{1}{3} y(1-y)(2y-1)$$

is an example of a type that arises repeatedly in the study of genetics. Find a relationship between y and x that does not involve derivatives.

Rewriting the equation with differentials, then separating variables, gives us

$$\frac{dy}{y(1-y)(2y-1)} = \frac{dx}{3}$$

Goal 50 | To Apply Separation of Variables

The integration of the left side, as in Case III earlier, depends on the use of an algebraic identity:*

$$\frac{1}{y(1-y)(2y-1)} = -\frac{1}{y} + \frac{1}{1-y} + \frac{4}{2y-1}$$

(Verify this!) So we have

$$\int \left(-\frac{1}{y} + \frac{1}{1-y} + \frac{4}{2y-1}\right) dy = \int \frac{dx}{3}$$

$$-\ln y - \ln(1-y) + 2\ln(2y-1) = \frac{x}{3} + C_1$$

provided y, $1-y$, and $2y-1$ are all positive. Then (justify each step):

$$\ln \frac{(2y-1)^2}{y(1-y)} = \frac{x}{3} + C_1$$

$$\frac{(2y-1)^2}{y(1-y)} = ke^{x/3}$$

Example 50-9

In psychology, one typical stimulus-response situation, known as *logarithmic response*, is that in which the response y changes at a rate inversely proportional to the stimulus x. Suppose $dy/dx = 2/x$, $x \geq x_0$, and that there is no response (that is, $y = 0$) when $x = x_0$. Determine y as a function of x.

$$\frac{dy}{dx} = \frac{2}{x} \qquad dy = \frac{2}{x} dx \qquad \int dy = \int \frac{2}{x} dx \qquad y = 2\ln x + C$$

Using the boundary condition we have

$$2\ln x_0 + C = 0 \qquad C = -2\ln x_0 \qquad y = 2\ln x - 2\ln x_0 = 2\ln \frac{x}{x_0}$$

---

*It is always possible theoretically to express a rational function $P(x)/Q(x)$, where the degree of P is less than that of Q, as a sum of so-called *partial fractions* whose denominators are distinct or repeated linear or quadratic factors. See, for example, George B. Thomas, Jr., *Calculus and Analytic Geometry* (Reading, Mass.: Addison-Wesley, 1972), pp. 371 ff.

HOMOGENEOUS FIRST-ORDER LINEAR DIFFERENTIAL EQUATIONS

A *first-order linear differential equation* is one of the form

$$\frac{dy}{dx} + a(x)\, y = b(x)$$

where $a(x)$ and $b(x)$ are functions of x. Some examples are

$$\frac{dy}{dx} + 3x^2 y = x^2 + 2 \qquad \frac{dy}{dx} + (5x - 1) y = e^x \qquad \frac{dy}{dx} + 5y = 2$$

A linear equation is *homogeneous* if the term $b(x)$ is identically zero. Some examples are

$$\frac{dy}{dx} + (x^2 - x + 1) y = 0 \qquad \frac{dy}{dx} + e^x y = 0 \qquad \frac{dy}{dx} + 2y = 0$$

Since the general form of a homogeneous first-order linear differential equation is

$$\frac{dy}{dx} + a(x)\, y = 0$$

it is separable and can be solved easily.

Example 50-10

Solve $dy/dx + y/x = 0$.

$$\frac{dy}{dx} = -\frac{y}{x} \qquad \frac{dy}{y} = -\frac{dx}{x} \qquad \int \frac{dy}{y} = -\int \frac{dx}{x}$$

$$\ln y = -\ln x + C \qquad \ln (xy) = C \qquad xy = k$$

(Explain the last two steps.)

EXERCISES

You may assume, whenever integrating du/u, that $u > 0$.

1. Find the general solution to each of the following separable differential equations.

Goal 50 | To Apply Separation of Variables

(a) $y^2 \, dy = (x + 1) \, dx$ (f) $\dfrac{dy}{dx} = y + 1$

(b) $e^y \, dy = dx$ (g) $y' = x + 1$

(c) $y' = \dfrac{x^2}{y}$ (h) $(x + 1) \, dy = e^y \, dx$

(d) $\dfrac{dy}{dx} = \dfrac{y - 1}{x}$ (i) $\dfrac{dy}{dx} + y = xy$

(e) $y' - \dfrac{e^x}{y} = 0$ (j) $\dfrac{dy}{dx} - x = 1 + xy + y$

2. Find the particular solution that satisfies the given side condition.
   (a) $y \, dy = (x + 1) \, dx$; $y = 2$ when $x = 0$
   (b) $y' = xy$; $y = 1$ when $x = 2$
   (c) $e^x y' = 2y$; $y = 1$ and $x = 0$
   (d) $\dfrac{dy}{dx} = x^2 y^2$; $y = 3$ when $x = 2$

3. Find the general solution to each of the following homogeneous first-order linear differential equations.

   (a) $\dfrac{dy}{dx} + (x^2 + 1) y = 0$ (c) $y' = y/x$

   (b) $\dfrac{dy}{dx} - (x^3 + x + 2) y = 0$ (d) $x^2 y' = -y$

4. Solve the differential equation
$$5\dfrac{dy}{dx} = (y + 3)(2 - y)$$
given that
$$\dfrac{5}{(y + 3)(2 - y)} = \dfrac{1}{y + 3} + \dfrac{1}{2 - y}$$

5. Do the following exercises in Goal 36: (a) 7, (b) 16, (c) 20, (d) 21, (e) 25

6. Suppose that the yearly marginal profit $P(t)$ is increasing at the constant rate of $100 per yr. If a resistance factor is introduced equal to 5% of the current marginal profit, then $dP/dt = 100 - 0.05P$. As $t \to \infty$, what is the limiting value of the marginal profit?

7. In certain stimulus-response situations the rate of change of the response y is both directly proportional to the current response level and inversely proportional to the stimulus x. Find the general solution

to the resulting differential equation. (Compare this resulting power formulation for stimulus-response with the logarithmic response of Example 50-9.)

8. Find the solution of the differential equation $dy/dx = -0.7y \ln y$ whose graph passes through the point $(0, 0.2)$. This is an example of a *Gompertz curve*. HINT: An antiderivative of $dy/(y \ln y)$ is $\ln \ln y$.

9. A common differential equation arising in the study of genetics is $y' = -0.01y^2 (1 - y)$, where y is the gene frequency, as a function of time, of a particular characteristic of an animal. Using the algebraic identity

$$\frac{1}{y^2(1-y)} = \frac{1}{y} + \frac{1}{y^2} + \frac{1}{1-y}$$

find the general solution of this equation.

10. Suppose that the population of a particular country, as a function of time, satisfies the differential equation $dP/dt = rP + I$, where r and I are constants measuring the birthrate and immigration rate respectively. Solve for P as a function of r, I, and an initial population P_0.

TEST A

Solve each of the following differential equations.

1. $y' = xy^2$
2. $y' = \dfrac{e^x}{y}$
3. $\dfrac{dy}{dx} = (x^2 - 3x + 2)y;\ y = 1$ when $x = -1$
4. $\dfrac{dy}{dx} = 2 - y;\ y = 1$ when $x = 2$

TEST B

Solve each of the following differential equations.

1. $y' - (x^5 + x)y = 0$
2. $dy/dx = e^x y^5$
3. $xy' = 2/y;\ y = 3$ when $x = e$
4. $dy/dx = -3(y - 4);\ y = 7$ when $x = 0$

Goal 50 | To Apply Separation of Variables

STUDY QUESTIONS

1. If dy/dx can be expressed as a function of y/x, then the substitution $u = y/x$, which yields $y = xu$ and $dy/dx = x(du/dx) + u$, changes the original equation into a separable one. For example, if

$$\frac{dy}{dx} = \frac{xy - y^2}{x^2}$$

then

$$\frac{dy}{dx} = \frac{y}{x} - \frac{y^2}{x^2}$$

and the substitutions described above give

$$x\frac{du}{dx} + u = u - u^2$$

$$x\frac{du}{dx} = -u^2$$

Complete this solution, and substitute "back"—that is, replace u by y/x—showing that the final solution is $x/y + C = \ln x$.

2. Apply the method of Study Question 1 to each of the following equations:

   (a) $\dfrac{dy}{dx} = \dfrac{y^2 - x^2}{xy}$ (c) $y' = e^{y/x} + \dfrac{y}{x}$

   (b) $\dfrac{dy}{dx} = \dfrac{x^2 y - y^3}{x^3}$

Summary of Goals 44–50

You should now be able to

(1) draw a rough sketch of the area between two given curves, express the area as a definite integral, and evaluate it;

(2) given the velocity or acceleration of a moving particle P and appropriate boundary conditions, find the position function of P;

(3) given the derivative of a function f and a boundary condition, find f;

(4) solve practical problems calling for a function or for its value at some specified time, given its rate of change and appropriate data;

(5) set up and evaluate an appropriate integral for the average (or mean) value of a continuous function over a specified interval;

(6) find the equilibrium point, the consumers' surplus, and the producer's surplus for a particular product, given the supply and demand functions;

(7) find the present value of

    (a) money worth \$$A$ in t yr, assuming it earns interest at a given rate, compounded continuously;

    (b) future periodic income;

    (c) future continuous income;

(8) determine whether a given function is a probability density function over a specified domain;

(9) find the probability that a variable x falls between a and b, given that $f(x)$ is a probability density function for x (or, equivalently, find the proportion of a specified population for which x is between a and b);

(10) state the order of a differential equation;

(11) verify that a given function is a solution of a given differential equation;

(12) find the general solution (or the particular solution for given side conditions) of a given differential equation;

(13) solve a given differential equation by the technique of separation of variables;

(14) apply the techniques presented for solving a differential equation arising from a practical problem.

Review Exercises

GOAL 44

1. Sketch the region bounded by the curves $y = 3x - x^2$ and $y = x^2 - 2x$. Find the area between the curves from $x = 0$ to $x = 2$.
2. Draw a sketch, then find the area between $y = x + 6$ and $y = x^2$.

GOAL 45

3. Find the position function $f(t)$ for a moving particle P if
   (a) its velocity is $v(t) = 3t^2 - 2t + 1$ and $f(1) = 5$.
   (b) its acceleration is $a(t) = 2e^{-t}$ and $f(0) = 3$, $f(-1) = e + 2$.
4. What initial velocity is needed in throwing a stone upward so that it will reach the height of 64 ft? HINT: $a(t) = -32$.
5. Suppose there are 50 units of a radioactive substance and it is decomposing at the rate of $5e^{-0.1t}$ units per day after t days. How much of the substance is left after 2 days?

GOAL 46

6. Find the average value of the function $f(x) = x^2 - x + 2$ over the interval $1 \leq x \leq 4$.
7. If the net value of an investment is $2000e^{0.05t}$ dollars after t yr, what is the average net value of the investment during the second year?

GOAL 47

8. Find the revenue function for a certain product if the manufacturer uses $m(x) = 14 - 0.006x$ as its marginal revenue function, in dollars per unit, at production level x, and if the manufacturer has no revenue at production level 0.
9. Find (a) the equilibrium point, (b) the consumers' surplus, and (c) the producer's surplus if the demand and supply functions (both in dollars per item) are given by $D(x) = -0.003x + 4$ and $S(x) = 0.0002x$.
10. The administrator of a new county medical clinic estimates that the clinic will receive 50 new patients per month for many years and that the fraction of patients who will continue to make monthly visits for

at least m months is $R(m) = e^{-m/25}$. (a) How many patients will the clinic have after 2 yr? (b) What limiting size will the patient load approach?

11. How much must the parents of a newborn child invest now at 6% per yr compounded continuously

   (a) to have $20,000 available on the child's eighteenth birthday?

   (b) if they want to give their child $5000 on his eighteenth, nineteenth, twentieth, and twenty-first birthdays?

12. If a new firm anticipates yearly profits of $3000, to be received continuously and at a constant rate, what is the present value of the first 5 yr profits, assuming an interest rate of 6% compounded continuously?

GOAL 48

13. Show that (a) $f(x) = \frac{1}{13}(3x^2 + 4x)$ is a probability density on the interval $1 \leq x \leq 2$; (b) $g(x) = 0.1e^{-0.1x}$ is a probability density on $0 \leq x < \infty$.

14. If a fast-food chain estimates that the amount x of money, in dollars, spent by customers per visit has a probability density $\frac{2}{(x+2)^2}$, (a) what proportion of customers spend no more than $5? (b) what is the probability that a customer chosen at random spends more than $1?

GOAL 49

15. What is the order of each differential equation?

   (a) $\left(\dfrac{dy}{dx}\right)^2 = 4x$ (b) $y'' - 4y' = x$

16. (a) Verify that $Ce^{2x} - e^{-2x}$ is a solution of $y'' = 4y$. (b) Find C in (a) if $y = 4$ when $x = 0$.

17. Find the general solution:

   (a) $y'' = e^{-x}$ (b) $\dfrac{dy}{dx} = x - \dfrac{1}{x}$

18. Find the particular solution of the differential equation

   $$y'' = 4x^2 - 3x$$

   for which $f(0) = -1$, $f(2) = \frac{1}{3}$.

GOAL 50

19. Find the general solution:

   (a) $\dfrac{dy}{dx} = \dfrac{4x}{2y+1}$ (b) $dy/dx + xy = 0$

20. Find the particular solution of the differential equation $\dfrac{dy}{dt} = -3y$ for which $y = 2$ when $t = 0$.

21. A colony of fruit flies is growing at a rate proportional to its size. If there are now 500 fruit flies and the growth rate is 30% per day, find an expression for the size P of the colony in t days.

Chapter Test A

1. Find the area between the curves $y = 2x - x^2$ and $y = -x$.
2. Set up, but do not evaluate, a definite integral for the area bounded by the x-axis, the lines $x = 1$ and $x = 5$, and the line $y = \tfrac{1}{2}x - 4$.
3. A particle moving on a line has acceleration $a(t) = 6t - 2$. Find its initial velocity and position if when $t = 1$ its velocity and position are 4 and -2 respectively.
4. An investment of $4000 grows at the rate of $320e^{0.08t}$ dollars per yr after t yr. What is it worth after 10 yr?
5. Find the average value of the function $y = e^{-x}$ on the interval $0 \leq x \leq 2$.
6. A company's maintenance cost for equipment and buildings is equal to $(400t + 9t^2 + 500)$ dollars per month after t months of operation. What is the average monthly cost during the first 2 yr?
7. Find (a) the equilibrium point, (b) the consumers' surplus, and (c) the producer's surplus for the demand and supply functions (in dollars per unit) $D(x) = -0.004x + 9$ and $S(x) = 0.0005x$.
8. A new national credit card company assumes that it can attract 200 new customers per month indefinitely and that the fraction of customers who will continue to use their cards for at least m months is $R(m) = e^{-(m/10)}$. (a) How many customers will the company have 6 months after starting? (b) If there is a limit to the number of customers, what is it?
9. If a father wants his college-age daughter to have an allowance of $2000 per yr for the next 4 yr, how much must he invest now if the bank will pay him 7% compounded continuously? Assume that the allowance is to be paid out continuously and at a constant rate.
10. How much must a 25-year-old invest now at 8% per yr compounded continuously to have $40,000 when he is 60?
11. Show that $f(x) = 2e^{-2x}$ is a probability density function over $0 \leq x < \infty$.
12. A TV-set manufacturer figures that the life expectancy of its picture tubes is between 2 and 6 yr and that the probability density function

for the life expectancy is $f(x) = \frac{1}{8}(x - 2)$ (for $2 \le x \le 6$). What is the probability that a tube chosen at random will fail during the third year?

13. Verify that Ce^{-2x} is a solution of $y'' = 4y$.

14. Find the general solution for each equation:

   (a) $\dfrac{d^2y}{dx^2} = 4x - 3$ (c) $\dfrac{dy}{dt} = ky$

   (b) $\dfrac{dy}{dx} = \dfrac{x}{y}$

15. Solve each equation, expressing y as a function of x:

   (a) $y'' = \dfrac{1}{x^2}$, $f(1) = 2$, $f(e) = 1$

   (b) $\dfrac{dy}{dx} = 1 - y$, $y = 0$ when $x = 2$

16. A substance is decaying at a rate proportional to the quantity present. If there were 10 oz initially and the decay rate (factor of proportionality) is 0.1% per yr, find the amount left after t yr.

Chapter 9

Multivariable Calculus

So far in this book all the theory and applications we have considered have involved functions of a single variable. Actually, however, almost everything with which we deal in the world is a function of many variables. In this chapter we extend some of the theory so far presented to functions of two or more variables. We begin with definitions, then graph some simple functions of two variables. To measure the rate of change of such a function or to find where it has a maximum or minimum value, we introduce its partial derivatives. Finally we present the method known as Lagrange multipliers for finding extreme values; it is used when we know how the independent variables are related.

Of special interest in this chapter is the way in which we generalize one-variable concepts, such as graphs, derivative, differential, extreme value, and the like. In both theory and applications involving functions of more than one variable, further insight is often attained by keeping in mind the single-variable parallels.

Goal 51. To understand functions of several variables
Goal 52. To find and apply partial derivatives
Goal 53. To find extreme values of functions of two variables
Goal 54. To find maxima or minima using Lagrange multipliers
Summary
Review Exercises
Chapter Test A

GOAL **51**

To understand functions of several variables

In this Goal we introduce functions of two or more variables, and their graphs. In particular, we consider, as an example, demand functions involving the prices of two related commodities. We begin with examples.

Example 51-1

When the Original Reproductions Company turned out x oil masterpieces at a profit of $500 each, their weekly profit function was $P(x) = 500x$. Now as a sideline they also whip off y finger paintings a week at a profit of $15 each. What is the new profit function?

The profit is now a function of the *two* variables x and y:

$$P(x, y) = 500x + 15y$$

For example, if 3 masterpieces and 20 finger paintings are fabricated in one week, then the profit is $P(3, 20) = 500(3) + 15(20) = \$1800$.

Example 51-2

The number M of matings per day among a certain population of dolphins is dependent both upon x, the number of dolphins, and y, the temperature (Celsius) of the water:

$$M(x, y) = \frac{1}{2500} xy(30 - y) \qquad x \geq 0, \ 0 < y < 30$$

Thus, in a colony of 100 dolphins living in 10° C water, the number of matings per day is $M(100, 10) = \dfrac{1}{2500} 100(10)(30 - 10) = 8$.

Example 51-3

The value A after n yr of an investment of P dollars at a yearly rate of interest r, compounded continuously, is a function of the three variables P, r, and n:

$$A(P, n, r) = Pe^{rn}$$

So $100 invested at 6% compounded continuously for 5 yr accumulates to

$$A(100, 5, 0.06) = 100e^{0.06(5)} \approx \$135$$

DEFINITIONS

We say that f is a (real-valued) function of two variables if f associates a unique real number with each pair (x, y) of real numbers in a set D; D is called the *domain* of the function.

Similarly, f is a function of three variables if f associates a real number with each triple (x, y, z) of real numbers in its domain.

Here are some functions of two variables:

$$f(x, y) = \frac{1}{xy} \qquad g(x, y) = e^x - y \qquad h(x, y) = 21$$

Here are some functions of three variables:

$$f(x, y, z) = xy + z \qquad g(x, y, z) = xyz \qquad h(x, y, z) = 3x + 1$$

If the domain of a function is not specified, we take it to be all pairs (or triples) of real numbers for which the function is defined.

Examples 51-4

(a) The domain of $f(x, y) = xy$ is the set of all pairs of real numbers (x, y).

(b) The domain of $g(x, y) = x/y$ is the set of all pairs for which $y \neq 0$.

(c) The domain of $h(x, y, z) = \dfrac{1}{y - 3} + \sqrt{z}$ is the set of all triples (x, y, z) for which both $y \neq 3$ and $z \geq 0$. For example, $(3, 4, 5)$ and $(0, -1, 0)$ are in D, but $(4, 3, 2)$ and $(1, -4, -4)$ are not.

(d) The domain of $p(x, y) = \sqrt{1 - x^2 - y^2}$ is the set of all pairs for which $x^2 + y^2 \leq 1$.

GRAPHS OF FUNCTIONS OF TWO VARIABLES

We defined the graph of $y = f(x)$ as the set of all points (x, y) in the plane that satisfy the equation $y = f(x)$. Similarly, the graph of $z = f(x, y)$ is the

Goal 51 | **To Understand Functions of Several Variables**

set of all points (x, y, z) in three-dimensional space that satisfy the equation $z = f(x, y)$.

One way of visualizing a three-dimensional coordinate system is to take a corner of a room, on the floor, as the origin, let the floor be the xy-plane, and let the two walls be the xz- and yz-planes. Then to locate, say, the point

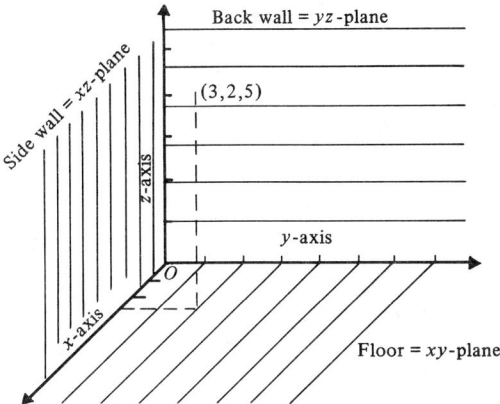

(3, 2, 5), we move 3 units (in the positive direction) along the x-axis, 2 units parallel to the y-axis, and 5 units up. This point lies in the *first octant*, in which every point has all three coordinates positive. To graph $z = f(x, y)$ we plot a point z units above (or below) each point (x, y) in the domain of f.

The graph of $y = f(x)$, as we know, is usually a curve in the xy-plane. The graph of $z = f(x, y)$ is most often a *curved surface* in three-dimensional space. Although it is generally quite difficult to graph a function of two variables, we will do two simple examples to provide some insight into the techniques used. In particular, it is frequently very helpful to determine the *traces* of a surface: its intersections with the coordinate planes. To find the trace of a surface in the yz-plane, for example, we set x equal to zero.

Example 51-5

Graph $z = f(x, y) = x^2 + y^2$.
The traces of the surface are as follows:

in the xy-plane ($z = 0$): the origin $(0, 0, 0)$;

in the yz-plane ($x = 0$): the parabola $z = y^2$;

in the xz-plane ($y = 0$): the parabola $z = x^2$.

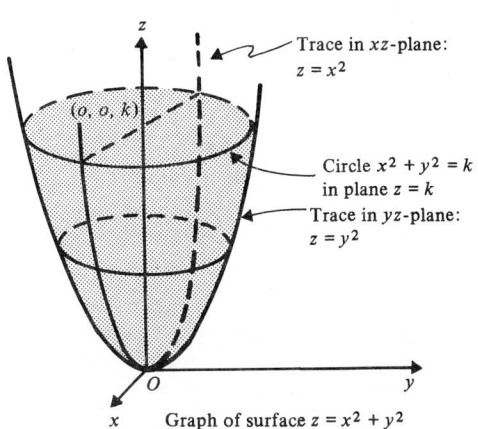

Graph of surface $z = x^2 + y^2$

Note also that, if z is a positive constant, say k, the curve in the plane $z = k$ is the circle $k = x^2 + y^2$. So every cross-section parallel to the xy-plane is a circle with center on the z-axis. There is no surface below the xy-plane (why?), but the surface extends infinitely upward (why?). This surface is called a *circular paraboloid*.

Example 51-6

Graph $z = f(x, y) = \sqrt{25 - x^2 - y^2}$.

The traces are as follows:

in the xy-plane, $0 = \sqrt{25 - x^2 - y^2}$ or $x^2 + y^2 = 25$: a circle of radius 5;

in the yz-plane, $z = \sqrt{25 - y^2}$, the upper half only of the circle $y^2 + z^2 = 25$ (why?);

in the xz-plane, the upper half only of the circle $x^2 + z^2 = 25$ (why?).

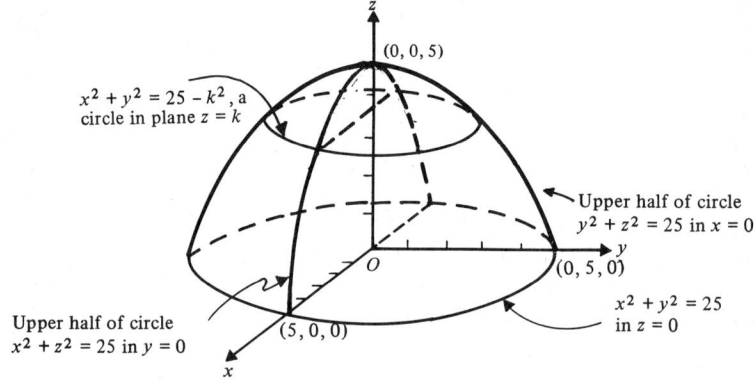

Graph of surface $z = \sqrt{25 - x^2 - y^2}$

Note that $0 \leq z \leq 5$ (why?) and that the surface intersects every plane $z = k$ ($0 \leq k < 5$) in a circle. The surface is a *hemisphere* with center at the origin and radius 5.

HOLDING ONE VARIABLE FIXED

Suppose, in the preceding example, that we hold x fixed at, say, 3. Then we can regard f as a function of the one variable y:

$$z = f(3, y) = \sqrt{25 - 3^2 - y^2} = \sqrt{16 - y^2}$$

Note that this curve is the intersection of the hemisphere and the plane $x = 3$ and is the upper half of the circle $y^2 + z^2 = 16$. On p. 497 we show only the part of the hemisphere that is in the first octant. We also show part of the semicircle whose equation is

Goal 51 | To Understand Functions of Several Variables

$$f(x, 2) = \sqrt{25 - x^2 - 4} = \sqrt{21 - x^2}$$

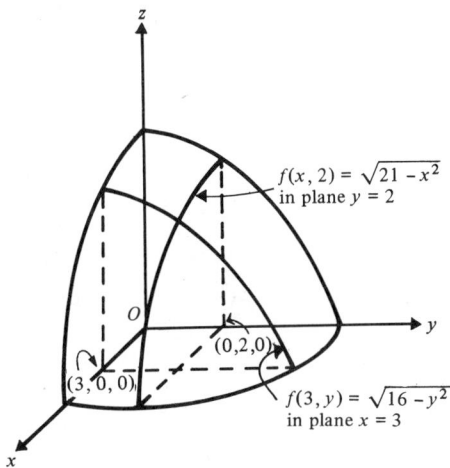

In the latter case we are viewing f as a function of the one variable x. Its graph is in the plane $y = 2$.

LEVEL CURVES

Although graphing a surface is often difficult, we can sometimes obtain useful information about a function of two variables rather easily by sketching its *level curves*.* These are obtained by setting z, or $f(x, y)$, equal to various constants.

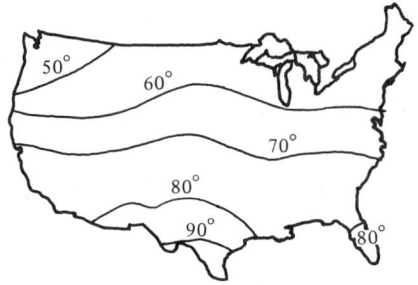

$f(x, y)$ = temperature at (x, y). An *isotherm* is a level curve of constant temperature.

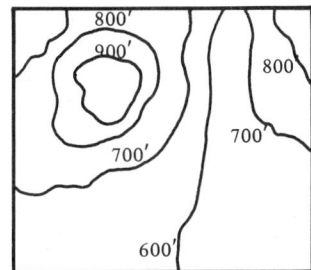

$f(x, y)$ = elevation in feet above sea level at (x, y). Along a level curve every point is at the same elevation.

---

*Also called *contour curves* or *contour lines*, and, in economics, *indifference curves*.

Example 51-7

Sketch some level curves for $z = f(x, y) = x^2 + y^2$.

When $z = 1$, we get $x^2 + y^2 = 1$, the circle of radius 1 centered at (0, 0).
When $z = 4$, we get $x^2 + y^2 = 4$, the circle of radius 2 with center at (0, 0).
When $z = c > 0$, we get the circle $x^2 + y^2 = c$, with center at the origin and radius $\sqrt{c}$.

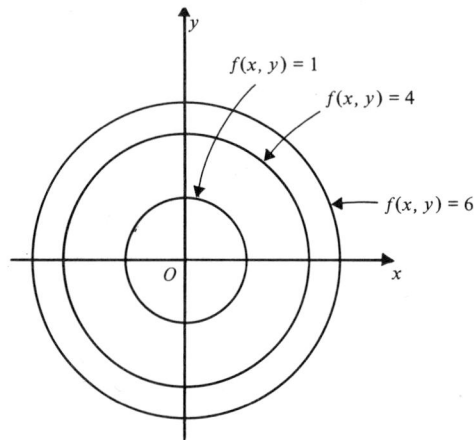

Level curves for $z = x^2 + y^2$

What does $z = 0$ yield? How about $z = -1$?

Note that these level curves are the *projections* in the *xy*-plane of the circular cross-sections of the paraboloid in Example 51-5.

Example 51-8

A new company plans to hire x skilled workers at \$5 per h and y unskilled laborers at \$2 per h. The hourly wages to be paid are therefore given by $W(x, y) = 5x + 2y$, $x \geq 0$, $y \geq 0$. What are the level curves and what do they represent?

Setting $W(x, y)$ equal to, say, \$500 corresponds to budgeting \$500 for hourly wages. The curve, in this case the straight line $5x + 2y = 500$, represents all possible combinations of skilled and unskilled laborers, each of which combinations will cost the company a total of \$500 per h.

On p. 499 we see three level curves for $W(x, y)$, equal to \$500, \$1000, and \$1500. For the last we observe that 300 skilled and no unskilled laborers, no skilled and 750 unskilled laborers, and 200 skilled and 250 unskilled laborers are all combinations that will cost the company \$1500 per h.

Goal 51 | To Understand Functions of Several Variables

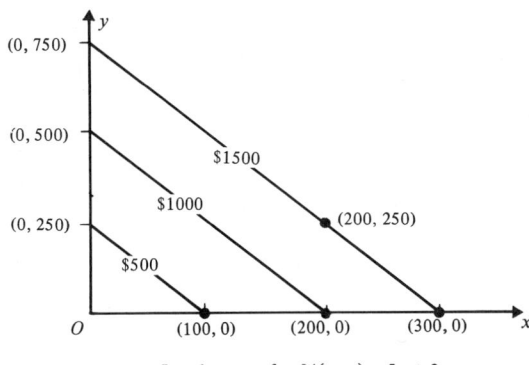

Level curves for $W(x, y) = 5x + 2y$

Example 51-9

Before bottling, the main processing plant of the Weece Queeze Company mixes x oz of orange juice costing 4¢ per oz with y oz of pure water costing 0.5¢ per oz. Thus the cost of this segment of production is $C(x, y) = 0.04x + 0.005y$ (dollars). What are the level curves and what do they represent?

Suppose $C(x, y) = 20$ (dollars). Then the level curve $0.04x + 0.005y = 20$ represents all combinations of orange juice and water that cost the company $20. For $20 the company can use, for example, 500 oz of orange juice and no water, or 4000 oz of water and no orange juice, or some such combination as 300 oz of juice and 1600 of water.

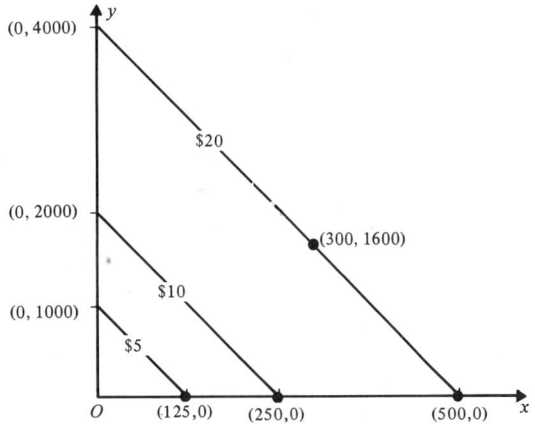

Level curves for $C(x, y) = 0.04x + 0.005y$

DEMAND FUNCTIONS OF TWO VARIABLES

We begin this topic by considering a town with two competing newspapers, the *Journal* and the *Times*. Although each paper has a number of faithful subscribers, there are also many readers whose choice of paper is influenced by price. Let p_J and p_T be the prices, respectively, of the *Journal* and the

Times; let D_J and D_T be the number of readers of each newspaper. We will regard both D_J and D_T as functions of price. It doesn't make sense to express D_J as a function only of p_J (or to express D_T as a function only of p_T) since an increase, say, in the *Journal* price not only causes a drop in the number of its readers but is likely also to bring about an increase in *Times* readers. We therefore conclude that D_J and D_T are both functions of the two variables p_J and p_T.

Example 51-10

Suppose the demand functions of the *Journal* and the *Times* are

$$D_J(p_J, p_T) = 5000 + 10{,}000 p_T - 20{,}000 p_J$$
$$D_T(p_J, p_T) = 4000 - 15{,}000 p_T + 12{,}000 p_J$$

(a) How many readers does each paper now have if the *Journal* sells for 10¢ and the *Times* for 15¢? (b) What is the effect of raising the price of the *Journal* to 12¢?

(a) We have

$$D_J(0.10, 0.15) = 5000 + 10{,}000(0.15) - 20{,}000(0.10) = 4500$$
$$D_T(0.10, 0.15) = 4000 - 15{,}000(0.15) + 12{,}000(0.10) = 2950$$

So at present the *Journal* has 4500 readers and the *Times* 2950.

(b) With the higher price of the *Journal*, we have

$$D_J(0.12, 0.15) = 5000 + 10{,}000(0.15) - 20{,}000(0.12) = 4100$$
$$D_T(0.12, 0.15) = 4000 - 15{,}000(0.15) + 12{,}000(0.12) = 3190$$

So the 2¢ increase in the *Journal* price would result in 400 fewer readers for the *Journal* and 240 more for the *Times*.

COMPETITIVE VERSUS COMPLEMENTARY PRODUCTS

In general, if we have demand functions $D_1(p_1, p_2)$ and $D_2(p_1, p_2)$ for two products, where p_1 and p_2 are their respective prices, then we note especially the following two cases:

CASE I

D_1 increases as p_2 increases, and D_2 increases as p_1 increases. The products involved are said to be *substitute* or *competitive products*. The newspapers in the previous example are competitive. So are radial snow tires manufactured by the two leading companies, or two popular electric razors, or the two best-selling portable color TV sets, and so on.

CASE II

D_1 decreases as p_2 increases, and D_2 decreases as p_1 increases. The products involved are said to be *complementary*. Note that in this case an increase in the price of either product causes the demand for both products to fall. Some examples are tonic water and gin, or skis and ski boots, or golf clubs and golf balls.

EXERCISES

1. In 1977 a discount gasoline station charged 61¢/gal for premium gas, 57¢/gal for regular, and 60¢/gal for no-lead. If x, y, and z represent the number of gallons sold of premium, regular, and no-lead gas respectively, express the revenue R in dollars as a function of the three variables x, y, and z. Calculate and give a physical interpretation of $R(100, 100, 150)$ and of $R(50, 100, 80)$.

2. The Spearman-Brown reliability formula hypothesizes that if the reliability of a test is r, $0 \leq r \leq 1$, and if the test is lengthened by a factor n, then the reliability of the new lengthened version is $f(n, r) = \dfrac{nr}{1 + (n-1)r}$. (a) Determine $f(6, \frac{1}{2})$, $f(4, \frac{1}{3})$, and $f(10, \frac{1}{5})$. (b) What is the effect of doubling the length of a test if the reliability is 0? if it is $\frac{1}{2}$? if it is 1? (c) What is the effect of tripling the length of a test if the reliability is 0? if it is $\frac{1}{2}$? if it is 1?

3. Professor Stewart ("Stu") Pitt figures that the number M of muggings in New York City on a given day is a function both of the highest temperature (x degrees Fahrenheit) reached that day and the number of days y until the next full moon:

$$M(x, y) = \frac{x}{50}(y^2 - 30y + 250)$$

   (a) Calculate $M(75, 6)$, $M(5, 10)$, and $M(90, 15)$. (b) How many muggings should be expected if the temperature reaches 100° F one day before a full moon? if it reaches only 25° F twelve days before a full moon?

4. Determine the domain of each of the following functions.

   (a) $f(x, y) = \dfrac{10x}{y}$
   (b) $f(x, y) = \sqrt{3x - y}$
   (c) $f(x, y) = \dfrac{1}{(x-3)(y-1)}$
   (d) $f(x, y, z) = \dfrac{1}{xyz}$
   (e) $f(x, y, z) = \sqrt{xyz}$
   (f) $f(x, y, z) = e^x \ln y - \sqrt[3]{z}$

5. (a) Find the traces of $y = x^2 + z^2$ and describe cross-sections in planes where $y = k$, $k > 0$. (b) Graph the surface.

6. Let $f(x, y) = y - x^2$. (a) Draw the level curves corresponding to $f(x, y) = 3$, $f(x, y) = 0$, and $f(x, y) = -2$. (b) Identify the curve $f(x, 4)$

and explain the geometric significance of this curve. (c) Evaluate $f(1, 4)$, $f(1 + h, 4)$, and $f(1 + h, 4) - f(1, 4)$. (d) Evaluate $f(2, 5)$, $f(2, 5 + k)$, and $f(2, 5 + k) - f(2, 5)$
(e) Evaluate $f(3 + h, 6 + k) - f(3, 6)$.

7. A power plant burns both coal and oil. Each ton of coal produces 70 lb of sulfur dioxide SO_2 pollution, while each ton of oil produces 180 lb of SO_2 pollution. If x tons of coal and y tons of oil are burned, express the amount of pollution P as a function of x and y. Sketch the level curves corresponding to $P(x, y) = 500$, $P(x, y) = 1260$, and $P(x, y) = 2520$.

8. A manufacturer determines that investing x dollars in labor and y dollars in materials will lead to the production of $10xy$ units of output. Sketch the level curves corresponding to 10,000 and to 50,000 units of output.

9. Tell whether each of the following pairs of demand functions describes products that are competitive, complementary, or neither.

   (a) $D_1(p_1, p_2) = 3000 - 1000p_1 - 300p_2$
       $D_2(p_1, p_2) = 3500 - 400p_1 - 250p_2$

   (b) $D_1(p_1, p_2) = 250 - 10p_1$
       $D_2(p_1, p_2) = 400 - 20p_2$

   (c) $D_1(p_1, p_2) = 10{,}000 - 300p_1 + 400p_2$
       $D_2(p_1, p_2) = 5{,}000 + 100p_1 - 100p_2$

   (d) $D_1(p_1, p_2) = 24 - 3p_1 - 2p_2$
       $D_2(p_1, p_2) = 18 - p_1 - p_2$

10. Classify each of the following pairs of products as to competitive, complementary, or neither.
    (a) chessmen; chessboards
    (b) cleansing tissues; dogfood
    (c) Mercedes-Benz; Cadillac
    (d) TV sets; vacuum cleaners
    (e) baseballs; bats
    (f) newspaper ads; TV commercials

TEST A

1. Suppose that the population P of foxes is a function both of x, the population of rabbits, and y, the number of hunters. If $P(x, y) = 0.05x - 2y$, determine the number of foxes when there are 1,000,000 rabbits and 2500 hunters.

2. What is the domain of $f(x, y, z) = \dfrac{\sqrt{z}}{xy}$

3. Let $f(x, y) = y^2 - x$. Draw the level curves corresponding to $f(x, y) = 1$, $f(x, y) = 0$, and $f(x, y) = -2$.
4. Give an example of two demand functions for products that are complementary.

TEST B

1. If $f(x, y, z) = x + y - 2$, determine $f(0, 2, 4)$ and $f(-1, 0, 500)$.
2. What is the domain of $f(x, y, z) = \dfrac{1}{x\sqrt{y}} - e^z$?
3. A sidewalk vendor sells hot dogs for $0.45 each and hamburgers for $0.80 each. If x and y are the number of hot dogs and hamburgers sold, respectively, express the revenue R as a function of x and y. Sketch the level curves corresponding to a revenue of $72 and to a revenue of $180.
4. Give an example of two demand functions for products that are competitive.

STUDY QUESTIONS

1. (a) Find and sketch the traces of the surface defined by $z = f(x, y) = 6 - 2x - 3y$. Note that the surface is a plane. Any equation of the form $ax + by + cz = d$ (a, b, c not all zero) is a plane.
   (b) Graph the planes: $y = 3$; $x = 1$; $z = 4$; $x + y = 4$; $x + y + z = 6$.

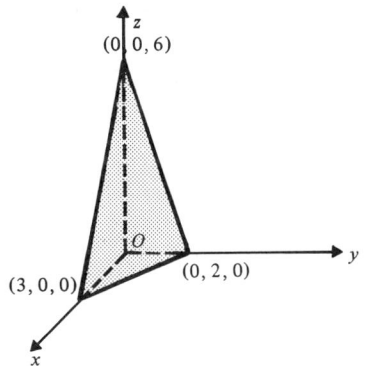

2. (a) Find the intersections of the graph of the three-dimensional surface $x^2 + y^2 = 4$ with the xy-plane, the plane $z = 1$, the plane $z = 5$, the plane $z = j$. What is the effect of the absent variable z?
   (b) Graph $x^2 + y^2 = 9$ in space.
   (c) Graph $y = x^2$ in space.

3. In Example 51-9 the company would also be interested in the production function $P(x, y) = x + y$. Sketch the level curves $P(x, y) = 650$ oz and $P(x, y) = 1900$ oz on the same graph on which are sketched the level curves $C(x, y) = \$5$ and $C(x, y) = \$20$. Determine and interpret the points of intersection.

GOAL 52

To find and apply partial derivatives

Measuring rates of change is just as important for functions of several variables as it is for those of a single variable. In this Goal we define partial derivatives and the total differential, both of which are generalizations of the corresponding one-variable concepts; we then consider some applications.

PARTIAL DERIVATIVES

Consider the function $z = f(x, y) = x - x^2y + y^3 - 5$. Suppose we hold y fixed at 2. Then, as noted in the preceding Goal, we can view f as a function of the one variable x:

$$f(x, 2) = x - x^2(2) + 2^3 - 5 = x - 2x^2 + 3$$

In the figure we see part of the graph of $f(x, 2)$, a parabola in the plane $y = 2$.

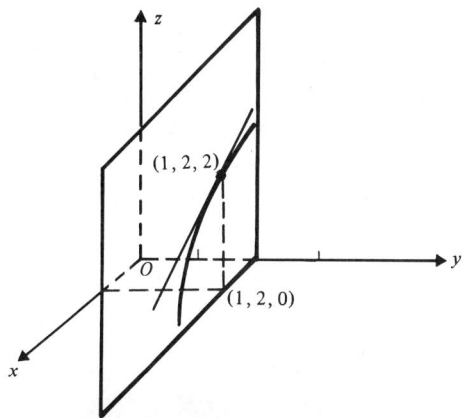

$f_x(1, 2)$ is the slope of the tangent to the curve in the plane $y = 2$ at the point where $x = 1$.

What is the rate of change of f with respect to x, if y is fixed at 2? The answer is just the derivative of the centered equation above with respect to x:

$$1 - 4x$$

When $x = 1$, for example, this rate of change equals $1 - 4(1)$, or -3. This derivative is denoted by any of the following:

$$\frac{\partial z}{\partial x}\bigg|_{(1,2)} \qquad \frac{\partial f}{\partial x}\bigg|_{(1,2)} \qquad f_x(1, 2)$$

which are read "the partial derivative of z (or f) with respect to x at the point (1, 2)." The figure shows that we can interpret $f_x(1, 2)$ geometrically as the slope of the tangent to the curve (the intersection of the surface with the plane $y = 2$) at the point where $x = 1$ (and $y = 2$). Note that when $x = 1$ and $y = 2$, $z = 2$.

In the preceding discussion we held y fixed at 2. We might have chosen y to be 3, or -2, or 17. In any case, we are interested in the rate of change of f with respect to x when y is held fixed. This is the *partial derivative of f with respect to x* and is obtained by differentiating f with respect to x while regarding y as a constant. The symbol we use, in the general case, is

$$\frac{\partial f}{\partial x} \quad \text{or} \quad \frac{\partial z}{\partial x} \quad \text{or} \quad f_x(x, y) \quad \text{or} \quad f_x$$

For the function considered above, $z = f(x, y) = x - x^2y + y^3 - 5$, we get

$$\frac{\partial f}{\partial x} = 1 - 2xy$$

To verify that this is correct, think of y as 7 or 68 or just some constant k. Note that y^3 is then also a constant; it therefore has derivative 0.

Since $z = f(x, y)$ is a function both of x and of y, we can also find $\partial f/\partial y$, the *partial derivative of f with respect to y*. To do this we merely change our perspective, hold x fixed (treat it as a constant), and differentiate with respect to y.

Example 52-1

For $z = f(x, y) = x - x^2y + y^3 - 5$, find

(a) $\dfrac{\partial f}{\partial y}$ (b) $f_y(0, -2)$ (c) $\dfrac{\partial z}{\partial y}\bigg|_{(4, -2)}$

Goal 52 | To Find and Apply Partial Derivatives

Solution:

(a) $\dfrac{\partial f}{\partial y} = -x^2 + 3y^2$ (Why?)

(b) $f_y(0, -2) = -0^2 + 3(-2)^2 = 12$

(c) $\dfrac{\partial z}{\partial y}\bigg|_{(4, -2)} = -4^2 + 3(-2)^2 = -4$

DEFINITIONS

In terms of limits, we define the partial derivatives of f as follows:

with respect to x: $\quad \dfrac{\partial f}{\partial x} = \lim\limits_{h \to 0} \dfrac{f(x + h, y) - f(x, y)}{h}$

with respect to y: $\quad \dfrac{\partial f}{\partial y} = \lim\limits_{k \to 0} \dfrac{f(x, y + k) - f(x, y)}{k}$

Note the similarity to the definition on page 88 of the derivative of a function of one variable.

Example 52-2

If $f(x, y) = x + y^2 - x^3 y^2$, determine

$$\dfrac{\partial f}{\partial x} \qquad \dfrac{\partial f}{\partial y} \qquad \dfrac{\partial f}{\partial x}\bigg|_{(1, 3)} \qquad \dfrac{\partial f}{\partial y}\bigg|_{(0, 1)}$$

We have

$\dfrac{\partial f}{\partial x} = 1 - 3x^2 y^2$ (remember that y^2 is viewed as a constant here!)

$\dfrac{\partial f}{\partial y} = 2y - 2x^3 y$ (both x and x^3 are regarded as constant)

$\dfrac{\partial f}{\partial x}\bigg|_{(1, 3)} = 1 - 3(1)^2 (3)^2 = -26 \qquad \dfrac{\partial f}{\partial y}\bigg|_{(0, 1)} = 2(1) - 2(0)^3(1) = 2$

Examples 52-3

Find $f_x(x, y)$ and $f_y(x, y)$ if

$$f(x, y) = \ln x + e^{xy} + (x^2 + 1)(y^3 - y)$$

We have

$$f_x(x, y) = \frac{1}{x} + ye^{xy} + 2x(y^3 - y)$$
$$f_y(x, y) = xe^{xy} + (x^2 + 1)(3y^2 - 1)$$

Example 52-4

The Madison Avenue Ad Agency knows that if a client spends x dollars on TV commercials and y dollars on newspaper ads, then this advertising will generate sales of

$$S(x, y) = (3x + 2y + 0.0001xy) \text{ dollars}$$

Suppose that currently $25,000 is spent on TV commercials and $10,000 on newspaper ads each month. What effect will a change in x or y have on sales?

To determine the effect on sales caused by a change in x (viewing y as fixed) we find $\partial S/\partial x$, then evaluate this partial derivative at $x = 25{,}000$ and $y = 10{,}000$.

$$\frac{\partial S}{\partial x} = 3 + 0.0001y \qquad \frac{\partial S}{\partial x}\bigg|_{(25\,000,10\,000)} = 3 + 0.0001(10{,}000) = 4$$

The procedure for finding the effect on sales caused by a change in y is similar:

$$\frac{\partial S}{\partial y} = 2 + 0.0001x \qquad \frac{\partial S}{\partial y}\bigg|_{(25\,000,10\,000)} = 2 + 0.0001(25{,}000) = 4.5$$

These answers may be interpreted as follows: An increase of $1 for TV commercials will produce an increase of approximately $4 in sales, whereas an increase of $1 for newspaper ads will produce an increase of approximately $4.50 in sales.

Example 52-5

Paul Douglas, formerly a senator from Illinois, was an economist by profession. He and his colleague Cobb developed a formula for measuring the productivity P of all manufacturing in the United States in terms of the investment in labor and the investment in capital (that is, in all capital goods, including machinery, equipment, buildings, and so on). Their formula (also called a *production* function) is

$$P(x, y) = 1.01x^{0.75}y^{0.25}$$

Goal 52 | **To Find and Apply Partial Derivatives**

where x and y are measures of the investments in labor and capital respectively. Find $\partial P/\partial x$, the *marginal productivity of labor*, and $\partial P/\partial y$, the *marginal productivity of capital*.

$$\frac{\partial P}{\partial x} = 1.01\,(0.75)x^{-0.25}y^{0.25} = 0.7575\left(\frac{y}{x}\right)^{0.25}$$

$$\frac{\partial P}{\partial y} = 1.01x^{0.75}\,(0.25)y^{-0.75} = 0.2525\left(\frac{x}{y}\right)^{0.75}$$

FUNCTIONS OF MORE THAN TWO VARIABLES

The partial derivatives of a function f of more than two variables are defined and calculated in the same manner as are those of a function of two variables. All the variables except one are held constant, f is viewed as a function of the single remaining variable, and f is then differentiated with respect to that variable.

Example 52-6

Let $f(x, y, z) = x + y^2 + xyz$. Find the partials of f with respect to x, y, and z.

$$f_x = \frac{\partial f}{\partial x} = 1 + yz \qquad f_y = \frac{\partial f}{\partial y} = 2y + xz \qquad f_z = \frac{\partial f}{\partial z} = xy$$

Example 52-7

An accepted measurement of the proper functioning of a person's kidneys involves the function $f(x, y, z) = xy/z$, where x = milligrams of urea per 100 ml of urine, y = milliliters of urine excreted per minute by the kidneys, and z = milligrams of urea per 100 ml of blood; f is the number of milliliters of urea removed from the blood per minute. Determine the three partial derivatives of f when $x = 300$, $y = 10$, and $z = 3$.

$$f_x(x, y, z) = \frac{y}{z} \qquad f_x(300, 10, 3) = \frac{10}{3}$$

$$f_y(x, y, z) = \frac{x}{z} \qquad f_y(300, 10, 3) = 100$$

$$f_z(x, y, z) = -\frac{xy}{z^2} \qquad f_z(300, 10, 3) = -\frac{1000}{3}$$

PARTIAL DERIVATIVES OF HIGHER ORDER

Let f be a function of the two variables x and y. Since $\dfrac{\partial f}{\partial x}$ and $\dfrac{\partial f}{\partial y}$, if they exist, are themselves functions of several variables, it makes sense to talk about the partial derivatives of $\dfrac{\partial f}{\partial x}$ and $\dfrac{\partial f}{\partial y}$. To parallel the single-variable case, we write

$$\frac{\partial^2 f}{\partial x^2} = \frac{\partial}{\partial x}\left(\frac{\partial f}{\partial x}\right) = f_{xx}$$

$$\frac{\partial^2 f}{\partial y^2} = \frac{\partial}{\partial y}\left(\frac{\partial f}{\partial y}\right) = f_{yy}$$

$$\frac{\partial^2 f}{\partial y \partial x} = \frac{\partial}{\partial y}\left(\frac{\partial f}{\partial x}\right) = f_{xy}$$

$$\frac{\partial^2 f}{\partial x \partial y} = \frac{\partial}{\partial x}\left(\frac{\partial f}{\partial y}\right) = f_{yx}$$

All the above are called *second-order* (or just *second*) *partial derivatives* of f. Note especially that

the notation in the first column above gives the order of differentiation reading from right to left:

$\dfrac{\partial^2 f}{\partial y \partial x}$ tells you to differentiate first with respect to x, then with respect to y;

$\dfrac{\partial^2 f}{\partial x \partial y}$ reverses the order of differentiation.

the notation in the right-hand column above gives the order of differentiation reading from left to right:

f_{xy} calls for a partial with respect to x followed by the partial with respect to y;

f_{yx} reverses that order of differentiation.

Example 52-8

Find all the second-order partial derivatives of $f(x,y) = x^3y + y^2$.

First we find the first-order partial derivatives:

$$\frac{\partial f}{\partial x} = 3x^2y \qquad \frac{\partial f}{\partial y} = x^3 + 2y$$

Then

$$\frac{\partial^2 f}{\partial x^2} = \frac{\partial}{\partial x}\left(\frac{\partial f}{\partial x}\right) = \frac{\partial}{\partial x}(3x^2y) = 6xy = f_{xx}$$

$$\frac{\partial^2 f}{\partial y \partial x} = \frac{\partial}{\partial y}\left(\frac{\partial f}{\partial x}\right) = \frac{\partial}{\partial y}(3x^2y) = 3x^2 = f_{xy}$$

$$\frac{\partial^2 f}{\partial y^2} = \frac{\partial}{\partial y}\left(\frac{\partial f}{\partial y}\right) = \frac{\partial}{\partial y}(x^3 + 2y) = 2 = f_{yy}$$

$$\frac{\partial^2 f}{\partial x \partial y} = \frac{\partial}{\partial x}\left(\frac{\partial f}{\partial y}\right) = \frac{\partial}{\partial x}(x^3 + 2y) = 3x^2 = f_{yx}$$

NOTE: For the particular function in Example 52-8, $f_{xy} = f_{yx}$. This is true for a certain class of functions. All the functions considered in this chapter and almost all practical applications fall into this class.

THE TOTAL DIFFERENTIAL

When f is a function of a single variable, the differential $f'(x)\,dx$* is extremely useful in approximating a change in f caused by a small change in x. Similarly, if f is a function of two variables, x and y, then the *total differential* given by

$$df = \frac{\partial f}{\partial x}\,dx + \frac{\partial f}{\partial y}\,dy = f_x\,dx + f_y\,dy$$

is an approximation to Δf, a change in f caused by small changes in both x and y. Earlier we noted that using differentials for approximating f when it is a function only of x is equivalent to approximating f by a tangent line to the curve $y = f(x)$. When f is a function of the two variables x and y, the analog is the approximation to f by a tangent plane to the surface $z = f(x, y)$. It will help at this time to review Goal 28 before proceeding to the examples that follow.

Example 52-9

Find the total differential of (a) $f(x, y) = x^2y + y^3$; (b) $g(x, y) = ye^x + \ln y$.

(a) $df = 2xy\,dx + (x^2 + 3y^2)\,dy$ (b) $dg = ye^x\,dx + \left(e^x + \dfrac{1}{y}\right)dy$

---

*In Goal 28 (p. 273) we defined dx to be equal to Δx. In this Goal, whenever we approximate Δf by df, with f as a function of x and y, we will assume that dx equals Δx and that dy equals Δy.

Example 52-10

Find df and Δf at $(2, 3)$ if $f(x, y) = xy - x^2$ and if x is increased by 0.1 and y is decreased by 0.2.

For the actual change in f we have

$$\Delta f = f(2 + \Delta x, 3 + \Delta y) - f(2, 3) = f(2.1, 2.8) - f(2, 3) = 1.47 - 2 = -0.53$$

To find df, we evalute its partial derivatives at $(2, 3)$:

$$f_x(x, y) = y - 2x \qquad f_x(2, 3) = -1$$
$$f_y(x, y) = x \qquad f_y(2, 3) = 2$$

So for the total differential we get

$$df = f_x(2, 3)\, dx + f_y(2, 3)\, dy = (-1)(0.1) + (2)(-0.2) = -0.5$$

Generally, the smaller the changes in the independent variables (x and y), the better is the approximation of df to Δf. This parallels the situation in the case of a single variable.

Example 52-11

The U. Toate TV Company, which manufactures two different models of portable TV sets, uses $P(x, y) = 5x + 2y + 0.001xy$ as its profit function, where x and y are the number of sets produced of each model. Find the approximate percentage change in profit caused by an increase in production of three sets of each model when the production level is $x = 500$ and $y = 300$.

The actual percentage change is $100\,\Delta P/P$. We will approximate this by evaluating $100\, dP/P$.

First we have

$$\frac{\partial P}{\partial x} = 5 + 0.001y \qquad \frac{\partial P}{\partial x}\bigg|_{(500,300)} = 5.3$$

$$\frac{\partial P}{\partial y} = 2 + 0.001x \qquad \frac{\partial P}{\partial y}\bigg|_{(500,300)} = 2.5$$

Next note that $dx = \Delta x = dy = \Delta y = 3$. Therefore

$$dP = \frac{\partial P}{\partial x}\, dx + \frac{\partial P}{\partial y}\, dy = (5.3)(3) + (2.5)(3) = 23.4$$

Since $P(500, 300) = 3250$,

$$100\,\frac{dP}{P} = 100 \cdot \frac{23.4}{3250} = 0.72\%$$

Goal 52 | **To Find and Apply Partial Derivatives**

Example 52-12

The Ritebedda Company manufactures writing paper of dimensions 8 in. by 10 in. with an error tolerance of 0.01 in. in each dimension. How close may they legitimately claim the area is to 80 sq in.?

Since the area is given by $f(x, y) = xy$, it follows that $f_x = y$ and $f_y = x$. Then $f_x(8, 10) = 10$ and $f_y(8, 10) = 8$. So

$$\Delta f \approx df = f_x(8, 10)\, dx + f_y(8, 10)\, dy = 10\,(0.01) + 8\,(0.01) = 0.18$$

Although we do not know what the exact error is in either dimension (we are told only that $|\Delta x| \leq 0.01$ and $|\Delta y| \leq 0.01$), by choosing dx and dy both equal to $+0.01$, we obtain an approximation to the *maximum* possible error in the area, namely, 0.18 sq in.

Example 52-13

Use the total differential to approximate $\sqrt{(2.98)^2 + (4.01)^2}$.

We let $f(x, y) = \sqrt{x^2 + y^2}$, $dx = \Delta x = -0.02$, $dy = \Delta y = 0.01$. Also, $x = 3$ and $y = 4$. We approximate $f(2.98, 4.01)$ by finding $f(3, 4) + df$, where $df = f_x(3, 4)\, dx + f_y(3, 4)\, dy$.

Since

$$f_x = \frac{x}{\sqrt{x^2 + y^2}} \qquad f_y = \frac{y}{\sqrt{x^2 + y^2}}$$

$$f_x(3, 4) = \frac{3}{5} \qquad f_y(3, 4) = \frac{4}{5}$$

we get

$$df = \frac{3}{5}\,(-0.02) + \frac{4}{5}\,(0.01) = -0.004$$

Thus

$$f(3, 4) + df = 5 - 0.004 = 4.996$$

This is an excellent approximation: to five decimal places, $\sqrt{(2.98)^2 + (4.01)^2}$ equals 4.99605.

COMPETITIVE VERSUS COMPLEMENTARY PRODUCTS

In Goal 51 (p. 500) we assumed that $D_1(p_1, p_2)$ and $D_2(p_1, p_2)$ were the demand functions for two products whose prices were p_1 and p_2 respectively. We considered two cases:

CASE I: The products are competitive (or are substitutes) if an increase in the price of either product causes an increase in the demand for the other. In terms of partial derivatives, we have, for this case,

$$\frac{\partial D_1}{\partial p_2} > 0 \qquad \text{and} \qquad \frac{\partial D_2}{\partial p_1} > 0$$

CASE II: The products are complementary if an increase in the price of either causes a decrease in the demand for the other. In this case, then,

$$\frac{\partial D_1}{\partial p_2} < 0 \quad \text{and} \quad \frac{\partial D_2}{\partial p_1} < 0$$

Example 52-14

If two products have demand functions $D_1(p_1, p_2) = 175 - 2p_1 + 5p_2^2$ and $D_2(p_1, p_2) = 200 + p_1^2 - 10p_2$, are they competitive, complementary, or neither?

We compute the partial derivatives:

$$\frac{\partial D_1}{\partial p_2} = 10p_2 \qquad \frac{\partial D_2}{\partial p_1} = 2p_1$$

Since p_1 and p_2 are both prices, they are both positive; and so are the two partial derivatives. This is therefore Case I: the products are competitive.

EXERCISES

1. Determine $f_x(x, y)$ and $f_y(x, y)$ for each of the following:
   (a) $f(x, y) = x + 2y - xy$
   (b) $f(x, y) = x^2 + y^3 + xy^2$
   (c) $f(x, y) = e^{3xy}$
   (d) $f(x, y) = \ln xy$
   (e) $f(x, y) = \dfrac{x}{y}$
   (f) $f(x, y) = (x^2 + 2)(y^3 - y)$
   (g) $f(x, y) = e^{xy^2+2}$
   (h) $f(x, y) = \ln(x^2 + y)$
   (i) $f(x, y) = \sqrt{x + y^2}$
   (j) $f(x, y) = \dfrac{y^2}{x^3}$

2. Certain medications, such as Imuran and propylthiouracil, are given in doses the size of which depends on body surface area. An accepted approximation to a person's surface area S, in square centimeters, is

$$S = 71.84 x^{0.425} y^{0.725}$$

where x is the person's weight in kilograms and y is the height in centimeters. Determine how surface area changes with respect to weight and with respect to height; that is, determine $\dfrac{\partial S}{\partial x}$ and $\dfrac{\partial S}{\partial y}$.

3. Suppose that the productivity function for a particular company is

$$P(x, y) = 5x^{1/3} y^{2/3}$$

where x is a measurement of the investment in labor and y is a measurement of the investment in capital. Find the marginal productivity of labor and the marginal productivity of capital when $x = 8$ and $y = 27$.

4. Find $\dfrac{\partial f}{\partial x}$, $\dfrac{\partial f}{\partial y}$, and $\dfrac{\partial f}{\partial z}$ and evaluate them at $(0, -1, 1)$ for each of the following:
   (a) $f(x, y, z) = x^2 + yz$
   (b) $f(x, y, z) = e^{xyz}$
   (c) $f(x, y, z) = \dfrac{xy}{z}$

5. Find the four second-order partial derivatives for each of the following functions:
   (a) $f(x, y) = xy + x^2 + y^3$
   (b) $f(x, y) = e^{xy}$
   (c) $f(x, y) = \dfrac{x}{y}$

6. Find the total differential of each of the following:
   (a) $f(x, y) = xy^2 - 5$ (b) $f(x, y) = e^{5xy}$ (c) $f(x, y) = \ln xy$

7. Find df and Δf at the point $(1, 4)$ if $f(x, y) = y^2 - xy$ and if x is increased by 0.1 and y by 0.2.

8. Find df and Δf at the point $(0, 1)$ if $f(x, y) = \dfrac{x}{y}$ and if x is decreased by 0.1 and y is increased by 0.05.

9. Consider the revenue function $R(x, y) = 3x + 2y - 0.001 xy$, where x and y are the number of units produced of two different products. If the current production level is $x = 250$ and $y = 200$, find the approximate percentage change in revenue resulting from an increase of 3 units of the first product and a decrease of 4 units of the second product.

10. A function developed in the study of proper performance of the kidney is $f(x, y, z) = \dfrac{xy}{z}$ (see Example 52-7). If x is measured to be 300 mg per 100 ml, with a possible error of 10, y is 10 ml/min with a possible error of 1, and z is 3 mg per 100 ml with a possible error of 0.1, use differentials to approximate the possible resulting error in f. HINT: $df = f_x \, dx + f_y \, dy + f_z \, dz$, where the partials are evaluated at an appropriate point.

11. The Reece-Eichel Tin Can Company manufactures cylindrical cans of radius 2 in. and height 5 in. If the manufacturing error tolerance is 0.1 in. for both the radius and the height, what is the possible resulting error in the intended volume of 20π cu in.? ($V = \pi r^2 h$.)

12. Use the total differential to approximate $\sqrt{(4.99)^2 + (12.02)^2}$.

13. Use the total differential to approximate $\dfrac{\sqrt{4.01}}{0.99}$.

14. Classify each pair of demand functions as competitive, complementary, or neither.
    (a) $D_1(p_1, p_2) = 1000 - p_1^2 + 3p_2$
        $D_2(p_1, p_2) = 500 + 2p_1^2 - 5p_2$
    (b) $D_1(p_1, p_2) = 100 - p_1$
        $D_2(p_1, p_2) = 300 - p_2^2$
    (c) $D_1(p_1, p_2) = 750 - p_1 - 4p_2$
        $D_2(p_1, p_2) = 1500 - 2p_1 - 0.1p_2^2$
    (d) $D_1(p_1, p_2) = 500 - 3p_1 + 2p_2 + 0.01p_2^2$
        $D_2(p_1, p_2) = 300 + 5p_1 - 0.09p_2^3$

15. Show that the *Journal* and the *Times* of Example 51-10 (p. 500) are competitive products.

16. Suppose $g(p_1, p_2)$ is a function giving the number of *Journal* readers, where p_1 is the price of the *Journal* and the p_2 the price of the *Times*. What sign do you expect $\dfrac{\partial g}{\partial p_2}$ to have? Why? What sign for $\dfrac{\partial g}{\partial p_1}$? Why?

TEST A

1. Determine $\dfrac{\partial f}{\partial x}$ and $\dfrac{\partial f}{\partial y}$ for each of the following functions:
   (a) $f(x, y) = 3x - 2y^2 + x^3 y^4$ (b) $f(x, y) = e^{x^2 - xy - 2}$

2. Consider the productivity function $P(x, y) = 3x^{1/4} y^{3/4}$, where x is a measure of the investment in labor and y is a measure of the investment in capital. Determine the marginal productivity of labor and the marginal productivity of capital when $x = 16$ and $y = 1$.

3. Find the total differential of each of the following:
   (a) $f(x, y) = xy^3 - 3$ (b) $f(x, y, z) = \ln xyz$

4. Consider the pollution index $P(x, y) = \sqrt{xy}$, where x and y are measurements of the quantity of two different pollutants in the air. Suppose $x = 40$ and $y = 10$. Use differentials to approximate the change in P due to increases in x and y of 0.1 and 0.05 respectively.

TEST B

1. Determine f_x and f_y for each of the following functions:
   (a) $f(x, y) = x^2 - y^3 + 2xy$ (b) $f(x, y) = \ln(2x + 3y - 1)$

2. Consider the productivity function $P(x, y, z) = 2x^{1/2}y^{1/3}z^{1/4}$, where x, y, and z are measures of expenditure on labor, equipment, and research respectively. Determine the marginal productivities of labor, equipment, and research when $x = 9$, $y = 8$, and $z = 1$.
3. Find the total differential of each of the following functions:
   (a) $f(x, y) = x^2 + x^2y$ (b) $f(x, y) = e^{x^2y^2}$
4. Consider the utility index $U(x, y) = e^{xy}$, which measures the amount of consumer satisfaction obtained from quantities of two products. Suppose $x = 0$ and $y = 3$. Use differentials to approximate the change in U due to increases in x and y of 0.2 and 0.1 respectively.

STUDY QUESTIONS

1. In this Goal we showed how partial derivatives are defined formally in terms of limits. We did not define or even discuss a limit of a function of two variables, nor will we do so here. This study question is, however, intended to provide insight into the complexity of the matter.

   Suppose $f(x, y) = 3x + 4y - 7$ and we investigate $\lim f(x, y)$ as $(x, y) \to (1, 2)$. It is clear that, as x gets closer and closer to 1 and y gets closer and closer to 2, $f(x, y)$ gets closer and closer to $3(1) + 4(2) - 7 = 4$. It is also clear that $f(x, y)$ approaches 4 regardless of what route (x, y) follows as it approaches $(1, 2)$—it can move along any line or curve whatsoever in the approach.

   In general, if

   $$\lim_{(x, y) \to (a, b)} f(x, y)$$

   is to exist and is to be equal, say, to L, then $f(x, y)$ must get closer and closer to L no matter how (x, y) approaches (a, b). Note the analogy with the single-variable case, where f must approach the same number as $x \to c$ from either the right or the left. In fact, to show that the limit in question does not exist we need only find two routes along which (x, y) can approach (a, b) for which f has different limits (or perhaps no limit at all for one of the routes).

   Let's examine

   $$\lim_{(x, y) \to (0, 0)} \frac{2x + y}{x + y}$$

   If we let $(x, y) \to (0, 0)$ along the x-axis (where $y = 0$), then

   $$\lim \frac{2x + y}{x + y} = \lim \frac{2x + 0}{x + 0} = 2$$

But if we let $(x, y) \to (0, 0)$ along the y-axis (where $x = 0$), then

$$\lim \frac{2x + y}{x + y} = \lim \frac{0 + y}{0 + y} = 1$$

Since these answers are different we conclude that $\lim f(x, y)$ as $(x, y) \to (0, 0)$ does not exist.

Which of the following limits exist?

(a) $\lim \dfrac{2x + y}{x + y}$ as $(x, y) \to (1, 0)$

(b) $\lim \dfrac{x + y}{x - y}$ as (i) $(x, y) \to (3, -1)$

       (ii) $(x, y) \to (2, 2)$

       (iii) $(x, y) \to (0, 0)$

2. The chain rule (Goal 15) can be generalized to functions of several variables. Suppose $f(x, y)$ is a real-valued function of two variables x and y which are themselves functions of t. Then f can be regarded as a function of the single variable t and

$$\frac{df}{dt} = \frac{\partial f}{\partial x} \frac{dx}{dt} + \frac{\partial f}{\partial y} \frac{dy}{dt}$$

For example, let $f(x, y) = x^2 y + x$, $x(t) = t^2$, and $y(t) = t + 2$. Determine $\dfrac{df}{dt}$ in two different ways: first by substituting to express f as a function of t and then differentiating; then by using the chain rule given above before substituting.

3. Consider the function $f(x, y)$ and the level curve $f(x, y) = c$. Suppose that the level curve is such that y can be expressed as a function of x. Then, taking the derivative with respect to x and applying the chain rule from Study Question 2, we obtain

$$\frac{\partial f}{\partial x} \frac{dx}{dx} + \frac{\partial f}{\partial y} \frac{dy}{dx} = 0 \quad \text{or} \quad \frac{\partial f}{\partial x} + \frac{\partial f}{\partial y} \frac{dy}{dx} = 0$$

Solving algebraically for $\dfrac{dy}{dx}$ results in

$$\frac{dy}{dx} = -\left(\frac{\partial f}{\partial x}\right) \bigg/ \left(\frac{\partial f}{\partial y}\right)$$

Goal 52 | To Find and Apply Partial Derivatives

For example, suppose $f(x, y) = y - x^2$ and we choose the level curve $f(x, y) = 3$. Find the slope of this level curve at the point (2, 7) in two different ways: first by finding the equation of the level curve, expressing y as a function of x, and finding the derivative when $x = 2$; then by using the formula above for $\dfrac{dy}{dx}$.

GOAL 53

To find extreme values of functions of two variables

As noted in Goal 25, a great many applications of the calculus involve finding extreme values of functions. This is as true for functions of two or more variables as for functions of a single variable. For the latter we used the first and second derivatives; to deal with a function of two or more variables we use its first and second partial derivatives. In this Goal we restrict our attention to functions of two variables; however, with appropriate modifications, the definitions and methods can be extended to three or more variables.

LOCAL MAXIMA AND MINIMA

We say that $f(x, y)$ has a *local minimum* at a point (a, b) in its domain if there is some circular region in the xy-plane with (a, b) as center such that $f(x, y) \geq f(a, b)$ for all (x, y) in the domain of f that are within the circular region. If $f(x, y)$ has a *local maximum* at (a, b), then $f(x, y) \leq f(a, b)$ for all (x, y) in the domain of f that are within the circular region. Geometrically, a surface "peaks" at a local maximum, "pits" at a local minimum.

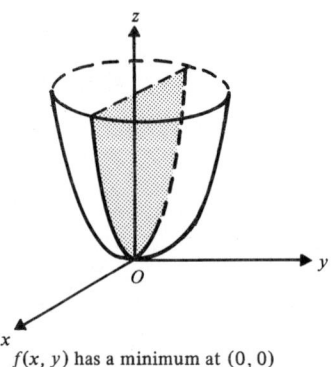

$f(x, y)$ has a minimum at $(0, 0)$

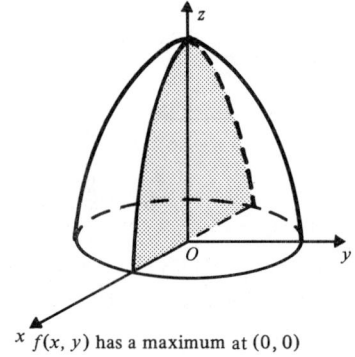

$f(x, y)$ has a maximum at $(0, 0)$

Goal 53 | To Find Extreme Values of Functions of Two Variables

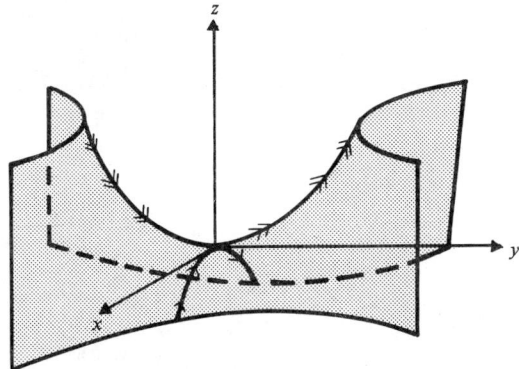

$f(x, y)$ has a saddle point at $(0, 0)$

In the figure on page 520, the surface at the left has a local min at $(0, 0)$, whereas the one at the right has a local max at $(0, 0)$. The surface shown above has neither a max nor a min at $(0, 0)$.

Note how the two definitions given earlier are satisfied respectively for the functions sketched on page 520. For the function sketched above, however, the curve in the xz-plane (shown by single arrowheads) has a local max at the origin, whereas the curve in the yz-plane (shown by double arrowheads) has a local min at the origin. The origin is called a *saddle point* for this surface. In any region of the xy-plane about the origin, we can find two points (x_1, y_1) and (x_2, y_2) such that $f(x_1, y_1) < 0$ but $f(x_2, y_2) > 0$.

CRITICAL POINTS

Suppose that $f(x, y)$ has a local max at (a, b) and that $f_x(a, b)$ and $f_y(a, b)$ both exist. Then the curve in which the surface intersects the plane $y = b$ must also have a local max at $x = a$, and so $f_x(a, b) = 0$. Also, the curve in which the graph of f intersects the plane $x = a$ must have a local max at

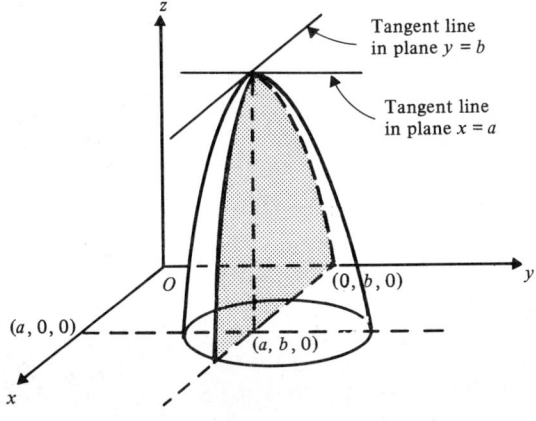

(The tangents shown are horizontal)

$y = b$, which implies that $f_y(a, b) = 0$. A similar argument holds if $f(x, y)$ has, instead, a local min at (a, b).

Any point at which both partial derivatives of f equal zero is called a *critical point*. We thus have:

If $f(x, y)$ has a local max or min at (a, b) and $f_x(a, b)$ and $f_y(a, b)$ both exist, then (a, b) is a critical point; that is,

$$f_x(a, b) = 0 \quad \text{and} \quad f_y(a, b) = 0$$

Example 53-1

Find all critical points of

(a) $f(x, y) = x^2 + y^2 + 2x - 4y + 1$
(b) $g(x, y) = x^2 + 3y^2 + xy - 5x - 8y + 2$
(c) $g(x, y) = y^3 - x^3 - 3y + 12x - 9$

(a) $f_x = 2x + 2; f_y = 2y - 4$. To find the critical values we set both f_x and f_y equal to zero; this yields $x = -1$ and $y = 2$. So f has only one critical point, at $(-1, 2)$.

(b) $g_x = 2x + y - 5; g_y = 6y + x - 8$. We want to solve simultaneously the pair of equations (see p. 555).

$$2x + y - 5 = 0$$
$$6y + x - 8 = 0$$

The solution is $x = 2$ and $y = 1$. So $(2, 1)$ is the only critical point of g.

(c) $h_x = -3x^2 + 12; h_y = 3y^2 - 3$. Since $h_x = 0$ if $x = \pm 2$ and $h_y = 0$ if $y = \pm 1$, h has four critical points; $(2, 1)$, $(2, -1)$, $(-2, 1)$, and $(-2, -1)$.

IDENTIFYING MAXIMUM OR MINIMUM POINTS

Suppose (a, b) is a critical point of f. Does f have a local max at (a, b)? a local min? neither? Often it is clear in a particular application that a max or min must exist. If we then find only one critical point, we may infer that this yields the expected extreme. Sometimes, however, the conclusion is not so obvious. Then either of two tests may be used:

TEST I

If (a, b) is a critical point of f, we compare $f(a+h, b+k)$, for small h and k, with $f(a, b)$.

If $f(a+h, b+k) < f(a, b)$ whenever h and k are sufficiently small, then (a, b) yields a local max.

Goal 53 | To Find Extreme Values of Functions of Two Variables

If $f(a+h, b+k) > f(a, b)$ whenever h and k are sufficiently small, then (a, b) yields a local min.

If for some small h and k, $f(a+h, b+k) < f(a, b)$, but for other small h and k, $f(a+h, b+k) > f(a, b)$, then f has neither a local max nor a local min at (a, b).

Note that Test I is a direct application of the definitions of local max and min of a function of two variables given earlier.

TEST II

Suppose f has a critical point at (a, b). Let

$$D = f_{xx}(a, b) f_{yy}(a, b) - [f_{xy}(a, b)]^2$$

Then

If $D > 0$ and $f_{xx}(a, b) < 0$, $f(a, b)$ is a local max.
If $D > 0$ and $f_{xx}(a, b) > 0$, $f(a, b)$ is a local min.
If $D < 0$, f has a saddle point at (a, b).
If $D = 0$, this test is inconclusive.

Test II is the analog of, albeit much more complicated than, the second-derivative test for max and min in the single-variable case. Test I is not efficient unless $f(x, y)$ is a simple function.

We now illustrate the use of both tests in identifying critical points.

Example 53-2

Find any local max or min of $f(x, y) = x^2 + y^2 - 4x - 2y + 3$ by using Tests I and II.

Since $f_x = 2x - 4$ and $f_y = 2y - 2$, f has a critical point at $(2,1)$.

Using Test I:

$$f(2+h, 1+k) = (2 + h)^2 + (1 + k)^2 - 4(2 + h) - 2(1 + k) + 3$$
$$= h^2 + k^2 - 2$$
$$f(2 + h, 1 + k) - f(2, 1) = (h^2 + k^2 - 2) - (-2) = h^2 + k^2$$

Since $h^2 + k^2$ is always positive (unless $h = k = 0$), $f(2,1)$ is a local min.

Using Test II: The first partials of f are given above. So $f_{xx} = 2$, $f_{yy} = 2$, $f_{xy} = 0$. Therefore

$$D = f_{xx}(2, 1) f_{yy}(2, 1) - [f_{xy}(2, 1)]^2$$
$$= 2 \cdot 2 \qquad\qquad - 0 \qquad\qquad = 4$$

Since $D > 0$ and $f_{xx}(2, 1) > 0$, $f(2, 1)$ by Test II is a local min, in agreement with the outcome of Test I.

Example 53-3

Use Test I to find any local max or min of $f(x, y) = x^2 - y^2 + 6y$.
Since $f_x = 2x$ and $f_y = -2y + 6$, f has a critical point at $(0, 3)$.

$$f(0+h, 3+k) = h^2 - (3 + k)^2 + 6(3 + k) = h^2 - k^2 + 9$$
$$f(0+h, 3+k) - f(0, 3) = h^2 - k^2 + 9 - (9) = h^2 - k^2$$

But $h^2 - k^2$ can be either positive or negative. For example, if $h = 0.2$ and $k = 0.1$, then $h^2 - k^2 > 0$, whereas if $h = 0.1$ and $k = 0.2$, then $h^2 - k^2 < 0$. We conclude that $(0, 3)$ yields neither a local max nor a local min. Verify that $f(0, 3)$ is a saddle point by applying Test II.

Example 53-4

Use Test II to identify any local max or min of $f(x, y) = 3x - x^3 - y^2 - 10y + 8$.

We have $f_x = 3 - 3x^2$, $f_y = -2y - 10$; $f_{xx} = -6x$; $f_{yy} = -2$; $f_{xy} = 0$. So $f_x = 0$ if $x = \pm 1$ and $f_y = 0$ if $y = -5$. Therefore f has two critical points, $(1, -5)$ and $(-1, -5)$.

First we test $(1, -5)$. $f_{xx}(1, -5) = -6$; $f_{yy}(1, -5) = -2$; $f_{xy}(1, -5) = 0$; $D = (-6)(-2) - 0^2 = 12 > 0$. Since $f_{xx}(1,-5) < 0$, $f(1, -5)$ is a local maximum.

Then we test $(-1, -5)$. $f_{xx}(-1,-5) = -6(-1) = 6$; $f_{yy}(-1, -5) = -2$; $f_{xy}(-1, -5) = 0$. $D = 6(-2) - 0^2 = -12 < 0$. Therefore $(-1, -5)$ is a saddle point of f.

Example 53-5

Find any local max or min of $f(x, y) = 4 + 2xy + 14x - 2x^2 - y^2 - 8y$ using Test II.

We have $f_x = 2y + 14 - 4x$, $f_y = 2x - 2y - 8$. Setting both partials equal to zero and solving simultaneously (p. 555) yields $x = 3$ and $y = -1$. Since $f_{xx} = -4$, $f_{yy} = -2$, and $f_{xy} = 2$, we have at $(3, -1)$ that $D = (-4)(-2) - 2^2 = 4 > 0$. Since $f_{xx}(3, -1) = -4 < 0$, $f(3, -1)$ is a local maximum.

AN APPLICATION TO THE METHOD OF LEAST SQUARES

An extremely important application of finding the minimum of a function of two variables is the problem of determining the straight line that best fits a set of data.

On p. 525 we see two graphs, one of data on sales of a product, one on demands for electricity, both plotted against time. The linear function or dotted line at the left can be used to predict the sales for year 6; the one at the right can be used to predict the demand for electricity in year 5.

The *best* straight line is usually defined to be that which minimizes the sum of the squares of the differences (the vertical distances) between the

Goal 53 | To Find Extreme Values of Functions of Two Variables

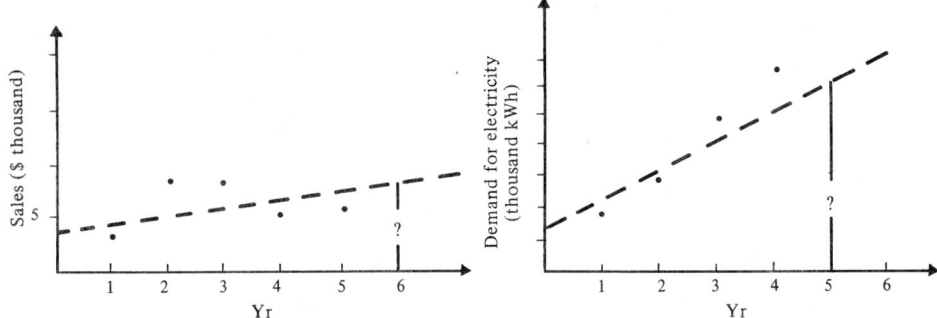

actual *y*-values and the corresponding *y*-values on the line. The line of best fit for a given set of data is often called the *regression line* for the data.

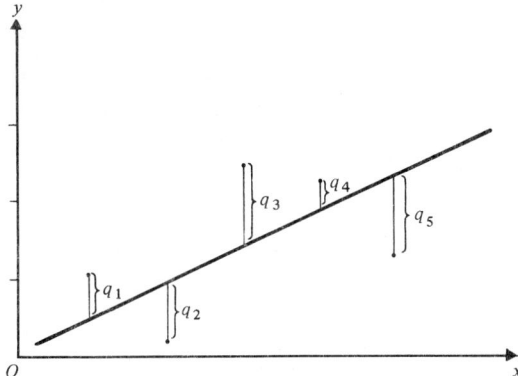

The dots are for the actual (or observed) *y*-values. The best straight line thus minimizes the sum

$$q_1^2 + q_2^2 + q_3^2 + q_4^2 + q_5^2$$

The method of least squares is demonstrated in the next two examples. The general case is the subject of Study Question 3.

Example 53-6

Find the line that best fits the points (1, 1), (2, 3), (3, 2), and (4, 3).

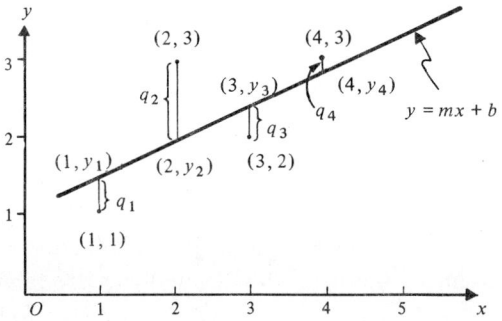

We see from the figure that we want to minimize the sum

$$Q = q_1^2 + q_2^2 + q_3^2 + q_4^2$$

If the best line has equation

$$y = mx + b$$

then we seek m and b for which Q is a minimum. Note first that y_1, y_2, y_3, and y_4 are the ordinates of points *on* the line. Thus $y_1 = m \cdot 1 + b$, $y_2 = m \cdot 2 + b$, $y_3 = m \cdot 3 + b$, $y_4 = m \cdot 4 + b$. Since

$$Q = (y_1 - 1)^2 + (y_2 - 3)^2 + (y_3 - 2)^2 + (y_4 - 3)^2$$

therefore

$$Q = (m \cdot 1 + b - 1)^2 + (m \cdot 2 + b - 3)^2 + (m \cdot 3 + b - 2)^2 + (m \cdot 4 + b - 3)^2$$

We can minimize Q by finding its critical point, where, as we know, $\partial Q/\partial m$ and $\partial Q/\partial b$ are both zero.

$$\frac{\partial Q}{\partial m} = 2(m + b - 1) + 2(2m + b - 3) \cdot 2 +$$
$$2(3m + b - 2) \cdot 3 + 2(4m + b - 3) \cdot 4$$
$$= 2(30m + 10b - 25) = 10(6m + 2b - 5)$$

$$\frac{\partial Q}{\partial b} = 2(m + b - 1) + 2(2m + b - 3) +$$
$$2(3m + b - 2) + 2(4m + b - 3)$$
$$= 2(10m + 4b - 9)$$

Now we set these partials equal to 0 and solve simultaneously for m and b:

$$\begin{cases} 10(6m + 2b - 5) = 0 \\ 2(10m + 4b - 9) = 0 \end{cases} \rightarrow \begin{cases} 12m + 4b - 10 = 0 \\ 10m + 4b - 9 = 0 \end{cases}$$

Subtracting the second equation from the first yields $2m - 1 = 0$, $m = \frac{1}{2}$. And substitution of this value into either equation on the left yields $b = 1$.

With Test II on page 523 we can quickly verify that Q has a minimum when $m = \frac{1}{2}$ and $b = 1$. The line of best fit is therefore $y = \frac{1}{2}x + 1$.

Example 53-7

A national chain, Pettable Used Pooches (PUP), sells second-hand stuffed animals. PUP's sales figures during the first 5 yr of operation are tabulated on page 527.

Goal 53 | **To Find Extreme Values of Functions of Two Variables**

Year	1	2	3	4	5
Sales ($)	25,000	32,000	38,000	46,000	54,000

Find the regression line for these data and use the resulting linear function to estimate PUP's sales in year 10.

We can express the given data as points in the plane as follows:

Year	$x_1 = 1$	$x_2 = 2$	$x_3 = 3$	$x_4 = 4$	$x_5 = 5$
Sales in Thousands of Dollars	$y_1 = 25$	$y_2 = 32$	$y_3 = 38$	$y_4 = 46$	$y_5 = 54$

Let $y = mx + b$ be the regression line sought. Then we want to minimize

$$Q = (mx_1 + b - y_1)^2 + (mx_2 + b - y_2)^2 + (mx_3 + b - y_3)^2 \\ + (mx_4 + b - y_4)^2 + (mx_5 + b - y_5)^2$$

So

$$Q = (m \cdot 1 + b - 25)^2 + (m \cdot 2 + b - 32)^2 + (m \cdot 3 + b - 38)^2 \\ + (m \cdot 4 + b - 46)^2 + (m \cdot 5 + b - 54)^2$$

$$\frac{\partial Q}{\partial m} = 2(m + b - 25) + 2(2m + b - 32) \cdot 2 + 2(3m + b - 38) \cdot 3 \\ + 2(4m + b - 46) \cdot 4 + 2(5m + b - 54) \cdot 5 \\ = 2(55m + 15b - 657)$$

$$\frac{\partial Q}{\partial b} = 2(m + b - 25) + 2(2m + b - 32) + 2(3m + b - 38) \\ + 2(4m + b - 46) + 2(5m + b - 54) \\ = 2(15m + 5b - 195) = 10(3m + b - 39)$$

Setting both partials equal to zero and solving simultaneously yields $m = 7.2$ and $b = 17.4$. Therefore the line of best fit (the regression line) for the data given is $y = 7.2x + 17.4$. An estimate of sales in year 10, in dollars, is

$$[7.2(10) + 17.4](1000) = \$89,400$$

EXERCISES

1. Find the critical points of each of the following functions.
   (a) $f(x, y) = x^2 + y^2 - 10x + 2y - 3$
   (b) $f(x, y) = -3x^2 - 2y^2 + 6x - 12y + 1$
   (c) $f(x, y) = 2x^2 - y^2 + 4x + 4y + 3$
   (d) $f(x, y) = x^2 + 2y^2 + 3xy - 8x - 11y + 1$
   (e) $f(x, y) = 3x^2 + y^2 - 2xy + 8x - 4y - 2$
   (f) $f(x, y) = -2x^2 - 3y^2 + 4xy - 4x + 6y - 1$

(g) $f(x, y) = x^3 + y^3 - 12x - 3y + 1$

(h) $f(x, y) = xy^2 - x$

2. For each critical point found in parts (a), (b), and (c) of Exercise 1, determine using Test I (p. 522) whether it yields a local max, a local min, or neither.

3. For each critical point found in Exercise 1, determine its nature by using Test II (p. 523).

4. Suppose the profits of Delicious Doggie Tidbits for the past 4 yr have been $10,000, $13,000, $15,000, and $18,000. Find the straight line that best approximates these data and use the resulting function to predict the profits of DDT for the next 2 yr.

5. The population figures for a particular animal species for the past 5 yr have been 180,000, 175,000, 168,000, 162,000, and 157,000. Find the straight line that best approximates these data and use the resulting function to predict how many years from now the species will become extinct.

TEST A

1. Find the critical points of the following functions and determine whether each yields a local maximum, a local minimum, or neither.

    (a) $f(x, y) = x^2 - y^2 + 4x - 2y + 1$

    (b) $f(x, y) = x^2 + 2y^2 - xy - 2x + y - 3$

2. Find the line that best fits the points (1, 0), (2, 2), (3, 2), and (4, 3).

TEST B

1. Find the critical points of the following functions and determine whether each yields a local maximum, a local minimum, or neither.

    (a) $f(x, y) = -3x^2 - y^2 + 6x - 4y + 2$

    (b) $f(x, y) = y^2 - 2x^2 + 8x - 1$

2. The profits of the Goode Appliance Service are shown in the table. Use the data to estimate the profits of GAS in 1983.

Year	1974	1975	1976	1977	1978
Profits in Thousands of Dollars	8	12	11	13	14

STUDY QUESTIONS

1. The function $f(x, y)$ has an *absolute minimum* at (a,b) if $f(a,b) \leq f(x, y)$ for all (x, y) in its domain. [$f(a, b)$ is an *absolute maximum*

if $f(a,b) \geq f(x, y)$.] If f is continuous on a *closed* domain (such as a region in the xy-plane bounded by a circle, triangle, or polygon), then f has both an abs max and abs min on its closed domain.

Now suppose f is also differentiable. To find absolute extremes of a function $f(x)$ of one variable on $[a,b]$, we evaluated f at each critical point and at the two endpoints a and b. Analogously, to find the absolute extremes of $f(x, y)$ on a closed domain $\mathcal{D}$ we evaluate f at each critical point, but now also at any critical points on the boundary of $\mathcal{D}$.

For example, to find the abs max and abs min of $f(x,y) = x^2 + y^2 - 2x$ over the circular region $\mathcal{D}$ given by $x^2 + y^2 \leq 4$, first we find the function's critical points: $\dfrac{\partial f}{\partial x} = 2x - 2$, $\dfrac{\partial f}{\partial y} = 2y$; these are zero at $(1,0)$, which is the only critical point. We note that $(1,0)$ is in $\mathcal{D}$. Now we look at f on the boundary of $\mathcal{D}$, the circle $x^2 + y^2 = 4$. Note that when $x^2 + y^2 = 4$, f reduces to the single-variable function $g(x) = 4 - 2x$, where $-2 \leq x \leq 2$. (Why?) Its curve is the intersection of the paraboloid $f(x,y) = x^2 + y^2 - 2x$ and the cylindrical surface $x^2 + y^2 = 4$. Since $g'(x) = -2$, $g'(x)$ never equals 0. g therefore attains its max and min at the endpoints of its domain, namely when $x = 2$ or -2. The corresponding points on the boundary of $\mathcal{D}$ are $(2, 0)$ and $(-2, 0)$. We now evaluate f at these two endpoints and at its one critical point $(1,0)$:

$$f(2,0) = 2^2 + 0^2 - 2 \cdot 2 = 0$$
$$f(-2,0) = (-2)^2 + 0^2 - 2(-2) = 8$$
$$f(1,0) = 1^2 + 0^2 - 2 \cdot 1 = -1$$

Therefore $f(1, 0)$ is the abs min and $f(-2, 0)$ is the abs max.

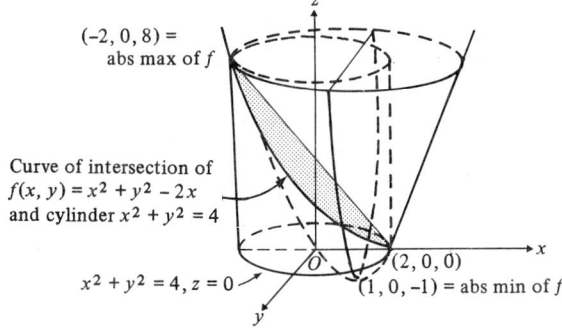

Find the abs max and abs min of $f(x, y) = x^3 + y^2 - 3x - y$ over the first-quadrant region bounded by the x- and y-axes and the line $x + y = 3$.

2. In an extremely useful applied procedure known as linear programming, we must find the extremes of $f(x,y) = ax + by + c$ on a closed polygonal region $\mathscr{D}$, where a, b, and c are constants (not all 0).

We note, since $\dfrac{\partial f}{\partial x}$ and $\dfrac{\partial f}{\partial y}$ are both constants, that f has no critical points inside the region. Furthermore, the boundary of $\mathscr{D}$ consists only of straightline segments, along each of which f reduces to a linear function (of one variable) that attains its max and min at the endpoints of the segment. So the maximum and minimum values achieved by f must occur at the *corner points* of its domain.

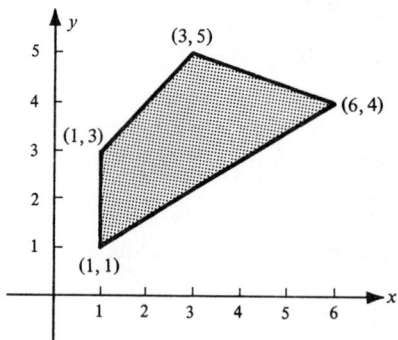

For example, to find the abs max and abs min of $f(x, y) = 3x + 4y - 10$ over the region shown, we need only recognize this as a problem in linear programming. We then evaluate f at each corner point. Since $f(1, 1) = -3$, $f(1, 3) = 5$, $f(3, 5) = 19$, and $f(6, 4) = 24$, we conclude that f has an abs min at $(1, 1)$ and abs max at $(6, 4)$.

Find the abs max and abs min of $f(x, y) = x + y + 3$ over the region of the xy-plane shown below.

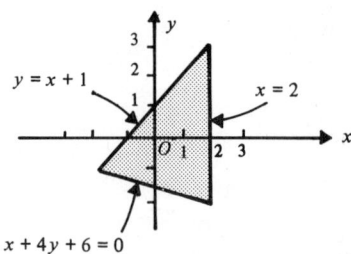

3. Formulas for m and b in the line of best fit $y = mx + b$ can be derived in the general case. The approach parallels exactly that used in Examples 53-6 and 53-7. If the data consist of the n points (x_1, y_1), (x_2, y_2), ..., (x_n, y_n), then we must minimize

Goal 53 | To Find Extreme Values of Functions of Two Variables

$$Q(m, b) = (mx_1 + b - y_1)^2 + (mx_2 + b - y_2)^2 + \ldots + (mx_n + b - y_n)^2$$

(a) Show that we can express $\dfrac{\partial Q}{\partial m}$ and $\dfrac{\partial Q}{\partial b}$ as follows:

$$\frac{\partial Q}{\partial m} = 2(m\Sigma x^2 + b\Sigma x - \Sigma xy) \qquad \frac{\partial Q}{\partial b} = 2(m\Sigma x + nb - \Sigma y)$$

where for succinctness of notation we have set

$$\Sigma x^2 = x_1^2 + x_2^2 + \ldots + x_n^2$$
$$\Sigma x = x_1 + x_2 + \ldots + x_n$$
$$\Sigma xy = x_1 y_1 + x_2 y_2 + \ldots + x_n y_n$$
$$\Sigma y = y_1 + y_2 + \ldots + y_n$$

(b) Show that if you set $\dfrac{\partial Q}{\partial m}$ and $\dfrac{\partial Q}{\partial b}$ equal to zero and solve simultaneously you get

$$m = \frac{n\Sigma xy - \Sigma x \cdot \Sigma y}{n\Sigma x^2 - (\Sigma x)^2} \qquad b = \frac{\Sigma y \cdot \Sigma x^2 - \Sigma xy \cdot \Sigma x}{n\Sigma x^2 - (\Sigma x)^2}$$

(c) Use these formulas to verify the answers obtained in Examples 53-6 and 53-7.

GOAL 54

To find maxima and minima using Lagrange multipliers

In this Goal we learn how to find the maximum or minimum of a function of several variables subject to a restriction known as a *side condition* or *constraint*. Here are some examples:

(1) Mrs. G. Thumm wants to put a fence around her rectangular garden to keep the rabbits out. She has 250 ft of fencing and decides to use all of it. What should the dimensions be for the largest garden? (The constraint is the fixed length of fencing.)

(2) Find the maximum and minimum of $f(x, y) = x + y$ subject to the side condition $x^2 + 3y^2 = 3$.

(3) The sum of the squares of two nonnegative numbers equals 200. Find their maximum and minimum product, if these exist. (The constraint is the fixed sum of the squares.)

In Goal 25 we solved (1) and (3) by using the given constraint to eliminate one variable, thus reducing each problem to finding the extreme of a function of a single variable. We will use this technique of substitution here, too, before introducing the generally more efficient method of Lagrange multipliers.

THE METHOD OF SUBSTITUTION

We begin with problems (1) and (3) above.

Example 54-1

Solve (1) by the method of substitution.

This is Example 25-1 of Goal 25 (p. 241). We noted there that, if x and y are the length and width respectively of the garden, we want to maximize the area A, where $A = xy$, subject to the constraint $2x + 2y = 250$, or

Goal 54 | To Find Maxima and Minima Using Lagrange Multipliers

$x + y = 125$. To express A as a function of only one variable we solved the constraint equation, getting $y = 125 - x$; so $A = x(125 - x) = 125x - x^2$; and $A' = 125 - 2x$, which is zero if $x = 62\frac{1}{2}$. With only one critical point, and with A equal to 0 at the endpoints of its domain (0 and 125), we concluded that A was a maximum for $x = 62\frac{1}{2}$. We found immediately that y, also, equals $62\frac{1}{2}$.

Example 54-2

Solve example (3) above by substitution.

This was Exercise 19 of Goal 25 (p. 248). Here we want to find the extremes of the product $P = xy$ subject to the constraint $x^2 + y^2 = 200$. Solving the constraint equation for y yields $y = \sqrt{200 - x^2}$ (the domain of x is $0 \leq x \leq \sqrt{200}$). So we can now express P as a function of the single variable x:

$$P = x\sqrt{200 - x^2}$$

$$P' = \frac{x \cdot (-2x)}{2\sqrt{200 - x^2}} + \sqrt{200 - x^2}$$

$$= \frac{-x^2}{\sqrt{200 - x^2}} + \sqrt{200 - x^2}$$

Setting P' equal to zero, we have

$$\frac{-x^2 + 200 - x^2}{\sqrt{200 - x^2}} = 0$$

or $2x^2 = 200$, $x = 10$. Thus $x = 10$ is a critical point. Since $P = 0$ at the endpoints of its domain, 0 and $\sqrt{200}$, and equals 100 when $x = 10$, the product is a maximum for the pair (10, 10) and a minimum for the pair $(0, \sqrt{200})$.

Example 54-3

Find the maximum and minimum of $f(x, y) = x^2 + y - 3$, subject to the constraint $y = 4x + 2$.

Among all the points (x, y) such that $y = 4x + 2$, we seek here those that maximize (or minimize) the quantity $x^2 + y - 3$. Direct substitution reduces f to a one-variable function which we denote by $h(x)$:

$$h(x) = x^2 + (4x + 2) - 3 = x^2 + 4x - 1$$

Since $h'(x) = 2x + 4$ is zero only if $x = -2$, and h has a minimum at $x = -2$, there is no maximum. Since $y = 4(-2) + 2 = -6$, we conclude that f has a minimum at $(-2, -6)$, but that it has no maximum, for the given side condition.

Example 54-4

Find the maximum and minimum, if they exist, of $f(x, y) = x^2 + y^2 + 6y - 3$ subject to the constraint $x^2 + 2y^2 = 2$.

It is easy to use the side condition to replace x^2 in $f(x, y)$ by $2 - 2y^2$, getting the one-variable function

$$h(y) = (2 - 2y^2) + y^2 + 6y - 3 = -y^2 + 6y - 1$$

where $-1 \leq y \leq 1$. (Why? It is important to keep track of the domain! Although $-y^2 + 6y - 1$ is defined for all real y, the side condition narrows the domain to $-1 \leq y \leq 1$.) We note that $f'(y) = -2y + 6$, which is zero when $y = 3$. But $y = 3$ is *not* in the domain of h! Therefore h has its max and min at the endpoints of its domain, namely, at $y = -1$ and $y = 1$. We substitute these in the constraint equation $x^2 + 2y^2 = 2$:

when $y = -1$, $x^2 + 2 = 2$ and $x = 0$
when $y = 1$, $x^2 + 2 = 2$ and $x = 0$

Since $f(0, 1) = 4$ and $f(0, -1) = -8$, we conclude that f, subject to the constraint, has a maximum at $(0, 1)$ and a minimum at $(0, -1)$.

LAGRANGE MULTIPLIERS

In all the examples done above it was easy to solve for one variable in the constraint equation and then substitute for it in $f(x, y)$. Often, however, this is difficult or even impossible. The method of Lagrange multipliers, named after the great French mathematician Joseph Louis Lagrange (1736–1813), is extremely powerful and does not depend on being able to solve a side-condition equation for one of the variables.

THE METHOD

To maximize or minimize $f(x, y)$ subject to the constraint equation $g(x, y) = 0$, we

(1) form the new function

$$F(x, y, \lambda) = f(x, y) + \lambda g(x, y)$$

The new variable, λ (Greek letter *lambda*), is called a *Lagrange multiplier*.

(2) solve the system of equations

$$\frac{\partial F}{\partial x} = 0 \qquad \frac{\partial F}{\partial y} = 0 \qquad g(x, y) = 0$$

Goal 54 | **To Find Maxima and Minima Using Lagrange Multipliers**

(3) evaluate $f(x, y)$ at each point (a, b) found in step (2). If f has a maximum or minimum it will occur at one of these points.

Example 54-5

Use Lagrange multipliers to find the maxima and minima of $f(x, y) = x + y$ subject to the constraint $x^2 + 3y^2 = 3$.

Note first that $g(x, y) = x^2 + 3y^2 - 3$. (Why?) We now follow the steps outlined above.

(1) $F(x, y, \lambda) = f(x, y) + \lambda g(x, y) = x + y + \lambda(x^2 + 3y^2 - 3)$
(2) $\partial F/\partial x = 1 + 2\lambda x = 0$; $\partial F/\partial y = 1 + 6\lambda y = 0$; $g(x, y) = x^2 + 3y^2 - 3 = 0$. From the first and second equations we get

$$\lambda = -\frac{1}{2x} \quad \text{and} \quad \lambda = -\frac{1}{6y}$$

Therefore

$$-\frac{1}{2x} = -\frac{1}{6y} \quad \text{and} \quad x = 3y$$

Now replace x by $3y$ in the third equation:

$$9y^2 + 3y^2 - 3 = 0; \quad 12y^2 = 3; \quad y = \pm\frac{1}{2}$$

Since $x = 3y$, we have two solutions:

$$\left(\frac{3}{2}, \frac{1}{2}\right) \quad \text{and} \quad \left(-\frac{3}{2}, -\frac{1}{2}\right)$$

3. Since $f(\frac{3}{2}, \frac{1}{2}) = 2$ and $f(-\frac{3}{2}, -\frac{1}{2}) = -2$, we conclude that, subject to the side condition, f has a maximum at $(\frac{3}{2}, \frac{1}{2})$ and a minimum at $(-\frac{3}{2}, -\frac{1}{2})$.

Example 54-6

Solve Example 54-1 above by the method of Lagrange multipliers.

We want to maximize the area $A = xy$ subject to the constraint $x + y = 125$, rewritten as $0 = x + y - 125 = g(x, y)$. We therefore have

(1) $F(x, y) = A(x, y) + \lambda g(x, y) = xy + \lambda(x + y - 125)$.
(2) $\partial F/\partial x = y + \lambda = 0$; $\partial F/\partial y = x + \lambda = 0$; $g(x, y) = x + y - 125 = 0$. From the first pair of equations we get immediately that $x = y$; from the third equation, therefore, we get $x = 62\frac{1}{2} = y$.
(3) We know from the given problem that, since x and y are the dimensions

of a rectangle, they are both positive. (Indeed, the area equals zero if $x = 0$ or 125!) Therefore we conclude that A is a maximum when the rectangle is a square.

NOTES ON THE METHOD

The steps outlined above make the method of Lagrange multipliers seem quite easy. However, it should be noted that solving three equations in three variables often entails very involved algebraic manipulations.

Although it is usually necessary to find the values only of x and y that satisfy the three equations, the Lagrange multiplier λ does have an interpretation. In an application to be considered shortly, we will discuss its economic significance.

No method has been given above for determining whether a solution found actually yields a maximum or minimum. Fortunately, in many practical applications the existence of a max or min can readily be inferred from the nature of the particular problem.

APPLICATION TO FUNCTIONS OF MORE THAN TWO VARIABLES

The following example illustrates the use of Lagrange multipliers in finding the extremes of a function of three variables.

Example 54-7

Find the maximum and minimum of $f(x, y, z) = x + 2y + 3z$ subject to the side condition $x^2 + y^2 + 2z^2 = 38$.

We follow the steps outlined earlier.

(1) $F(x, y, z, \lambda) = x + 2y + 3z + \lambda(x^2 + y^2 + 2z^2 - 38)$.

(2) $\partial F/\partial x = 1 + 2x\lambda = 0$; $\quad \partial F/\partial y = 2 + 2y\lambda = 0$; $\partial F/\partial z = 3 + 4z\lambda = 0$; and $g(x, y, z) = x^2 + y^2 + 2z^2 - 38 = 0$. From the first three equations we get

$$\lambda = -\frac{1}{2x} = -\frac{1}{y} = -\frac{3}{4z}$$

So

$$y = 2x \text{ and } z = \frac{3x}{2}$$

Now we substitute in the fourth equation:

$$x^2 + (2x)^2 + 2\left(\frac{3x}{2}\right)^2 - 38 = 0; \qquad \frac{19}{2}x^2 = 38; \qquad x = \pm 2$$

Goal 54 | **To Find Maxima and Minima Using Lagrange Multipliers**

Using the expressions for y and z in terms of x, we get the two points $(2, 4, 3)$ and $(-2, -4, -3)$.

(3) Since $f(2, 4, 3) = 19$ and $f(-2, -4, -3) = -19$, we conclude that, subject to the side condition, f has a maximum at $(2, 4, 3)$ and a minimum at $(-2, -4, -3)$.

AN APPLICATION TO ECONOMICS

We begin with a specific problem.

Example 54-8

Maximize the productivity function

$$P(x, y) = x^{3/4} y^{1/4}$$

subject to the budget restriction $5x + 20y = 80,000$. (Here x and y are the number of units invested, respectively, in labor and in capital, and \$5 and \$20 are the respective costs per unit. The side condition tells us that the total budget for these two items is \$80,000.)

We have

(1) $F(x, y, \lambda) = x^{3/4} y^{1/4} + \lambda (5x + 20y - 80,000)$

(2) $\dfrac{\partial F}{\partial x} = \dfrac{3}{4} \left(\dfrac{y}{x} \right)^{1/4} + 5\lambda = 0; \quad \dfrac{\partial F}{\partial y} = \dfrac{1}{4} \left(\dfrac{x}{y} \right)^{3/4} + 20\lambda = 0;$ and $g(x, y) = 5x + 20y - 80,000 = 0$. From the first two equations we get

$$\lambda = -\dfrac{3}{20} \left(\dfrac{y}{x} \right)^{1/4} = -\dfrac{1}{80} \left(\dfrac{x}{y} \right)^{3/4}$$

This yields the following:

$$12 \left(\dfrac{y}{x} \right)^{1/4} = \left(\dfrac{x}{y} \right)^{3/4}; \quad 12 \left(\dfrac{y}{x} \right)^{1/4} \left(\dfrac{x}{y} \right)^{1/4} = \left(\dfrac{x}{y} \right)^{3/4} \left(\dfrac{x}{y} \right)^{1/4};$$

$$12 = \left(\dfrac{x}{y} \right)^{4/4} = \dfrac{x}{y}$$

So $x = 12y$. When we substitute in the third equation, we get $5(12y) + 20y = 80,000$, or $y = 1000$. So $x = 12,000$.

(3) The productivity is maximized when 12,000 units are invested in labor and 1000 units in capital.

ECONOMIC SIGNIFICANCE OF λ

Suppose that the total amount budgeted in Example 54-8, instead of $80,000, is B dollars, so that the constraint equation is $5x + 20y = B$. Let us express the maximum production as a function of B.

To do this we proceed precisely as in Example 54-8 until we get to $x = 12y$. But now we substitute in the equation $5x + 20y = B$, obtaining $y = \frac{1}{80}B$. If we denote the solution for (x, y) by (a, b), then the maximum production is

$$P(a,b) = x^{3/4}y^{1/4} \text{ when } x = \frac{12B}{80} \text{ and } y = \frac{B}{80}$$

Substituting yields

$$P(a, b) = \left(\frac{12B}{80}\right)^{3/4}\left(\frac{B}{80}\right)^{1/4} = 12^{3/4}\left(\frac{B}{80}\right)^{4/4} = 12^{3/4} \cdot \frac{B}{80}$$

From Example 54-8 we have an expression for λ in terms of x and y:

$$\lambda = -\frac{1}{80}\left(\frac{x}{y}\right)^{3/4}$$

If we replace (x, y) by (a, b) we have

$$\lambda = -\frac{1}{80}\left(\frac{12B}{80 \cdot \frac{B}{80}}\right)^{3/4} = -\frac{1}{80} \cdot 12^{3/4}$$

So we see that the maximum productivity can be written

$$P(a, b) = -\lambda B$$

But this means that

$$\frac{dP(a, b)}{dB} = -\lambda$$

That is: the rate of change of the maximum productivity with respect to the total amount budgeted is equal to minus the Lagrange multiplier λ!

With a calculator we can obtain the numerical value of λ just above: it is about -0.08. We thus conclude that an increase of, say, $100 in the total budget results in increased productivity of about 8 units. The Lagrange multiplier λ in this context is called the *marginal productivity* (or *utility*) *of money*.

Goal 54 | To Find Maxima and Minima Using Lagrange Multipliers

EXERCISES

1. Use the substitution technique to find the maximum and minimum of $f(x, y)$, if they exist, subject to the given side condition for each of the following.
   (a) $f(x, y) = x^2 - x + y - 2$ subject to $y = x^2 - 3x + 1$
   (b) $f(x, y) = x + y^2 + 3y + 1$ subject to $y^2 + 5y = x - 2$
   (c) $f(x, y) = x^2 + 2y - 2x$ subject to $y = -x^2 + 3x$
   (d) $f(x, y) = 4y^2 - x^2 + 6x - 5$ subject to $x^2 + y^2 = 1$
   (e) $f(x, y) = 2x^2 - y^2 - 18x + 2$ subject to $x^2 + y^2 = 25$

2. Using Lagrange multipliers, find the maximum and minimum of f, if they exist, subject to the given side condition for each of the following.
   (a) $f(x, y) = x + 2y + 1$ subject to $2x^2 + y^2 = 8$
   (b) $f(x, y) = 2x - y - 1$ subject to $x^2 + 3y^2 = 39$
   (c) $f(x, y) = 2x^2 + 3y^2 + 2$ subject to $x - y - 5 = 0$.
   (d) $f(x, y) = -y^2 - x^2 - 3$ subject to $2x + y = 5$
   (e) $f(x, y) = xy + 1$ subject to $x^2 + y^2 = 2$
   (f) $f(x, y, z) = 2x - y + 2z$ subject to $x^2 + y^2 + 2z^2 = 7$
   (g) $f(x, y, z) = 3x + y - z$ subject to $3x^2 + 2y^2 + z^2 = 2$

3. Suppose that labor and capital cost $4 and $16 per unit, respectively. If $48,000 is budgeted for these two items, and if the production function is

$$P(x, y) = x^{2/3}y^{1/3}$$

where x and y are the numbers of units invested respectively in labor and capital, how many units should be invested in labor and in capital to maximize production?

4. In Exercise 3, let the amount budgeted be B dollars instead of $48,000. Determine the maximum production as a function of B. What is λ when production is maximized? How can maximum production be expressed in terms of B and λ? For each $1 increase in the budget B, how does the maximum production change?

5. Suppose that labor and capital cost $20 and $25 per unit respectively, that $100,000 is budgeted for these items, and that the production function is $P(x, y) = xy$. How many units should be invested in labor and in capital to maximize production? What is the maximum production?

TEST A

1. Use the substitution technique to find the maximum and minimum of $f(x, y)$, if they exist, subject to the given side condition.
   (a) $f(x, y) = x + y^2 - 4y + 3$ subject to $y^2 + 8y - 2 = x$
   (b) $f(x, y) = x^2 + 3y^2 + 8y - 1$ subject to $x^2 + y^2 = 9$
2. Using Lagrange multipliers, find the maximum and minimum of f, if they exist, subject to the given side condition for each of the following.
   (a) $f(x, y) = 3x + y - 2$ subject to $x^2 + 2y^2 = 38$
   (b) $f(x, y) = x^2 + 2y^2 - 1$ subject to $2x - y = 9$

TEST B

1. Use the substitution technique to find the maximum and minimum of $f(x, y)$, if they exist, subject to the given side condition.
   (a) $f(x, y) = y + x^2 + x + 1$ subject to $y = x^2 - 5x + 1$
   (b) $f(x, y) = x^2 + 2y^2 + x - 1$ subject to $x^2 + y^2 = 1$
2. Using Lagrange multipliers, find the maximum and minimum of f, if they exist, subject to the given side condition for each of the following.
   (a) $f(x, y) = 2x - y + 1$ subject to $3x^2 + y^2 = 21$
   (b) $f(x, y) = 3x^2 + y^2 - 2$ subject to $x + y = 8$

STUDY QUESTIONS

1. Many writers of mathematics textbooks use the following alternative procedure with Lagrange multipliers:
   To find extreme values of $f(x, y)$ subject to the constraint $g(x, y) = 0$, the new function

   $$F(x, y) = f(x, y) + \lambda g(x, y)$$

   is formed. Then the following system of equations is solved:

   $$\frac{\partial F}{\partial x} = 0; \quad \frac{\partial F}{\partial y} = 0; \quad \frac{\partial F}{\partial \lambda} = 0$$

   Show that this procedure is equivalent to the one outlined on page 534.
2. Consider the problem of finding the maximum or minimum of $f(x, y, z)$ subject to the *two* side conditions $L(x, y, z) = 0$ and $M(x, y, z) = 0$. This can be solved using *two* Lagrange multipliers, λ and μ. We set

   $$F(x, y, z, L, M) = f(x, y, z) + \lambda L(x, y, z) + \mu M(x, y, z)$$

Goal 54 | To Find Maxima and Minima Using Lagrange Multipliers

If the maximum or minimum exists, it will be among the solutions to the system of equations

$$\frac{\partial F}{\partial x} = 0; \qquad \frac{\partial F}{\partial y} = 0; \qquad \frac{\partial F}{\partial z} = 0; \qquad L = 0; \qquad M = 0$$

Find the maximum of $f(x, y, z) = xy + xz$ subject to the side conditions $xy = 1$ and $x^2 + z^2 = 1$.

3. Suppose that the production function $P(x, y)$ is subject to the budget restriction $c_1 x + c_2 y = B$, where c_1 and c_2 are the respective unit costs. Using the Lagrange method for maximizing P, solve for λ, and show that

$$\frac{\dfrac{\partial P}{\partial x}}{\dfrac{\partial P}{\partial y}} = \frac{c_1}{c_2}$$

This equation says: When the total productivity is maximized, then the marginal productivities of x and y are in the same ratio as their unit costs.

Summary of Goals 51–54

You should now be able to

(1) evaluate $f(a, b)$ and find the domain of f, given $z = f(x, y)$;
(2) find the traces of a surface;
(3) graph a surface such as $z = x^2 + y^2$ or $z = \sqrt{9 - x^2 - y^2}$;
(4) find and sketch some level curves of functions like those in (3);
(5) state whether a given pair of demand functions describes products that are complementary, competitive, or neither;
(6) give an example of demand functions for a pair of complementary or competitive products;
(7) given a function of two or more variables, find and evaluate its partial derivatives and its second partial derivatives;
(8) determine a rate of change of $z = f(x, y)$ with respect to x or y in a practical problem; e.g., find the marginal productivity of labor or of capital;
(9) find the total differential of a function of two or more variables;
(10) given $z = f(x, y)$, use the total differential df as an approximation for Δf;
(11) use the partial derivatives of demand functions for two products to determine whether the products are competitive, complementary, or neither;
(12) find the critical points of a function of two variables;
(13) determine whether a critical point yields a local max, a local min, or neither;
(14) find the straight line that best fits a set of data (the regression line for the data) using the method of least squares;
(15) use the substitution technique to find a max or min of $f(x, y)$, given a constraint or side condition involving x and y;
(16) use Lagrange multipliers rather than substitution in (15);
(17) use Lagrange multipliers to find a max or min of $f(x, y, z)$ in a practical problem.

Review Exercises

GOAL 51

1. (a) The material for a box costs $2 a sq ft for the bottom, $1.50 a sq ft for the sides, and $1 a sq ft for the top. If its dimensions are l, w, and h (with h the height), express the cost C of the material as a function of l, w, and h. (b) How much does the material cost for a box that has length 5 ft, width 4 ft, and height 3 ft?

2. An instructor in a business calculus course told his class that the course grade would be the average of the grades in three hour-long tests, with the quiz average counted as equivalent to two additional tests. If x denotes the sum of a student's three test scores and y is her quiz average, express her course grade G as a function of x and y. What course grade is given a student whose test scores are 55, 70, and 75 if her quiz average is 82?

3. Find the domain of each function:
   (a) $f(x, y) = x^2 + 4y^2$
   (b) $f(x, y) = \dfrac{4}{xy - x}$
   (c) $f(x, y) = \sqrt{y - x}$
   (d) $f(x, y, z) = x \ln y + \sqrt{z}$

4. (a) Find the traces of $z = 4x^2 + 4y^2$. (b) Describe the curve in which this surface intersects the plane $z = 4$; the plane $z = 16$. (c) Sketch the surface.

5. Sketch the level curves of the surface in Exercise 4 for which $z = 0$, $z = 4$, $z = 16$.

6. Suppose a manufacturer makes a profit of $3, $4, and $5 per item when he produces x, y, and z items respectively of three different products. Express the total profit P in terms of x, y, and z and explain the significance of the level curves for this profit function.

7. What are competitive (or substitute) products? complementary ones?

8. Tell whether each pair of demand functions is for products that are competitive, complementary, or neither:
   (a) $D_1(p_1, p_2) = 10 - 2p_1 + 5p_2$
   $D_2(p_1, p_2) = 8 + p_2 - 3p_2$
   (b) $D_1(p_1, p_2) = 700 - 30p_1 - 10p_2$
   $D_2(p_1, p_2) = 500 - 10p_1 - 4p_2$

GOAL 52

9. Find $f_x(x, y)$ and $f_y(x, y)$ for each function:
   (a) $f(x, y) = 2x^2 - 3y - 5xy^2$
   (b) $f(x, y) = e^{-xy} - x$
   (c) $f(x, y) = \ln \dfrac{x}{y} + \dfrac{x}{y}$

10. Find the four second-order partial derivatives for each function in Exercise 9.

11. A company produces $f(x, y) = 60x^{2/3}y^{1/3}$ units of a product, where x denotes the units of labor, in man-hours per production run, and y denotes the units of capital, in thousands of dollars. (a) Find the marginal productivities of labor and capital if a production run uses 125 man-hours and if the capital invested is $64,000. (b) Interpret the results obtained in part (a).

12. Find the total differential of each function:
    (a) $f(x, y) = x^2y + 1$ (b) $f(x, y, z) = \ln \dfrac{x}{yz}$ (c) $f(x, y) = e^{-xy}$

13. Use differentials to approximate the change in productivity in Exercise 11 when labor is increased from 125 to 130 man-hours and the capital investment is increased from $64,000 to $66,000.

14. Find the percentage change in productivity in Exercise 11 when the capital invested is decreased by $1000 and the labor is increased by 5 man-hours.

15. If a corrugated box is designed to be 3 in. by 3 in. by 5 in. and the manufacturing error tolerance is 0.05 in. in each dimension, what is the maximum possible error in the intended volume of 45 cubic in.

16. Use the total differential to approximate $(3.1)/\sqrt{1 + 7.8}$.

17. Use partial derivatives to explain why the demand functions

$$D_2(p_1, p_2) = 800 - 3p_1^2 - 4p_2$$
$$D_2(p_1, p_2) = 1200 - 5p_1 - p_2$$

are for complementary products.

GOAL 53

18. Find the critical points for each function and determine whether a critical point yields a local max, a local min, or neither:
    (a) $f(x, y) = x^2 + y^2 + 6x - 2y + 5$
    (b) $f(x, y) = -2x^2 + y^2 - xy + 7x - 5y + 2$

19. Find the line that best fits the points (1, 2), (2, 3), (3, 5), and (4, 6).
20. A homeowner used the amount of gas indicated from March through June:

Month	March	April	May	June
Amount of Gas Used, in Therms	230	180	140	110

Use the method of least squares to estimate the amount of gas used in July.

GOAL 54

21. Use the substitution technique to find the local max and min, if they exist, of $f(x, y) = x + y^2 + y - 2$ subject to $x = y^2 + 3y + 1$.
22. Use Lagrange multipliers to find the max and min, if they exist, of $f(x, y) = 2y - x + 3$ subject to $x^2 + 3y^2 = 21$.
23. Find the dimensions of the rectangle of largest area that can be inscribed in a circle of radius 2.

Chapter Test A

1. A seashell seller in Hawaii sells a murex for $0.40, a cowry for $0.30, and a small cone for $0.65. If the numbers he sells of these shells are denoted by x, y, and z respectively, express his revenue R as a function of x, y, and z.
2. For Exercise 1, explain the significance of the level curve for which $R = 200$.
3. If $f(x, y) = y^2 \sqrt{5 - x}$, (a) find $f(1, 3)$; (b) determine the domain of f; (c) identify the curve $f(4, y)$ and interpret it geometrically.
4. (a) Find the traces of $z = \sqrt{9 - x^2 - y^2}$. (b) Tell why $0 \leq z \leq 3$. (c) Describe the cross-sections in the plane $z = k$ where $|k| \leq 3$. (d) Identify the surface.
5. Explain why products whose demand functions are

$$D_1(p_1, p_2) = 25 - 5p_1 + p_2$$
$$D_2(p_1, p_2) = 18 + 3p_1 - 2p_2$$

are called *competitive* or *substitute* products.

6. Evaluate $\dfrac{\partial f}{\partial x}$ and $\dfrac{\partial f}{\partial y}$ at (2, 0) if

   (a) $f(x, y) = xe^y - y \ln x$ (b) $f(x, y) = \dfrac{y^2}{x} + 3x$

7. Find the four second partial derivatives of $f(x, y) = xy^2 - 3x + 4y^3$.

8. A manufacturer's profit when he produces x and y units respectively of two products is equal, in dollars, to

$$P(x, y) = 90x + 40y - 0.2x^2 - 0.1y^2 - 0.05xy - 5000$$

   Find the marginal profits of x and y when $x = 200$ and $y = 100$.

9. Use differentials to approximate the change in profit in Exercise 8 when the production of each product is increased by 5 units. What is the resulting percentage change in the profit?

10. Use the total differential to approximate $60(126)^{2/3}(62)^{1/3}$.

11. Use partial derivatives to characterize demand functions of competitive products.

12. Find the critical point of $f(x, y) = 2x^2 + y^2 - xy + x - 2y - 3$ and determine whether it yields a local max, a local min, or neither.

13. Find the line that best fits the points (0, 3), (1, 3), (2, 4), and (3, 6).

14. Find the extreme values, if any, of $f(x, y) = 25 - x^2 - y^2$ subject to the condition $x + y = 4$, using (a) substitution; (b) Lagrange multipliers.

Appendix

Review of Basic Mathematics

Table I   Natural Logarithms

Table II   Exponentials

Review of Basic Mathematics

Numbers

Integers: ..., −3, −2, −1, 0, 1, 2, 3, ...
Positive integers: 1, 2, 3, ...
Zero: 0
Positive integers: 1, 2, 3, ...
Negative integers: −1, −2, −3, ...
Nonnegative integers: 0, 1, 2, 3, ...
Rational numbers: those that can be expressed in the form p/q, p and q integers, $q \neq 0$. Examples:

$$\tfrac{3}{2} \qquad -\tfrac{1}{5} \qquad 6 = \tfrac{6}{1} \qquad 0.37 = \tfrac{37}{100}$$

Every rational number can be expressed as a terminating or repeating decimal.

Irrational numbers: numbers that are not rational. Examples:

$$\sqrt{2} \qquad \sqrt[3]{3} \qquad \pi$$

Every irrational number corresponds to an infinite nonrepeating decimal.

Real numbers: the set, denoted by $\mathscr{R}$, composed of the rationals and the irrationals. Every real number corresponds to exactly one point on the number line, and conversely.

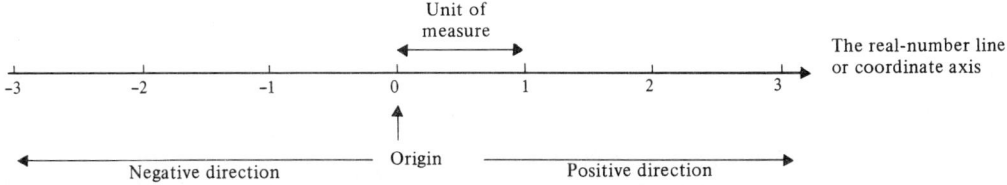

Square-root convention: If b is positive, then $\sqrt{b}$ is the *positive* number c such that $c^2 = b$. Examples:

$$\sqrt{9} = 3 \qquad \sqrt{(-4)^2} = 4 \qquad \sqrt{8} = 2\sqrt{2}$$

Inequalities

Definitions: a is *less than* b if a is to the left of b on the number line; a is *greater than* b if a is to the right of b on the number line.

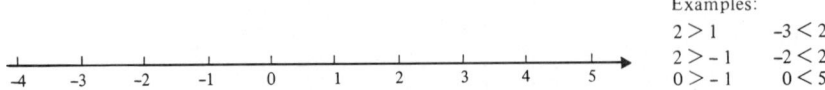

Examples:
$2 > 1 \qquad -3 < 2$
$2 > -1 \qquad -2 < 2$
$0 > -1 \qquad 0 < 5$

Properties:

(a) If $a < b$, then $a + c < b + c$ for any number c.
(b) If $a < b$ and $c > 0$, then $ac < bc$.
(c) If $a < b$ and $c < 0$, then $ac > bc$.
(d) If $a < b$ and $b < c$, then $a < c$.

Note that property (c) says that the sense of an inequality is *reversed* if we multiply or divide by a *negative* number.

Examples:

(a) $3 < 5$ and $3 + 9 < 5 + 9$ (that is, $12 < 14$);
$3 < 5$ and $3 - 4 < 5 - 4$ (that is, $-1 < 1$).
(b) $-1 < 4$ and $5(-1) < 5(4)$ (that is, $-5 < 20$).
(c) $-1 < 4$ but $(-3)(-1) > (-3)(4)$ (that is, $3 > -12$).
(d) $-5 < -1$ and $-1 < 2$; also $-5 < 2$.

Note also that "$a \leq b$" denotes that a is less than or equal to b. If x is nonnegative, then $x \geq 0$; if x is positive, then $x > 0$; if x is negative, then $x < 0$.

Intervals

An *open interval* (a, b) consists of all numbers x such that $a < x < b$; an open interval does not include the endpoints.

A *closed interval* $[a, b]$ consists of all x such that $a \leq x \leq b$; a closed interval includes both endpoints.

A *half-open interval* (a, b] or [a, b), denoted respectively also by

$a < x \leq b$
$a \leq x < b$

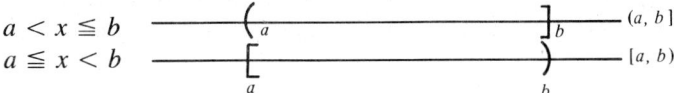

— (a, b]
— [a, b)

includes only one endpoint.

Absolute Value

The absolute value of x, denoted by $|x|$, is defined for any real number x by

$$|x| = \begin{cases} x & \text{if } x \geq 0 \\ -x & \text{if } x < 0 \end{cases}$$

$|x|$ is the distance of the point x on a number line from the origin and is always nonnegative.

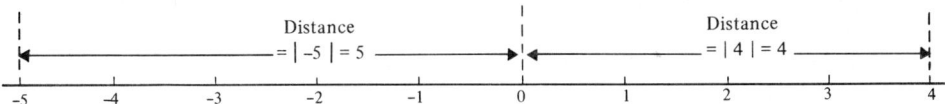

Note, for example, that $|-5| = |5| = 5$; in general, $|x| = |-x|$. Since distance is nonnegative, the distance between two points a and b on a number line is $|a - b|$ or $|b - a|$.

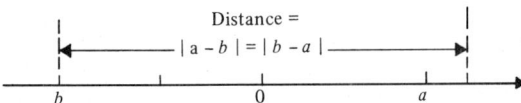

Example: The distance between the points -2 and 4 is $|4 - (-2)| = 6$ or $|-2 - 4| = |-6| = 6$.

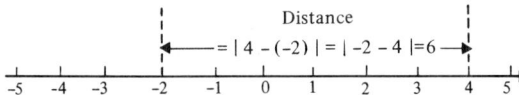

The Pythagorean Theorem

If the lengths of the sides of a right triangle are as shown at the right, then

$c^2 = a^2 + b^2$

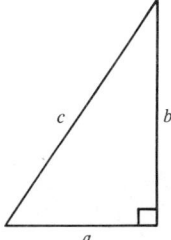

Rectangular Coordinates and Graphing

Rectangular coordinate system: To set one up we draw two number lines perpendicular to each other. They are usually called the *x*- and *y*-axes. The point where they meet is the origin, *O*. They divide the plane into four *quadrants*, numbered as shown. With each point *P* in the plane we associate a pair of coordinates, *x* (the *abscissa*) and *y* (the *ordinate*), for the respective distances of *P* horizontally and vertically from the *y*- and *x*-axes. The figure at the right shows how different points in the plane are associated with pairs of real numbers.

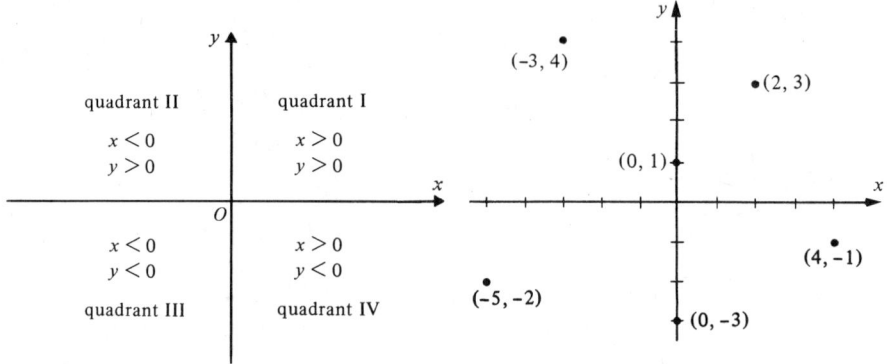

The distance between two points, $P_1(x_1, y_1)$ and $P_2(x_2, y_2)$, can be obtained immediately from the Pythagorean Theorem:

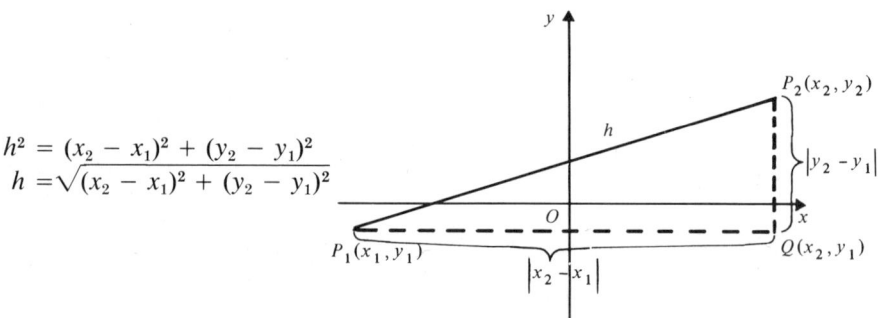

$$h^2 = (x_2 - x_1)^2 + (y_2 - y_1)^2$$
$$h = \sqrt{(x_2 - x_1)^2 + (y_2 - y_1)^2}$$

Example: If $P_1 = (-2, 3)$ and $P_2 = (4, -1)$, then

$$P_1P_2 = \sqrt{[(4 - (-2)]^2 + [-1 - 3]^2} = \sqrt{36 + 16} = \sqrt{52} = 2\sqrt{13}$$

The *graph* of an equation in *x* and *y* is the set of all points whose coordinates satisfy the equation.

The *equation* of a circle is derived easily from the distance formula, using the fact that any point (*x*, *y*) on the circle is at a fixed distance (the radius) from the center.

Review of Basic Mathematics

If the center of the circle is (h, k) and the radius is r, then

$\sqrt{(x - h)^2 + (y - k)^2} = r$

$(x - h)^2 + (y - k)^2 = r^2$

The equation of a circle with center at the origin $(0, 0)$ is

$$x^2 + y^2 = r^2$$

Examples:

(a) $(x - 4)^2 + (y + 1)^2 = 9$ is the equation of the circle with center at $(4, -1)$ and radius 3.

(b) $x^2 + y^2 = 16$ is the equation of the circle with center at O and radius 4.

Products of Binomials

PRODUCT OF THE SUM AND DIFFERENCE OF TWO QUANTITIES
$$(a + b)(a - b) = a^2 - b^2$$

Examples:

$$(x + 2)(x - 2) = x^2 - 4$$
$$(3s - 2t)(3s + 2t) = 9s^2 - 4t^2$$

SQUARE OF A BINOMIAL
$(a + b)^2 = a^2 + 2ab + b^2$
$(a - b)^2 = a^2 - 2ab + b^2$

Examples:
$(2x + 5)^2 = 4x^2 + 20x + 25$
$(3y - 4z)^2 = 9y^2 - 24yz + 16z^2$

FOIL METHOD

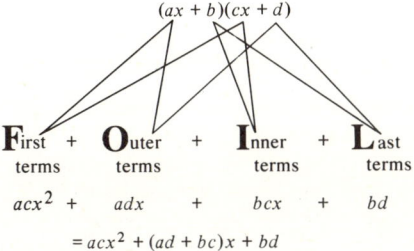

$acx^2 + adx + bcx + bd$

$= acx^2 + (ad + bc)x + bd$

Examples:

$$(x+5)(x-2) = x^2 - 2x + 5x - 10 = x^2 + 3x - 10$$

$$(3p-q)(4p-5q) = 12p^2 - 15pq - 4pq + 5q^2 = 12p^2 - 19pq + 5q^2$$

Factoring

REMOVING COMMON FACTORS
Examples:

$$3x + 3xy = 3x(1+y)$$
$$5(a+b) - (a+b)^2 = (a+b)(5-a-b)$$

DIFFERENCE OF TWO SQUARES
Examples:

$$a^2 - b^2 = (a-b)(a+b)$$
$$9x^2 - 16 = (3x-4)(3x+4)$$
$$y^4 - 81 = (y^2 + 9)(y^2 - 9)$$
$$= (y^2 + 9)(y+3)(y-3)$$

TRINOMIALS

	Examples:
$a^2 + 2ab + b^2 = (a+b)^2$	$x^2 + 16x + 64 = (x+8)^2$
$a^2 - 2ab + b^2 = (a-b)^2$	$y^2 - 6xy + 9x^2 = (y-3x)^2$
$x^2 + (c+d)x + cd = (x+c)(x+d)$	$x^2 - 4x + 3 = (x-3)(x-1)$
	$y^2 - 2y - 3 = (y-3)(y+1)$
	$5s^2 + 19s - 4 = (5s-1)(s+4)$

Solving Equations

LINEAR EQUATIONS
$ax + b = 0$ has the solution $x = -b/a$.
Examples:

(a) $4x + 8 = 0$
    $4x = -8$ (subtracting 8 from both sides)
    $x = -2$ (dividing by 4)

(b) $5x - 2 = 3x + 4$
    $2x - 2 = 4$ (subtracting $3x$)
    $2x = 6$ (adding 2)
    $x = 3$ (dividing by 2)

QUADRATIC EQUATIONS
$x^2 = b$, where $b = 0$, yields $x = \pm\sqrt{b}$.

Review of Basic Mathematics

$(x - a)(x - b) = 0$ yields $x = a$ or $x = b$.
$(ax - c)(bx - d) = 0$ yields $x = c/a$ or $x = d/b$.
$ax^2 + bx + c = 0$ yields the *quadratic formula*

$$x = \frac{-b \pm \sqrt{b^2 - 4ac}}{2a}$$

NOTE: If possible, always factor a quadratic equation into linear factors rather than using the quadratic formula.

Examples:

(a) $x^2 = 4 \to x = \pm 2$.
(b) $x^2 - 8 = 0 \to x^2 = 8 \to x = \pm\sqrt{8} \to x = \pm 2\sqrt{2}$.
(c) $x^2 - 5x - 6 = 0 \to (x - 6)(x + 1) = 0 \to x = 6$ or $x = -1$.
(d) $4x^2 - 12x + 9 = 0 \to (2x - 3)^2 = 0 \to 2x = 3 \to x = \frac{3}{2}$.
(e) $6x^2 - x - 2 = 0 \to (2x + 1)(3x - 2) = 0 \to x = -\frac{1}{2}$ or $x = \frac{2}{3}$.
(f) $x^2 - x - 3 = 0 \to$

$$x = \frac{1 \pm \sqrt{1 + 12}}{2} = \frac{1 \pm \sqrt{13}}{2}$$

Equations in Two Unknowns

(i) Both linear in x and y: For example,

$$2x - 3y = 5 \quad \text{and} \quad x - 5y + 1 = 0$$

By *substitution*. Solve the second equation for x and substitute in the first:

$$x = 5y - 1 \quad\quad 2(5y - 1) - 3y = 5$$
$$10y - 2 - 3y = 5$$
$$7y = 7 \text{ and } y = 1$$

Then $x = 5(1) - 1 = 4$. So the solution is $(4, 1)$.

By *elimination*. We write

$$2x - 3y = 5 \quad\quad\quad \text{(E1)}$$
$$x - 5y = -1 \quad\quad\quad \text{(E2)}$$

Multiply Equation (E2) by 2 and subtract from (E1):

$$\begin{array}{r} 2x - 3y = 5 \\ 2x - 10y = -2 \\ \hline 7y = 7 \text{ and } y = 1 \end{array}$$

Substitute $y = 1$ in (E1) to get $x = 4$, as above.
  (ii) One linear, one quadratic: For example,

$$x + y = 4 \quad \text{and} \quad y = x^2 - 3x - 4$$

Solve the first equation for y and substitute in the second:

$$y = 4 - x \qquad 4 - x = x^2 - 3x - 4$$
$$0 = x^2 - 2x - 8$$
$$0 = (x - 4)(x + 2)$$
$$x = 4 \text{ or } -2$$

Substituting these values for x in the first equation, we get: when $x = 4$, $y = 0$, and when $x = -2$, $y = 6$. So the common solutions (or points of intersection of the graphs) are $(4, 0)$ and $(-2, 6)$.

FRACTIONAL EQUATIONS
Example (a):

$$\frac{1}{3} + \frac{1}{x} = -\frac{1}{2}$$

Multiply through by $6x$, the least common denominator (lcd):

$$(6x)\left(\frac{1}{3} + \frac{1}{x}\right) = (6x)\left(-\frac{1}{2}\right)$$
$$2x + 6 = -3x$$
$$5x = -6 \text{ and } x = -6/5$$

Note that this value of x checks when substituted in the original equation:

$$\frac{1}{3} + \frac{1}{-\left(\frac{6}{5}\right)} = \frac{1}{3} - \frac{5}{6} = \frac{2-5}{6} = -\frac{3}{6} = -\frac{1}{2}$$

Example (b):
$$1 + \frac{2}{x-1} = \frac{4}{x^2 - 1}$$

Multiply by $x^2 - 1$, the lcd:

$$(x^2 - 1)\left(1 + \frac{2}{x-1}\right) = (x^2 - 1)\frac{4}{x^2 - 1}$$
$$x^2 - 1 + 2(x + 1) = 4$$
$$x^2 + 2x - 3 = 0$$
$$(x + 3)(x - 1) = 0$$
$$x = -3 \text{ or } 1$$

First we check $x = -3$; the left side equals

$$1 + \frac{2}{-3-1} = 1 - \frac{2}{4} = \frac{1}{2}$$

The right side equals

$$\frac{4}{(-3)^2 - 1} = \frac{4}{8} = \frac{1}{2}$$

So $x = -3$ is a solution. Now try $x = 1$. If we replace x by 1 in the original equation, we get

$$1 + \frac{2}{0} = \frac{4}{0}$$

Since division by zero is not allowed, we reject $x = 1$. The only solution to the given equation is 3. Possible solutions for fractional equations with variable denominators must *always* be checked.

Solving Inequalities

We use the properties of inequalities noted on page 550. Examples:

(a) $x - 5 \leq 2$
$\quad\quad x \leq 7$ (adding 5 to both sides)

(b) $5x + 8 \geq 3$
$\quad\quad 5x \geq -5$ (adding -8 to both sides)
$\quad\quad x \geq -1$ (dividing by 5)

(c) $2 - 3x \leq 6$
$\quad\quad -3x \leq 4$ (adding -2)
$\quad\quad x \geq -\frac{4}{3}$ (dividing by -3, which *reverses* the sense of the inequality)

Table I Natural Logarithms of Numbers

n	$\ln n$	n	$\ln n$	n	$\ln n$
0.0		4.5	1.5041	9.0	2.1972
0.1	−2.3026	4.6	1.5261	9.1	2.2083
0.2	−1.6094	4.7	1.5476	9.2	2.2192
0.3	−1.2040	4.8	1.5686	9.3	2.2300
0.4	−0.9163	4.9	1.5892	9.4	2.2407
0.5	−0.6932	5.0	1.6094	9.5	2.2513
0.6	−0.5108	5.1	1.6292	9.6	2.2618
0.7	−0.3567	5.2	1.6487	9.7	2.2721
0.8	−0.2231	5.3	1.6677	9.8	2.2824
0.9	−0.1054	5.4	1.6864	9.9	2.2925
1.0	0.0000	5.5	1.7047	10	2.3026
1.1	0.0953	5.6	1.7228	11	2.3979
1.2	0.1823	5.7	1.7405	12	2.4849
1.3	0.2624	5.8	1.7579	13	2.5649
1.4	0.3365	5.9	1.7750	14	2.6391
1.5	0.4055	6.0	1.7918	15	2.7081
1.6	0.4700	6.1	1.8083	16	2.7726
1.7	0.5306	6.2	1.8245	17	2.8332
1.8	0.5878	6.3	1.8405	18	2.8904
1.9	0.6419	6.4	1.8563	19	2.9444
2.0	0.6931	6.5	1.8718	20	2.9957
2.1	0.7419	6.6	1.8871	25	3.2189
2.2	0.7885	6.7	1.9021	30	3.4012
2.3	0.8329	6.8	1.9169	35	3.5553
2.4	0.8755	6.9	1.9315	40	3.6889
2.5	0.9163	7.0	1.9459	45	3.8067
2.6	0.9555	7.1	1.9601	50	3.9120
2.7	0.9933	7.2	1.9741	55	4.0073
2.8	1.0296	7.3	1.9879	60	4.0943
2.9	1.0647	7.4	2.0015	65	4.1744
3.0	1.0986	7.5	2.0149	70	4.2485
3.1	1.1314	7.6	2.0281	75	4.3175
3.2	1.1632	7.7	2.0412	80	4.3820
3.3	1.1939	7.8	2.0541	85	4.4427
3.4	1.2238	7.9	2.0669	90	4.4998
3.5	1.2528	8.0	2.0794	95	4.5539
3.6	1.2809	8.1	2.0919	100	4.6052
3.7	1.3083	8.2	2.1041		
3.8	1.3350	8.3	2.1163		
3.9	1.3610	8.4	2.1282		
4.0	1.3863	8.5	2.1401		
4.1	1.4110	8.6	2.1518		
4.2	1.4351	8.7	2.1633		
4.3	1.4586	8.8	2.1748		
4.4	1.4816	8.9	2.1861		

Table II The Exponential Function 0.00–1.50

x	e^x	e^{-x}	x	e^x	e^{-x}	x	e^x	e^{-x}
0.00	1.0000	1.00 000	**0.50**	1.6487	.60 653	**1.00**	2.7183	.36 788
0.01	1.0101	0.99 005	0.51	1.6653	.60 050	1.01	2.7456	.36 422
0.02	1.0202	.98 020	0.52	1.6820	.59 452	1.02	2.7732	.36 059
0.03	1.0305	.97 045	0.53	1.6989	.58 860	1.03	2.8011	.35 701
0.04	1.0408	.96 079	0.54	1.7160	.58 275	1.04	2.8292	.35 345
0.05	1.0513	.95 123	**0.55**	1.7333	.57 695	**1.05**	2.8577	.34 994
0.06	1.0618	.94 176	0.56	1.7507	.57 121	1.06	2.8864	.34 646
0.07	1.0725	.93 239	0.57	1.7683	.56 553	1.07	2.9154	.34 301
0.08	1.0833	.92 312	0.58	1.7860	.55 990	1.08	2.9447	.33 960
0.09	1.0942	.91 393	0.59	1.8040	.55 433	1.09	2.9743	.33 622
0.10	1.1052	.90 484	**0.60**	1.8221	.54 881	**1.10**	3.0042	.33 287
0.11	1.1163	.89 583	0.61	1.8404	.54 335	1.11	3.0344	.32 956
0.12	1.1275	.88 692	0.62	1.8589	.53 794	1.12	3.0649	.32 628
0.13	1.1388	.87 810	0.63	1.8776	.53 259	1.13	3.0957	.32 303
0.14	1.1503	.86 936	0.64	1.8965	.52 729	1.14	3.1268	.31 982
0.15	1.1618	.86 071	**0.65**	1.9155	.52 205	**1.15**	3.1582	.31 664
0.16	1.1735	.85 214	0.66	1.9348	.51 685	1.16	3.1899	.31 349
0.17	1.1853	.84 366	0.67	1.9542	.51 171	1.17	3.2220	.31 037
0.18	1.1972	.83 527	0.68	1.9739	.50 662	1.18	3.2544	.30 728
0.19	1.2092	.82 696	0.69	1.9937	.50 158	1.19	3.2871	.30 422
0.20	1.2214	.81 873	**0.70**	2.0138	.49 659	**1.20**	3.3201	.30 119
0.21	1.2337	.81 058	0.71	2.0340	.49 164	1.21	3.3535	.29 820
0.22	1.2461	.80 252	0.72	2.0544	.48 675	1.22	3.3872	.29 523
0.23	1.2586	.79 453	0.73	2.0751	.48 191	1.23	3.4212	.29 229
0.24	1.2712	.78 663	0.74	2.0959	.47 711	1.24	3.4556	.28 938
0.25	1.2840	.77 880	**0.75**	2.1170	.47 237	**1.25**	3.4903	.28 650
0.26	1.2969	.77 105	0.76	2.1383	.46 767	1.26	3.5254	.28 365
0.27	1.3100	.76 338	0.77	2.1598	.46 301	1.27	3.5609	.28 083
0.28	1.3231	.75 578	0.78	2.1815	.45 841	1.28	3.5966	.27 804
0.29	1.3364	.74 826	0.79	2.2034	.45 384	1.29	3.6328	.27 527
0.30	1.3499	.74 082	**0.80**	2.2255	.44 933	**1.30**	3.6693	.27 253
0.31	1.3634	.73 345	0.81	2.2479	.44 486	1.31	3.7062	.26 982
0.32	1.3771	.72 615	0.82	2.2705	.44 043	1.32	3.7434	.26 714
0.33	1.3910	.71 892	0.83	2.2933	.43 605	1.33	3.7810	.26 448
0.34	1.4049	.71 177	0.84	2.3164	.43 171	1.34	3.8190	.26 185
0.35	1.4191	.70 469	**0.85**	2.3396	.42 741	**1.35**	3.8574	.25 924
0.36	1.4333	.69 768	0.86	2.3632	.42 316	1.36	3.8962	.25 666
0.37	1.4477	.69 073	0.87	2.3869	.41 895	1.37	3.9354	.25 411
0.38	1.4623	.68 386	0.88	2.4109	.41 478	1.38	3.9749	.25 158
0.39	1.4770	.67 706	0.89	2.4351	.41 066	1.39	4.0149	.24 908
0.40	1.4918	.67 032	**0.90**	2.4596	.40 657	**1.40**	4.0552	.24 660
0.41	1.5068	.66 365	0.91	2.4843	.40 252	1.41	4.0960	.24 414
0.42	1.5220	.65 705	0.92	2.5093	.39 852	1.42	4.1371	.24 171
0.43	1.5373	.65 051	0.93	2.5345	.39 455	1.43	4.1787	.23 931
0.44	1.5527	.64 404	0.94	2.5600	.39 063	1.44	4.2207	.23 693
0.45	1.5683	.63 763	**0.95**	2.5857	.38 674	**1.45**	4.2631	.23 457
0.46	1.5841	.63 128	0.96	2.6117	.38 289	1.46	4.3060	.23 224
0.47	1.6000	.62 500	0.97	2.6379	.37 908	1.47	4.3492	.22 993
0.48	1.6161	.61 878	0.98	2.6645	.37 531	1.48	4.3929	.22 764
0.49	1.6323	.61 263	0.99	2.6912	.37 158	1.49	4.4371	.22 537
0.50	1.6487	.60 653	**1.00**	2.7183	.36 788	**1.50**	4.4817	.22 313

Table II (continued)   The Exponential Function 1.5–10.0

x	e^x	e^{-x}	x	e^x	e^{-x}	x	e^x	e^{-x}
1.5	4.4817	.22 313	3.0	20.086	.04 9787	4.5	90.017	.011 109
1.6	4.9530	.20 190	3.1	22.198	.04 5049	4.6	99.484	.010 052
1.7	5.4739	.18 268	3.2	24.533	.04 0762	4.7	109.95	.009 0953
1.8	6.0496	.16 530	3.3	27.113	.03 6883	4.8	121.51	.008 2297
1.9	6.6859	.14 957	3.4	29.964	.03 3373	4.9	134.29	.007 4466
2.0	7.3891	.13 534	3.5	33.115	.030 197	5.0	148.41	.006 7379
2.1	8.1662	.12 246	3.6	36.598	.027 324	5.5	244.69	.00 40868
2.2	9.0250	.11 080	3.7	40.447	.024 724	6.0	403.43	.00 24788
2.3	9.9742	.10 026	3.8	44.701	.022 371	6.5	665.14	.00 15034
2.4	11.023	.09 0718	3.9	49.402	.020 242	7.0	1 096.6	.00 09119
2.5	12.182	.082 085	4.0	54.598	.01 8316	7.5	1 808.0	.00 05531
2.6	13.464	.074 274	4.1	60.340	.01 6573	8.0	2 981.0	.000 3355
2.7	14.880	.067 206	4.2	66.686	.01 4996	8.5	4.914.8	.000 2035
2.8	16.445	.060 810	4.3	73.700	.01 3569	9.0	8 103.1	.000 1234
2.9	18.174	.055 023	4.4	81.451	.01 2277	9.5	13 360	.000 0749
3.0	20.086	.04 9787	4.5	90.017	.011 109	10.0	22 026	.000 0454

Selected Answers

Answers to Goal Exercises

Answers to Goal Tests A

Answers to Chapter Review Exercises

Answers to Chapter Tests A

Answers to Goal Exercises

GOAL 1

1, 2, 3. The descriptions define functions from S into T in 1, 2, 3, but from T into S only in 2.

4. $\mathcal{R}$ **10.** $\mathcal{R}$
5. $x \neq 0$ **11.** $\mathcal{R}$
7. $x \geqq 0$ **12.** $x \neq 0, 5$

15. $f(-1) = f(3) = f(75) = f(a) = 5$. The domain of f is $\mathcal{R}$; its range is the single number 5.

17. $f(2) = f(-2) = 7$; $f(0) = 3$; $f(10) = 103$; $f(x + h) = x^2 + 2hx + h^2 + 3$; $f(x + h) - f(x) = 2hx + h^2$.

19. $g(1) = 3$; $g(-1) = -1$; $g(0) = -3$; $g(1 + h) = \dfrac{3}{1 + 2h}$.

	$f(x)$:	$f(x + h) - f(x)$:
20.	x^2	$x^2 + 2hx + h^2 - x^2 = 2hx + h^2$
22.	$\dfrac{2}{3x + 1}$	$\dfrac{2}{3x + 3h + 1} - \dfrac{2}{3x + 1} = \dfrac{-6h}{(3x + 3h + 1)(3x + 1)}$
24.	$2x - 1$	$2(x + h) - 1 - (2x - 1) = 2h$

25. F   **27.** T   **29.** F

GOAL 2

1.
  (a) 1   (c) no slope   (f) 0   (g) 2

2.
  (a) $y = 3x + 2$ (c) $y = 2x + 1$ (f) $y = -1$

3.

(a) $y = x - 2$ (c) $x = -2$ (f) $y = 1$ (g) $y = 2x$

4. Several lines are graphed on each set of axes (note that the graph of $y = 0$ is the x-axis and that the graph of $x = 0$ is the y-axis).

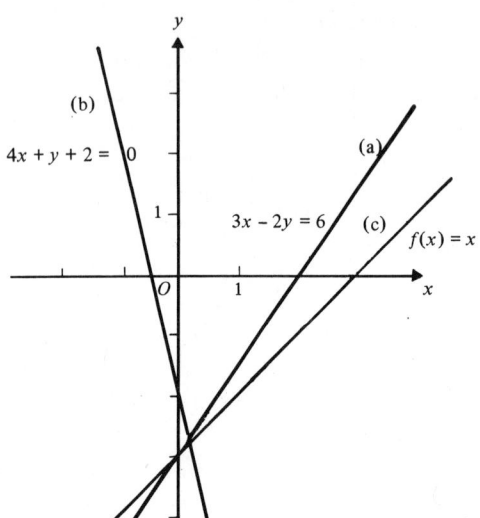

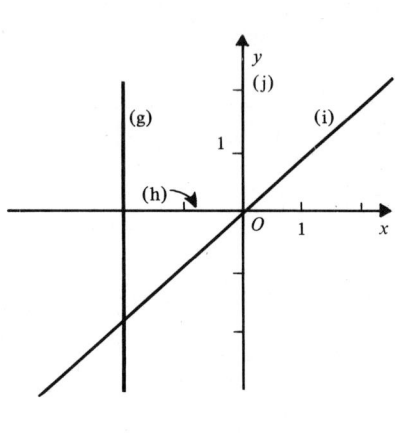

5.

(a)

(d)

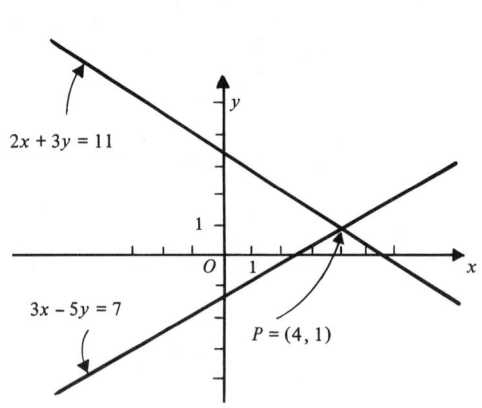

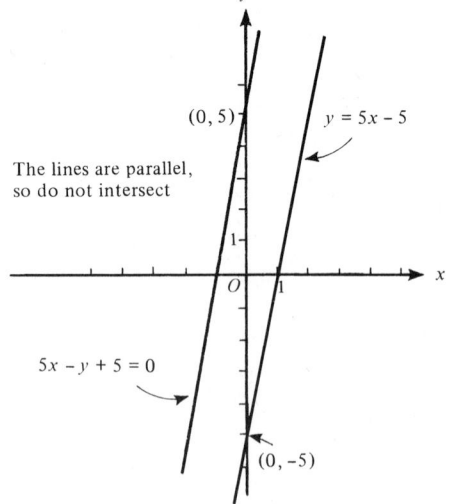

6.

(a) $y = -3$ (c) $y = x + 5$ (f) $4x + 3y = 14$

Answers to Goal Exercises

7. (a) and (d) are increasing; (c) and (f) are decreasing; (b) and (e) are constant functions.

8.
   (a) $C(x) = 200 + 4x$ (b) $C(500) = \$2200$
   (c) $P(x) = 6x - (200 + 4x) = 2x - 200$
   (d) We take the domain to be $x \geq 0$ for both functions, even though practically only whole boxes are produced and there is an upper limit to the number that can be produced.

12.
   (a) $19.25 (b) $450

13. 480 g. $w = 120b$, where w is the weight of the baby's brain in grams and b is its body weight in kilograms (1 kg = 1000 g).

15.
   (a) This is the set of parallel lines with slope 1.
   (b) Each member of this family has y-intercept 3.
   (c) This is a set of parallel lines, each with slope -2.
   (d) Every member of this family goes through the point $(-1,2)$.

17.
   (a) $k = -10$ (b) $k = 12$ (c) $k = 1$

GOAL 3

To save space, we have not sketched all the graphs in the exercises. For those omitted we give crucial facts, such as the location of the vertex and whether the curve opens up or down.

1. Each parabola opens up with vertex $(0,0)$; $y = 0.8x^2$ is least steep, $y = 2x^2$ most.

3. $y = x^2 + 1$ has vertex at $(0,1)$; $y = x^2 - 5$ has vertex at $(0,-5)$. All three parabolas open up.

5. $y = -(x-3)^2$ has vertex at $(3, 0)$; $y = -(x+1)^2$ has vertex at $(-1, 0)$. All three parabolas are concave down.

7. $y = 2(x+1)^2 - 1.5$ has vertex at $(-1, -1.5)$ and opens up.

9.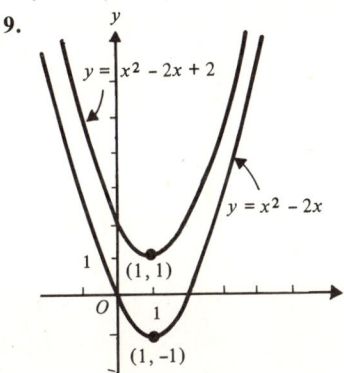

10.

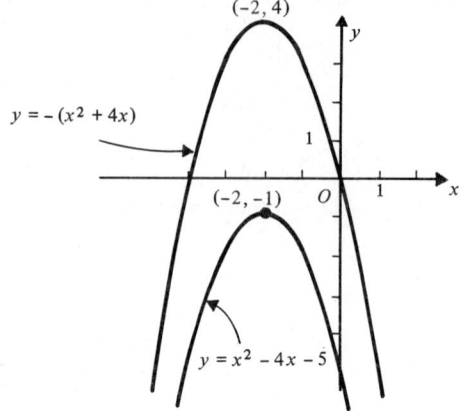

12. $f(x) = 500 + 10x - \dfrac{x^2}{2}$ has vertex at (10,550) and opens down.

14. $f(x) = 150x - 2.5x^2$ has vertex at (30,2250) and opens down.

15. (0, 5); $x = 0$

17. (1, −4); $x = 1$

19. (10,550); $x = 10$

21. 40 sec; maximum height is 1600 ft

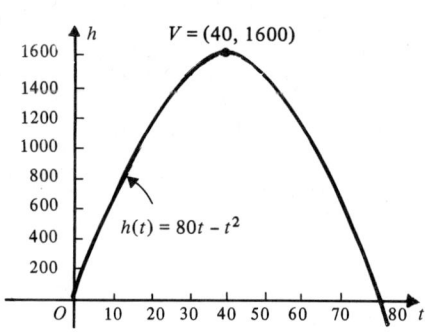

23. 40 mph

26. $y = \sqrt{x}$ is a function on the domain $x \geq 0$. It is the upper half of the parabola $y^2 = x$.

28. $C(x) = 0.002x^2 - 0.4x + 1000$. Producing 100 gadgets yields minimum cost.

Answers to Goal Exercises

GOAL 4

1.
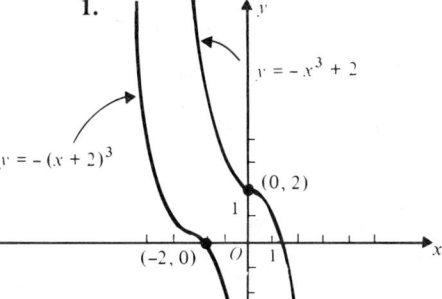

3.
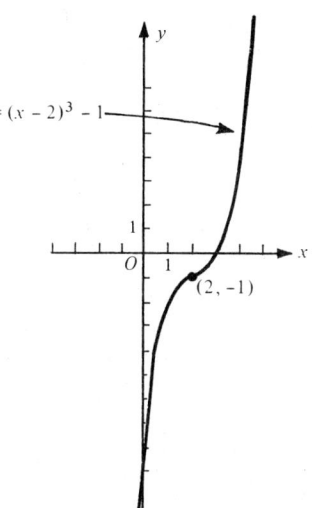

Exercise:	**5.**	**7.**	**9.**
Point of Inflection:	$(0, 0)$	$(0, 6)$	$(-2, -100)$
Concave Up:	$x > 0$	$x > 0$	$x > -2$

11.
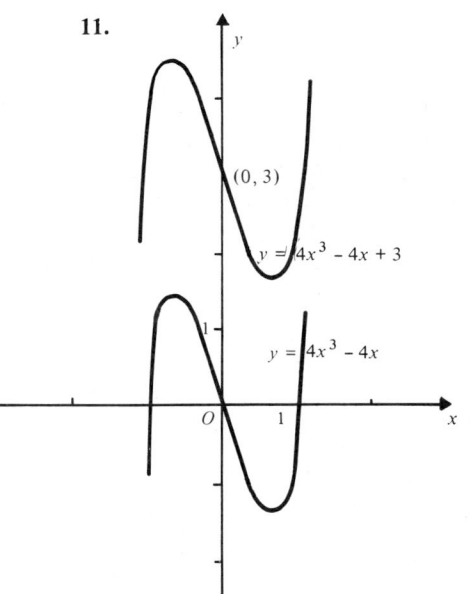

13.
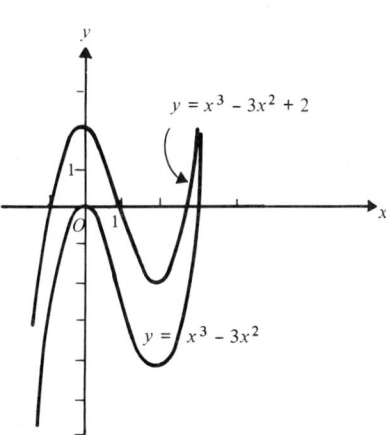

15.

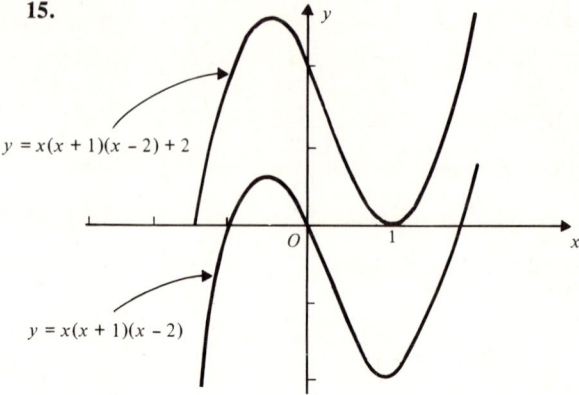

17.

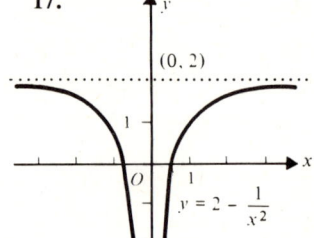

18. $x = 2, y = 0$

20. $x = 3, y = 2$

22. Only (c) is odd; (a), (e), and (g) are even.

24.

(a) $h(-3) = h(-\frac{1}{2}) = -1$; $h(0) = 0$; $h(1) = h(\sqrt{2}) = h(\pi) = 1$. (b) The range of h is $\{-1, 0, 1\}$. h is a function since every real number in the domain of h is associated with only one number.

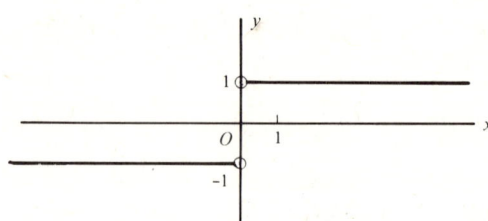

25. The graph is identical with that of $f(x) = |x|$.

Answers to Goal Exercises

27.

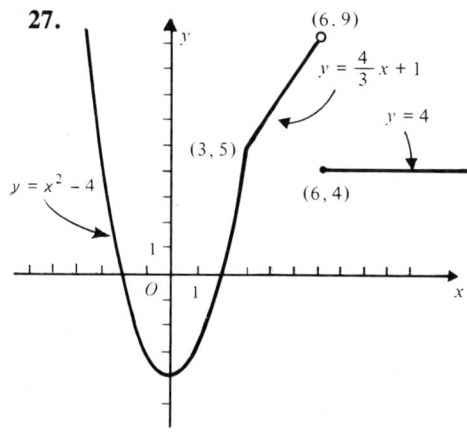

29.

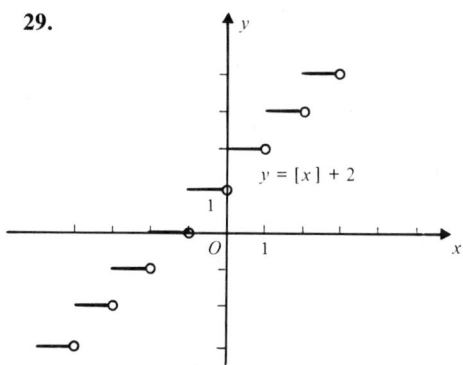

31. If x is the number of students on the tour, then

$$f(x) = \begin{cases} 500x & \text{if } 0 < x < 60 \\ x[500 - (x-60)2] = x(620 - 2x) & \text{if } 60 \leq x \leq 100 \\ 420x & \text{if } x > 100 \end{cases}$$

GOAL 5

	(a) $f(x) + g(x)$:	(b) $f(x)g(x)$:	(c) $3f(x) - g(x)$:	(d) $\dfrac{g(x)}{f(x)}$:
1.	$2x - 1$	$-3x^2 + 5x - 2$	$10x - 7$	$\dfrac{1-x}{3x-2}$
3.	$\dfrac{1}{1-x} + 1 - x^2$	$1 + x \ (x \neq 1)$	$\dfrac{3}{1-x} + x^2 - 1$	$(1-x^2)(1-x) \ (x \neq 1)$

5. $f(g(x))$: $(x+2)^3$ $g(f(x))$: $x^3 + 2$

7. x x

9. $\sqrt{x^2 + 4}$ $x + 4$

For Exercises 11 through 16 we give one choice of f and g; others are possible.

11. Let $f(x) = x^6$, $g(x) = 3x + 2$.

13. Let $f(x) = \dfrac{4}{\sqrt{x}}$, $g(x) = 2 - x$.

15. Let $f(x) = \dfrac{100}{x}$, $g(x) = 4 - 3x$

Of Exercises 17 through 26, numbers 19, 22, 24, and 25 are not one-to-one.

17. $g(x) = \dfrac{x-1}{4}$

21. $g(x) = \dfrac{1}{x} - 2$

26. $g(x) = \dfrac{1}{x-1}$

27. (a)

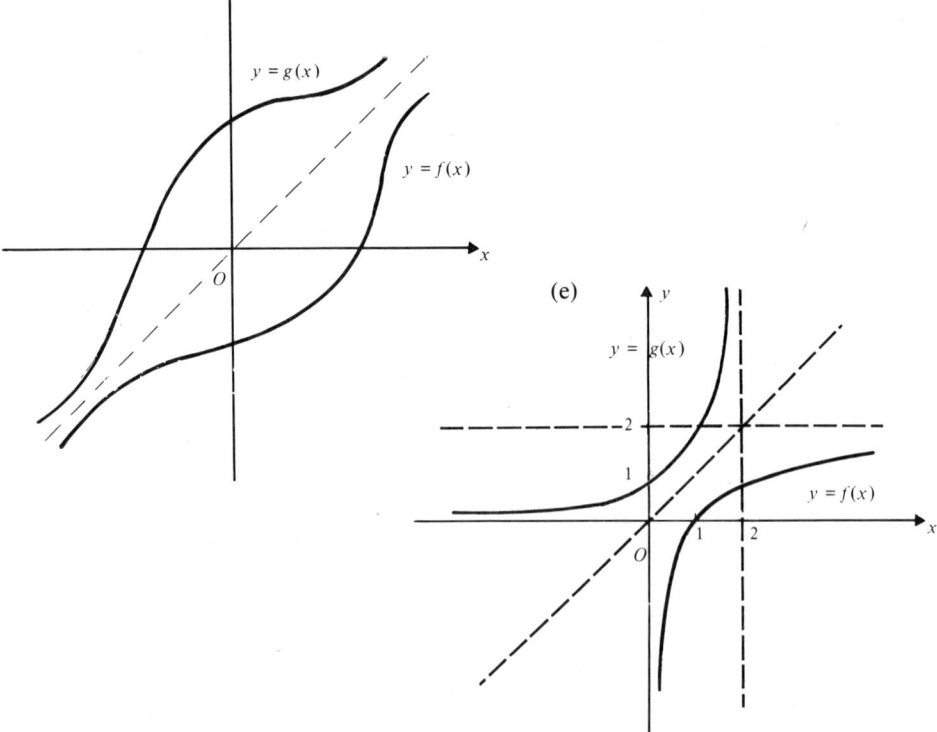

29. $D(p(t)) = \dfrac{4000}{0.1t + 35}$. In 12 months consumers will buy about 110 toasters daily.

GOAL 6

1. $xy = 300$

   (a) $c(x) = 4\left(2x + \dfrac{600}{x}\right)$

   (b) $c(x) = 4\left(x + \dfrac{600}{x}\right)$

   (c) $c(x) = 9x + \dfrac{2400}{x}$

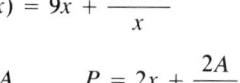

3. $xy = A$ $P = 2x + \dfrac{2A}{x}$

4. $P(t) = \tfrac{25}{29}t + 20$

5. $V = x^2 y = 24$

   $C(x) = \left(0.6x^2 + \dfrac{28.8}{x} + 2\right)$ dollars

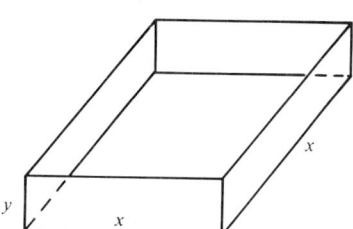

7. If N is the number of oranges, then

   (a) $N(x) = x[\,360 - (x - 20)\ 15] = x(660 - 15x)$

   (b) $N(y) = (360 - 15y)(20 + y)$

9.
   (a) If x is the number on the trip, then

   $$R(x) = x\left(35 - (x - 100)\dfrac{1}{4}\right) = \dfrac{x(240 - x)}{4}$$

   (b) If y is the number of passengers in excess of 100, then

   $$R(y) = (100 + y)\left(35 - \dfrac{y}{4}\right) = \dfrac{(100 + y)(140 - y)}{4}$$

11. If p is the admission price, then we are given that $p(2000) = 8$ and $p(3000) = 6$. Since p is a linear function of x, this yields

    $$p(x) = -\dfrac{x}{500} + 12$$

13. Let x be the number of $2 increases in the weekly rental. Then $R(x) = (50 + 2x)(150 - x)$.

14. If x is the number of tickets sold, then
$$p(x) = -\frac{3}{10,000}x + 27$$

16. $p(x) = -0.004x + 50$.

17. The box has dimensions $9 - 2x$ by $12 - 2x$ by x. Its volume, $V(x)$, is therefore $x(9 - 2x)(12 - 2x)$.

19. $TQ = \sqrt{x^2 + 4}$
$QF = 4 - x$
With x in miles, $t(x)$ in hours, we have

$$t(x) = \begin{cases} \dfrac{\sqrt{x^2+4}}{20} + \dfrac{4-x}{40} + \dfrac{1}{12} & \text{if } x < 4 \\ \dfrac{\sqrt{5}}{10} & \text{if } x = 4 \end{cases}$$

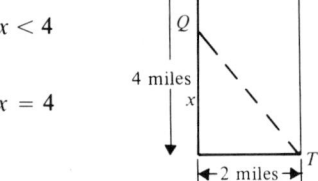

21.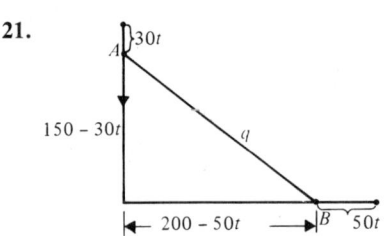

$q^2 = (150 - 30t)^2 + (200 - 50t)^2$
$q = \sqrt{(150 - 30t)^2 + (200 - 50t)^2}$

23. Let y be the length of the box, x that of the side of the square end. The combined length and girth is $4x + y$; so $4x + y = 84$. The volume is x^2y or $x^2(84 - 4x)$. So $V(x) = x^2(84 - 4x)$.

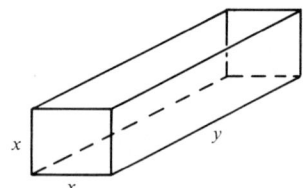

25. Use x for the square. The area of the square is $\dfrac{x^2}{16}$; the area of the triangle is $\dfrac{y^2\sqrt{3}}{36}$. Since $x + y = 20$,
$x = 20 - y$ and $A(y) = \dfrac{(20-y)^2}{16} + \dfrac{y^2\sqrt{3}}{36}$

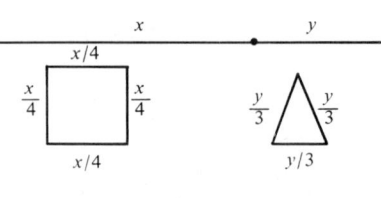

26. $R = 0.0005x(150 - x)$

Answers to Goal Exercises   573

GOAL 7

Note that a negative rate of change at a particular instant (as in Exercises 3 and 5) indicates that the function is decreasing at that instant.

1. Using t from 60 to 61:   2.42
   from 60 to 60.1:   2.402
   from 59.9 to 60:   2.398
3. Using x from 100 to 105: -0.095
   from 100 to 101: -0.099
   from 99 to 100: -0.101
5. Using x from -1.01 to -1: -1.01
   from -1 to -0.99: -0.99
7. Using c from 3 to 4:   7
   from 3 to 3.1:   6.1
   from 2.99 to 3:   5.99

GOAL 8

Below are given, first, complete solutions, including the pictures called for, for a few of the exercises. After that are listed the slopes of the curves in the odd-numbered exercises, at the designated points.

1. $P = (1, 2)$, $Q = (1 + h, 2(1 + h)^2)$.

$$\text{Slope of line } PQ = \frac{2(1 + h)^2 - 2}{(1 + h) - 1} = \frac{2(1 + 2h + h^2) - 2}{h}$$
$$= \frac{4h + 2h^2}{h} = \frac{h(4 + 2h)}{h} = 4 + 2h \text{ if } h \neq 0$$

The slope of the curve $y = 2x^2$ at $(1, 2)$ is 4.

11. $P = (0, 1)$, $Q = \left(h, \dfrac{1}{h + 1}\right)$.

$$\text{Slope of line } PQ = \frac{\dfrac{1}{h+1} - 1}{0 + h - 0} = \frac{\dfrac{1}{h+1} - \dfrac{h+1}{h+1}}{h}$$
$$= \frac{\dfrac{1 - (h+1)}{h+1}}{h} = -\frac{1}{h+1} \text{ if } h \neq 0$$

The slope of $y = \dfrac{1}{x + 1}$ at $(0, 1)$ is -1.

16. $P = (3, 19)$, $Q = (3 + h, 19)$.

Slope of line $PQ = \dfrac{19 - 19}{3 + h - 3} = \dfrac{0}{h} = 0$ if $h \neq 0$

The slope of the line $y = 19$ at $(3, 19)$ is zero.

Slopes of the curves in the odd-numbered exercises at the designated points:

1. 4 **7.** 5 **13.** 2
3. 4 **9.** $-\frac{2}{9}$ **15.** 0
5. 36 **11.** -1 **17.** 1

GOAL 9

1. $6x$ **5.** 8 **9.** $-\dfrac{1}{(x+1)^2}$
3. $3x^2$ **7.** $-\dfrac{2}{x^2}$

GOAL 10

1. 7 **11.** 4 **21.** ∞ **31.** 1
3. 7 **13.** $2x$ **23.** $+\infty$ **33.** No limit
5. 17 **15.** $2x^2 - 3x + 1$ **25.** $-\infty$ **35.** T
7. 0 **17.** 4 **27.** None **37.** F
9. 3 **19.** 5 **29.** 0 **39.** T

GOAL 11

1. The function jumps every quarter mile.
2. The function jumps each instant a person is born or dies.
4. The function jumps whenever a car enters or leaves the tollway.
6. $g(1)$ is not defined.
8. $\lim\limits_{y \to 1} g(y)$ does not exist.
10. $f(6)$ is not defined.
11. $\lim\limits_{x \to 6} f(x) = 12$ but $f(6) = 0$.
13. $\lim\limits_{z \to 1} g(z) = 4$ but $g(1) = 5$.

In problems 16 through 20, the functions in 16, 18, and 20 are continuous everywhere. In 17, g is discontinuous at $x = 1$. In 19, f has points of discontinuity at 0 and -1.

Answers to Goal Exercises

GOAL 12

1. 0
3. 0
5. 3
7. −5
9. $87x^{86}$
11. $8y^7 - 15y^4$
13. $4\frac{2}{3}$
15. $12x^{11} - 10x^9 + 1$
17. $12(3x^2 - 2x + 3)$
19. $-18\frac{3}{4}$

GOAL 13

1. 10 mph
3. −1
5. 10,050 bacteria per hr
7. $59.10 per unit
9. 9000 gal per day
11. 200 people per day
13. $12 per handbick
15. 2 mm per g
17. 89.2 units per h
19. 3 ft per yr

GOAL 14

1.
  (a) $f'(x) = 3x^2 + 2x - 1$
  (c) $h'(z) = \dfrac{1}{(z+1)^2}$
  (e) $c'(x) = (x^3 - 5x^2 + x + 1)(6x - 1) + (3x^2 - x + 2)(3x^2 - 10x + 1)$
  (g) $f'(x) = -10/x^{11}$
  (h) $g'(x) = -\dfrac{19x + 20}{x^{21}}$

2.
  (a) $\frac{1}{2}$ (c) 2   (e) −14

3. $-\frac{3}{4}$

4. (1, 2) and (−1, −2)

7.
  (a) $C'(x) = 20 + 0.2x$
  (b) $A(x) = \dfrac{40 + 20x + 0.1x^2}{x}$
  (c) $A'(x) = \dfrac{0.1x^2 - 40}{x^2}$
  (d) 30, 25.80, $\dfrac{21}{250}$

8. If y is the number of trees per acre in excess of 20 then
  (a) $N(y) = (360 - 15y)(20 + y)$
  (b) $N'(y) = 60 - 30y$

11.

(a) $T'(x) = 2ax - 3bx^2$.

(b) $T'(x) = x(2a - 3bx)$; this equals 0 if $x = 0$ or $\dfrac{2a}{3b}$.

GOAL 15

1. $33x^{32}$
3. $\frac{4}{3}x^{1/3}$
5. $-\frac{3}{4}x^{-7/4}$
7. $\frac{1}{4}x^{-3/4}$
9. $10(x + 1)^9$
11. $15(5x + 2)^2$
12. $-16(8x - 4)^{-3}$
14. $-\dfrac{10}{(5x + 1)^3}$
16. $-2(5x^4 - x + 3)^{-3}(20x^3 - 1)$
18. $-\dfrac{10x}{(x^2 + 1)^6}$
19. $3x(x + 1)^{1/2}$
21. $\dfrac{2x^3}{\sqrt{x^4 - 1}}$
23. $20x^2(x^2 + 1)^9 + (x^2 + 1)^{10}$
25. $\dfrac{x}{2\sqrt{x + 1}} + \sqrt{x + 1}$
26. $(x^2 - 1)^{-5} - 10x^2(x^2 - 1)^{-6}$
28. $12x^2(x^2 + x)^2(x^3 - 1)^3 + 2(x^2 + x)(2x + 1)(x^3 - 1)^4$
30. $-\dfrac{1}{2x\sqrt{x}}$
31. $5\left(\dfrac{x + 1}{x - 1}\right)^4 \left(\dfrac{-2}{(x - 1)^2}\right)$
33. 150
35. -9

GOAL 16

1. $\dfrac{x}{y}$
3. $2xy^2$
5. $\dfrac{6 - 2xy}{x^2}$
7. $-\dfrac{2xy^2}{2x^2y + 1}$
8. $\frac{1}{5}$
10. $\frac{27}{32}$
11. $-\frac{3}{5}$
13. $-\frac{3}{4}$
15. Each unit increase in x results in a decrease of about $\frac{11}{4}$ units in the population y.
17. 1 km

GOAL 17

1. (a) $56x^6$ (c) $6x + 4$ (e) $\dfrac{2}{(x + 1)^3}$

2. (a) $-\frac{1}{108}$ (c) $-\frac{3}{4}$

Answers to Goal Exercises

3.

(a) $120x^3$ (c) $-\dfrac{6}{x^4}$

4. $n = 8$

5.

	Velocity $v(t) = f'(t)$	Acceleration $a(t) = f''(t)$:
(a)	$32t - 5$	32
(d)	$-\dfrac{2}{t^3}$	$\dfrac{6}{t^4}$

6.

(a) all t (c) $t > \tfrac{1}{3}$

8. $n''(t) = 0$ if $t = 8\tfrac{1}{3}$.

GOAL 18

	Equation of Tangent:	Equation of Normal:
1.	$y = 4x + 2$	$y = -\tfrac{1}{4}x + \tfrac{25}{4}$
3.	$y = -6x + 11$	$y = \tfrac{1}{6}x - \tfrac{15}{2}$
5.	$y = 1$	$x = 0$
7.	$y = -\tfrac{1}{9}x + \tfrac{2}{3}$	$y = 9x - \tfrac{80}{3}$
9.	$y = \tfrac{1}{4}x + 1$	$y = -4x + 18$
11.	$y = -\tfrac{3}{4}x + \tfrac{11}{4}$	$y = \tfrac{4}{3}x + \tfrac{2}{3}$
13.	$y = -\tfrac{1}{5}x + \tfrac{8}{5}$	$y = 5x - 14$
15.	$y = -2$	$x = 1$

GOAL 19

1. f decreases if $x < -2$ and if $x > 1$; it increases if $-2 < x < 1$.

3. f decreases if $x < 0$ and if $0 < x < 2$; it increases if $2 < x < 4$ and if $x > 4$.

	Critical Points:	Intervals of Decrease:	Intervals of Increase:
5.	$(2, -3)$	$x < 2$	$x > 2$
7.	$(\tfrac{2}{3}, \tfrac{10}{3})$	$x > \tfrac{2}{3}$	$x < \tfrac{2}{3}$
9.	$(1, 6), (2, 5)$	$1 < t < 2$	$t < 1; t > 2$
11.	$(-2, -16), (2, 16)$	$x < -2; x > 2$	$-2 < x < 2$
13.	$(1, 2)$	—	$x < 1; x > 1$
15.	$(0, 0)$	$x < 0$	$x > 0$
17.	$(-2, -30), (0, 2), (1, -3)$	$x < -2; 0 < x < 1$	$-2 < x < 0; x > 1$
19.	$(0, 0), (2, 48), (3, 27)$	$2 < t < 3$	$t < 0; 0 < t < 2; t > 3$

21. $f(x) = x^{2/3}$ is continuous everywhere, but $f'(x) = \dfrac{2}{3x^{1/3}}$ is not defined at 0. If $x < 0$, f decreases; if $x > 0$, f increases.

23–30. Many different functions have the properties specified in each of these exercises. Below we show the graph of one acceptable function.

23.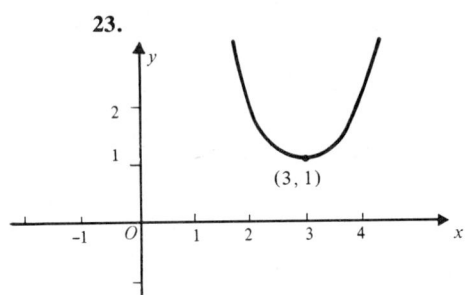

25.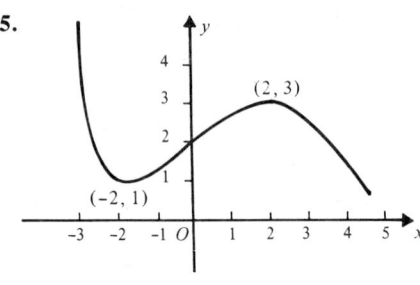

27.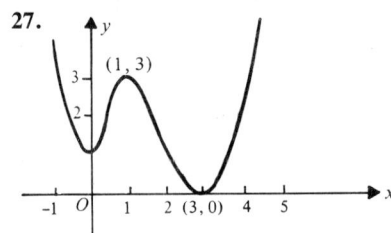

29.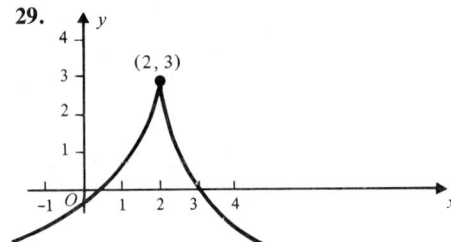

GOAL 20

1. f has a local min at $x = -2$, a local max at $x = 1$.
3. f has a local min at $x = 2$.

	Local Min:	Local Max:
5.	$(2, -3)$	none
7.	none	$(\frac{2}{3}, \frac{10}{3})$
9.	$(2, 5)$	$(1, 6)$
11.	$(-2, -16)$	$(2, 16)$
13.	none	none
15.	$(0, 0)$	none
17.	$(-2, -30), (1, -3)$	$(0, 2)$
19.	$(3, 27)$	$(2, 48)$
21.	$(0, 0)$	none

Answers to Goal Exercises

GOAL 21

	Abs Max:	Abs Min:
1.	(2, 4)	(0, 0)
3.	(4, 23)	(0, −1)
5.	(−1, 7)	(2, −20)
7.	(3, 30)	(−1, −2)
9.	(2, 17)	(−1, −1)
11a.	(1, 1)	(5, $\tfrac{1}{5}$)
11b.	none	(1, 1)
13.	(0, 1)	(−2, $\tfrac{1}{5}$)
17.	(0, 0)	(2, −48)
19.	(5, $\tfrac{2}{3}$)	(0, −1)

20. abs max at (−1, 10) and (3, 10); abs min at (0, 1) and (2, 1)
21. (a) 6 ft; (b) 5 ft
22. (a) 2875; (b) 2500

GOAL 22

1.

	Points of Inflection:	Concave Up:	Concave Down:
(a)	($\tfrac{1}{3}$, $-\tfrac{11}{27}$)	$x > \tfrac{1}{3}$	$x < \tfrac{1}{3}$
(c)	(2, −7)	$x > 2$	$x < 2$
(e)	(−1, −1), (1, −1)	$x < -1, x > 1$	$-1 < x < 1$
(h)	(−1, 0), (1, 0)	$x < -1, 1 < x$	$-1 < x < 1$

HINT: $|x^2 - 1| = x^2 - 1$ if $|x| \geq 1$ but $|x^2 - 1| = 1 - x^2$ if $|x| < 1$

(i) none $\qquad\qquad\qquad\qquad\qquad x > 0 \qquad\qquad x < 0$

(a) $f'(x) > 0$:   (b) $f''(x) > 0$:   (c) Both $f'(x) > 0$ and $f''(x) > 0$:

2. $1 < x < 3 \qquad x < 2 \qquad 1 < x < 2$
5. $x < 2 \qquad x < 2, x > 2 \qquad x < 2$

6. Below we sketch one of many possible functions for the examples shown:

(a) $\qquad\qquad\qquad$ (b) $\qquad\qquad\qquad$ (d)

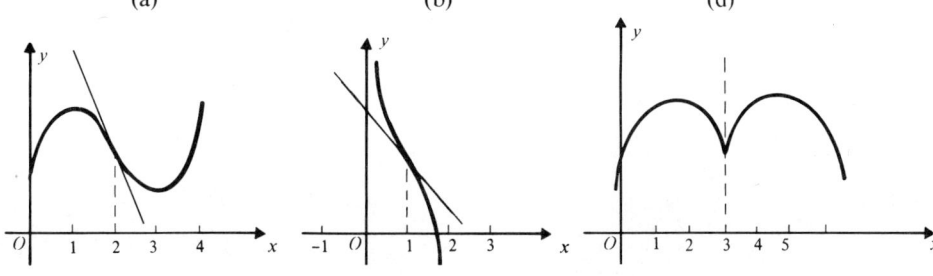

7. See the answers (p. 577) to the exercises for Goal 19.

GOAL 23

1.

(a) 0   (f) ∞ (no limit)   (k) $-\frac{1}{2}$
(c) $-\frac{1}{2}$ (g) $\frac{1}{3}$ (l) ∞ (no limit)
(e) 0   (i) 5

2.

Vertical Asymptotes: Horizontal Asymptotes:
(a) $x = 0$ $y = 0$
(c) $x = 1$ $y = 1$
(e) $x = -1, x = 2$ $y = 1$
(g) $x = 0, x = 1, x = -1$ $y = 5$
(i) $x = 3$ $y = 0$

HINT: Factor the denominator and note that $\lim\limits_{x \to 0} y = \frac{1}{3}$.

3. 0

5. 2750

GOAL 24

1.

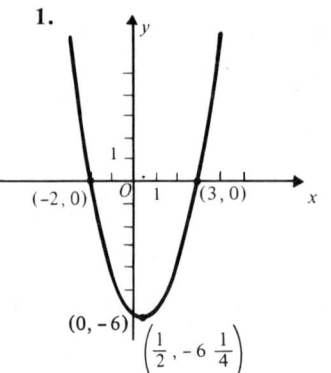

4.

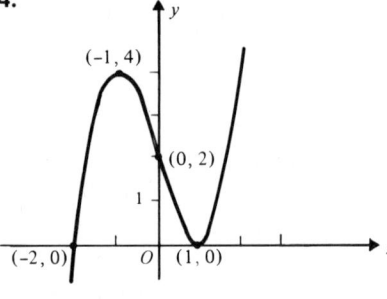

6.

Answers to Goal Exercises

7.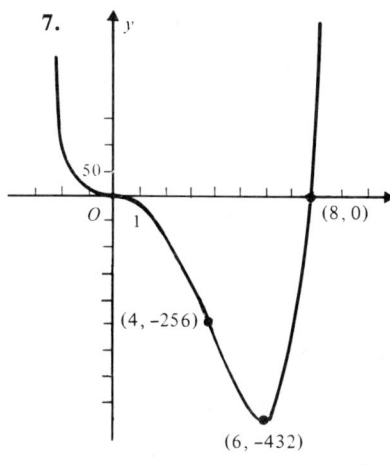

9.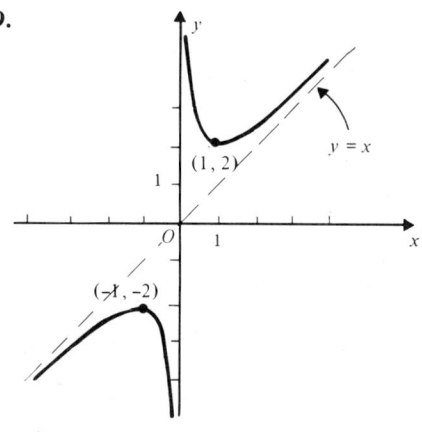

GOAL 25

For exercises based on those in Goal 6 it may be helpful to refer to the answers for Goal 6 to check the functions being maximized or minimized.

1. We have

$$R(r) = r\left(60 - \frac{r - 200}{5}\right) = 60r - \frac{r^2}{5} + 40r = 100r - \frac{r^2}{5}$$

$$R'(r) = 100 - \frac{2r}{5}$$

And this equals zero when $r = 250$. Also

$$R''(r) = -\frac{2}{5}$$

which is less than zero for all r; therefore $r = 250$ yields a maximum.

2. 20 m by 20 m
5. 22 trees per acre
7. 120 passengers; $3600
9. 44,000 tickets
11. Yes. His total time is a minimum when $x = 900/\sqrt{3} \approx 520$ ft; this time is just under 118 sec.
12. The company should provide ferry service along a straight line from the town to the factory. The minimum time (by this route) is under $13\frac{1}{2}$ min.
15. After $\frac{145}{34} \approx 4.3$ h
16. 14 in. × 14 in. × 28 in.
17. 23
18. 25
20. Use all of the wire for the square. (The lone critical point yields a minimum, not a maximum.)

21. 3 ft × 3 ft × 3 ft
22. 4 ft × 4 ft × 3 ft
24. 2.5 sec; 100 ft
25. 40 mph
27. when $t = 10$
28. 14
29. $65; $8450
31. $r = \dfrac{2r_0}{3}$
33. $2.5g$

GOAL 26

1. $\dfrac{8}{5\pi} \approx \dfrac{1}{2}$ cm per h
3. approaching at 3 knots
5. increasing, one per month
7. $1.20 per yr
9. 16,800 km per h per h

GOAL 27

1. 3
3. 1
5. 2
7. -1 and $\frac{4}{3}$
8. The MVT does not apply; the function is not continuous at $x = 1$, which is in the given interval.
10. $\frac{8}{27}$
13. $f(x) = x^3 + 1$
15. $f(x) = \dfrac{1}{x} - 1$
17. $f(x) = \dfrac{x^3}{3} - 1$

GOAL 28

1.
   (a) $\Delta y = 7$
       $dy = 6$
   (c) $\Delta y = -0.59$
       $dy = -0.6$
   (e) $\Delta y = -0.00059999$
       $dy = -0.0006$
2. (a) 8.0625   (c) 2.01667   (e) 9.35259
3. $43
5. $200\pi \approx 628$ ft
7. 5025, 5050, 5075, 5100, 5125
9. ± 0.0083 in.
11. $11,600; $330; 2.84%
13. When $n = 100$ and $p = \$25$, a $1 increase causes a decrease in the demand of approximately 3 items.
14.
   (a) $dy = (6x^2 - 5)\, dx$
   (c) $ds = \dfrac{-t}{\sqrt{4 - t^2}}\, dt$
   (f) $du = 3(x^2 + 3x + 8)^2(2x + 3)\, dx$

Answers to Goal Exercises

GOAL 29

1.

(a)
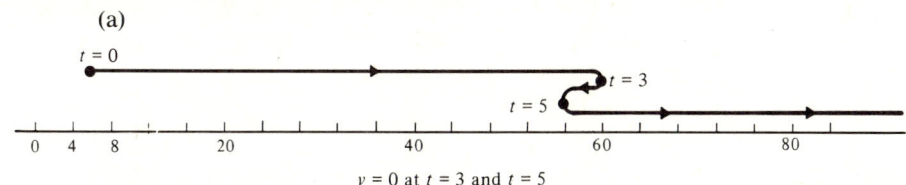
$v = 0$ at $t = 3$ and $t = 5$

(e)
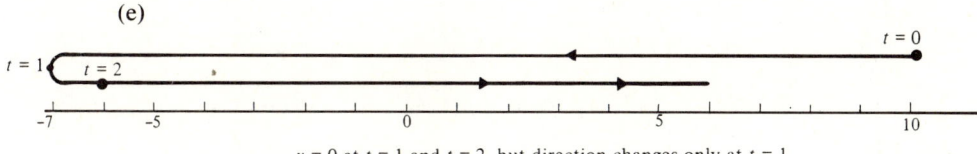
$v = 0$ at $t = 1$ and $t = 2$, but direction changes only at $t = 1$

2. (b) The object is speeding up when $1 < t < 2\frac{1}{2}$ and when $t > 4$, slowing down when $0 < t < 1$ and $2\frac{1}{2} < t < 4$.

3.

	Height of Building (ft)	Initial Velocity (ft/sec)	Maximum Height (ft)	Velocity at Impact (ft/sec)
(a)	320	128	576	-192
(e)	425	-5*	425	-165

4.

(a) Same position at $t = 1$ and $t = 7$; same velocity at $t = 4$.

(b) Same position at $t = \frac{1}{2}$; never have same velocity.

5. Superman will catch Mr. Wright when they are 6 ft above the ground.

GOAL 30

1.

(b) $R(x) = 8x - 0.00002x^2$ $R'(x) = 8 - 0.00004x$ 200,000 units

(d) $R(x) = 10,000$ $R'(x) = 0$ revenue constant

2.

(b) $C'(x) = 3 + 0.00012x$ $A(x) = \dfrac{6000}{x} + 3 + 0.00006x$ 10,000 units

(d) $C'(x) = -10 + 0.002x$ $A(x) = \dfrac{30,250}{x} - 10 + 0.001x$ 5500 units

---

*The negative sign here indicates that the stone was thrown *down*.

3.
(b) $P(x) = -6000 + 5x - 0.00008x^2$ $P'(x) = 5 - 0.00016x$
31,250 units; $72,125
(d) $P(x) = -20,250 + 10x - 0.001x^2$ $P'(x) = 10 - 0.002x$
5000 units; $4750

4.

		Demand Inelastic:	Demand Elastic:
(a) $E(p) = \dfrac{p}{3-p}$	$0 < p < 3$	$p < \tfrac{3}{2}$	$p > \tfrac{3}{2}$
(d) $E(p) = 1$	$0 < p$	never	never
(e) $E(p) = \dfrac{2p}{60-p}$	$0 < p < 60$	$p < 20$	$p > 20$

5.
(a) $1.50   (d) revenue is constant   (e) $20

6. 12,000 boxes at $6.80 per box

8. Not at all. The cost of the units themselves—item (1) on page 291—is a constant regardless of the unit cost. The term has no effect on $C'(x)$.

9. Shipments of 100 clarinets each five times a year; $C(100) = \$20,150$.

GOAL 31

1.
(a) 1   (c) $\tfrac{1}{36}$ (e) 4   (g) 7   (i) $\tfrac{1}{16}$

2.
(a) $7^2 = 49$ (c) $9^0 = 1$ (d) $10^{-2} = \tfrac{1}{100}$

3.
(a) $\log_5 625 = 4$ (c) $\log_2 \tfrac{1}{32} = -5$ (e) $\log_7 7 = 1$

4.
(a) 3   (c) -1 (e) $-\tfrac{1}{2}$ (g) 25   (i) 3

5.
(a) x^9 (b) t^{20} (c) y^5 (d) t^{35} (e) x^2

6.
(a) $\log_b 50$ (c) $\log_b 2$ (e) 2

7.
(a) -2 (d) 5

8.
(a) 1.0792   (c) 0.301   (e) -0.4771

Answers to Goal Exercises

9.

(a)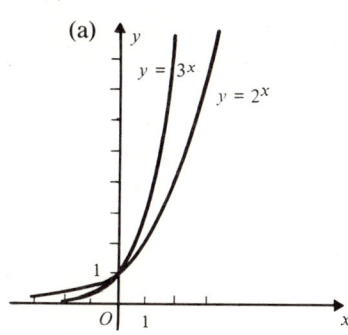

(c) (graph showing $y = 3^x$ and $y = \log_3 x$)

11.

(a) $\log_b b^x = x \log_b b$ [by log law (3)] $= x \cdot 1 = x$.

12.

(a) T (b) F (c) T (d) F (e) T (f) F

13. 10^{-6} watt per sq m

14.

(a) $10^5 I_0$ (b) 100 times more intense

GOAL 32

1.

(a) e^5 (b) e^6 (d) $\dfrac{1}{e^8}$ (g) $t^2 + 2te^t + e^{2t}$

2.

(a) 6.0496 (c) 20.086 (d) 0.14957 (f) 1.0000

3.

(a)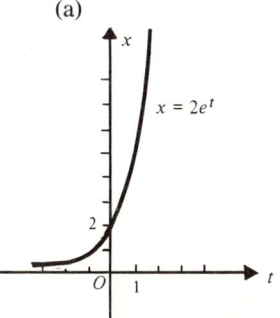

(c)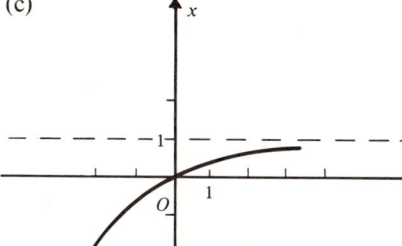

$x = 1 - e^{-(t/2)}$

GOAL 33

1.
(a) 1.792   (c) −1.099   (e) 3.584

2.
(a) 5   (c) 64   (e) $-x + \ln x$ (g) $\dfrac{e^x}{x}$ (h) 0

3.
(a) 1.1632   (c) −1.6094   (e) 4.3175
(g) 4.7875   (i) 18.4208

4.
(a) $y = \ln \dfrac{1}{x} = -\ln x$; the graph is the reflection of $y = \ln x$ in the x-axis.

5.
(a) 0.683   (c) 277.2   (e) 69,320

6. $k = 0.0094$

7.
(a) 11

GOAL 34

1.
(a) $\dfrac{8}{x}$ (g) $\dfrac{2x}{x^2 + 1} + \dfrac{6x}{3x^2 - 2}$

(c) $\dfrac{2}{x}$ (j) $\dfrac{1}{2x\sqrt{1 + \ln x}}$

(d) $\dfrac{2 \ln x}{x}$ (k) $\dfrac{1}{x \ln 2x}$

(f) $x + 2x \ln x$ (l) 0

2.
(a) $y = x - 1$ (c) $y = 0$

3.
(a) $f'(x) = \dfrac{1}{x} \log_{10} e$

5. $\dfrac{dM}{dI} = \dfrac{1}{I} \log_{10} e$ or $\dfrac{1}{(\ln 10) \cdot I}$

6. $f'(x) = \dfrac{1}{x} > 0$ for all $x > 0$, so the graph of f rises; $f''(x) = -\dfrac{1}{x^2} < 0$ for all $x > 0$, so the graph is everywhere concave down.

Answers to Goal Exercises

GOAL 35

1.

(a) $5e^x$ (h) $x^2e^x + 2xe^x$ (n) 0

(d) $-8e^{-8t}$ (j) $\dfrac{2t^2 e^{t^2} - 2e^{t^2}}{t^3}$ (o) 1

(f) $-\dfrac{1}{t^2} e^{1/t}$ (k) $\dfrac{-e^x}{(1+e^x)^2}$ (p) 1

(g) $\dfrac{e^{\sqrt{x}}}{2\sqrt{x}}$ (m) $\dfrac{e^x}{x} + e^x \ln x$

2.

(a) $y = x + 1$ (c) $y = 1$

3.

(a) $f'(t) = 10e^{10t} = 10f(t); k = 10$

(b) $f'(t) = 5(\tfrac{1}{2} e^{t/2}) = \tfrac{1}{2}(5 e^{t/2}) = \tfrac{1}{2} f(t); k = \tfrac{1}{2}$

4.

(a) $f'(t) = -(-e^{-t}) = 7 - (7 - e^{-t}) = 7 - f(t); k = 1, A = 7$

(c) $f'(t) = -2(-4e^{-4t}) = 4[3 - (3 - 2e^{-4t})] = 4[3 - f(t)]; k = 4, A = 3$

6.

(a) $f(t) = 3e^{4t}$ (b) $f(t) = 8e^{-3t}$

7.

(a) $f(t) = 8 - 2e^{-3t}$

8. $f(t) = \dfrac{5}{1 + 4e^{-15t}}$

GOAL 36

NOTE: Many of the numerical results below are rounded off as seems reasonable.

1. $N(t) = 1000e^{0.03t}$, t in hours
3. $2P_0 = P_0 e^{0.04t} \rightarrow 0.04t = \ln 2 \rightarrow t \approx 17.3$ yr
4. about 23,100 yr
7. about 22 yr
10. about $1500
11. about $2019
13. $Q(t) = Q_0 e^{-(9.76 \times 10^{-10} t)}$, t in years
14. about 1860 yr ago
16. 0.005 mg/ml

19. about 25°
21. about 43,200 words
22. about 10 days
24. about $7\tfrac{2}{3}$ h
25. $N(t) = \dfrac{50{,}000}{1 + 499e^{-0.16t}}$, t in days
27. about 507 people
29. 307 cm $\approx$ 3 m

GOAL 37

1. $3\frac{1}{8}$; $3\frac{3}{10}$ miles
3. $101
4. in sq units:
   (a) 39.375 (b) 65 (c) 74.375

GOAL 38

2.
(a) $\lim_{n\to\infty} \left[3\left(\frac{1}{n}\right)^2 \left(\frac{1}{n}\right) + 3\left(\frac{2}{n}\right)^2 \left(\frac{1}{n}\right) + \ldots + 3\left(\frac{n}{n}\right)^2 \left(\frac{1}{n}\right) \right] = 1$

4. The final answer is $\frac{5}{2}$ sq units.
6. The final answer is $\frac{17}{2}$.
7. $\frac{5}{6}$ sq units.
8. $\lim_{n\to\infty} \left\{ \left[\left(\frac{5}{n}\right)^2 + \frac{5}{n}\right]\left(\frac{5}{n}\right) + \left[\left(\frac{10}{n}\right)^2 + \frac{10}{n}\right]\left(\frac{5}{n}\right) + \ldots \right.$
$\left. + \left[\left(\frac{5n}{n}\right)^2 + \frac{5n}{n}\right]\left(\frac{5}{n}\right) \right\} = \lim_{n\to\infty} \frac{5}{n}\left[\left(\frac{5}{n}\right)^2 (1^2 + 2^2 + \ldots + n^2) + \right.$
$\left. \frac{5}{n}(1 + 2 + \ldots + n) \right] = \frac{125}{3} + \frac{25}{2} = \frac{325}{6}$

10. 39 sq units.

GOAL 39

1. $\int_1^3 (5x^3 + x^2 + 2)\, dx$
3. $\int_a^b f(x)\, dx$
6. $\int_{14}^{21} 15(2^t)\, dt$
7. $\int_0^1 W(t)\, dt$
8. $\int_4^6 100e^{-2t}\, dt$
10. $\int_8^{16} R(t)\, dt$

GOAL 40

1.
   (a) 16 (c) 8 (d) $\ln 2 \approx 0.693$
3. $0.41
5. 26 sq units

Answers to Goal Exercises

GOAL 41

1.
(a) $x^7 + C$
(c) $\frac{3}{2}x^4 + C$
(e) $\frac{x^{11}}{11} - \frac{x^{10}}{10} + \frac{7x^5}{5} + 2x + C$
(g) $\frac{x^6}{6} + \frac{x^2}{2} + \frac{1}{x} - \frac{1}{2x^2} + C$
(i) $\frac{x^3}{3} + x + \ln x + \frac{2}{x} + C$
(j) $\frac{x^2}{2} - x + 2\ln x + C$
(l) $2\sqrt{x} + C$
(n) $\frac{4}{5}x^{5/4} + C$
(o) $\frac{2}{5}x^{5/2} + \frac{2}{3}x^{3/2} + 2x^{1/2} + C$
(p) $18e^x + C$
(q) $3e^x - \ln x + C$

2.
(a) $-\frac{17}{12}$ (c) $-\frac{4}{3}$ (e) 3 (g) $-\frac{1}{100}$ (i) $e - 1$

3.
(a) $\frac{338}{3}$ (c) $\$130.31$ (e) 1176 tons

4.
(a) $\frac{1}{2}$ (b) diverges (c) diverges

5.
(a) $\frac{1}{2}$ sq unit (b) The area is infinite since $\int_2^\infty \frac{1}{x}\,dx$ diverges.

GOAL 42

1. $\dfrac{(x^4 - x^2)^5}{5} + C$

4. $\dfrac{(\ln x)^8}{8} + C$

7. $e^{(5x^2 - x + 1)} + C$

8. $e^{(x^3 + 8x)} + C$

9. $\ln(x^3 + x^2 - x) + C$

10. $\ln(4x^2 - 5) + C$

12. $\dfrac{(x^3 + 3x^2 - 3x)^6}{18} + C$

14. $2e^{(x^5 - 5x)} + C$

15. $\frac{1}{3}\ln(3x^4 + x^3 + 1) + C$

17. $\frac{2}{3}(5x^2 + 3x - 2)^{3/2} + C$

19. $\frac{2}{21}(x^7 - 7x^2)^{3/2} + C$

21. $-\dfrac{1}{4(x^2 + 2x - 1)^4} + C$

23. $-\dfrac{5}{x^2 + x} + C$

25. $2e^{\sqrt{x+1}} + C$

26. $\frac{1}{18}(3x + 4)^6 + C$

27. $\frac{1}{5}\ln(5x + 2) + C$

28. $\frac{25}{4}x^4 + 10x^3 + \frac{9}{2}x^2 + C$

31. $\frac{1}{2}$

33. $\frac{8}{3}$

35. $1 - \dfrac{1}{e^2}$

GOAL 43

1. $\frac{1}{3} xe^{3x} - \frac{1}{9}e^{3x} + C$
3. $-x^2 e^{-x} - 2xe^{-x} - 2e^{-x} + C$ or $-e^{-x}(x^2 + 2x + 2) + C$
4. $\frac{1}{6}x^6 \ln x - \frac{1}{36}x^6 + C$
6. $\dfrac{x(x + 3)^{11}}{11} - \dfrac{(x + 3)^{12}}{132} + C$
8. 1
10. $\frac{1}{3}[8 \ln 2 - \frac{7}{3}]$

GOAL 44

1.
  (a) $\int_1^4 [2x - (x^2 - 4x)] \, dx = \int_1^4 (6x - x^2) \, dx = 24$

  (c) $\int_{-1}^0 (4 - x^2 - x) \, dx = \dfrac{25}{6}$

  (f) $\int_{-1}^0 [2 - (-x^3 + x^2 + 6x)] \, dx = \int_{-1}^0 (2 + x^3 - x^2 - 6x) \, dx = \dfrac{53}{12}$

  (h) $\int_0^4 (\sqrt{x} + 2 - x) \, dx = 5\dfrac{1}{3}$

  (j) $\int_0^1 (e^x - x) \, dx = e - \dfrac{3}{2}$

2.
  (a) $\int_1^3 (x + 1) - (x^2 - 3x + 4) \, dx = \int_1^3 (4x - x^2 - 3) \, dx = \dfrac{4}{3}$

  (c) $\int_2^3 (-x^2 + 5x - 6) \, dx = \dfrac{1}{6}$

  (d) $\int_0^1 (\sqrt{x} - x^2) \, dx = \dfrac{1}{3}$

GOAL 45

1.
  (a) $a(t) = 12t^2 - 2$
      $f(t) = t^4 - t^2 + 1$

  (c) $a(t) = 12e^{2t}$
      $f(t) = 3e^{2t} + 5t + 2$

2.
  (a) $v(t) = 3t^2 + 2$
      $f(t) = t^3 + 2t - 3$

  (e) $v(t) = 3e^{3t} + 2$
      $f(t) = e^{3t} + 2t + 1$

  (d) $v(t) = -\dfrac{2}{(t + 1)^2}$
      $f(t) = \dfrac{2}{t + 1} + 1$

  (g) $v(t) = \dfrac{1}{t + 1}$
      $f(t) = \ln(t + 1)$

Answers to Goal Exercises

3. (a) 69 ft (b) 600 ft
4. 100 ft
5. 440 ft
8. $W(t) = 1500e^{0.06t} \approx \$2025$
10. about 22 hr

GOAL 46

1.
   (a) $9\frac{1}{3}$ (d) 32 (e) ≈ 1750 (g) $\ln 2 \approx 0.693$
2. average = $68\frac{1}{3}$; maximum = $72\frac{1}{2}$; minimum = 60
3. ≈ 5.8 lb

GOAL 47

1.
   (a) $15x - 0.001x^2$ (b) $10x - 0.0005x^2$
2.
   (a) $4x + 0.001x^2 + 6000$
   (c) $25x + 0.4x^2 + 0.004x^3 + 25{,}000$
3.
   (a) (5000, \$1.50); \$3750; \$3750
   (c) (3000, \$4.86); \$1980; \$7290
5.
   (a) ≈ 1430; $3750(1 - e^{-q/25})$; 3750
6. $\approx \$5488$
7. $\approx \$9692$
8.
   (a) \$33,180   (c) \$53,000
9.
   (a) about 111 (b) $500(1 - e^{-q/20})$ (c) 500
10. $\approx \$7870$

GOAL 48

1. All except (e), for which $f(x)$ is not nonnegative on its domain
2. $\frac{1}{9}$; $\frac{5}{9}$
4. $\frac{3}{20}$; $\frac{17}{60}$; $\frac{17}{30}$
5. 0.117
7. $\frac{7}{16}$; $\frac{3}{4}$
10. $\frac{1}{6}$

GOAL 49

1.
(a) 1   (c) 1   (d) 2   (f) 3

2.
(a) $f'(x) = 3x^2 + 1$
(b) $f'(x) = e^x + 2x$
(d) $f'(x) = 5(ce^{5x}) = 5f(x)$
(f) $f'(x) = 4x^3 - 3x^2 + C, f''(x) = 12x^2 - 6x$
(h) $f'(x) = \dfrac{C}{x}, f''(x) = -\dfrac{C}{x^2} = -\dfrac{1}{x}\left(\dfrac{C}{x}\right) = -\dfrac{1}{x}f'(x)$

3.
(a) $f(x) = x^3 + x + 3$
(c) $f(x) = x^2 + x$
(e) $f(x) = \dfrac{e^{2x}}{x}$
(g) $f(x) = 2e^{2x}$

4.
(a) $f(x) = \dfrac{x^3}{3} + 2x^2 - x + c$
(e) $f(x) = \dfrac{1}{4}e^{2x} + cx + k$
(c) $f(x) = \ln x - \dfrac{x^2}{2} + c$

5.
(a) $f(x) = x^3 + 3x^2 - x - 1$
(d) $f(x) = \dfrac{x^4}{12} - \dfrac{x^3}{3} - \dfrac{7}{4}x + 2$

GOAL 50

1.
(a) $\dfrac{y^3}{3} = \dfrac{x^2}{2} + x + C$
(g) $y = \dfrac{x^2}{2} + x + C$
(c) $\dfrac{y^2}{2} = \dfrac{x^3}{3} + C$
(h) $-e^{-y} = \ln(x+1) + C$
(e) $\dfrac{y^2}{2} = e^x + C$

2.
(a) $y^2 = x^2 + 2x + 4$
(c) $\ln y = 2 - 2e^{-x}$

3.
(a) $y = ke^{-(x^3/3)-x}$
(d) $y = ke^{1/x}$

4. $\dfrac{y+3}{2-y} = ke^{x/5}$

5.
(a) about 22 yr   (b) 0.005 mg/ml   (d) about 43,200 words

6. $2000

Answers to Goal Exercises

8. $y = 0.2e^{-0.7x}$

10. $P = \dfrac{1}{r}[(rP_0 + I)e^{rt} - I]$

GOAL 51

2.
  (a) $f(6, \tfrac{1}{2}) = \tfrac{6}{7}; f(4, \tfrac{1}{3}) = \tfrac{2}{3}; f(10, \tfrac{1}{5}) = \tfrac{5}{7}$
  (c) $f(3, 0) = 0; f(3, \tfrac{1}{2}) = \tfrac{3}{4}; f(3, 1) = 1$

3.
  (a) $M(75, 6) = 159; M(5, 10) = 5; M(90, 15) = 45$
  (b) $M(100, 1) = 442; M(25, 12) = 17$

4.
  (a) all (x, y) for which $y \neq 0$
  (d) all (x, y, z) for which $x \neq 0$, $y \neq 0$, and $z \neq 0$
  (e) all (x, y, z) for which $xyz \geq 0$
  (f) all (x, y, z) for which $y > 0$

7. $P(x, y) = 70x + 180y, \ x \geq 0, \ y \geq 0$

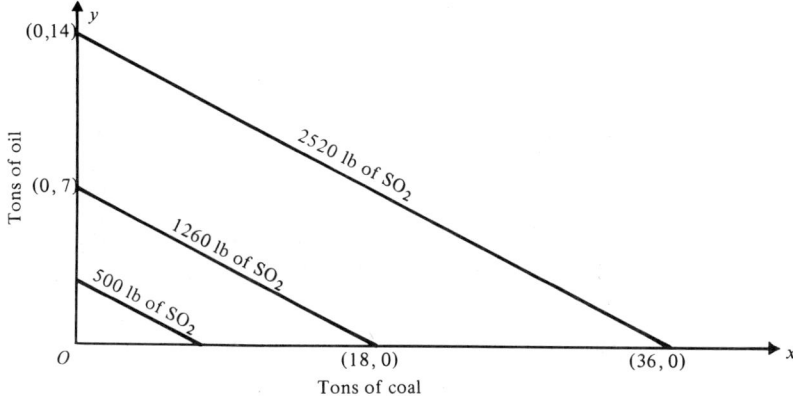

8. $P(x, y) = 10xy, \ x \geq 0, \ y \geq 0$

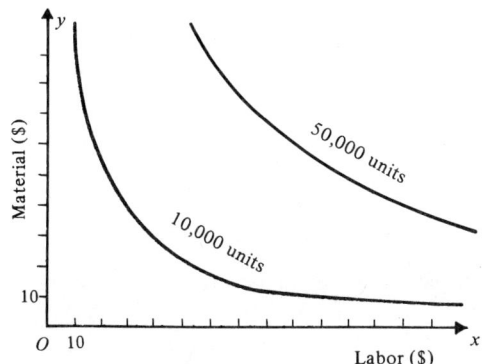

9.
(a) complementary (c) competitive
(b) neither (d) complementary

10.
(a) complementary (d) neither (f) competitive

GOAL 52

1.
(a) $f_x = 1 - y$; $f_y = 2 - x$ (g) $f_x = y^2 e^{xy^2+2}$; $f_y = 2xye^{xy^2+2}$

(c) $f_x = 3ye^{3xy}$; $f_y = 3xe^{3xy}$ (i) $f_x = 2 \dfrac{1}{\sqrt{x+y^2}}$; $f_y = \dfrac{y}{\sqrt{x+y^2}}$

(e) $f_x = \dfrac{1}{y}$; $f_y = -\dfrac{x}{y^2}$

2. $\dfrac{\partial A}{\partial x} = 30.532 \dfrac{y^{0.725}}{x^{0.575}}$; $\dfrac{\partial A}{\partial y} = 52.084 \dfrac{x^{0.425}}{y^{0.275}}$

4.
(a) $0; 1; -1$ (b) $-1; 0; 0$

5.
(a) $\dfrac{\partial^2 f}{\partial x^2} = 2$ $\dfrac{\partial^2 f}{\partial x \partial y} = \dfrac{\partial^2 f}{\partial y \partial x} = 1$ $\dfrac{\partial^2 f}{\partial y^2} = 6$

(c) $\dfrac{\partial^2 f}{\partial x^2} = 0$ $\dfrac{\partial^2 f}{\partial x \partial y} = \dfrac{\partial^2 f}{\partial y \partial x} = -\dfrac{1}{y^2}$ $\dfrac{\partial^2 f}{\partial y^2} = \dfrac{2x}{y^3}$

6.
(a) $df = y^2 \, dx + 2xy \, dy$ (b) $df = 5ye^{5xy} \, dx + 5xe^{5xy} \, dy$

7. $df = 1$; $\Delta f = 1.02$

9. $\approx 0.13\%$

10. 300 ml/min

11. 2.4π

13. 2.0225

14.
(a) competitive   (b) neither   (c) complementary
(d) competitive

16. We expect $\dfrac{\partial g}{\partial p_2}$ to be positive, since if the *Journal* price is held fixed then the number who read it should increase if the price of the *Times* increases. But $\dfrac{\partial g}{\partial p_1}$ should be negative, since if the price of the *Times* is unchanged and that of the *Journal* is raised, then the number of *Journal* readers should decrease.

Answers to Goal Exercises

GOAL 53

1.
   - (a) $(5, -1)$
   - (c) $(-1, 2)$
   - (f) $(0, 1)$
   - (h) $(0, 1), (0, -1)$

2.
   - (a) $f(5+h,-1+k) = h^2 + k^2 - 29; f(5,-1) = -29$, a local min
   - (c) $f(-1+h,2+k) = 2h^2 - k^2 + 5; f(-1,2) = 5$, a saddle point

3.
   - (a) $D = (2)(2) - 0 = 4 > 0$. Since $f_{xx} > 0$, $f(5,-1)$ is a local min.
   - (c) $D = (4)(-2) - 0 = -8 < 0$; f has a saddle point at $(4,-2)$.
   - (f) $D = (-4)(-6) - 16 = 8 > 0$. Since $f_{xx} < 0$, $f(-4,-6)$ is a local max.
   - (h) $(0,1)$: $D = (0)(0) - 4 = -4 < 0$; f has a saddle point at $(0,1)$.
     $(0,-1)$: $D = (0)(0) - 4 = -4 < 0$; f has a saddle point at $(0,-1)$.

4. The regression line is $y = 2.6x + 7.5$, where x is the year (year 1 was 4 yr ago) and y the profits in thousands of dollars. According to this function, profits next year will be approximately $20,500; the year after, approximately $23,100.

GOAL 54

1.
   - (a) no maximum; minimum -3 at $(1,-1)$
   - (c) no minimum; maximum 4 at $(2,2)$
   - (e) maximum 142 at $(-5,0)$; minimum -50 at $(3,\pm 4)$

2.
   - (a) maximum 7 at $(\frac{2}{3}, \frac{8}{3})$; minimum -5 at $(-\frac{2}{3}, -\frac{8}{3})$
   - (c) no maximum; minimum 32 at $(3,-2)$. Note, for example, that $f(5,0) > f(3,-2)$.
   - (e) maximum 2 at $(1,1)$ and at $(-1,-1)$; minimum 0 at $(1,-1)$ and $(-1,1)$
   - (g) maximum 3 at $(\frac{2}{3}, \frac{1}{3}, -\frac{2}{3})$; minimum -3 at $(-\frac{2}{3}, -\frac{1}{3}, \frac{2}{3})$

3. $x = 8000; y = 1000$

5. Labor 2500 units; capital 2000 units; maximum productivity 5,000,000 units.

Answers to Goal Tests A

GOAL 1

1.
   (a) R (b) $x \neq 2$ (c) $x \geq 4$ (d) $x \neq 0, 3$

2.

	$f(x)$:	$f(1)$:	$f(-1)$:	$f(0)$:	$f(-5)$:
(a)	$2x + 4$	6	2	4	-6
(b)	$3x^2 - 1$	2	2	-1	74
(c)	$\dfrac{2}{x-3}$	-1	$-\frac{1}{2}$	$-\frac{2}{3}$	$-\frac{1}{4}$
(d)	$\sqrt{1-x}$	0	$\sqrt{2}$	1	$\sqrt{6}$

3.

	$g(x)$:	$g(x+h) - g(x)$:
(a)	$3x + 1$	$3h$
(b)	$x^2 - 1$	$2hx + h^2$
(c)	$\dfrac{1}{x}$	$\dfrac{-h}{x(x+h)}$
(d)	$\sqrt{x-4}$	$\sqrt{x+h-4} - \sqrt{x-4}$

GOAL 2

1.
   (a) -4 (b) 2   (c) 0   (d) 3

2.
   (a) $y = 2x$ (b) $y = -x + 3$ (c) $x - 3y = 3$

Answers to Goal Tests A

3. Note that two lines are graphed on each pair of coordinate axes (and note that the graph of (c) coincides with the x-axis):

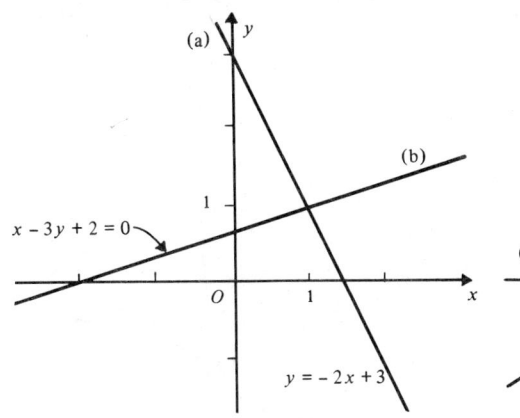
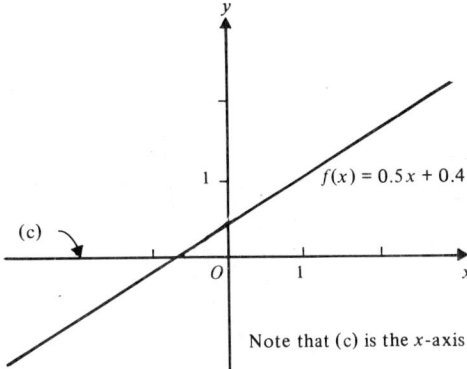

4. $f(x) = -3x + 3$

GOAL 3

1.
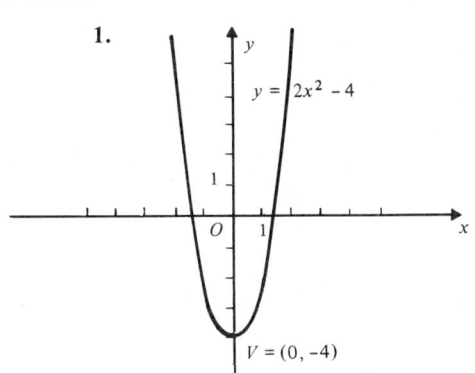

2.
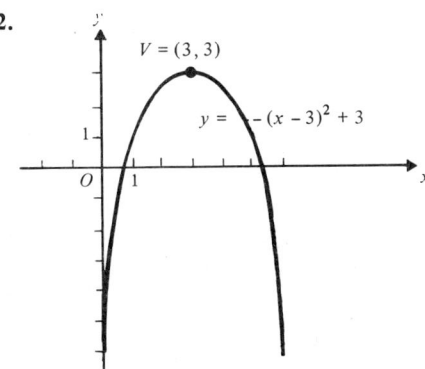

3.
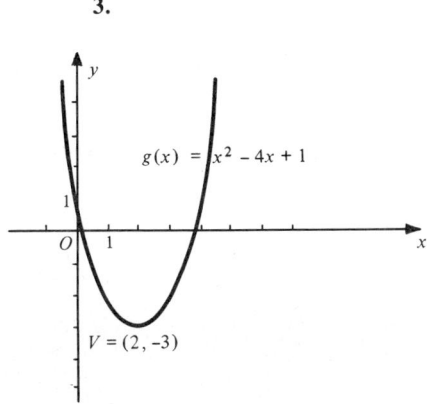

4.
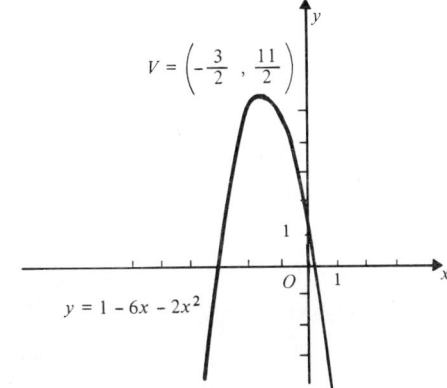

5. $V = (20, -4)$ is a minimum.

6. $V = (3, 104\frac{1}{2})$ is a maximum.

GOAL 4

1.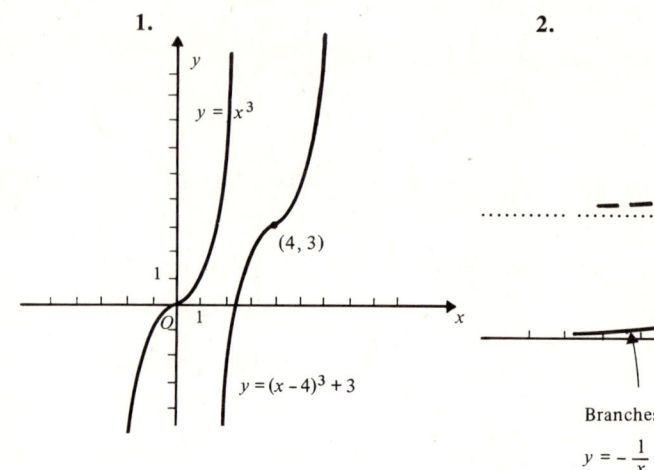

2.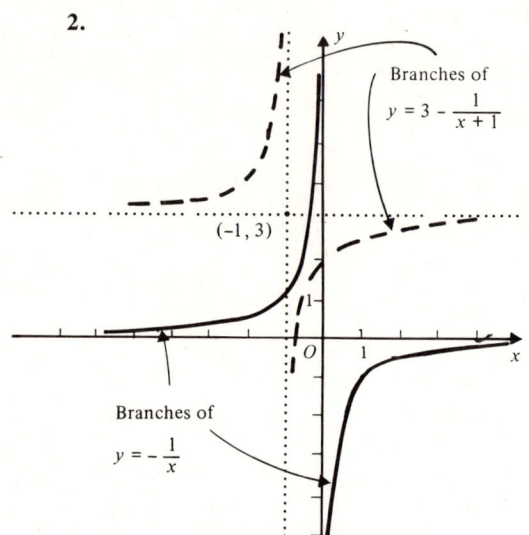

3. The point of inflection of $y = 4 - (x - 2)^3$ is at $(2, 4)$; it is concave up if $x < 2$.

4.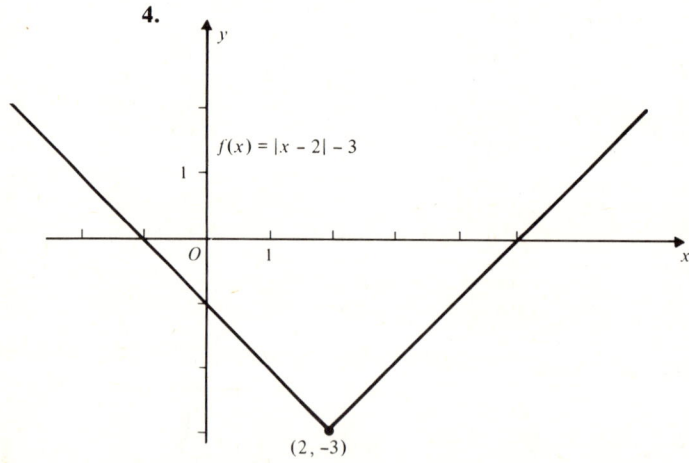

GOAL 5

1. $f(x) + g(x) = 2x^2 - 1 + \dfrac{1}{x + 1}$; $f(x)g(x) = \dfrac{2x^2 - 1}{x + 1}$;

$f(g(x)) = \dfrac{1 - 2x - x^2}{(x + 1)^2}$; $g(f(x)) = \dfrac{1}{2x^2}$.

Answers to Goal Tests A

2. $f(x)$ can be $\dfrac{3}{x^2}$ and $g(x)$ can be $1 - x - x^2$.

3. Only (a) and (d) are one-to-one functions.

4. $g(x) = (x + 4)^{1/3}$ is the inverse of $f(x)$.

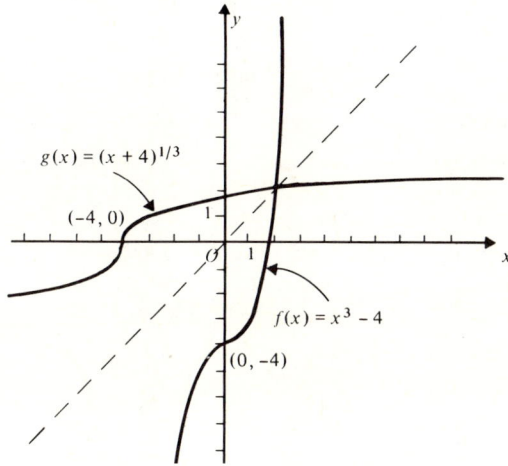

GOAL 6

1.
$2x + y = 100$
Area $= A = xy$
$A(y) = \dfrac{y(100 - y)}{2}$

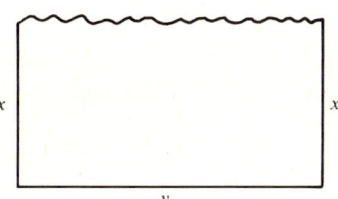

2.
(a) Let x be the new monthly rate. Then
$$R(x) = x\left(25 - \dfrac{x - 95}{5}\right) = \dfrac{x(220 - x)}{5} \qquad (x \geqq 95)$$

(b) Let y be the number of $5 increases in the monthly rate. Then
$$R(y) = (25 - y)(95 + 5y) \qquad (y \geqq 0)$$

3.
(a) If p is the price and x the number of watches he sells then
$$p - 15 = -\dfrac{5}{2000}(x - 10{,}000)$$

So
$$p(x) = -\frac{x}{400} + 40$$

(b) We have
$$C(x) = 5000 + 8x$$
$$R(x) = x\left(-\frac{x}{400} + 40\right) = -0.0025x^2 + 40x$$
$$P(x) = -0.0025x^2 + 32x - 5000$$

GOAL 7

1. Using x from 2 to 2.5: 6.5
   from 2 to 2.1: 6.1
   from 1.99 to 2: 5.99

2. Using t from 4 to 4.2: 830
   from 4 to 4.1: 820
   from 3.99 to 4: 809

GOAL 8

1. $P = (4,8)$, $Q = \left(4 + h, \dfrac{(4+h)^2}{2}\right)$

$$\frac{\dfrac{(4+h)^2}{2} - 8}{h} = \frac{\dfrac{16 + 8h + h^2}{2} - 8}{h}$$

$$= \frac{4h + \dfrac{h^2}{2}}{h} = 4 + \frac{h}{2} \qquad \text{if } h \neq 0$$

Since $4 + \dfrac{h}{2}$ becomes closer and closer to 4 as h becomes closer and closer to 0, the slope of $y = \dfrac{x^2}{2}$ at (4,8) is 4.

2. $P = (2,9)$, $Q = (2 + h, (2 + h)^3 + 1)$.

$$\frac{(2+h)^3 + 1 - 9}{h} = \frac{8 + 12h + 6h^2 + h^3 + 1 - 9}{h}$$

$$= \frac{12h + 6h^2 + h^3}{h} = 12 + 6h + h^2 \qquad \text{if } h \neq 0$$

Since $12 + 6h + h^2$ becomes closer and closer to 12 as h becomes closer and closer to zero, the slope of $y = x^3 + 1$ at (2,9) is 12.

3. $P = \left(3, \dfrac{1}{3}\right)$, $Q = \left(3+h, \dfrac{1}{3+h}\right)$.

$$\dfrac{\dfrac{1}{3+h} - \dfrac{1}{3}}{h} = \dfrac{\dfrac{3}{3(3+h)} - \dfrac{3+h}{3(3+h)}}{h} = \dfrac{\dfrac{-h}{3(3+h)}}{h}$$

$$= \dfrac{-1}{3(3+h)} \quad \text{if } h \neq 0$$

Since $\dfrac{-1}{3(3+h)}$ becomes closer and closer to $-\dfrac{1}{9}$ as h becomes closer and closer to zero, the slope of $y = \dfrac{1}{x}$ at $\left(3, \dfrac{1}{3}\right)$ is $-\dfrac{1}{9}$.

GOAL 9

1. $\dfrac{f(x+h) - f(x)}{h} = \dfrac{[(x+h)^2 + 3] - [x^2 + 3]}{h}$

$= \dfrac{x^2 + 2xh + h^2 + 3 - x^2 - 3}{h} = 2x + h$ if $h \neq 0$.

As $h \to 0$, $2x + h \to 2x$; so $f'(x) = 2x$.

2. $\dfrac{g(3+h) - g(3)}{h} = \dfrac{[2(3+h)^3 + (3+h)] - [2(3)^3 + 3]}{h}$

$= \dfrac{2(27 + 27h + 9h^2 + h^3) + 3 + h - 54 - 3}{h}$

$= \dfrac{54 + 54h + 18h^2 + 2h^3 + 3 + h - 57}{h} = \dfrac{55h + 18h^2 + 2h^3}{h} =$

$55 + 18h + 2h^2$ if $h \neq 0$. As $h \to 0$, $55 + 18h + 2h^2 \to 55$; so $g'(3) = 55$.

3. $\dfrac{g(y+h) - g(y)}{h} = \dfrac{\dfrac{4}{y+h} - \dfrac{4}{y}}{h} = \dfrac{\dfrac{4y}{y(y+h)} - \dfrac{4(y+h)}{y(y+h)}}{h}$

$= \dfrac{4y - 4y - 4h}{y(y+h)} \times \dfrac{1}{h} = \dfrac{-4h}{y(y+h)} \times \dfrac{1}{h} = \dfrac{-4}{y(y+h)}$ if $h \neq 0$.

As $h \to 0$, $\dfrac{-4}{y(y+h)} \to \dfrac{-4}{y(y+0)} = -\dfrac{4}{y^2}$; so $g'(y) = -\dfrac{4}{y^2}$.

GOAL 10

1. 5
2. -2
3. 6
4. $6x$
5. $-\infty$
6. $+\infty$
7. 0
8. No limit

GOAL 11

1. f is not continuous at $x = 3$ because lim $f(x)$ as $x \to 3$ is 10 but $f(3)$ is 8.
2. g is discontinuous at $y = 0$ because lim $g(y)$ as $y \to 0$ does not exist.
3. h is continuous at $z = 2$ (and everywhere else, too!).
4. g is discontinuous at $x = 1$ because $g(1)$ is not defined.
5. f has points of discontinuity at $x = -4$ and $x = 3$.

GOAL 12

1. 5
2. 0
3. $10t^9 - 12t^3 - 2$
4. $\frac{1}{2}$

GOAL 13

1. 26 tons/day
2. 145 people/day
3. $6/unit
4. 44.2 tons/h

GOAL 14

1. $f'(x) = (x^2 + x + 2)(1) + (x - 1)(2x + 1) = 3x^2 + 1$
2. $\dfrac{dy}{dx} = \dfrac{(2 - x)(-1) - (1 - x)(-1)}{(2 - x)^2} = \dfrac{-1}{(2 - x)^2}$
3. $\dfrac{dy}{dx} = \dfrac{-3}{(x + 2)^2}$ and when $x = 1$, $\dfrac{dy}{dx} = \dfrac{-3}{(1 + 2)^2} = -\dfrac{1}{3}$

GOAL 15

1. $\frac{5}{2}x^{3/2}$
2. $-5(3x^2 - x)^{-6}(6x - 1)$
3. $\dfrac{1}{2\sqrt{x + 1}}$
4. $20x(2x + 1)^9 + (2x + 1)^{10}$

GOAL 16

1. $6y \dfrac{dy}{dx} = 2x + 1$ so $\dfrac{dy}{dx} = \dfrac{2x + 1}{6y}$
2. $x \dfrac{dy}{dx} + y \cdot 1 = 5$ so $\dfrac{dy}{dx} = \dfrac{5 - y}{x}$
3. $2x^2y \dfrac{dy}{dx} + 2xy^2 = \dfrac{dy}{dx}$ so $\dfrac{dy}{dx} = \dfrac{2xy^2}{1 - 2x^2y}$
4. $2x + 8yy' = 0$ so $y' = -\dfrac{x}{4y}$; at $(3, -1)$, $y' = \dfrac{3}{4}$

Answers to Goal Tests A

GOAL 17

1. $f'(x) = 15x^2 - 6x + 1$ and $f''(x) = 30x - 6$
2. $y = 2x^{-1}$; $\dfrac{dy}{dx} = -2x^{-2}$; $\dfrac{d^2y}{dx^2} = 4x^{-3} = \dfrac{4}{x^3}$
3. $f'(x) = 7x^6 - 4x$; $f''(x) = 42x^5 - 4$; $f'''(x) = 210x^4$; $f'(1) = 3$; $f''(1) = 38$; $f'''(1) = 210$.
4.
   (a) $v(t) = f'(t) = 4t^3 - 12t^2$
   (b) $a(t) = f''(t) = 12t^2 - 24t = 12t(t - 2)$; $a(t) > 0$ when $t > 2$.

GOAL 18

1. $\dfrac{dy}{dx} = 4x + 1$ and $\dfrac{dy}{dx}$ at $(1, 0)$ is 5. The slope of the tangent is 5; the slope of the normal is $-\tfrac{1}{5}$.

   The equation of the tangent is $\dfrac{y - 0}{x - 1} = 5$, or $y = 5x - 5$.

   The equation of the normal is $\dfrac{y - 0}{x - 1} = -\dfrac{1}{5}$, or $y = -\dfrac{1}{5}x + \dfrac{1}{5}$.

2. $\dfrac{dy}{dx} + 2xy\dfrac{dy}{dx} + y^2 = 0$; $\dfrac{dy}{dx} = -\dfrac{y^2}{1 + 2xy}$; at $(2, 1)$,
   $\dfrac{dy}{dx} = -\dfrac{1}{1 + 4} = -\tfrac{1}{5}$.

   The slope of the tangent is $-\tfrac{1}{5}$; the slope of the normal is 5.

   The equation of the tangent is $\dfrac{y - 1}{x - 2} = -\dfrac{1}{5}$, or $y - 1 = -\dfrac{1}{5}x + \dfrac{2}{5}$, or

   $$y = -\dfrac{1}{5}x + \dfrac{7}{5}$$

   The equation of the normal is $\dfrac{y - 1}{x - 2} = 5$, or $y - 1 = 5x - 10$, or

   $$y = 5x - 9$$

GOAL 19

1.
   (a) $f'(x) = 6x^2 + 30x + 36 = 6(x^2 + 5x + 6) = 6(x + 2)(x + 3)$; this is zero if $x = -2$ or -3; these are the critical points.

Interval	x_0	Sign of $f'(x_0)$
$x < -3$	-4	$6(-2)(-1) > 0$
$-3 < x < -2$	$-2\frac{1}{2}$	$6(-\frac{1}{2})(\frac{1}{2}) > 0$
$-2 < x$	0	$6(2)(3) > 0$

Therefore

if $x > -3$	f increases
if $-3 < x < -2$	f decreases
if $-2 < x$	f increases

(b) $g'(x) = 12x^3 - 12x^2 = 12x^2(x-1)$; this is zero if $x = 0$ or 1, the critical points.

Interval	x_0	Sign of $g'(x_0)$
$x < 0$	-1	$12(-1)^2(-2) < 0$
$0 < x < 1$	$\frac{1}{2}$	$12(\frac{1}{2})^2(-\frac{1}{2}) < 0$
$1 < x$	2	$12(2)^2(1) > 0$

Therefore

if $x < 0$	f decreases
if $0 < x < 1$	f decreases
if $1 < x$	f increases

2.

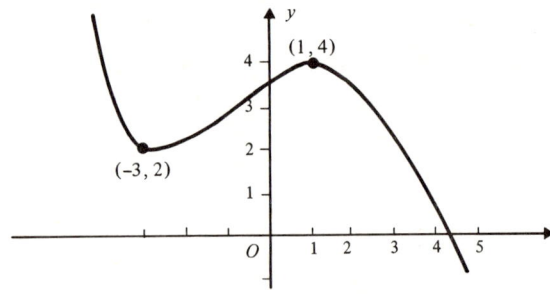

(Many answers are possible.)

GOAL 20

1. $f'(x) = 12x^2 + 6x - 6 = 6(2x^2 + x - 1) = 6(2x-1)(x+1)$; this is zero if $x = \frac{1}{2}$ or -1. The intervals to consider are $x < -1$, $-1 < x < \frac{1}{2}$, and $x > \frac{1}{2}$. For the first, choose $x_0 = -2$; then $f'(x_0) = 6(-5)(-1) > 0$. For the second, choose $x_0 = 0$; then $f'(x_0) = 6(-1)(1) < 0$. And for the third, choose $x_0 = 1$, giving $f'(x_0) = 6(1)(2) > 0$.

A rough sketch ignoring the functional values may be helpful. f has a local max at $(-1, 8)$ and a local min at $(\frac{1}{2}, \frac{5}{4})$. See figure at the left on p. 605.

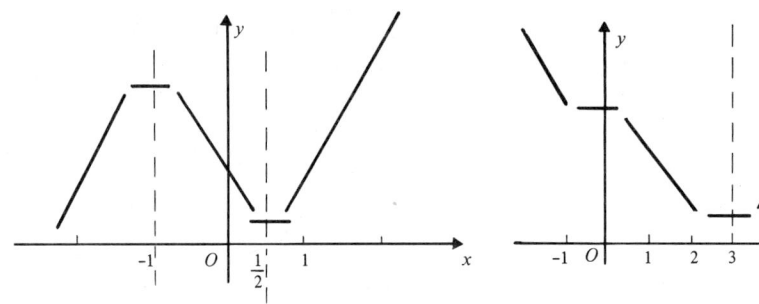

2. $g'(x) = 4x^3 - 12x^2 = 4x^2(x - 3)$; this is zero at $x = 0$ and $x = 3$. The intervals are $x < 0$, $0 < x < 3$, $3 < x$. For the first, choosing $x_0 = -1$, we have $g'(x_0) = 4(-1)^2(-4) < 0$. For the second, with $x_0 = 1$, $g'(x_0) = 4(1)^2(-2) < 0$. For the third, with $x_0 = 4$, $g'(x_0) = 4(4)^2(1) > 0$.

g has a local minimum at $(3, 3)$. The point $(0, 30)$ is neither a local maximum nor a local minimum. See figure at the right above.

GOAL 21

1. $f'(x) = 2x - 4$, which is zero if $x = 2$; $x = 2$ is in the interval $0 \leq x \leq 3$. Since $f(0) = 2$, $f(2) = -2$, and $f(3) = -1$, the abs max is at $(0, 2)$, the abs min at $(2, -2)$.

2. $f'(x) = -2x + 2$, which is zero when $x = 1$, but $x = 1$ is not in $[2, 4]$. Since $f(2) = 1$ and $f(4) = -7$, the abs max is at $(2, 1)$, the abs min at $(4, -7)$.

3. $f'(x) = 3x^2 - 12x + 9 = 3(x^2 - 4x + 3) = 3(x - 1)(x - 3)$; this is zero when $x = 1$ or 3. Only $x = 1$ is in the interval $[0, 2]$. Since $f(0) = -1$, $f(1) = 3$, and $f(2) = 1$, the abs max is at $(1, 3)$, the abs min at $(0, -1)$.

4. $f'(x) = 2x - \dfrac{2}{x^2} = \dfrac{2x^3 - 2}{x^3} = \dfrac{2(x^3 - 1)}{x^3}$, which is zero if $x = 1$; $x = 1$ is in $[\tfrac{1}{2}, 2]$. Since $f[\tfrac{1}{2}] = 4\tfrac{1}{4}$, $f(1) = 3$, and $f(2) = 5$, the abs min is at $(1, 3)$, the abs max at $(2, 5)$.

GOAL 22

1. $f(x) = -x^4 + 2x^3 + 12x^2 - 10$; $f'(x) = -4x^3 + 6x^2 + 24x$; $f''(x) = -12x^2 + 12x + 24 = -12(x^2 - x - 2) = -12(x - 2)(x + 1)$. This is zero if $x = -1$ or 2, which may yield points of inflection. The intervals to consider are $x < -1$, $-1 < x < 2$, and $2 < x$. We examine the sign of $f''(x)$ on these intervals:

Interval	x_0	Sign of $f''(x_0)$
$x < -1$	-2	$-12(-4)(-1) < 0$
$-1 < x < 2$	0	$-12(-2)(1) > 0$
$2 < x$	3	$-12(1)(4) < 0$

So f has inflection points at -1 and at 2; the curve is concave down if $x < -1$ or if $x > 2$, concave up if $-1 < x < 2$.

2. (a) $f'(x) > 0$ if $x < 0$. (b) $f''(x) > 0$ if $x < 0$, $0 < x < 2$, or $x > 4$.
(c) $f'(x) > 0$ and $f''(x) > 0$ if $x < 0$.

3. $f(x) = (x^2 - 1)^2$; $f'(x) = 2(x^2 - 1) \cdot 2x = 4x(x^2 - 1)$: this is zero if $x = 0, 1$, or -1, the critical points. Since $f'(x) = 4(x^3 - x)$, $f''(x) = 4(3x^2 - 1)$.
$f''(0) = -4$, so is negative; $f''(-1) = 4(2)$, so is positive; $f''(1) = 4(2)$, so is positive. Therefore f has a local max at $(0, 1)$ and local minima at $(-1, 0)$ and $(1, 0)$.

GOAL 23

1.
  (a) $-\frac{3}{2}$ (b) 0   (c) ∞ (no limit)   (d) 7

2.

	Vertical Asymptotes:	Horizontal Asymptotes:
(a)	$x = 0$	$y = 0$
(b)	$x = 1, x = -3$	$y = 0$
(c)	$x = \frac{5}{2}$	$y = -\frac{3}{2}$
(d)	$x = -1, x = 2$	$y = 1$

GOAL 24

1. $f(x) = -x^2 + 2x + 3 = -(x + 1)(x - 3)$. Intercepts: $(0, 3)$, $(-1, 0)$, $(3, 0)$. $f'(x) = -2x + 2 = -2(x - 1)$; $f''(x) = -2$. f has a local max at $(1, 4)$; the curve is concave down everywhere. See figure at left below.

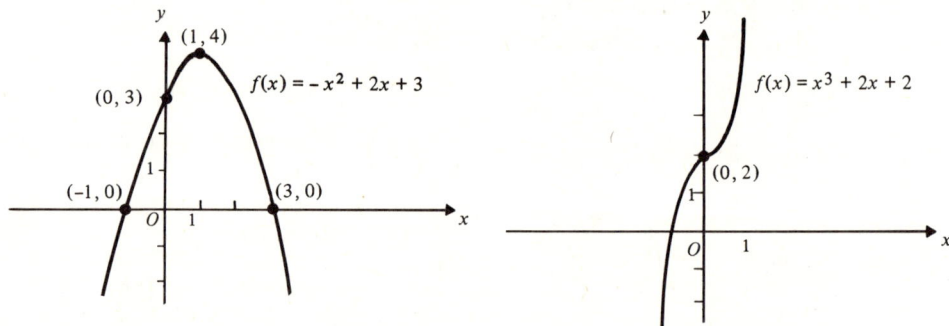

2. $f(x) = x^3 + 2x + 2$; $f(0) = 2$; $f'(x) = 3x^2 + 2$; $f''(x) = 6x$. f has no critical points; there is a point of inflection at $(0, 2)$. Concave up if $x > 0$, down if $x < 0$. See figure at right above.

3. $f(x) = \dfrac{4x - 4}{2x + 3}$. Asymptotes: $x = -\frac{3}{2}$, $y = 2$. Intercepts: $\left[0, -\frac{4}{3}\right]$, $(1, 0)$.
$f'(x) = \dfrac{20}{(2x + 3)^2}$; $f''(x) = \dfrac{-80}{(2x + 3)^3}$. No critical points or inflection points.
Curve concave up if $x < -\frac{3}{2}$, down if $x > -\frac{3}{2}$. The figure is on page 607.

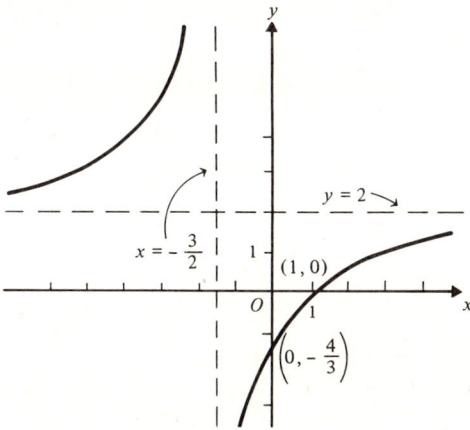

GOAL 25

1. We want to maximize $A = xy$. Since $x + 2y = 100$, $y = \frac{1}{2}(100 - x)$ and $A(x) = \frac{1}{2}x(100 - x) = \frac{1}{2}(100x - x^2)$; then $A'(x) = \frac{1}{2}(100 - 2x)$, which is zero if $x = 50$. Since the domain of $A(x)$ is $x > 0$, we check $A''(x)$, which equals $(\frac{1}{2})(-2)$ for all x. Since $A''(x) < 0$, the graph of $A(x)$ is concave down and $A(x)$ is a maximum at $x = 50$. The dimensions are therefore 50 ft by 25 ft for maximum area.

2. Let x be the number of participants in excess of 60. Then the fare for each will be $(800 - 8x)$ dollars. The revenue is $R(x) = (60 + x)(800 - 8x)$ and its domain is $[0, 15]$. Since $R'(x) = (60 + x)(-8) + (800 - 8x)(1) = -480 - 8x + 800 - 8x = 320 - 16x$, this is zero if $x = 20$. Since 20 is not in the interval $[0, 15]$, the domain of $R(x)$, we evaluate $R(x)$ only at 0 and at 15. $R(0) = (60)(800) = \$48,000$; $R(15) = (75)(800 - 8 \cdot 15) = \$51,000$. Therefore 75 passengers yield maximum revenue.

3. The cost per box in cents is

   cost of bottom + cost of top + cost of four sides

   $= x^2 \cdot \frac{3}{2} + x^2 \cdot \frac{1}{2} + 4xy \cdot \frac{9}{8}$

   $= 2x^2 + \frac{9}{2}xy$

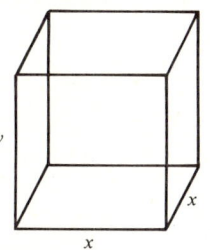

Since $x^2 y = 24$, $y = \dfrac{24}{x^2}$ and the cost for 100,000 boxes is

$$C(x) = 100,000 \left(2x^2 + \frac{9x}{2} \times \frac{24}{x^2} \right) = 100,000 \left(2x^2 + \frac{108}{x} \right)$$

Then

$$C'(x) = 100,000 \left(4x - \frac{108}{x^2} \right) = \frac{100,000}{x^2}(4x^3 - 108)$$

and this is zero if $x^3 = 27$, $x = 3$. Since the domain is $x > 0$, we find $C''(x)$:

$$C''(x) = 100{,}000 \left(4 + \frac{216}{x^3} \right)$$

and this is positive for $x > 0$. Therefore the cost is a minimum when $x = 3$. The dimensions for minimum cost are 3 in. × 3 in. × $\frac{8}{3}$ in.

GOAL 26

1. If x, y, and z represent the distances, respectively, between the first car and the city, the second car and the city, and the two cars, at any time, then $x^2 + y^2 = z^2$ and we must determine $\frac{dz}{dt}$ when $x = 30$ and $y = 40$. By implicit differentiation with respect to t and the chain rule, we have

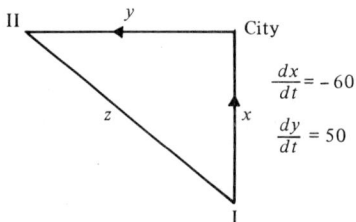

$$2x \frac{dx}{dt} + 2y \frac{dy}{dt} = 2z \frac{dz}{dt} \text{ or } x \frac{dx}{dt} + y \frac{dy}{dt} = z \frac{dz}{dt}$$

At any time

$$\frac{dz}{dt} = \frac{x \frac{dx}{dt} + y \frac{dy}{dt}}{z} = \frac{x(-60) + y(50)}{z}$$

When $x = 30$ and $y = 40$, $z = 50$. Therefore at that instant

$$\frac{dz}{dt} = \frac{(30)(-60) + (40)(50)}{50} = +4$$

At that time, then, the cars are separating at 4 km/h.

2. We want $\frac{dy}{dt}$ when $x = 1000$ and $\frac{dx}{dt} = 4$. Since $y^2 = x^2 - 750x$,

$$2y \frac{dy}{dt} = 2x \frac{dx}{dt} - 750 \frac{dx}{dt} \text{ and } \frac{dy}{dt} = \frac{x \frac{dx}{dt} - 375 \frac{dx}{dt}}{y}$$

at any time. At the instant in question we note that $y = 500$. Then

Answers to Goal Tests A

$$\frac{dy}{dt} = \frac{1000(4) - 375(4)}{500} = 5$$

So the y-population is increasing at the rate of five organisms per hour.

GOAL 27

1.

(a) $\dfrac{f(2) - f(-1)}{2 - (-1)} = \dfrac{0 - (-6)}{3} = 2$; $f'(x) = -2x + 3$; $-2x + 3 = 2$ gives $x = \frac{1}{2}$. Since $\frac{1}{2}$ is in the given interval, it serves as a c.

(b) $\dfrac{f(5) - f(0)}{5 - 0} = \dfrac{79 - (-1)}{5} = 16$; $f'(x) = 3x^2 - 4x + 1$.
$3x^2 - 4x + 1 = 16$; $3x^2 - 4x - 15 = 0$; $(3x + 5)(x - 3) = 0$ yields $x = -1\frac{2}{3}$ or $x = 3$. Only $x = 3$ is in the interval $0 \leq x \leq 5$, so 3 serves as a c.

(c) $\dfrac{f(8) - f(2)}{8 - 2} = \dfrac{\frac{1}{8} - \frac{1}{2}}{8 - 2} = -\dfrac{1}{16}$; $f'(x) = -\dfrac{1}{x^2}$; $-\dfrac{1}{x^2} = -\dfrac{1}{16}$ gives $x = \pm 4$; $x = 4$ is in the interval $2 \leq x \leq 8$, so 4 serves as a c.

2.

(a) $f(x) = x^4 + C$; $0 + C = 3$, $C = 3$, and so $f(x) = x^4 + 3$.

(b) $f(x) = x + C$; $1 + C = -1$, $C = -2$, and so $f(x) = x - 2$.

GOAL 28

1. $\Delta y = (2.1^3 - 2.1) - (2^3 - 2) = 7.161 - 6 = 1.161$; $dy = f'(2)\Delta x = 11(0.1) = 1.1$

2. $f(x) = \sqrt{x}$, $c = 4$, $\Delta x = 0.2$; $dy = f'(c)\Delta x = \dfrac{1}{2\sqrt{4}}(0.2) = 0.05$; $\sqrt{4.2} \approx f(c) + dy = 2.05$

3. $P(x) = -10{,}000 + 5000x - 400x^2$, $c = 6.5$, $\Delta x = 0.25$. $dy = P'(c)\Delta x = (-200)(0.25) = -50$. A \$0.25 increase from \$6.50 to \$6.75 produces approximately a \$50 drop in profit.

4. $x = 700A = 700\pi r^2$, $\dfrac{dx}{dr} = 1400\pi r$, $\Delta x = 100$. $\Delta x \approx dx = 1400\pi(2\frac{1}{4})\Delta r$; $100 = 1400\pi(2\frac{1}{4})\Delta r$; $\Delta r \approx 0.01$. The error in the measurement of the radius must not exceed 0.01 in.

GOAL 29

1. $f(t) = 2t^3 - 21t^2 + 60t + 5$; $v(t) = 6t^2 - 42t + 60$. The initial position is $f(0) = 5$. Direction changes when $v(t) = 6(t - 5)(t - 2)$ changes sign; that is, when $t = 2$ and $t = 5$. $f(2) = 57$ and $f(5) = 30$. The object is speeding up when $2 < t < 3\frac{1}{2}$ and again when $t > 5$.

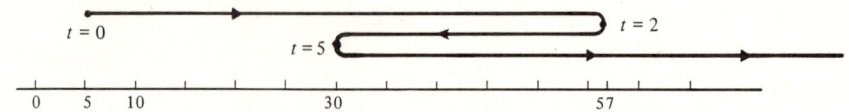

2. $f(t) = -16t^2 + 80t + 96$; $v(t) = -32t + 80$. Maximum height attained when $v(t) = 0$; that is, when $t = 2\frac{1}{2}$. At that time the distance is $f(2\frac{1}{2}) = 196$ ft. The ball hits the ground when $f(t) = 0$; that is, when $t = 6$. At that time the velocity is $v(6) = -112$ ft/sec.

GOAL 30

1. $R(x) = x(7 - 0.001x) = 7x - 0.001x^2$; $R'(x) = 7 - 0.002x$, and this is zero when $x = 3500$. Revenue is maximized at a production level of 3500 units.
$A(x) = \dfrac{C(x)}{x} = \dfrac{1000}{x} + 2 + 0.00025x$; $A'(x) = -\dfrac{1000}{x^2} + 0.00025$, and this is zero when $x = 2000$. Average cost is minimized at a production level of 2000 units.
$P(x) = R(x) - C(x) = (7x - 0.001x^2) - (1000 + 2x + 0.00025x^2) = -1000 + 5x - 0.00125x^2$; $P'(x) = 5 - 0.0025x$, and this is zero when $x = 2000$. Profit is maximized at a production level of 2000 units.
$P(2000) = -1000 + 5(2000) - 0.00125(2000)^2 = \$4000$.
$p(2000) = 7 - 0.001(2000) = \$5$.

2. $E(p) = -\dfrac{p}{D(p)} D'(p) = -\dfrac{p}{1500(5-p)} (1500)(-1) = \dfrac{p}{5-p}$
$E(3) = \dfrac{3}{5-3} = \dfrac{3}{2}$; this is greater than 1, so the revenue will decrease with an increase in price.

3. We have, in dollars,

$C(x)$	= cost of 1000 radios	+ yearly storage costs	+ yearly shipping costs
	$= 1000(8)$	$+ \dfrac{x}{2}(1)$	$+ \dfrac{1000}{x}(20)$
	$= 8000$	$+ \dfrac{x}{2}$	$+ \dfrac{20{,}000}{x}$

So
$$C'(x) = \dfrac{1}{2} - \dfrac{20{,}000}{x^2}$$

and this is zero if $x^2 = 40{,}000$, $x = 200$. Since $C''(x) > 0$ for all x, ordering 200 radios per shipment minimizes total annual cost.

GOAL 31

1.
    (a) 1   (b) 8   (c) 9   (d) t^{12} (e) $\log_b 2$

Answers to Goal Tests A

2.
(a) $2^6 = 64$ (b) $\log_3 \frac{1}{27} = -3$ (c) $9^{3/2} = 27$ (d) $\log_{10} 1 = 0$

3.
(a) 13   (b) −3   (c) 50

4.
(a) $\log_6 (6^{2t-4}) = \log_6 36$ or $6^{2t-4} = 6^2$
$2t - 4 = 2$ $2t - 4 = 2$
$t = 3$ $t = 3$

(b) $\log_8 2 = \log_8 (8^{t-4})$ or $2^1 = (2^3)^{t-4} = 2^{3t-12}$
$\frac{1}{3} = t - 4$ $3t - 12 = 1$
$t = 4\frac{1}{3}$ $t = 4\frac{1}{3}$

5.

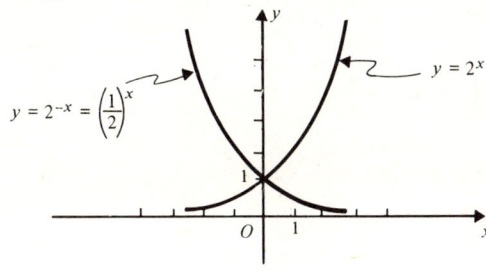

GOAL 32

1. $(e^5)^2 = e^{5 \cdot 2} = e^{10}$; $e^t e^t = e^{t+t} = e^{2t}$; $\dfrac{e^5}{e^2} = e^{5-2} = e^3$

2. $e^{-2.5} \approx 0.0821$; $e^{0.2} \approx 1.2214$; $e^{1.1} \approx 3.0042$

3.

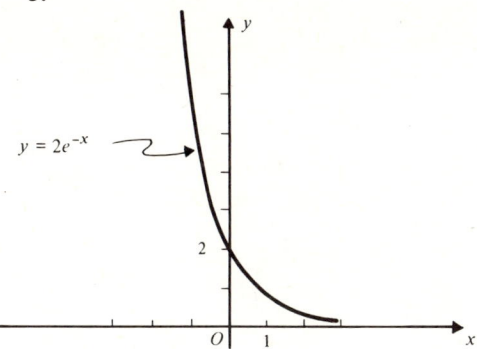

GOAL 33

1.
(a) $\ln \dfrac{1}{e} = \ln e^{-1} = -1$ (c) $\ln \sqrt{e} = \ln e^{1/2} = \dfrac{1}{2}$

(b) $e^{\ln 2} = 2$ (d) $e^{-5 \ln 3} = (e^{\ln 3})^{-5} = 3^{-5} = \dfrac{1}{3^5} = \dfrac{1}{243}$

2.
   (a) $3t = \ln 51 = 1.6292 + 2.3026 = 3.9318$; so $t = 1.3106$.
   (b) $-0.2t = \ln 0.1 = -2.3206$; so $t = 11.513$.
   (c) $-0.0005t = \ln \frac{1}{2} = -0.6932$; so $t = 1386.4$.

3. The graph of $y = \ln x$ is the reflection of $y = e^x$ in the line $y = x$.

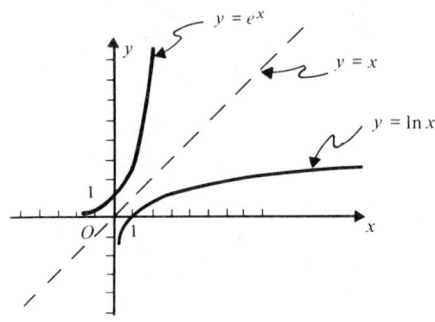

4. If $e^{36} = 10^k$, then $\ln e^{36} = \ln 10^k$ and $36 = k \ln 10$,

$$k = \frac{36}{\ln 10} \approx \frac{36}{2.3026} \approx 16$$

to the nearest integer.

GOAL 34

1.
   (a) $\dfrac{d}{dx} 4 \ln 3x = 4 \cdot \dfrac{3}{3x} = \dfrac{4}{x}$

   (b) $\dfrac{d}{dx} \ln (x^3 + x^2 - 1) = \dfrac{3x^2 + 2x}{x^3 + x^2 - 1}$

   (c) $\dfrac{d}{dx} \ln \dfrac{1-x}{x} = \dfrac{d}{dx} [\ln(1-x) - \ln x] = \dfrac{-1}{1-x} - \dfrac{1}{x}$
   $= \dfrac{1}{x-1} - \dfrac{1}{x}$, or $\dfrac{1}{x(x-1)}$

   (d) $\dfrac{d}{dx} x \ln(1-x) = x \dfrac{d}{dx} \ln(1-x) + \ln(1-x) \dfrac{d}{dx} x$
   $= \dfrac{x \cdot -1}{1-x} + \ln(1-x) \cdot 1 = \dfrac{x}{x-1} + \ln(1-x)$

   (e) $\dfrac{d}{dx}[\ln(2x+1)]^2 = 2\ln(2x+1) \cdot \dfrac{2}{2x+1} = \dfrac{4\ln(2x+1)}{2x+1}$

   (f) $\dfrac{d}{dx} \ln(\ln x) = \dfrac{1}{x \ln x}$

2. $f(x) = \ln(x+1)$, $f'(x) = \dfrac{1}{x+1}$, $f'(0) = 1$. The equation of the tangent is $\dfrac{y-0}{x-0} = 1$, or $y = x$.

3. $f(x) = \dfrac{\ln x}{x}$; $f'(x) = \dfrac{x \cdot \dfrac{1}{x} - \ln x \cdot 1}{x^2} = \dfrac{1 - \ln x}{x^2}$; this is zero if $\ln x = 1$; that is, if $x = e$. Note that $f'(x)$ is positive if $x < e$, negative if $x > e$, so that $(e, f(e))$ is a local max. Since there is only the one critical value (at $x = e$), and $f(e) = \dfrac{\ln e}{e} = \dfrac{1}{e}$, the point $\left(e, \dfrac{1}{e}\right)$ is an absolute max.

GOAL 35

1.

(a) $\dfrac{d}{dx} e^{x^2 - 5x + 2} = (2x - 5)e^{x^2 - 5x + 2}$

(b) $\dfrac{d}{dt} te^{3t} = e^{3t} + 3te^{3t}$

(c) $\dfrac{d}{dx} \dfrac{x}{e^x} = \dfrac{e^x(1) - xe^x}{(e^x)^2} = \dfrac{1-x}{e^x}$

(d) $\dfrac{d}{dt} \dfrac{3}{1 + e^{-t}} = \dfrac{-3(-e^{-t})}{(1 + e^{-t})^2} = \dfrac{3e^{-t}}{(1 + e^{-t})^2}$

2. $y = 2e^{2x}$; $\dfrac{dy}{dx} = 4e^{2x}$; tangent: $\dfrac{y-2}{x-0} = 4$ and $y = 4x + 2$

3. $f'(t) = -5(-2e^{-2t}) = -2(-5e^{-2t}) = -2f(t)$; $k = -2$

4. $f(t) = ce^{kt}$ and $k = 2$ since $f'(t) = 2f(t)$. Since $f(0) = 8$, $ce^{2(0)} = ce^0 = c = 8$, and $f(t) = 8e^{2t}$.

GOAL 36

1. The population satisfies $dP/dt = 0.03P$, whose solution is $P = P_0 e^{0.03t}$. We are asked to find t when $3P_0$. $3P_0 = P_0 e^{0.03t}$ gives $0.03t = \ln 3 = 1.0986$ and $t = 36.6$ yr.

2. The quantity of C^{14} present satisfies $dQ/dt = -0.00012Q$, whose solution is $Q(t) = Q_0 e^{-0.00012t}$. We seek t when $Q(t) = 0.2Q_0$:

$$0.2Q_0 = Q_0 e^{-0.00012t} \qquad 0.2 = e^{-0.00012t}$$
$$-0.00012t = \ln 0.2 = -1.6094 \qquad t \approx 13{,}400$$

The animal died about 13,400 yr ago.

3. The amount of money A satisfies $dA/dt = kA$, whose solution is $A = A_0 e^{kt}$. Since $A = 1000$ when $t = 0$, $A_0 = 1000$; since $A = 1500$ when $t = 10$,

$$1500 = 1000e^{10k} \qquad 10k = \ln 1.5 = 0.4055 \qquad k = 0.04055$$

So $A = 1000e^{0.04055t}$. When $t = 15$, $A = 1000e^{0.60825} \approx 1000e^{0.61} \approx \$1840$.

4. The water temperature satisfies $dW/dt = -k(W - 20)$, whose solution is $W(t) = 20 + ce^{-kt}$. Since $W(0) = 70$, $70 = 20 + c$ and $c = 50$. Since $W(1) = 50$, $50 = 20 + 50e^{-k}$, $e^{-k} = 0.6$, $k = -\ln 0.6 = 0.5108$. So $W(t) = 20 + 50 e^{-0.5108t}$.

We seek the t for which $W(t) = 32$.

$$32 = 20 + 50e^{-0.5108t} \qquad 0.24 = e^{-0.5108t}$$
$$-0.5108t = \ln 0.24 = -1.4271 \qquad t = 2.79$$

In approximately $2\frac{3}{4}$ h the water will turn to ice.

5. The number who have been infected by time t satisfies $dn/dt = kn(15{,}000 - n)$ with solution

$$n(t) = \frac{15{,}000}{1 + ce^{-15{,}000kt}}$$

From $n(0) = 100$ we get

$$100 = \frac{15{,}000}{1 + c} \qquad c = 149$$

From $n(1) = 500$ we get

$$500 = \frac{15{,}000}{1 + 149e^{-15{,}000k}} \qquad 1 + 149e^{-15{,}000k} = 30$$

$$e^{-15{,}000k} = \frac{29}{149}$$

We seek $n(2)$:

$$n(2) = \frac{15{,}000}{1 + 149(e^{-15{,}000k})^2} = \frac{15{,}000}{1 + 149\left(\frac{29}{149}\right)^2} \approx 2260$$

About 2260 people will have caught the disease by the end of next week.

GOAL 37

1. $[(2\frac{1}{4})^2 + 2(2\frac{1}{4}) - 3](\frac{1}{2}) + [(2\frac{3}{4})^2 + 2(2\frac{3}{4}) - 3](\frac{1}{2}) + [(3\frac{1}{4})^2 + 2(3\frac{1}{4}) - 3](\frac{1}{2}) + [(3\frac{3}{4})^2 + 2(3\frac{3}{4}) - 3](\frac{1}{2}) = 24.625$ sq. units

2. Factoring out the constant subinterval length (0.2) and the constant factor (100) yields
(0.2) (100) $[0.1 + 2(3) + 0.1 + 2(3.2) + 0.1 + 2(3.4) + 0.1 + 2(3.6) + 0.1 + 2(3.8)] = 690$ bacteria

Answers to Goal Tests A

GOAL 38

$$\text{Area} = \lim_{n\to\infty} \left\{ \left[2\left(\frac{1}{n}\right)^2 + 1\right]\left(\frac{1}{n}\right) + \left[2\left(\frac{2}{n}\right)^2 + 1\right]\left(\frac{1}{n}\right) + \cdots \right.$$
$$\left. + \left[2\left(\frac{n}{n}\right)^2 + 1\right]\left(\frac{1}{n}\right) \right\}$$

$$= \lim_{n\to\infty} \frac{1}{n}\left[2\left(\frac{1}{n}\right)^2 (1^2 + 2^2 + \ldots n^2) + \overbrace{(1 + 1 + \cdots + 1)}^{n \text{ ones}}\right]$$

$$= \lim_{n\to\infty} \left[\frac{2}{n^3}\left(\frac{n^3}{3} + \frac{n^2}{2} + \frac{n}{6}\right) + \frac{1}{n} \cdot n\right] = \frac{2}{3} + 1 = \frac{5}{3} \text{ sq units}$$

GOAL 39

1. $\int_1^3 \left(\frac{1}{x} + 2\right) dx$ **2.** $\int_0^3 10 e^{-t} dt$

GOAL 40

1.

(a) $\int_0^1 (5x^4 - 3x^2 + 1)\, dx = \left[x^5 - x^3 + x\right]_0^1 = (1 - 1 + 1) - (0 - 0 + 0) = 1$

(b) $\int_1^2 \left(2x + \frac{1}{x}\right) dx = \left[x^2 + \ln x\right]_1^2 = (4 + \ln 2) - (1 + 0) = 3 + \ln 2$

2. 16 sq units

GOAL 41

1.

(a) $\dfrac{x^9}{9} - \dfrac{7x^4}{4} + 3x + C$

(b) $\int \dfrac{3 + x^2 - x^3}{x^4} dx = \int \left(\dfrac{3}{x^4} + \dfrac{1}{x^2} - \dfrac{1}{x}\right) dx = -\dfrac{1}{x^3} - \dfrac{1}{x} - \ln x + C$

(c) $2e^x - \dfrac{3x^2}{2} + C$

2.

(a) $\int_0^1 (x^9 + 3x^4)\, dx = \left[\dfrac{x^{10}}{10} + \dfrac{x^5}{5}\right]_0^1 = \dfrac{1}{10} + \dfrac{1}{5} = \dfrac{3}{10}$

(b) $\int_0^4 (3\sqrt{x} - x^{3/2})\, dx = \left[2x^{3/2} - \dfrac{2}{5}x^{5/2}\right]_0^4 = 16 - \dfrac{2}{5}(32) = \dfrac{16}{5}$

3. $\int_3^\infty \dfrac{1}{x^2}\, dx = \lim_{n\to\infty} \int_3^n x^{-2}\, dx = \lim_{n\to\infty} \left[-\dfrac{1}{x}\right]_3^n = \lim_{n\to\infty} -\left[\dfrac{1}{n} - \dfrac{1}{3}\right] = \dfrac{1}{3}$

GOAL 42

1. $\dfrac{(x^3 + 3x^2 - 1)}{11} + C$

2. $e^{(5x^2+2)} + C$

3. In formula I, $u = x^6 + 2x^3 + 3x^2$, $n = 4$, $du = (6x^5 + 6x^2 + 6x)\,dx = 6(x^5 + x^2 + x)\,dx$. So for the given integral we have

$$\frac{1}{6}\int u^4\,du = \frac{1}{6}\left(\frac{u^5}{5}\right) + C = \frac{1}{30}(x^6 + 2x^3 + 3x^2)^5 + C$$

4. In formula I, $u = 8x - 5$, $n = 20$, $du = 8\,dx$. So

$$\int (8x - 5)^{20}\,dx = \frac{1}{8}\int (8x - 5)^{20}\, 8\,dx$$
$$= \frac{1}{8}\cdot\frac{(8x-5)^{21}}{21} + C = \frac{1}{168}(8x - 5)^{21} + C$$

5. In formula II, $u = x^2 + 2$, $du = 2x\,dx$:

$$\int \frac{1}{x^2 + 2}\, x\,dx = \frac{1}{2}\int \frac{1}{x^2 + 2}(2x\,dx) = \frac{1}{2}\ln(x^2 + 2) + C$$

6. $\displaystyle\int_0^1 (x^2 + 1)^3 x\,dx = \frac{1}{2}\int_0^1 (x^2 + 1)^3\, 2x\,dx = \frac{1}{2}\left[\frac{(x^2 + 1)^4}{4}\right]_0^1$

$$= \frac{1}{8}(2^4 - 1^4) = \frac{15}{8}$$

GOAL 43

1. $\displaystyle\int \overbrace{x}^{u}\,\overbrace{e^{2x}\,dx}^{dv} = \overbrace{x}^{u}\cdot\overbrace{\tfrac{1}{2}e^{2x}}^{v} - \int \overbrace{\tfrac{1}{2}e^{2x}}^{v}\cdot\overbrace{1\,dx}^{du}$

$\quad = \tfrac{1}{2}xe^{2x} - \tfrac{1}{2}\int e^{2x}\,dx = \tfrac{1}{2}xe^{2x} - \tfrac{1}{4}e^{2x} + C$

2. $\displaystyle\int \overbrace{x}^{u}\,\overbrace{(x + 9)^5\,dx}^{dv} = \overbrace{x}^{u}\cdot\overbrace{\tfrac{1}{6}(x + 9)^6}^{v} - \int \overbrace{\tfrac{1}{6}(x + 9)^6}^{v}\cdot\overbrace{1\,dx}^{du}$

$\quad = \tfrac{1}{6}x(x + 9)^6 - \tfrac{1}{42}(x + 9)^7 + C$

3. $\displaystyle\int x \ln x\,dx = \int \overbrace{\ln x}^{u}\cdot\overbrace{x\,dx}^{dv} = \overbrace{\ln x}^{u}\cdot\overbrace{\tfrac{x^2}{2}}^{v} - \int \overbrace{\tfrac{x^2}{2}}^{v}\cdot\overbrace{\tfrac{1}{x}\,dx}^{du}$

$\quad = \tfrac{1}{2}x^2 \ln x - \tfrac{1}{2}\int x\,dx = \tfrac{1}{2}x^2 \ln x - \dfrac{x^2}{4} + C$

4. $\displaystyle\int \overbrace{(x + 1)}^{u}\,\overbrace{e^{-x}\,dx}^{dv} = \overbrace{(x + 1)}^{u}\,\overbrace{(-e^{-x})}^{v} - \int \overbrace{-e^{-x}}^{v}\cdot\overbrace{1\,dx}^{du}$

$\quad = -(x + 1)e^{-x} + \int e^{-x}\,dx = -(x + 1)e^{-x} - e^{-x} + C$

Answers to Goal Tests A

5. $\int_1^3 x \ln x \, dx = \left[\frac{1}{2} x^2 \ln x - \frac{x^2}{4} \right]_1^3$

$= \left(\frac{1}{2} \cdot 9 \cdot \ln 3 - \frac{9}{4} \right) - \left(\frac{1}{2} \cdot 1 \cdot 0 - \frac{1}{4} \right)$

$= \frac{9}{2} \ln 3 - \frac{8}{4} = \frac{9}{2} \ln 3 - 2$

GOAL 44

1. $\int_0^2 [x^3 - (x^2 - 3x)] \, dx = \int_0^2 (x^3 - x^2 + 3x) \, dx$

$= \left[\frac{x^4}{4} - \frac{x^3}{3} + \frac{3x^2}{2} \right]_0^2 = 7 \frac{1}{3}$

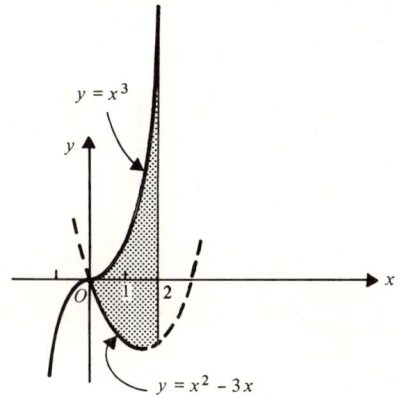

2. $\int_{-1}^0 [(1 - x^2) - (-2)] \, dx = \int_{-1}^0 (3 - x^2) \, dx = \left[3x - \frac{x^3}{3} \right]_{-1}^0 = 2\frac{2}{3}$

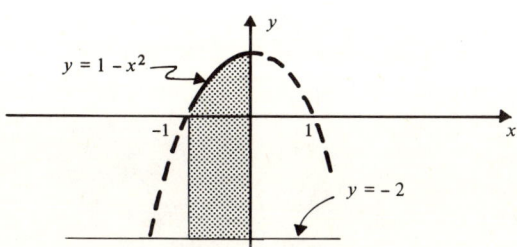

Selected Answers

3. $\int_0^4 [x - (x^2 - 3x)] \, dx = \int_0^4 (4x - x^2) \, dx = \left[2x^2 - \frac{x^3}{3} \right]_0^4 = \frac{32}{3}$

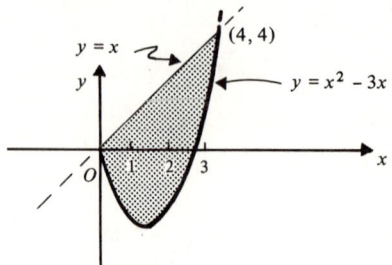

GOAL 45

1. $a(t) = v'(t) = 2$
   $f(t) = \int v(t) \, dt = t^2 + 3t + C$
   $4 = f(0) = (0)^2 + 3(0) + C$ and $C = 4$
   $f(t) = t^2 + 3t + 4$

2. $v(t) = \int a(t) \, dt = 6t^2 - 8t + C$
   $f(t) = \int v(t) \, dt = 2t^3 - 4t^2 + Ct + K$
   $-2 = f(1) = 2(1)^3 - 4(1)^2 + C(1) + K = -2 + C + K$ } solving simultaneo
   $0 = f(2) = 2(2)^3 - 4(2)^2 + C(2) + K = 2C + K$ } gives $C = 0$, $K = 0$
   Thus $v(t) = 6t^2 - 8t$
   $f(t) = 2t^3 - 4t^2$

3. If $A(t)$ is the amount left after t yr, then

$$A(t) = \int -0.375 e^{-0.025t} \, dt = -\frac{0.375}{-0.025} \int e^{-0.025t} (-0.025 \, dt) = 15 e^{-0.025t} + C$$

Since $A(0) = 15$, $15 = 15 e^0 + C$ and $C = 0$. So

$$A(100) = 15 e^{-2.5} \approx 1.2 \text{ g}$$

GOAL 46

1.
   (a) $\frac{1}{2-(-1)} \int_{-1}^{2} (x^2 + x + 1) \, dx = \frac{1}{3} \left[\frac{x^3}{3} + \frac{x^2}{2} + x \right]_{-1}^{2} = \frac{5}{2}$

   (b) $\frac{1}{4-1} \int_1^4 \frac{1}{x^2} \, dx = \frac{1}{3} \left[-\frac{1}{x} \right]_1^4 = \frac{1}{4}$

2. $\frac{1}{1-0} \int_0^1 500 e^{0.1t} \, dt = \left[5000 e^{0.1t} \right]_0^1 \approx \$526$

Answers to Goal Tests A

GOAL 47

1. Profit $P(x) = \int(-0.01x + 6.5)\,dx = -0.005x^2 + 6.5x + k$
   $200 = P(100) = -0.005(100)^2 + 6.5(100) + k = 600 + k;\ k = -400$
   $P(x) = -0.005x^2 + 6.5x - 400$ dollars

2. To find the equilibrium point we set $D(x) = S(x)$:

$$-0.004x + 3 = 0.000004x^2$$
$$0.000004x^2 + 0.004x - 3 = 0$$
$$4x^2 + 4000x - 3{,}000{,}000 = 0$$
$$(2x - 1000)(2x + 3000) = 0$$

So $x = 500$ is the only positive answer. Since $D(500) = S(500) = 1$, 500 items at \$1 each is the equilibrium point.

$$\text{Consumers' surplus} = \int_0^{500} [D(x) - 1]\,dx$$

$$= \int_0^{500} (-0.004x + 2)\,dx = \Big[-0.002x^2 + 2x\Big]_0^{500} = \$500$$

$$\text{Producer's surplus} = \int_0^{500} [1 - S(x)]\,dx$$

$$= \int_0^{500} (1 - 0.000004x^2)\,dx = \Big[x - \frac{0.000004}{3}x^3\Big]_0^{500} \approx \$333$$

3. $\int_0^{30} 150e^{-(30-t)/15}\,dt = \Big[2250\,e^{-(30-t)/15}\Big]_0^{30} = 2250(1 - e^{-2}) \approx 1945$

4. $\int_0^5 3600e^{-0.06t}\,dt = \Big[-60{,}000e^{-0.06t}\Big]_0^5 = 60{,}000(1 - e^{-0.3}) \approx \$15{,}550$

GOAL 48

1. (a) and (c); (b) is not nonnegative throughout $-1 \le x \le 2$

2. $\int_0^5 \frac{1}{20} e^{-x/20}\,dx = \Big[-e^{-x/20}\Big]_0^5 = 1 - 0.779 = 0.221$

3. $\int_0^{25} \frac{1}{100}\,dx = \Big[\frac{1}{100}x\Big]_0^{25} = \frac{1}{4}$

GOAL 49

1. It is of the third order.

2. Substituting $f(x) = x^3 + Cx$ for y and $f'(x) = 3x^2 + C$ for y' gives

$$3x^2 + C = 2x^2 + \frac{x^3 + Cx}{x}$$

de equals $2x^2 + x^2 + C = 3x^2 + C$.

$) \, dx = x^6 - \dfrac{x^3}{3} + C;$

$\left(-\dfrac{x^3}{3} + C\right) dx = \dfrac{x^7}{7} - \dfrac{x^4}{12} + Cx + k$

$(- 6x + 1) \, dx = \frac{1}{3} e^{3x} - 3x^2 + x + C$

$\frac{1}{3} + C$ so $C = \frac{5}{3}$

$^{3x} - 3x^2 + x + \frac{5}{3}$

$= xy^2 \qquad \dfrac{dy}{dx} = xy^2 \qquad \dfrac{dy}{y^2} = x \, dx$

$\displaystyle\int \dfrac{dy}{y^2} = \int x \, dx \qquad -\dfrac{1}{y} = \dfrac{x^2}{2} + C$

2. $y' = \dfrac{e^x}{y} \qquad y \, dy = e^x \, dx \qquad \displaystyle\int y \, dy = \int e^x \, dx \qquad \dfrac{1}{2} y^2 = e^x + C$

3. $\dfrac{dy}{dx} = (x^2 - 3x + 2)y \qquad \dfrac{dy}{y} = (x^2 - 3x + 2) \, dx$

$\ln y = \dfrac{x^3}{3} - \dfrac{3x^2}{2} + 2x + C$

Since $y = 1$ when $x = -1$, we have

$$\ln 1 = 0 = -\dfrac{1}{3} - \dfrac{3}{2} - 2 + C \qquad \text{and} \qquad C = \dfrac{23}{6}$$

so

$$\ln y = \dfrac{x^3}{3} - \dfrac{3x^2}{2} + 2x + \dfrac{23}{6}$$

and

$$y = e^{(x^3/3) - (3x^2/2) + 2x + (23/6)}$$

4. $\dfrac{dy}{dx} = 2 - y \qquad \dfrac{dy}{2-y} = dx \qquad \displaystyle\int \dfrac{dy}{2-y} = \int dx$

$- \ln(2 - y) = x + C$

From the fact that $y = 1$ when $x = 2$, we get

$$-\ln 1 = 0 = 2 + C \qquad \text{so} \qquad C = -2$$

Therefore (reading across),

$-\ln(2-y) = x - 2 \qquad \ln(2-y) = 2 - x$

$2 - y = e^{2-x} \qquad y = 2 - e^{2-x}$

GOAL 51

1. $P(1{,}000{,}000,\ 2500) = 0.05(1{,}000{,}000) - 2(2500) = 50{,}000 - 5{,}000 = 45{,}000$
2. The domain is the set of all (x, y, z) for which $x \ne 0$, $y \ne 0$, and $z \ge 0$.

Answers to Goal Tests A

3. $f(x, y) = 1$ gives $y^2 - x = 1$ or $x = y^2 - 1$
$f(x, y) = 0$ gives $y^2 - x = 0$ or $x = y^2$
$f(x, y) = -2$ gives $y^2 - x = -2$ or $x = y^2 + 2$

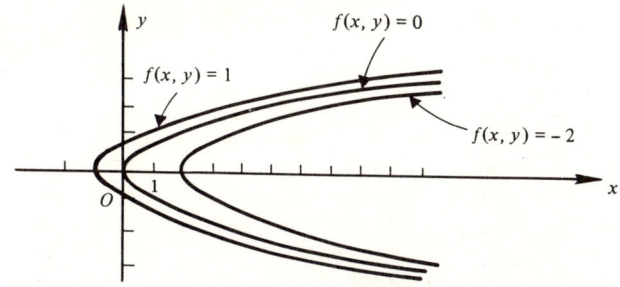

4. Something like the following; note that in each equation the signs of the second and third terms on the right-hand side are all negative:

$$D_1(p_1, p_2) = 300 - 10p_1 - 5p_2$$
$$D_2(p_1, p_2) = 100 - 2p_1 - 3p_2$$

GOAL 52

1.

(a) $\dfrac{\partial f}{\partial x} = 3 + 3x^2 y^4 \qquad \dfrac{\partial f}{\partial y} = -4y + 4x^3 y^3$

(b) $\dfrac{\partial f}{\partial x} = (2x - y)e^{x^2 - xy - 2} \qquad \dfrac{\partial f}{\partial y} = -xe^{x^2 - xy - 2}$

2.

$\dfrac{\partial P}{\partial x} = \dfrac{3}{4}\left(\dfrac{y}{x}\right)^{3/4}, \dfrac{\partial P}{\partial y} = \dfrac{9}{4}\left(\dfrac{x}{y}\right)^{1/4}$; when $x = 16, \ y = 1$,

$\dfrac{\partial P}{\partial x} = \dfrac{3}{32}, \dfrac{\partial P}{\partial y} = \dfrac{9}{2}$

3.

(a) $df = f_x\, dx + f_y\, dy = y^3\, dx + 3xy^2\, dy$

(b) $df = \dfrac{1}{xyz} yz\, dx + \dfrac{1}{xyz} xz\, dy + \dfrac{1}{xyz} xy\, dz = \dfrac{1}{x} dx + \dfrac{1}{y} dy + \dfrac{1}{z} dz$

4. $f_x = \dfrac{y}{2\sqrt{xy}} \qquad f_x(40, 10) = \dfrac{1}{4}$

$f_y = \dfrac{x}{2\sqrt{xy}} \qquad f_y(40, 10) = 1$

$\Delta P \approx dP = f_x(40, 10)\, dx + f_y(40, 10)\, dy$

$= \dfrac{1}{4}(0.1) + (1)(0.05) = 0.075$

$f_y = -2y - 2$. The critical point is $(-2, -1)$. $f_{xx} = 2$;
$= 0$. $D = (2)(-2) - 0 = -4 < 0$; f has a saddle point at
ither a local max nor local min.

$y - 2$; $f_y = 4y - x + 1$.

$$\begin{cases} 2x - y - 2 = 0 \\ -x + 4y + 1 = 0 \end{cases} \rightarrow \begin{cases} 2x - y - 2 = 0 \\ -2x + 8y + 2 = 0 \end{cases}$$

which $y = 0$, $x = 1$. So f has one critical point, at $(1, 0)$. Then $f_{xx} = 2$;
$= 4$; $f_{xy} = -1$; $D = (2)(4) - 1 = 7 > 0$. Since $f_{xx} > 0$, $f(1, 0)$ is a local
.l.

$= mx + b$ is the line of best fit, we minimize

$= (m \cdot 1 + b - 0)^2 + (m \cdot 2 + b - 2)^2 + (m \cdot 3 + b - 2)^2 +$
$m \cdot 4 + b - 3)^2$

$$\frac{\partial Q}{\partial m} = 2(m + b) + 2(2m + b - 2) \cdot 2 + 2(3m + b - 2) \cdot 3 +$$

$$2(4m + b - 3) \cdot 4$$

$$= 2(30m + 10b - 22)$$

$$\frac{\partial Q}{\partial b} = 2(m + b) + 2(2m + b - 2) + 2(3m + b - 2) +$$

$$2(4m + b - 3)$$

$$= 2(10m + 4b - 7)$$

We set these equal to zero and solve simultaneously.

$$\begin{cases} 30m + 10b - 22 = 0 \\ 10m + 4b - 7 = 0 \end{cases} \rightarrow \begin{cases} 30m + 10b - 22 = 0 \\ 30m + 12b - 21 = 0 \end{cases}$$

from which $b = -\frac{1}{2}$, $m = 0.9$. So the line that best fits the data is $y = 0.9x - 0.5$.

GOAL 54

1.
   (a) $f(x, y) = x + y^2 - 4y + 3$ reduces to $g(y) = (y^2 + 8y - 2) + y^2 - 4y + 3 = 2y^2 + 4y + 1$. So $g'(y) = 4y + 4$, and g has a minimum at $y = -1$. Then $x = (-1)^2 + 8(-1) - 2 = -9$. So f has an absolute minimum at $(-9, -1)$, and no maximum.
   (b) $f(x, y) = x^2 + 3y^2 + 8y - 1$ reduces to $g(y) = (9 - y^2) + 3y^2 + 8y - 1 = 2y^2 + 8y + 8$, where $-3 \le y \le 3$. So $g'(y) = 4y + 8$, which is zero when $y = -2$. When $y = -2$, $x = \pm\sqrt{9 - y^2} = \pm\sqrt{5}$. There are thus

four points to check: $(\sqrt{5}, -2)$, $(-\sqrt{5}, -2)$, $(0, 3)$, and $(0, -3)$. $f(\sqrt{5}, -2) = f(-\sqrt{5}, -2) = 0$, absolute minima. $f(0, 3) = 50$, absolute maximum. $f(0, -3) = 2$, not an extreme value.

2.

(a) $F(x, y, \lambda) = 3x + y - 2 + \lambda(x^2 + 2y^2 - 38)$

$\dfrac{\partial F}{\partial x} = 3 + 2\lambda x \qquad \dfrac{\partial F}{\partial y} = 1 + 4\lambda y$

$\dfrac{\partial F}{\partial x} = 0$ and $\dfrac{\partial F}{\partial y} = 0$ gives $\lambda = -\dfrac{3}{2x}$

$1 + 4\left(-\dfrac{3}{2x}\right) y = 0; \ x = 6y$

$(6y)^2 + 2y^2 = 38; \ 38y^2 = 38; \ y = \pm 1$

$f(6, 1) = 17$, the absolute maximum
$f(-6, -1) = -21$, the absolute minimum

(b) $F(x, y, \lambda) = x^2 + 2y^2 - 1 + \lambda(2x - y - 9)$

$\dfrac{\partial F}{\partial x} = 2x + 2\lambda \qquad \dfrac{\partial F}{\partial y} = 4y - \lambda$

$\dfrac{\partial F}{\partial x} = 0$ gives $\lambda = -x$; $\dfrac{\partial F}{\partial y} = 0$ gives $\lambda = 4y$

So $x = -4y$ and since $2x - y - 9 = 0$ we get $x = 4, y = -1$
f has an absolute minimum at $(4, -1)$
[Note, for example, that $f(3, -3) > f(4, -1)$.]

Answers to Chapter Review Exercises

CHAPTER 1

1.
   (a) is a function.

2.
   (a) $\mathscr{R}$ (c) $\mathscr{R}$

3.

	$f(1)$:	$f(-2)$:	$f(q)$:	$f(x+h) - f(x)$:
(a)	3	−3	$2q + 1$	$2h$
(c)	5	8	$q^2 + 4$	$2xh + h^2$

4.
   (a) $\frac{4}{3}$ (c) $-\dfrac{2h}{h} = -2$ if $h \neq 0$

5.
   (a) $= \frac{1}{2}x - 4$ (b) $y = -2x + 2$

7. $k = \frac{2}{3}$

9. 3 sec; 144 ft

10. $V = (\frac{1}{2}, 4\frac{1}{4})$; opens down

11.

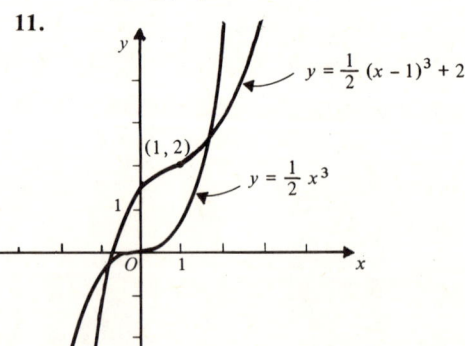

Answers to Chapter Review Exercises

12. Inflection point at $(-2, -1)$; concave down if $x < -2$

14. $f(x)g(x) = \dfrac{x^2 + 3}{x - 1}$ $\qquad \dfrac{f(x)}{g(x)} = (x^2 + 3)(x - 1)$

$f(g(x)) = \dfrac{1}{(x - 1)^2} + 3 \qquad g(f(x)) = \dfrac{1}{x^2 + 2}$

16.

(b) $y = \left(\dfrac{x + 5}{2}\right)^{1/3}$

(c) $y = 3 - \dfrac{1}{x}$ or $y = \dfrac{3x - 1}{x}$

18. If x is the number of passengers, then

$$R(x) = \begin{cases} 15x & x \leq 100 \\ x[15 - (x - 100)\,0.05] \\ \quad = 20x - 0.05x^2 & x > 100 \end{cases}$$

20. If p is the admission price in dollars and n is the number who attend at price p, then $n = -100p + 1100$. The cost (in dollars) $C(p) = 1000 + 1.50n$; the revenue (in dollars) $R(p) = np = -100p^2 + 1100p$; the profit (in dollars)

$$P(p) = R(p) - C(p) = -100p^2 + 1100p - [1000 + 1.50(-100p + 1100)]$$
$$= -100p^2 + 1250p - 2650$$

CHAPTER 2

1. 3.2; 3.1; 2.9

3. 8

5.
 (a) 8 (c) $-2x$ (e) 1

6.
 (a) 8 (c) 0 (e) 16 (g) $-6x$

7.
 (a) ∞ (c) 1 (e) $-\infty$

8.
 (a) No limit (b) 0

11.
 (a) none (c) $y = \pm 2$ (f) the set of all integers

12.
 (a) 0 (c) 1 (e) $7 - 5t^4 - 6t^5$ (g) $-x^9$ (i) 1.8

13. 16

14. -4

17. $2000; $6000

18. 8 gal/sec

CHAPTER 3

1.
  (a) $-2x(x^2 + 3x - 2) + (1 - x^2)(2x + 3)$ (b) $\frac{8}{25}$

3.
  (a) $A(x) = 3x - 5 + \dfrac{200}{x}$ (b) $2.50

4.
  (c) $\dfrac{x}{\sqrt{x^2 + 1}}$
  (e) $9x^2 \sqrt{2x^3 - 1}$
  (f) $\dfrac{4}{(x^2 + 4)^{3/2}}$
  (g) $2x(3 - 5x^2)(3 - x^2)^3$

5.
  (b) 60

6.
  (a) $\dfrac{dy}{dx} = \dfrac{2y - 3x^2}{2y - 2x}$

7. $-\frac{1}{6}$

8. $-\frac{35}{63} \approx -56¢$

9.
  (a) $6x - 6$

10.
  (a) $\frac{3}{4}$ (c) 16

12. $t = 11\frac{1}{9}$

CHAPTER 4

1.
  (a) tangent $y = -x + 8$; normal $y = x + 4$

2.
  (a) $(-1, 5)$ (d) $(-1, -1)$ and $(1, 3)$
  (c) none   (f) $(-3, 0)$

3.

	Increasing	Decreasing
(a)	$x < -1$	$x > -1$
(c)	all x	
(e)	$x > 0$	$x < 0$

4.

	Increasing	Decreasing
(a)		$x < 2, x > 2$
(c)	$x > 0$	$x < 0$

5.

(b)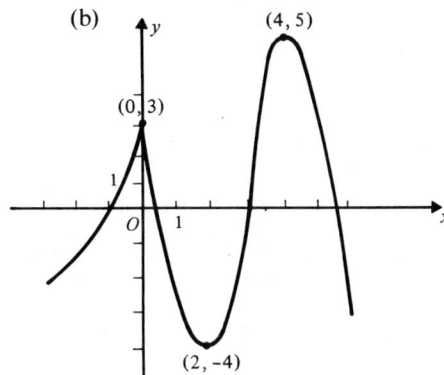

6.
(a) (−1, 4) and (3, −28)   (c) (0, 4)

7.

	Abs Max	Abs Min
(a)	(1,6)	(−2, −12)
(c)	(1, 1)	(−1, −11)
(e)	none	(4,4)

8. 30

9. 25

10.

	Concave Up	Concave Down	Points of Inflection
(a)	$x > 1$	$x < 1$	(1, −12)
(c)	$x < 0, x > 0$	nowhere	none

12.
(b) $g'(x) = -4(1 - x)^3$, which equals zero if $x = 1$; $g''(x) = 12(1 - x)^2$, which is zero if $x = 1$. Since $g'(x) < 0$ if $x > 1$, $g'(x) > 0$ if $x > 1$, g has a local min at $x = 1$.

13.
(a) 0   (b) 2   (c) ∞   (d) −3

14.

	Horizontal Asymptotes	Vertical Asymptotes
(a)	$y = 0$	$x = \pm 2$
(c)	$y = 0$	none
(f)	none	$x = 0$

15. If this formula holds, his vocabulary approaches (but never reaches) 50,000 words.

CHAPTER 5

1. 50 ft × 100 ft
4. 540 mph
6. (1, 2)
7. $f(x) = x^3 + 2$
8. 2.983
9. about $50
10. (b) $-\dfrac{x}{\sqrt{1-x^2}}\,dx$
13. 1500 units; 1000 units; 2000 units
14. $E(p) = \dfrac{p}{6-p}$. Since $E(4) > 1$, a small increase in price results in decreased revenue.
15. 6 shipments

CHAPTER 6

1.
   (a) 64   (d) t^{20}
2.
   (a) 1   (c) 4
3. $-\tfrac{1}{3}$
5.
   (a) e^5 (c) 1
6.

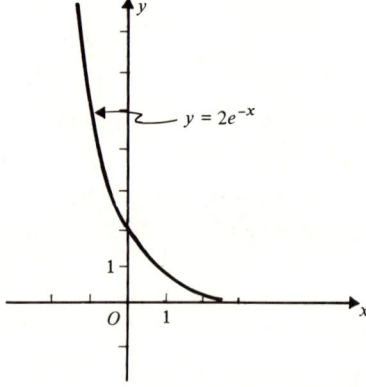

7.
   (a) 3   (c) -2
8. 2.66
10. 17
11. 7.1×10^8 yr

Answers to Chapter Review Exercises

12.
  (a) $\dfrac{3}{x}$
  (c) $\dfrac{x^2 - 1}{x} + 2x \ln x$
  (e) $\dfrac{1}{x \ln 2x}$

13.
  (a) $4e^{4x}$ (c) $xe^x + e^x$ (e) $e^x \ln x + \dfrac{e^x}{x}$

14. $y = 1 - 2x$
15. 3
16. $f(t) = 4e^{-t}$
18. $\approx \$598.60$
20. ≈ 1833 animals
21. about 312 people

CHAPTER 7

1. 18 ft
3. $2440
4. $\displaystyle\lim_{n\to\infty} \dfrac{1}{n}\left[2 \cdot \dfrac{1}{n^2} + 2\left(\dfrac{2}{n}\right)^2 + 2\left(\dfrac{3}{n}\right)^2 + \ldots + 2\left(\dfrac{n}{n}\right)^2 \right]$
   $= \dfrac{2}{n^3}(1^2 + 2^2 + 3^2 + \ldots + n^2) = \dfrac{2}{n^3}\left(\dfrac{n^3}{3} + \dfrac{n^2}{2} + \dfrac{n}{6}\right) = \dfrac{2}{3}$

6. $\displaystyle\int_0^2 (4x^3 + x - 3)\, dx$

8. $\displaystyle\int_0^4 -5e^{-0.1t}\, dt$

9. (a) 10   (b) $e^2 - 1$ 10. $(\ln 3)$ ft   11. 26

12.
  (a) $x^3 + x^2 + 4x + C$
  (b) $\dfrac{x^3}{3} + \ln x + C$
  (d) $2e^x - 3x + C$

13.
  (a) $\tfrac{3}{2} - \ln 2$ (b) $e - 1$

14.
  (b) does not exist

15.
- (a) $\frac{1}{7}(x^2 + 1)^7 + C$
- (c) $\frac{1}{2}\ln(2x^3 + 2x^2 + 5) + C$
- (e) $\frac{2}{9}(3x - 1)^{3/2} + C$

16.
- (a) $\frac{123}{4}$
- (c) $\frac{1}{2}\ln 5$ or $\ln\sqrt{5}$

17.
- (a) $-xe^{-x} - e^{-x} + C$
- (b) $\dfrac{x^5}{5}\ln x - \dfrac{1}{25}x^5 + C$

CHAPTER 8

1. $\frac{14}{3}$
2. $\frac{125}{6}$
4. 64 ft/sec
7. $\approx \$2156$
8. $M(x) = 14x - 0.003x^2$.
9.
- (a) (1250, $0.25)
- (b) $2343.75
- (c) $156.25

10.
- (a) about 771 patients
- (b) 1250 patients

12. $12,960

13.
- (a) $\dfrac{1}{13}\displaystyle\int_1^2 (3x^2 + 4x)\,dx = \dfrac{1}{13}\Big[x^3 + 2x^2\Big]_1^2 = 1$

14.
- (a) $\frac{5}{7}$
- (b) $\frac{2}{3}$

17.
- (a) $y = e^{-x} + C_1 x + C_2$

18. $y = f(x) = \dfrac{x^4}{3} - \dfrac{x^3}{2} - 1$

19.
- (a) $y^2 + y = 2x^2 + C$

21. $P = 500 e^{0.30t}$

CHAPTER 9

1.
- (a) $C(l, w, h) = 3lw + 3lh + 3wh = 3(w + lh + wh)$; C in dollars
- (b) $141

Answers to Chapter Review Exercises

3.
(a) all pairs of real numbers (c) $y \geq x$

4.
(a) The origin, in the xy-plane; the parabola $z = 4x^2$ in the xz-plane; the parabola $z = 4y^2$ in the yz-plane.

(b) $4x^2 + 4y^2 = 4$ is the circle $x^2 + y^2 = 1$; $4x^2 + 4y^2 = 16$ is the circle $x^2 + y^2 = 4$.

(c)

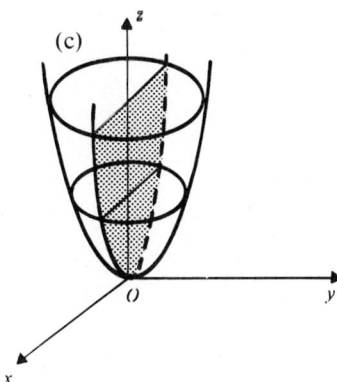

Surface of $z = 4x^2 + 4y^2$

7. If two products are *competitive*, then an increase in the price of either product results in increased demand for the other. If the products are *complementary*, then an increase in the price of either product causes a decrease in the demand for the other.

8.
(a) competitive (b) complementary

9.
	$f_x(x, y)$:	$f_y(x, y)$:
(a)	$4x - 5y^2$	$-3 - 10xy$
(c)	$\dfrac{1}{x} + \dfrac{1}{y}$	$\dfrac{1}{y} - \dfrac{x}{y^2}$

10.
	f_{xx}:	f_{yy}:	$f_{xy} = f_{yx}$:
(a)	4	$-10x$	$-10y$
(c)	$-\dfrac{1}{x^2}$	$\dfrac{1}{y^2} + \dfrac{2x}{y^3}$	$-\dfrac{1}{y^2}$

11.
(a) labor 32; capital 125/4

(b) An increase of 1 man-hour per production run, from $x = 125$ to 126, when y is held fixed at 64, results in an increase in production of 32 units. An increase of \$1000 in capital invested, from $= 64$ to 65, when x is held fixed at 125, causes an increase in production of approximately 31 units.

12.
(a) $df = 2xy\, dx + x^2\, dy$
(c) $df = -ye^{-xy}\, dx - xe^{-xy}\, dy$

14. 2.1%

16. ≈ 1.044

17. $\dfrac{\partial D_1}{\partial p_2} = -4 < 0;\ \dfrac{\partial D_2}{\partial p_1} = -5 < 0;$ so an increase in the price of either product causes a decrease in demand for the other. This characterizes complementary products.

18.
(a) $f(-3, 1)$ is a local minimum

19. $y = 1.4x + 0.5$

20. ≈ 65 therms

22. $f(3, -2)$ equal to -4 is an absolute minimum; $f(-3, 2)$ equal to 10 is an absolute maximum

Answers to Chapter Tests A

CHAPTER 1

1. $x \neq 0, 1$
2. $f(0) = -\dfrac{1}{2}$; $f(-1) = -\dfrac{1}{3}$;
3. $\dfrac{3h}{h} = 3$ if $h \neq 0$
4. $y = x + 4$
5.

$m = 2$
y-intercept $= -3$

6. $k = 2$
7.
(a)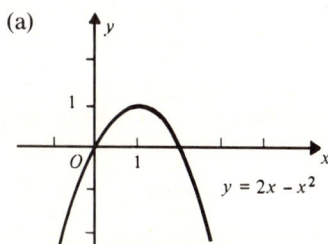
$y = 2x - x^2$

(b)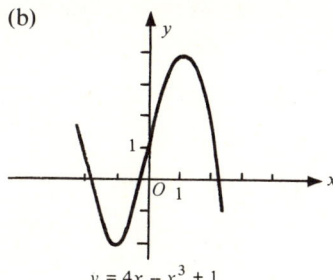
$y = 4x - x^3 + 1$

(c)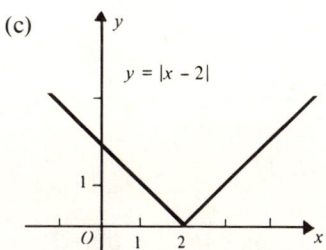
$y = |x - 2|$

633

8. $f(g(x)) = 20 - 13x + 2x^2$;
   $g(f(x)) = 4 - x - 2x^2$.
9. Let $f(x) = \sqrt{x}$ and $g(x) = 5x + 1$.
10. If $g(x)$ is the inverse of $f(x)$, then $g(x) = (x - 1)^{1/3}$

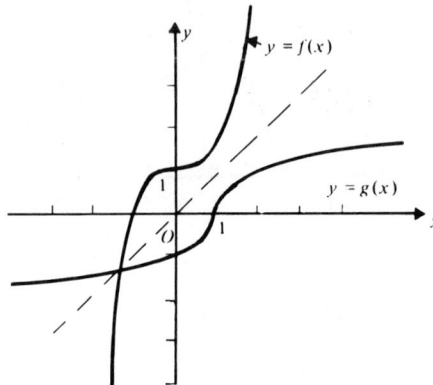

11. $2x^2 y = \dfrac{64}{3}$, where x is the width in feet

The cost $C = 1.50(4x^2 + 2xy + 4xy)$
$= 3(2x^2 + 3xy)$

$C(x) = 3\left[2x^2 + 3x \cdot \dfrac{64}{3(2x^2)}\right] = 3\left[2x^2 + \dfrac{32}{x}\right]$

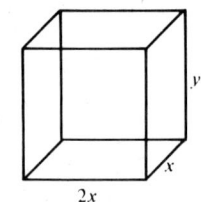

12.
(a) If n is the number he can sell and p is the price per frisbee, then

$$p(n) = -\dfrac{n}{40,000} + \dfrac{9}{2}$$

(b) $C(n) = 3000 + \dfrac{1}{2}n$

$R(n) = p \cdot n = -\dfrac{n^2}{40,000} + \dfrac{9}{2}n$

$P(n) = R(n) - C(n) = -\dfrac{n^2}{40,000} + 4n - 3000$

CHAPTER 2

1. 12.2; 11.8
2. 12
3.
   (a) $-4x$ (b) -3

4.

(a) 2 (e) 0
(b) −4 (f) no limit
(c) ½ (g) 0
(d) no limit (h) no limit

5. One example (with $c = 0$) is

$$f(x) = \begin{cases} \dfrac{x^2 + x}{x} & x \neq 0 \\ 2 & x = 0 \end{cases}$$

6. $x = 0$ and -2

7. Yes; $\lim f$ as $x \to 2^- = \lim f$ as $x \to 2^+ = f(2) = 5$.

8.
  (a) $3x^3 - 21x^2 + 1$ (b) 28   (c) 0   (d) −9

9. $x = 0, 3$

10. decreasing by ≈ 167 people per yr

11. $12.50

CHAPTER 3

1.
  (a) $(3x + 2)(1 - 2x) + 3(4 + x - x^2) = 14 + 2x - 9x^2$
  (b) $-\frac{1}{18}$

2. $(0, 4)$ and $(-4, -4)$

3.
  (a) $-\dfrac{x}{\sqrt{4 - x^2}}$ (b) $\dfrac{2 + x}{(2 - x)^3}$

4.

  (a) $xy' + y = 2x$; $y' = \dfrac{2x - y}{x}$

  (b) $y = x + \dfrac{1}{x}$; $y' = 1 - \dfrac{1}{x^2}$

  (c) Replace y in (a) by its value taken from (b):

$$y' = \frac{2x - \left(x + \dfrac{1}{x}\right)}{x} = \frac{x - \dfrac{1}{x}}{x} = 1 - \frac{1}{x^2}$$

5. 2

6.

(a) $24x^2 - 6$ (b) $\dfrac{3}{\sqrt{2x}}$

7. $t > {}^{td}$

CHAPTER 4

1. tangent $y = 1$; normal $x = 1$

2.

	Local Max	Local Min	Inflection Points
(a)	$\dfrac{2}{\sqrt{3}}, \approx 2.1$	$\dfrac{2}{\sqrt{3}}, \approx 4.1$	$(0, -1)$
(b)	$(0, 4)$	$(\pm 1, 3)$	$\left(\pm \dfrac{1}{\sqrt{3}}, 3\dfrac{4}{9}\right)$

3.
 (a) See page 35, Example 4-4.
 (b)

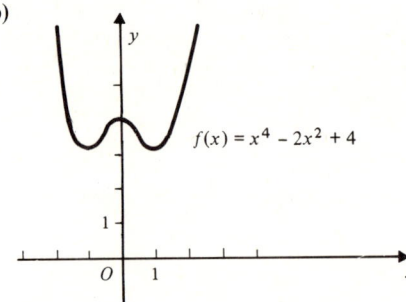

$f(x) = x^4 - 2x^2 + 4$

4. See page 200, Example 20-2.

5.

	Abs Max	Abs Min
(a)	$(3, 14)$	$(1, -2)$
(b)	$[\tfrac{1}{2}, 2]$	$[4, \tfrac{1}{4}]$
(c)	none	$[8, \tfrac{1}{8}]$

6.

	Horizontal Asymptotes	Vertical Asymptotes
(a)	none	none
(b)	$y = -1$	$x = \pm 2$
(c)	none	none
(d)	$y = 0$	none

Answers to Chapter Tests A

7. See page 568, Exercise 17, with equivalent equation.
8.
   (a) $N(t) \to 30$ as $t \to \infty$.
   (b) No, there is no abs max on the open interval $t > 0$.

CHAPTER 5

1. 2 in. by 2 in. by 3 in.
2. $10
3. ≈ 0.01 ft/min
4. $\approx \$257$ per h
5.
   (a) $\left(\sqrt{3}, \dfrac{\sqrt{3}}{3}\right)$ (b) $(1, 1)$
6. 0.075
7. $\dfrac{x}{\sqrt{x^2 + 5}}\, dx$
8.
   (a) only at $t = 2$ (b) 1
9. $3.20
10. 400

CHAPTER 6

1.
   (a) 3   (b) 2   (c) $\tfrac{1}{8}$ (d) x^6 (e) $\sqrt{x}$
2. $-\tfrac{1}{4}$
3. See figure.
4. 0.46
5. 26
6.
   (a) $2e^{2x} + e^{-x}$
   (b) $2x(1 + 2x \ln x)$
   (c) $x^2 e^x + 2xe^x$
   (d) $\dfrac{1}{2x\sqrt{1 + \ln 2x}}$
   (e) $\dfrac{-x}{(4 - x^2)}$
   (f) $\dfrac{1}{2} e^{-(x/2)}$
7. $y = x - 1$
8. $f'(t) = ce^{kt} \cdot k = k(ce^{kt}) = kf(t)$

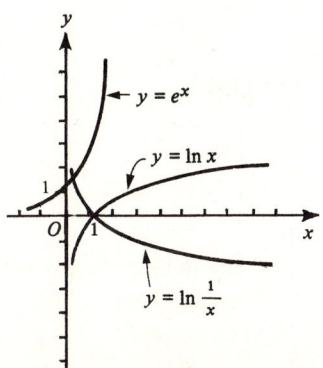

9. in about 6.9 yr
10. 7.0%
11. 25° F
12. about 550 people

CHAPTER 7

1. 65

2. $\lim_{n \to \infty} \frac{1}{n}\left[4\left(\frac{1}{n}\right)^2 + 4\left(\frac{2}{n}\right)^2 + \ldots + 4\left(\frac{n}{n}\right)^2\right] = \lim_{n \to \infty} \frac{4}{n^3}(1^2 + 2^2 + \ldots + n^2) = \lim_{n \to \infty} \frac{4}{n^3}\left(\frac{n^3}{3} + \frac{n^2}{2} + \frac{n}{6}\right) = \frac{4}{3}$

3. $\int_0^1 4x^2\, dx = \left[4\frac{x^3}{3}\right]_0^1 = \frac{4}{3}(1 - 0) = \frac{4}{3}$

4. The definite integral represents the growth in population between times t_1 and t_2.

5. $\int_{100}^{120} (5 + 0.04x - 0.0003x^2)\, dx$

6. $115.20

7.
   (a) $\frac{x^4}{2} - \frac{x^2}{2} + 5 \ln x + C$

   (b) $\frac{x^3}{3} - 2x - \frac{1}{x} + C$

   (c) $-\frac{1}{3}(1 - x^2)^{3/2} + C$

   (d) $\frac{x^2}{2} \ln x - \frac{x^2}{4} + C$

   (e) $\frac{1}{4} \ln (4x + 1) + C$

   (f) $\frac{1}{2}\sqrt{4x + 1} + C$

8.
   (a) 24.2   (b) 1   (c) $\frac{1}{2} \ln 6$

CHAPTER 8

1. $\frac{9}{2}$

2. $\int_1^5 -\left(\frac{1}{2}x - 4\right) dx = \int_1^5 \left(4 - \frac{1}{2}x\right) dx$

3. Initial velocity = 3; initial position = −5.

4. ≈ $8902

5. ≈ 0.43

6. $7028

7.
   (a) (2000, $1)   (b) $8000   (c) $1000

Answers to Chapter Tests A

8.
   (a) about 902 customers (b) 2000
9. $\approx \$6978$
10. $\approx \$2432$
11. $\int_0^\infty 2e^{-2x}\, dx = \lim_{n\to\infty} \int_0^n 2e^{-2x}\, dx = \lim_{n\to\infty} \left[-e^{-2x}\right]_0^n = \lim_{n\to\infty} -\left[\dfrac{1}{e^{2n}} - \dfrac{1}{e^0}\right] = 1$
12. $\frac{1}{16}$
13. $y = Ce^{-2x} \to y' = -2Ce^{-2x} \to y'' = 4Ce^{-2x} = 4y$
14.
   (a) $y = \dfrac{2}{3}x^3 - \dfrac{3}{2}x^2 + C_1 x + C_2$
   (b) $y^2 = x^2 + C$
   (c) $y = ce^{kt}$
15.
   (a) $y = -\ln x + 2$ (b) $y = 1 - e^{2-x}$
16. $Q = 10e^{-0.001t}$

CHAPTER 9

1. In dollars, $R(x, y, z) = 0.40x + 0.30y + 0.65z$.
2. Any triple (a, b, c) for which $R(a, b, c) = 200$ is on the level curve. His revenue when he sells a murex shells, b cowries, and c cones is \$200.
3.
   (a) 18.
   (b) The domain of f is $x \leq 5$, y any real.
   (c) $f(4, y) = y^2$; this is the parabola in which the surface intersects the plane $x = 4$.
4.
   (a) In the xy-plane, the circle $x^2 + y^2 = 9$; in the yz-plane, the upper half of the circle $y^2 + z^2 = 9$; in the xz-plane, the upper half of the circle $x^2 + z^2 = 9$.
   (b) $z = \sqrt{9 - (x^2 + y^2)}$. Since $x^2 + y^2 \geq 0$, $z \leq \sqrt{9} = 3$; also $z \geq 0$, so $0 \leq z \leq 3$.
   (c) The curve in the plane $z = k$ ($|k| \leq 3$) is the circle $x^3 + y^2 = 9 - k^2$.
   (d) The surface is a hemisphere, center at O, radius 3.
5. In the D_1 function, the coefficient of p_2 is positive; in the D_2 function the coefficient of p_1 is positive. These facts imply that the demand for either product increases if the price of the other increases. This defines competitive products.

6.

	$f_x (2, 0)$:	$f_y(2, 0)$:
(a)	1	$2 - \ln 2$
(b)	3	0

7. $f_{xx} = 0$; $f_{xy} = 2y = f_{yx}$; $f_{yy} = 2x + 24y$
8. $P_x(200,100) = 5$; $P_y(200,100) = 10$
9. $75; 1%
10. ≈ 5970
11. If $D_1(p_1, p_2)$ and $D_2(p_1, p_2)$ are the demand functions for competitive products at prices p_1 and p_2, then $\partial D_1/\partial p_2$ and $\partial D_2/\partial p_1$ are both positive.
12. $f(0, 1)$ which equals -4 is a local minimum
13. $y = x + \frac{5}{2}$
14. $f(2, 2)$ which equals 17 is an absolute maximum

Index

Numerals not in parentheses refer to pages. Those in parentheses refer to numbered exercises or problems, with letter abbreviations as follows: A = Goal Test A; B = Goal Test B; ChT = Chapter Test; SQ = Study Question.

abscissa, 552
absolute maxima and minima, 203–7, 528(SQ1)
absolute value, 551
absolute-value function, 91, 114, 194
acceleration, 165, 277, 432–4
acoustics, 310, 312(13), 333(4), 437(4)
actuarial applications, 465(5)
advertising, 349, 352, 355(22), 508
antiderivative, 265, 393, 395–8
antidifferential, 395, 405–6
arbitrary constant, 395, 470–1, 475, 479
area between curves, 425–8, 429(3, SQ1)
area under a curve, 368–9, 373–9, 390–1, 400
asymptote, 37–8, 99–100, 220–4, 226
autocatalytic reaction, 63(26), 249(32), 352, 355(28)
average cost, 143, 144(6,7), 146(SQ3,4), 171(3), 286–7, 292(2), 293(A1), 294(B1,SQ3–5), 298(13)
average inventory, 290–1
average rate of change, 74–6
average speed, 439–40
average value of a function, 440–443, 445(SQ3)
average velocity, 73–4, 439–40, 444(4), 445(SQ4)

base, 303–6, 309
binomial, 553–4
blood pressure, 115(SQ5), 127, 128(15), 250(33), 274(10)
boundary condition, 470
business, simplifying assumptions, 244, 283–4
business, staffing for, 259 (B2)

capital value, 454
chain rule, 150–3, 155, 405–6, 518(SQ2)
change of variable in integration, 404–9
chemical decomposition, 387(A2)

chemicals: purification of, 225(4); transformation of, 128(18)
circle, 552–3
closed domain, 528(SQ1)
closed interval, 203, 550
Cobb–Douglas formula, 508
communications, 458–62
competitive vs. complementary products, 500–1, 513–4
composition of functions, 44–7
compound interest, 104(SQ1), 315–6, 322(SQ2), 346–7, 354(7–11), 356(A3,B3), 360(18), 361(ChT10), 453–4
concavity, 23, 32–3, 37, 210–1
constant function, 13
constant of integration, 395, 431–4, 476
constraint, 532
consumers' surplus, 447–9, 455(3), 456(A2, B2), 487(9), 489(ChT7)
continuous function, 106–11, 114, 116
contour curve or line, 497 fn
convergent improper integral, 400
coordinates, 552
corner point, 530(SQ2)
cost, 4, 8(14), 16, 19(8), 29(23, 28), 45–6, 51(28), 75–6, 79(3), 107, 120, 128(7,8,13), 144(6,7), 206, 208(21), 220, 273(3), 285–7, 290–2, 367–8, 371(3,B2), 384, 386(4), 389, 393(3), 420(3), 434, 455(2), 499, 543(1), and numerous exercises in Goals 6, 25, and 30
cost, fixed, 16
crime, 46, 501(3)
critical point, 187–91, 227–8, 521–4
crop yield, 45, 60(7), 144(8), 247(5)
cube root, approximation of, 270

decay: exponential, 344–8, 353; rate of, 344
decreasing function, 13, 30, 184–91

641

definite integral: applications of, 382–6, 390–1, 426, 461; definition, 381–2; evaluating, 389–93, 398–9, 409–10, 415–6; vs. indefinite, 395
demand function, 19, 19(9,10), 51(29), 79(8), 160, 162(6), 172(8), 258(7), 274(13), 288–90, 446, 463–4, 499–501, 514, 516(14), and numerous exercises in Goals 6 and 30
demography, 134(ChT10)
density function (probability), 459
derivative(s): definition, 87–9; of exponential functions, 335–7; higher order, 164–8; of logarithmic functions, 330–2; partial, 505–11; of polynomials, 121; of products and quotients, 139–42, 145–6; of sums, 120–1; undefined at a point, 91, 187, 191; also 91, 117–20, 164, 186–7
differentiability, 114, 187, 191, 204–5
differential, 269–73; total, 511–3
differential equations: classification of, 469; linear, 473(SQ1,2), 482; order of, 469; separable, 474–82; also 468–71
differentiation, 117; implicit, 157–61; logarithmic, 334 (SQ2)
diffusion, 167; see also disease, drugs, rumor
discontinuity, 108, 113–4, 191
distance between points, 552
distance problems, 62(21,22), 248(15), 252(SQ3), 256–7(3), 258(A1), 297(4)
divergent improper integral, 400
divorces, number of, 385
dolphins, matings of, 493
domain, 3, 6–7, 244, 494
dosage determinations (medicine), 514(2)
dropout expectancy, 451–3
drug: concentration in bloodstream, 225(3), 354(16), 355(24); diffusion of, 479; reaction to, 168; see also sensitivity

e, 315–8
earthquake, magnitude of, 312(14), 330(5)
ecology, 465(A2), 528(5); see also pollution
economics, 283–92, 446–54
elastic demand, 288–90, 292(4,5), 293(A2), 294(B2, SQ6), 298(14)
endpoints of an interval, 203
equations of a line, 15–7, 21(SQ1)
equilibrium point, 447–9, 455(3), 456(A2,B2), 487(9), 489(ChT7)
error tolerance, 104(SQ2), 272–3, 274(9), 275(B4),513, 515(11), 544(15)
even vs. odd function, 9, 30, 37, 42
e^x, 318–21, 326, 328, 335, 337, 398; values of, 559–60

explicit vs. implicit functions, 156–7
exponent, 303–6
exponential equations, 306–9, 327
exponential functions, 305–6, 319–21, 336–40
extinction of a species, 528(5)
extreme values, 196–200, 203–6, 215–7, 500–24, 528(SQ1), 532–7, 540(SQ2)

factor of proportionality, 16
factoring, 554
family of lines, 20
FOIL, 553
fractional equations, 556–7
function(s): absolute value, 38–9; composition of, 44–7; constant, 13; continuous or discontinuous, 106–11; definition and notation, 3, 6–7, 9; domain of, 3, 6–7; exponential, 305–6; inverse, 10, 47–9, 309–10, 326; linear, 12–18; logarithmic, 309–10; one-to-one, 10; polynomial, 31, 227; range of, 3, 7; rational, 35–8; of several variables, 493–501, 522–4
fundamental theorem of the calculus, 391–2, 394(SQ1,2)
future income: continuous, 454, 457(SQ1,4); periodic, 453–4
future value, 456 (SQ1)

general solution of differential equation, 469–71
genetics, 484(9)
geology, 274(8), 443
geometry and mensuration, 29(22), 54, 59(2–4), 61(17), 62(20), 63(23–25, A1), 64(SQ1), 145(9,10), 257(4), 258(8), 271–2, 274(4), 297(1,5), 545(23), and numerous exercises in Goal 25
glass blowing, 258(6)
Gompertz curve, 484(8)
graphing and graphs, 192, 552; of functions of one variable, 227–32, 319–21, 325–6; of functions of two variables, 494–6; of inverse functions, 42, 48–9; straight line, 12–6
greatest-integer function, 39
growth: of an animal, 128(6); of a colony (bacteria, fruit flies, etc.), 74–5, 80(A2), 128(5), 129(B2), 166, 167(5), 253–4, 257(1), 275(A4), 352, 353(1,2), 356(B1), 366–7, 371(A2), 379, 386(6), 393(4), 438(10), 489(21); exponential, 344–8, 353; of a tree, 129(19); unrestricted vs. restricted, 350–1; see also population

Index

half-life, 327, 436
half-open interval, 551
hemisphere, 496
highway engineering, 465(7,A3)
homogeneous differential equation, 473(SQ1,2), 482
hyperbola, 37

implicit differentiation, 157–61, 251–2(SQ1)
implicitly defined function, 156–7
improper integral, 399–400
increasing function, 13, 30, 184–91
indefinite integral, 395
independent vs. dependent variable, 7
indifference curve, 497 fn
inelastic demand, 289
inequalities, 550, 557
infinite limits, 98–9; of integration, 399–400
inflection, point of, 32, 166, 211–5, 228
initial amount, 345
initial condition, 470
instantaneous rate of change, 77–9
integrand, 382
integration, 395–8; by parts, 413–7; by substitution, 404–10; see also definite integral
intensity: of light, 269, 356(29,30); of sound, 310
intercepts: of a graph, 227; of a line, 12, 13, 21
interest, 4, 19(12); see also compound interest
intermediate value theorem, 34, 114–5
intervals, closed and open, 203, 551–2
inventory control (optimal order size), 283, 290–2, 293(8–10,A3), 294(B3), 298(15), 299(ChT10)
inverse functions, 10, 42, 47–9, 309–10, 326
investments, 385, 435, 437(8), 442, 444(A2), 453–4, 455(6–8), 456(10,A4,B4,SQ1), 457(SQ4), 487(7), 488(11,12), 489(ChT4,9,10), 494
irrational exponent, 314(SQ4)
irrational number, 549

kidney, functioning of, 509, 515(10)

Lagrange multipliers, 534–8, 540(SQ1)
learning, memorizing, 166, 169(8,B4), 173(12), 236(15), 238(ChT8), 349, 355(21)
least squares, 524–7
level curves, 497–9, 518(SQ3)
life expectancy of a commodity, 462–3, 464(2), 489(ChT12)
limits, 85, 94–101, 104–5; of function of two variables, 517(SQ1); "at infinity," 220–2;

of integration, 382; left- and right-hand, 94–5, 98–101
linear approximation, 269
linear equations: forms of, 15, 20, 21; solving, 554
linear functions, 12–8
linear programming, 530(SQ2)
lines: parallel, 16–7; perpendicular, 16–7; of best fit, 524–5, 530(SQ3)
ln x (natural log of x), 324–6, 330; values of, 558
local maxima and minima, 196; of function of two variables, 520–1
logarithm, 306–9
logarithmic differentiation, 334(SQ2)
logarithmic equation, 306–7
logarithmic function, 309–10, 330–2
logarithmic response, 481
logistic curve, 351, 357(SQ1), 480

machining, see error tolerance
maintenance costs, 450–1, 455(4), 489(ChT6)
manufacturing, see production
mapping, 4
marginal average cost, 143
marginal cost, 78, 287
marginal productivity: of capital, 509; of labor, 509; of money, 538
marginal revenue, 287
marginal utility, 167, 538
marriages, number of, 384, 399
maximum sustainable population, 351
maxima and minima, 24–8, 197–201, 215–6, 244, 520–4, 532–8; absolute, 203–6, 528–30
mean of probability density function, 466(SQ4)
mean value of a function, see average value
Mean Value Theorem (MVT), 260–5; for Integrals, 444(SQ1)
mechanics, 79(1,7), 115(SQ4), 161, 162(17), 254–6, 257(2), 258(9), 274(5), 297(5), 298(ChT3), 349, 354(12), 371(1); see also motion problems
median of probability density function, 467(SQ5)
meteorology, 195, 444(B2)
model, mathematical, 53
motion problems, 3, 29(21), 66(9), 73–4, 80(B2,SQ1), 127(1,2), 130(2), 133(13), 150, 249(24), 365–6, 383, 389(2), 393(2), 394(B2), 419(1), 420(10), 487(4), and numerous exercises in Goals 17, 20, 45
MVT, see Mean Value Theorem

natural logarithm, *see* ln x
nonnegative integers, 549
normal lines, 180–1
numbers, kinds of, 549
number-theory applications, 248(19), 532–3

oblique asymptote, 226
octant, first, 495
odd vs. even function, 9, 30, 37, 42
one-to-one function, 10, 47–8, 52
open interval, 203, 550
optimal order size, *see* inventory control
order of differential equation, 469
ordinate, 552

parabola, 22–8, 30
paraboloid, circular, 495–6
partial derivative, 505–11
particular solution of differential equation, 469–71
parts formula, 413, 416–7
percentage change, 271, 288, 294(SQ6), 512
percentage error, 274(11)
physiology, 249(31), 387(SQ1), 403(SQ4)
point of tangency, 177
pollution, 56, 59(4), 126, 128(9,10), 129(A1,B1), 387(9), 502(7), 516(A4)
polynomials, 31, 95, 111, 227
population: human, 345, 350–1, 353(3), 355(26), 356(A1), 360(17,20), 385, 390, 422(ChT4), 480, 484(10); nonhuman, 79(2), 119, 225(5), 354(17), 357(B5), 371(2), 399, 442, 502(A1), 528(5)
position function, 165, 277, 431–4
postage function, 41(30), 100–1, 107
power functions: derivative of, 119, 148–9, 155; antiderivative of, 396–7
predator-prey relations, 159, 162(15), 258(5,A2), 274(12)
predicting group characteristics, 464(3,4), 465(6,8,10), 466(B2,3), 488(14)
predicting group size, 451–3, 455(5,9), 456(A3,B3), 457(SQ3), 487(10), 489(ChT8)
predicting sales, 451–3, 455(5), 456(A3,B3)
present value, 345, 453–4, 455(6–8), 456(10,A4,B4), 457(SQ4), 488(12), 489(ChT9,10)
probability, 317, 458–64, 466(SQ1–3), 488(14), 489(ChT12)
producer's surplus, 449–50, 453(3), 456(A2,B2), 487(9), 489(ChT7)
production, 128(17), 129(A4), 249(28), 269, 502(8), 539(3–5), 541(SQ3), and numerous exercises in Goal 30

productivity, 508–9, 514(3), 516(A2), 517(B2), 537–8, 544(11,13,14); *see also* marginal productivity
profit, 19(8), 127, 128(13,14), 129(A3,B3), 133(17), 150–2, 208(22), 271, 274(6,11), 275(A3,B3), 384, 386(5), 420(7), 422(ChT5), 456(A1), 483(6), 493, 512, 528(4,B2), 543(6), 546(8,9), and numerous exercises in Goals 6, 25, 30
projections, 498
proportionality, 16, 338, 345
puzzle-solving, 462–3, 465(9)
Pythagorean Theorem, 551

quadrants, 552
quadratic equations, 554–5
quadratic formula, 555
quadratic functions, 22–8

radioactive decay, 327, 329(6), 347, 353(4,5), 354(6,14,15), 356(A2,B2), 359(11), 360(19), 361(ChT9), 420(8), 435–6, 438(A3), 444(3), 487(5), 490(16)
radioisotope dating, 347–8, 354(6,14,15), 356(A2)
range of function, 3
rates of change, 124–6, 338–40; relative, 334(SQ1)
rational functions, 35–8, 111, 221–22
rational numbers, 549
real numbers, 6, 549
reflection of curve, 33
regression line, 525
related-rate problems, 253–7
relative maxima and minima, 197
reliability of a test, 501(2)
removable discontinuity, 113–4
restricted (inhibited) growth model, 350–1, 477–9
revenue, 42(31), 45–6, 54–6, 133(17), 135(ChT11), 140–1, 236(8,9), 275(SQ2), 384, 387(B2), 437(7), 454(1), 487(8), 501(1), 503(B3), 515(9), 545(ChT1), and numerous exercises in Goals 6, 25, 30
Richter scale (for earthquakes), 312(14)
Riemann integral and sum, 382
right- and left-hand limits, 94–5, 98–9
Rolle's Theorem, 268(SQ3)
routing problems, 57–8, 61(18,19), 246–7, 248(11,12)

saddle point, 521
sales, 352, 451–3, 455(5), 456(A3,B3), 508, 526
secant line, 76–7
second derivative, 164–70, 211, 215–6

sensitivity to a drug, 127, 128(15,16), 145(11), 250(33), 274(10)
separable differential equations, 474–82
shipping costs, 290–1
side condition, 470, 532, 540(SQ2)
simultaneous equations, 555–6
slope: of a curve, 76–8, 81–4; of a line, 12–3
solution of differential equation, 469–71
Spearman–Brown reliability formula, 501(2)
speed: increasing or decreasing, 278; most economical, 29(3), 206–7, 249(25)
spread: of disease, 126–7, 128(11,12), 129(A2), 133(16), 167, 249(27), 274(7), 351, 355(25), 356(A5), 360(21), 386, 436; of rumor, 64(SQ2), 249(30), 352, 355(27), 361(ChT12), 385, 387(8), 399, 438(9)
square root, 29, 550
step function, 39
stimulus-response, 481, 483(7)
stock market, 3
storage costs, 290–1
subinterval, 366
substitution in integration, 404–10
sum formulas, first n integers, squares, cubes, 380
supply and demand, 446–50, 455(3), 456(A2,B2), 487(9), 489(ChT7)
supply function, 19(11), 66(6)
surfaces, 495–6
symmetry, 23, 30, 32, 36, 48

tangent line, 85, 177–81, 187, 269–70
tangent plane, 511
temperature, 5, 19(14), 129(B4), 167, 259(B1), 269; human, 128(16), 145(11), 195, 387(10); law of cooling, 348–50, 354(18,19), 355(20), 356(A4,B4), 361(ChT11); in meteorology, 65(1), 106, 150, 442, 444(2)
therapy, 105(SQ3), 356(B2)
total differential, 511–3
trace of a surface, 495
translation of curve, 24–8, 32–5
transportation, 61(19), 248(12)
tumor, growth of, 387(7)

undefined at a point, 6, 91, 187, 191
unrestricted (or uninhibited) growth, 350–1, 476
utility, 167, 169(7), 517(B4)

vary directly, 16
value of machinery, 443, 444(5)
velocity, 73–4, 165, 277–80, 431–2; limiting, 349, 355(23), 479
vertical tangent, 177

wages, 498–9
Weber–Fechner law, 167
weight: of a baby's brain, 19(12); of an animal, 385, 435, 438(B3)

y-intercept, 12